I0044425

The Multi-Agent Transport Simulation MATSim

Edited by
Andreas Horni, Kai Nagel, Kay W. Axhausen

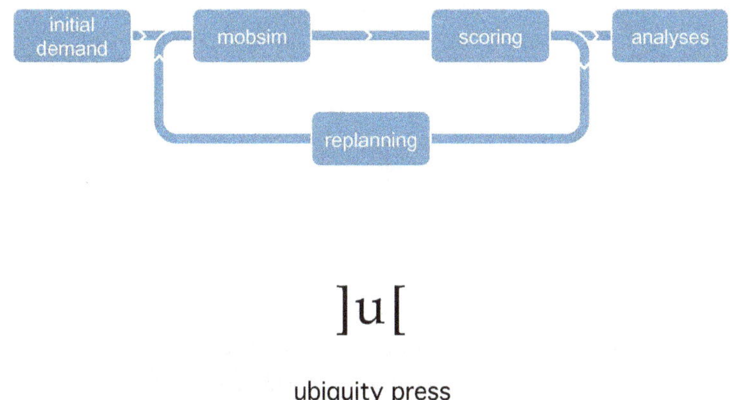

]u[

ubiquity press
London

Published by
Ubiquity Press Ltd.
6 Windmill Street
London W1T 2JB
www.ubiquitypress.com

First published 2016

Cover Illustration by Dr. Marcel Rieser, Senozon AG

Print and digital versions typeset by diacriTech.

ISBN (Hardback): 978-1-909188-75-4
ISBN (PDF): 978-1-909188-76-1
ISBN (EPUB): 978-1-909188-77-8
ISBN (Mobi/Kindle): 978-1-909188-78-5

DOI: http://dx.doi.org/10.5334/baw

The full text of this book has been peer-reviewed to ensure high academic standards. For full review policies, see http://www.ubiquitypress.com/

Suggested citation:
Horni, A, Nagel, K and Axhausen, K W (eds.) 2016 *The Multi-Agent Transport Simulation MATSim*. London: Ubiquity Press. DOI: http://dx.doi.org/10.5334/baw.
License: CC-BY 4.0

To read the free, open access version of this book online, visit http://dx.doi.org/10.5334/baw or scan this QR code with your mobile device:

Contents

Cover and Title Photos

The following cover and title photos have been provided by Dr. Marcel Rieser, Senozon AG.

©Dr. Marcel Rieser, Senozon AG

Portland, Oregon. View from the south to the city center, from the Portland Aerial Tram. June 2008. ©Dr. Marcel Rieser, Senozon AG

Zürich, Switzerland. Tracks at Zürich Main Station. May 2011. ©Dr. Marcel Rieser, Senozon AG

Berne, Switzerland. Car and bike park at Berne Main Station. June 2011. ©Dr. Marcel Rieser, Senozon AG

Gotthard railway model at the Swiss Museum of Transport, Lucerne, Switzerland. February 2004. ©Dr. Marcel Rieser, Senozon AG

Preface

Developing complex software for over a decade with a heterogeneous group of engineers and scientists, each with widely different skill levels and expertise across multiple locations around the world, requires dedication and mechanisms unusual for a university environment.

This book is one of these mechanisms. It allows us, collectively, to take stock and present a coherent state-of-the-system: for us and anyone interested in this approach. It highlights basics for the student who wants to undertake a small first research project as part of his or her degree, provides a description of the main functionalities, in detail, for the engineer setting up MATSim (Multi-Agent Transport Simulation) to conduct a policy analysis and, finally, fits the approach into the theoretical background of complex systems in computer science and physics.

The choice of the additional e-book format is an advantage, as it allows us to keep the book up-to-date with future chapters, revisions and, if necessary, errata. Equally importantly it allows you, the readers, to select those sections relevant to your needs.

The book comes at an important time for the system; for most of the first decade, its use was limited to the original developers and users in Berlin and Zürich. It is now much more widely consulted around the world, as we document in the chapter summarizing contributions on scenarios so far.

Scenario: This term will occur again and again. In MATSim context, it is defined as the combination of specific agent populations, their initial plans and activity locations (home, work, education), the network and facilities where, and on which, they compete in time-space for their slots and modules, i.e., behavioral dimensions, which they can adjust during their search for equilibrium. Within these scenarios, the user can experiment and explore with behavioral utility function parameters, with the sampling rate of the population between 1 % and 100 %, with algorithm parameters, e.g., the share of the sample engaged in replanning in any iteration, or behavioral dimensions or exact settings necessary to avoid gridlock due to the traffic flow dynamics. The creation of a scenario is a substantial effort, and the framework makes a number of tools available to accelerate it: population synthesizers, network editors, network converters between popular formats and the MATSim representation, e.g., OSM (OpenStreetMap) or GTFS (General Transit Feed Specification), semi-automatic network matching to join information, among others.

A large group of colleagues has been involved and many of them are contributors to this book; this is a list of those involved, other than ourselves, in Berlin, Singapore and Zürich.

Amit Agarwal
Milos Balac
Dr. Michael Balmer
Henrik Becker
Joschka Bischoff
Patrick Bösch
Dr. David Charypar
Dr. Nurhan Cetin
Dr. Artem Chakirov
Dr. Yu Chen
Dr. Francesco Ciari
Dr. Christoph Dobler
Thibaut Dubernet
Dr. Alexander Erath
Dr. Matthias Feil
Prof. Dr. Gunnar Flötteröd
Pieter J. Fourie
Dr. Christian Gloor
Dr. Dominik Grether
Dr. Jeremy K. Hackney
Dr. Johannes Illenberger
Prof. Dr. Johan W. Joubert
Ihab Kaddoura
Dr. Benjamin Kickhöfer
Dr. Gregor Lämmel
Nicolas Lefebvre
Dr. Michal Maciejewski
Dr. Fabrice Marchal
Alejandro Marmolejo
Dr. Konrad Meister
Dr. Manuel Moyo Oliveros
Kirill Müller
Dr. Andreas Neumann
Dr. Thomas Nicolai
Sergio A. Ordóñez Medina
Dr. Bryan Raney
Dr. Marcel Rieser
Dr. Nadine Rieser-Schüssler
Daniel Röder
Mohit Shah
Dr. Lijun Sun
Alexander Stahel
Prof. Dr. David Strippgen
Theresa Thunig
Dr. Basil Vitins
Michael Van Eggermond
Dr. Rashid Waraich
Dominik Ziemke
Michael Zilske

Additional contributors are mentioned as authors of their respective chapters in this book. We hope to acknowledge the contributions of more colleagues from other groups in future versions of this book and in the software.

Special thanks go to a number of people who greatly helped improving this book beyond their own chapters. Benjamin Kickhöfer's deep knowledge of MATSim's mathematical base, particularly its interpretation within the discrete choice framework, made the discussions accompanying the writing of this book very fruitful. Thibaut Dubernet's, Marcel Rieser's and Michael Zilske's outstanding expertise on software core development helped us very much and also improved the software structure during the writing of this book. Marcel Rieser's layout and illustrations greatly improved the book's appearance. Joschka Bischoff's effort to document basic information about every module will greatly help readers make a quick step into respective functionality.

The efficient and productive copy editing by Karen Ettlin is gratefully acknowledged.

The reported effort was funded and supported over the years by numerous agencies. Several particularly important sources are: ETH (Eidgenössische Technische Hochschule) Zürich and TU (Technische Universität) Berlin, the DFG (Deutsche Forschungsgemeinschaft), the SNF (Schweizerischer Nationalfonds), the Swiss ASTRA (BundesAmt für STRAssen), and the NRF (Singaporean National Research Foundation), through their repeated grants and projects supporting different dissertations over the years. A more complete list is provided on pages xxi ff. This support is gratefully acknowledged by all researchers.

The publication of this book was funded by the following institutions. The publisher services are funded by the EU (European Union) FP7 post-grant Open Access Pilot (OpenAIRE) and by DFG. The book's copy-editing is funded by the SNF under B-0010_166808. The support is highly appreciated.

We hope this book captures the interest of more researchers and engineers and encourages them to get involved in this joint effort. This would enable us to provide this framework, which has to be continuously adapted to our policy needs, together and ensure that it stays at the forefront of travel behavior modeling.

The editors
Andreas Horni, Kai Nagel, Kay W. Axhausen
Zürich and Berlin, February 2016

Acknowledgments

A project this dispersed and as long as the MATSim (Multi-Agent Transport Simulation) project draws on many sources for its support. We hope that we have not forgotten any institution here. We are grateful to all of them that they have made this open-source effort possible and we hope that they will continue to do so in the spirit of intellectual discovery and sharing.

In every case, we have to thank our home institutions for providing the basic intellectual and computing infrastructure for our work. ETH (Eidgenössische Technische Hochschule) Zürich was home to Prof. Nagel and his group when he started the project and continues to be the basis for Prof. Axhausen and his team. TU (Technische Universität) Berlin became Prof. Nagel's new platform after his move. Both institutions provided support through base funding for staff, servers and data access, which allow us to provide ongoing support to the overall project.

The following projects and sponsors funded particular persons and implementations:

TU Berlin (Kai Nagel, Amit Agarwal, Ulrike Beuck, Joschka Bischoff, Yu Chen, Gunnar Flötteröd, Dominik Grether, Johannes Illenberger, Ihab Kaddoura, Benjamin Kickhöfer, Gregor Lämmel, Michal Maciejewski, Manuel Moyo Oliveros, Andreas Neumann, Thomas Nicolai, Marcel Rieser, David Strippgen, Theresa Thunig, Jakub Wilk, Dominik Ziemke, Michael Zilske) undertook this work in the framework of the following projects: "COOPERS (Co-Operative Networks for Intelligent Road Safety) (EU (European Union) 026814); "Modelling and simulation approaches for livable cities" (Volvo Research and Education Foundation SP-2004-49); "Travel impacts of social networks and networking tools" (Volkswagen Stiftung I/82 714); "Numerical Last-mile Tsunami Early Warning and Evacuation Information System" (BMBF (Bundesministerium für Bildung und Forschung/Federal Ministry of Education and Research) 03FG0666E); "Adaptive Traffic Control" (BMBF 03NAPAI4); "State Estimation for traffic simulations as coarse grained systems" (DFG (Deutsche Forschungsgemeinschaft) NA 682/1-1); "Detailed assessment of transport measures using micro-simulation" (DFG NA 682/3-1); "Simulation of Multidestination Pedestrian Crowds" (DFG NA 682/5-1); "SustainCity: Micro-simulation for the prospective of sustainable cities in Europe (EU 7th Framework 244557); "Contributions of transport towards the realization of a 2000 W city" (DFG NA 682/6-1); "GRIPS (GIS-based Risk analysis, Information, and Planning System for the evacuation of areas)" (BMBF 13N11382); "MINTE (MItigating Negative Transport Externalities in industrialized and newly industrializing countries)" (DAAD (Deutscher Akademischer Austauschdienst – German Academic Exchange Service) scholarship for doctoral students), "eCab: Simulation-based system for the sustainable management of electrically powered taxi fleets" (Einstein Stiftung Berlin A-2012-132); "Optimization and network wide analysis of traffic signal control" (DFG NA 682/7-1); "MAXess: Measuring accessibilities for policy evaluation" (ERA (European Research Action – Country consortia), ERAfrica, BMBF 01DG14008); "An agent-based evolutionary approach for the user-oriented optimization of complex public transit systems" (DFG NA682/11-1).

ETH Zürich (Kay Axhausen, Milos Balac, David Charypar, Francesco Ciari, Christoph Dobler, Thibout Dubernet, Andreas Horni, Nadine Rieser, Rashid Waraich) could also draw on the following grants: "A generalized approach to population synthesis" (SNF (Schweizerischer Nationalfonds) 205121_138270 25); "Agent-based modelling of retailers and their reactions to road pricing" (ETH TH-19042); "Agent-based simulation for location-based services" (KTI (Kommission für Technologie und Innovation) 8443.1 ESPP-ES); "An investigation of strategies leading to a 2000 W City using a bottom-up model of urban energy flows" (SNF 105218-122632 1); "Assessment of the impacts of the Westumfahrung Zürich (Kanton Zürich)"; "Autonomous Cars—The next revolution in mobility" (SNF 200021_159234 43); "Choice models for transport modelling: Accounting for similarities between alternatives in large scale choice sets" (SNF 205120-121889 14); "Deriving and assessing strategies for limiting the spread of airborne diseases using a social contact model: The case of influenza" (SNF); "Destination Choice Modeling for Discretionary Activities: Fundamentals of Choice Set Formation and Impacts of Spatial Competition" (SNF 205121_132086 20); "Dynamic Traffic Self-organization in China: Network Spatial-temporal Methodology and MATSim Simulation" (SNF IZ69Z0_13113917); "Integrated modelling and analysis of energy and transport systems" (ETH TH-22 07-03); "Large-scale multi-agent simulation of travel behaviour and traffic flow" (ETH TH-7959); "Large-scale stochastic optimization for agent-based traffic simulations" (ETH TH-18951); "MAXess: Measuring accessibility in policy evaluation" (ERA, ERAfrica IZEAZ0_154310 37); "Models without (personal) data?" (SNF 200021_144134 29); "Optimising public transport: Making smart cards more useful" (SNF IZKSZ2_162185 44); "Post Car World" (SNF CRSII1_147687 21); "SCCER (Swiss Competence Center for Energy Research) Energy and Mobility" (KTI 33290); "Sharing is Saving: how collaborative mobility can reduce the impact of energy consumption for transportation" (NFP (Nationales Forschungsprogramm) 407140_153807 41); "Simulation evacuation scenarios and Schwingerfest: Evacuation study" (BABS (Bundesamt für Bevölkerungsschutz, Switzerland)); "SURPRICE (Sustainable mobility through Road User Charging)" (ERA, ERA.net); "SustainCity: Micro-simulation for the prospective of sustainable cities in Europe" (EU 7th Framework 244557); "THELMA (Technology-centered ELectric Mobility Assessment)" (CCEM (Competence Center Energy and Mobility)); "ToPDAd (Tool supported Policy Development for regional Adaptation)" (EU 7th Framework 308620); "Travel behaviour in a dynamic spatial and social context: Modelling the Interdependence of Social Network Interactions and spatial choices" (SNF 105212-112482 10) and "Travel impacts of social networks and networking tools" (Volkswagen Stiftung I/82 714).

The NRF (Singaporean National Research Foundation) together with ETH Zürich supported the work of Alexander Erath, Pieter Fourie, Sergio Ordonez Medina, Artem Chakirov and Michael Van Eggermond as part of FCL (Future Cities Laboratory).

The co-operation which funded Lun Zhang's work (Tongji University) was based on two grants (EG01-032010, NIP02-092010) of the Sino-Swiss Cooperation Project Program funded by ETH Zürich.

The work reported by Senozon AG (Michael Balmer, Marcel Rieser, Daniel Röder, Christoph Dobler and Andreas Neumann) is based on projects undertaken since it was set up in 2010, especially noteworthy are the following clients: BVG (Berliner Verkehrsbetriebe), BfS (Bundesamt für Statistik – Federal Statistical Office), Peter Vovsha, Parsons Brinckerhoff, NY, Prof. Ulrich Weidmann, Transport Systems Group (VS) of the IVT (Institut für Verkehrsplanung und Transportsysteme – Institute for Transport Planning and Systems).

University of Pretoria (Johan Joubert) was supported by grants of the South African National Treasury and the National Research Foundation grant FA2007051100019.

At RMIT (Royal Melbourne Institute of Technology) Lin Padgham and Dhirendra Singh were supported by the ARC (Australian Research Council) Discovery DP1093290, ARC Linkage LP130100008 and Telematics Trust grants. They would like to thank Agent Oriented Software for the use of the JACK BDI (Belief Desire Intention) platform.

The work of Seungjae Lee and Atizaz Ali at the University of Seoul was supported by a grant (11 High-Tech Urban G06) from High-tech Urban Development Program funded by Ministry of Land, Infrastructure and Transport of Korean government.

At the National Institute for Environmental Studies, the research of Daisuke Murakami was supported by the Environment Research and Technology Development Fund (S-10) of Japan's Ministry of the Environment.

The work on the Trondheim scenario by Stefan Flügel, Julia Kern and Frederik Bockemühl was supported by the Research Council of Norway with "Future Sustainable Transport for Industry and Trade in Norway" (208420/F40).

The work on the Santiago de Chile scenario by Benjamin Kickhöfer and Alejandro Tirachini has been supported by Chile's CONICYT (Comisión Nacional de Investigación Científica y Tecnológica – National Commission for Scientific and Technological Research) through the FONDECYT (Fondo Nacional de Desarrollo Científico y Tecnológico) Grant 11130227.

The research presented by the University of Poznan (Michal Maciejewski, Waldemar Walerjanczyk) was partially supported by the grants PBS1/A6/11/2012 and ERA-NET-TRANSPORT-III/2/2014 from the National Centre for Research and Development (Poland).

At the Universite de Liege (Mario Cools, Jacques Teller, Ismail Saadi) the work was supported by the ARC grant for Concerted Research Actions, financed by the Wallonia-Brussels Federation on "Landuse change and future flood risk: influence of micro-scale spatial patterns (FLOODLAND)".

Oleg Saprykin, Olga Saprykina and Tatyana Mikheeva were supported by the Ministry of Education and Science of the Russian Federation at Samara State Aerospace University.

Chengxiang (Tony) Zhuge (Zhejiang University, Beijing Jiaotong University) and Chunfu Shao's project "Evolution Mechanism, Regulation and Control Methods of Urban Transportation Supply and Demand Structure" was funded by the National Natural Science Foundation of China (51338008).

Sashikanth Gurram, Abdul R. Pinjari and Amy L. Stuart work at the University of South Florida and benefited from a grant by the National Science Foundation (0846342) on "Tampa, Florida: High Resolution Simulation of Urban Travel and Network Performance for Estimating Mobile Source Emissions".

The work of Maxime Lenormand at UIB (Universitat Autónoma de Barcelona) and Miguel Picornell at Nommon was in the context of a EU 7th Framework grant (EUNOIA (Evolutive User-centric Networks fOr Intraurban Accessibility), 318367).

The work for Toronto (Adam Weis, Khandker Nurul Habib, Peter Kucirek, Eric Miller, CF Shao) was funded in part by an Natural Sciences and Engineering Research Council (Canada) Discovery Grant and by the sponsors of the University of Toronto Travel Modelling Group: Metrolinx, the Ontario Ministry of Transportation, the Cities of Toronto, Hamilton, Mississauga and Brampton, and the Regional Municipalities of Durham, Halton, Peel and York.

The work at Shinshu University (Rolando Armas) is supported by the Ecudoran National Secretariat of Higher Education, Science, Technology and Innovation.

National University of Ireland Maynooth and Dublin (Gavin McArdle, Aonghus Lawlor, Eoghan Furey) were supported by the Science Foundation Ireland by a Strategic Research Cluster grant (07/SRC/I1168) under the National Development Plan.

The work at the University of Melbourne (Nicole Roland) was based on an Australian Research Council grant on "Integrating Mobility on Demand" (Linkage Project LP120200130).

Daisuke Fukuda's work at Tokyo Tech was supported by a Grant-in-Aid for Scientific Research from the Japan Society for the Promotion of Science (B) number 25289160 and by the CART (Committee on Advanced Road Technology), Ministry of Land, Infrastructure, Transport, and Tourism, Japan.

The results from Erasmus University Rotterdam (Paul Bouman, Milan Lovric) were made possible by a grant of the NYBPM (Nederlandse Organisatie voor Wetenschappelijk Onderzoek – Netherlands Organization for Scientific Research) funding the ComPuTr (Complexity in Public Transport) project.

The research leading to the results reported by UCL (University College London) (Camilo Ruiz, Joan Serras, Mike Batty, Melanie Bosredon, Vassilis Zachariadis) has received funding from Engineering and Physical Sciences Research Council of UK (United Kingdom) under grant agreement number EP/G057737/1 (SCALE project; 2009–2013), the European Union 7th Framework Programme FP7/2007–2013 under grant agreement number 318367 (EUNOIA project) and the European Research Council under grant agreement number 249393 (MECHANICITY project; 2010–2015).

The past and ongoing work at KTH (Kungliga Tekniska Högskolan – Royal Institute of Technology) Stockholm (Gunnar Flötteröd) was based on the following grants: "IHOP2: Flexible coupling of disaggregate travel demand models and network simulation packages" (TRV (Trafikverket – Swedish Transport Administration) 2015/2950); "SMART-PT: Smart public Transport" (ERA, Eranet Transport III—Future traveling, VINNOVA 2014-03976) and "PETRA (PErsonal TRansport Advisor): an integrated platform of mobility patterns for Smart Cities to enable demand-adaptive transportation system" (EU 7th Framework Program 609042). He is supported by the KTH strategic research program in transport TRENoP (Transport REsearch with Novel Perspectives).

The data sources and support which the authors obtained are too numerous to list here. Please see the original papers, theses and reports as cited in the various chapters. Special thanks go to OSM (OpenStreetMap) and their contributors, who have made the procurement of high-quality highly detailed network data much easier than it was before.

Selected Sponsors

BABS	Bundesamt für Bevölkerungsschutz – Federal Office for Civil Protection	Switzerland
BMBF	Bundesministerium für Bildung und Forschung/Federal Ministry of Education and Research	Germany
DFG	Deutsche Forschungsgemeinschaft – German Research Foundation	Germany
ERA	European Research Action	Country consortia
EU	European Union	European countries
KTI	Kommission für Technologie und Innovation/ Commission for Technology and Innovation	Switzerland
NFP	Nationales Forschungsprogramm – National Research Program	Switzerland
NRF	National Research Foundation	Singapore
NSF	National Science Foundation	USA
SNF	Schweizerischer Nationalfonds – Swiss National Research Foundation	Switzerland

Contributors

Editors

Andreas Horni
Institute for Transport Planning and Systems (IVT)
ETH Zürich
horni@senozon.com

Kay W. Axhausen
Institute for Transport Planning and Systems (IVT)
ETH Zürich
axhausen@ivt.baug.ethz.ch

Kai Nagel
Transport Systems Planning and Transport Telematics (VSP)
TU Berlin
nagel@vsp.tu-berlin.de

Authors (alphabetically)

Amit Agarwal
Transport Systems Planning and Transport Telematics (VSP)
TU Berlin
amit.agarwal.iitd@gmail.com

Hernan Aguirre
Faculty of Engineering
Shinshu University, Japan
ahernan@shinshu-u.ac.jp

Atizaz Ali
Departement of Transportation Engineering
University of Seoul
atizaz.ali@uos.ac.kr

Yalcin Alver
Department of Civil Engineering
Ege University, 35100 Bornova, Izmir, Turkey
yalcin.alver@ege.edu.tr

Rolando Armas
Faculty of Engineering
Shinshu University, Japan
rolando.armas@iplab.shinshu-u.ac.jp

Milos Balac
Institute for Transport Planning and Systems (IVT)
ETH Zürich
milos.balac@ivt.baug.ethz.ch

Michael Balmer
Senozon AG
balmer@senozon.com

Mike Batty
Centre for Advanced Spatial Analysis (CASA)
University College London
m.batty@ucl.ac.uk

Gian Ricardo Berkenbrock
Software/Hardware Integration Lab (LISHA)
Universidade Federal de Santa Catarina (UFSC) Joinville
gian.rb@ufsc.br

Davi Guggisberg Bicudo
Universidade Federal de Santa Catarina (UFSC) Joinville
davi.bicudo@me.com

Joschka Bischoff
Transport Systems Planning and Transport Telematics (VSP)
TU Berlin
bischoff@vsp.tu-berlin.de

Frederik Bockemühl
Master's student at Hasselt University
frederik.bockemuhl@student.uhasselt.be

Patrick M. Bösch
Institute for Transport Planning and Systems (IVT)
ETH Zürich
boesch@ivt.baug.ethz.ch

Melanie Bosredon
Centre for Advanced Spatial Analysis (CASA)
University College London
m.bosredon.11@ucl.ac.uk

Paul Bouman
Department of Technology and Operations Management
Rotterdam School of Management (RSM)
Erasmus University Rotterdam
research@pcbouman.nl

Andrew Campbell
CEE Systems and Transportation
University of California, Berkeley
andrew.campbell@berkeley.edu

Artem Chakirov
Future Cities Laboratory
Singapore-ETH Centre
chakirov@ivt.baug.ethz.ch

David Charypar
Institute for Transport Planning and Systems (IVT)
ETH Zürich
dcharypar@gmail.com

Francesco Ciari
Institute for Transport Planning and Systems (IVT)
ETH Zürich
ciari@ivt.baug.ethz.ch

Mario Cools
Local Environment Management & Analysis (LEMA)
University of Liège
mario.cools@ulg.ac.be

Dhirendra Singh
School of Computer Science and I.T.
RMIT University, Melbourne, Australia
dhirendra.singh@rmit.edu.au

Christoph Dobler
Senozon AG
dobler@senozon.com

Thibaut Dubernet
Institute for Transport Planning and Systems (IVT)
ETH Zürich
thibaut.dubernet@ivt.baug.ethz.ch

Alexander Erath
Future Cities Laboratory
Singapore-ETH Centre
erath@ivt.baug.ethz.ch

Sidney Feygin
CEE Systems and Transportation
University of California, Berkeley
sid.feygin@berkeley.edu

Gunnar Flötteröd
Department of Transport Science
KTH Royal Institute of Technology
gunnar.floetteroed@abe.kth.se

Stefan Flügel
Institute of Transport Economics
Norwegian Centre for Transport Research
stefan.flugel@toi.no

Pieter Fourie
Future Cities Laboratory
Singapore-ETH Centre
fourie@ivt.baug.ethz.ch

Daisuke Fukuda
Department of Civil Engineering
Tokyo Institute of Technology
fukuda@plan.cv.titech.ac.jp

Eoghan Furey
National Centre for Geocomputation
NUI Maynooth
eoghan.furey@nuim.ie

Dominik Grether
Transport Systems Planning and Transport Telematics (VSP)
TU Berlin
dominik.grether@alumni.tu-berlin.de

Sashikanth Gurram
Department of Civil & Environmental Engineering
University of South Florida
sgurram@mail.usf.edu

Khandker M. Nurul Habib
Department of Civil Engineering
University of Toronto
khandker.nurulhabib@utoronto.ca

Walter J. Hernández B.
Centro de Computación Gráfica
Universidad Central de Venezuela, Caracas
walter.hernandez@ciens.ucv.ve

Johannes Illenberger
Transport Network Development and Transport Models (GSV)
DB Mobility Logistics AG
johannes.illenberger@deutschebahn.com

Johan W. Joubert
Department of Industrial and Systems Engineering
University of Pretoria
johan.joubert@up.ac.za

Julia Kern
Mathematical Optimization and Scientific Information
Zuse Institute Berlin
kern@zib.de

Benjamin Kickhöfer
Transport Systems Planning and Transport Telematics (VSP)
TU Berlin
kickhoefer@vsp.tu-berlin.de

Hubert Klüpfel
Maleto
hubert@maleto.de

Peter Kucirek
TMG Travel Modelling Group, Toronto
peter.kucirek@alum.utoronto.ca

Gregor Lämmel
Institute for Advanced Simulation (IAS)
Forschungszentrum Jülich GmbH
g.laemmel@fz-juelich.de

Aonghus Lawlor
Insight Centre for Data Analytics
University College Dublin
aonghus.lawlor@insight-centre.org

Seungjae Lee
Departement of Transportation Engineering
University of Seoul
sjlee@uos.ac.kr

Maxime Lenormand
Instituto de Fisica Interdisciplinar y Sistemas Complejos (IFISC)
Campus Universitat de les Illes Balears
maxime@ifisc.uib-csic.es

Milan Lovric
Department of Technology and Operations Management
Rotterdam School of Management (RSM) Erasmus University Rotterdam
lovric.milan@gmail.com

Jiangshan Ma
Shanghai Maritime University
tonny.achilles@gmail.com

Michal Maciejewski
Division of Transport Systems
Poznan University of Technology
michal.maciejewski@put.poznan.pl

Gavin McArdle
National Centre for Geocomputation
Maynooth University
Gavin.McArdle@nuim.ie

Tatyana Mikheeva
Department of Transportation Organization and Management
Samara State Aerospace University, Samara, Russia
mikheevati@its-spc.ru

Sudatta Mohanty
CEE Systems and Transportation
University of California, Berkeley
sudatta.mohanty@berkeley.edu

Daisuke Murakami
Center for Global Environmental Research
National Institute for Environmental Studies, 16-2, Onogawa, Tsukuba, Ibaraki, 305-8506, Japan
murakami.daisuke@nies.go.jp

Mehmet Metin Mutlu
Department of Civil Engineering
Ege University, 35100 Bornova, Izmir, Turkey
mmetinm@gmail.com

Héctor E. Navarro U.
Centro de Computación Gráfica
Universidad Central de Venezuela, Caracas
hector.navarro@ciens.ucv.ve

Andreas Neumann
Senozon Deutschland GmbH
earlier: Transport Systems Planning and Transport Telematics (VSP)
TU Berlin
neumann@senozon.de

Pelin Onelcin
Department of Civil Engineering
Ege University, 35100 Bornova, Izmir, Turkey
pelin.onelcin@ege.edu.tr

Sergio Arturo Ordóñez Medina
Future Cities Laboratory
Singapore-ETH Centre
ordonez@ivt.baug.ethz.ch

Lin Padgham
School of Computer Science and I.T.
RMIT University, Melbourne, Australia
lin.padgham@rmit.edu.au

Miguel Picornell
Nommon Solutions and Technologies
miguel.picornell@nommon.es

Abdul R. Pinjari
Department of Civil & Environmental Engineering
University of South Florida
apinjari@usf.edu

Alexei Pozdnoukhov
CEE Systems and Transportation
University of California, Berkeley
alexeip@berkeley.edu

Marcel Rieser
Senozon AG
rieser@senozon.com

Nadine Rieser-Schüssler
Ernst Basler + Partner AG
earlier: Institute for Transport Planning and Systems (IVT), ETH Zürich
nadine.rieser@ebp.ch

Daniel Röder
Senozon Deutschland GmbH
roeder@senozon.de

Nicole Ronald
Department of Infrastructure Engineering
University of Melbourne
nicole.ronald@unimelb.edu.au

Ismaïl Saadi
Local Environment Management & Analysis (LEMA)
University of Liège
ismail.saadi@ulg.ac.be

Oleg Saprykin
Department of Transportation Organization and Management
Samara State Aerospace University, Samara, Russia
saprykinon@gmail.com

Olga Saprykina
Department of Transportation Organization and Management
Samara State Aerospace University, Samara, Russia
olga_grineva_@mail.ru

Joan Serras
Centre for Advanced Spatial Analysis (CASA)
University College London
j.serras@ucl.ac.uk

Hajime Seya
Graduate School for International Development and Cooperation
Hiroshima University
hseya@hiroshima-u.ac.jp

Chunfu Shao
School of Traffic and Transportation
Beijing Jiaotong University, Beijing, China
cfshao@bjtu.edu.cn

Norihito Shinkai
Regional Futures Research Center Co. Ltd.
shinkai@refrec.jp

David Strippgen
Interactive Systems & Game Technologies
Hochschule für Technik und Wirtschaft (HTW)
david.strippgen@htw-berlin.de

Amy L. Stuart
Department of Civil & Environmental Engineering and Department of Environmental & Occupational Health
University of South Florida
astuart@health.usf.edu

Jacques Teller
Local Environment Management & Analysis (LEMA)
University of Liège
Jacques.Teller@ulg.ac.be

Theresa Thunig
Transport Systems Planning and Transport Telematics (VSP)
TU Berlin
thunig@vsp.tu-berlin.de

Alejandro Tirachini
Transport Engineering Division, Civil Engineering Department
Universidad de Chile
alejandro.tirachini@ing.uchile.cl

Camilo Vargas-Ruiz
Centre for Advanced Spatial Analysis (CASA)
University College London
camilo.ruiz@ucl.ac.uk

Waldemar Walerjanczyk
Division of Transport Systems
Poznan University of Technology
waldemar.walerjanczyk@put.poznan.pl

Rashid A. Waraich
Institute for Transport Planning and Systems (IVT)
ETH Zürich
waraich@ivt.baug.ethz.ch

Adam Weiss
Department of Civil Engineering
University of Toronto
adam.weiss@utoronto.ca

Kaoru Yamada
Oriental Consultants Global Co. Ltd.
yamada-kr@oriconsul.com

Yoshiki Yamagata
Center for Global Environmental Research
National Institute for Environmental Studies, 16-2, Onogawa, Tsukuba, Ibaraki, 305-8506, Japan
yamagata@nies.go.jp

Elvira B. Yaneza
College of Computer Studies
Xavier University-Ateneo de Cagayan de Oro City, Philippines
eyaneza@xu.edu.ph

Mogeng Yin
CEE Systems and Transportation
University of California, Berkeley
mogengyin@berkeley.edu

Vassilis Zachariadis
Centre for Advanced Spatial Analysis (CASA)
University College London
v.zachariadis@ucl.ac.uk

Lun Zhang
Transport Information Engineering
Tongji University Shanghai, China
lun_zhang@tongji.edu.cn

Chengxiang Zhuge
Department of Geography
University of Cambridge
earlier: School of Traffic and Transportation, Beijing Jiaotong University, Beijing, China
cz293@cam.ac.uk

Dominik Ziemke
Transport Systems Planning and Transport Telematics (VSP)
TU Berlin
ziemke@vsp.tu-berlin.de

Michael Zilske
Transport Systems Planning and Transport Telematics (VSP)
TU Berlin
zilske@vsp.tu-berlin.de

Copy-Editing

Karen Ettlin
karen.ettlin@datazug.ch

Introduction

The book is intended to give new MATSim users a quick start in running MATSim. It also provides more experienced MATSim users and MATSim developers with information on how to extend MATSim by plugging in available modules (e.g., the contributions), or by programming against the MATSim API (Application Programming Interface) to implement their own MATSim extensions. Another of this book's goals is to contextualize the methods used in MATSim within a broader theoretical background. By compiling our conceptual insights on MATSim gained over the years, the book also contributes to methodological discussions on joint microsimulation of travel demand and traffic flow, a relatively new field, or, more generally, spatial demand and its congestion generation.

The book is divided into four parts, focused on *using* (Part I), *extending* (Part II), and *understanding* (Part III) MATSim, while simultaneously providing practical, technical, and methodological information. The last part of the book (Part IV) then presents an array of MATSim scenarios that have been created around the world.

Part I: Using MATSim

This part enables users to run MATSim with only the config file, a population and a network. They are given general information to assess whether MATSim is a suitable tool and method for their specific research question.

Chapter 1 introduces the MATSim basics, including its underlying co-evolutionary principle and its traffic flow model. Chapter 2 shows the MATSim novice how to set up and run a basic MATSim scenario. Scoring is central to MATSim; a full chapter, Chapter 3, scrutinizes scoring. Chapter 4 lists the config file options available for basic scenarios containing config file, a population and a network.

Part II: Extending MATSim

This part presents technical information on how to extend the base functionality of MATSim by additional input data beyond config file, population and network, as well as by programming against the API.

Chapter 5 introduces MATSim's modular architecture. It also explains how to use the available modules introduced in Chapters 6 through 42. Chapter 43 describes modules that were important in the past but whose development was discontinued. Chapter 44 briefly describes MATSim organization, i.e., its development process, code structure, the team and the community, and summarizes their development tools. Chapter 45 goes one step further and explains to readers how to write their own MATSim extensions, and how to then contribute them to MATSim, including details about points where MATSim can be extended; it also digs a bit deeper and provides details about the very central MATSim concept of events. Explanations about how to inject alternative or additional modules and how in general to write MATSim scripts in Java is also found here.

Part III: Understanding MATSim

This part presents theoretical aspects underlying the previous two parts. For example, the MATSim score is no longer simply denoted by S without interpretation, but is here contextualized within the discrete choice framework (Chapter 49) and becomes related to utility, commonly denoted by U. The first chapter, Chapter 46 starts with a summary of MATSim's history, written by Kai Nagel and Kay W. Axhausen. Chapter 47 then elaborates on agent-based traffic assignment and qualitatively contextualizes MATSim within classical concepts. Here, the focus is on development from static to dynamic traffic assignment and, finally, agent-based traffic assignment. Chapter 48 quantitatively contextualizes MATSim within classical concepts by presenting it as a fundamentally stochastic tool, based on random distributions and understandable as a Monte Carlo engine. Chapter 50 analyzes MATSim's traffic flow model in relation to kinematic waves, while Chapter 51 provides an economic view on MATSim.

Part IV: Scenarios

At this point, when readers have a complete picture of MATSim and are ready to set up their own real-world MATSim scenario, Chapters 52 through 96 show them the numerous and highly varied scenarios that have been implemented around the world.

The book concludes with a discussion of promising research avenues (Chapter 97).

Related Material

The book concentrates on the more stable aspects of MATSim application and development. In the future, revisions of Chapters 1 to 5 will be presented once a year. Additional material is referenced from `http://matsim.org`, for example under `http://matsim.org/docs`, `http://matsim.org/javadoc`, `http://matsim.org/extensions`, `http://matsim.org/faq`, or `http://matsim.org/issuetracker`.

PART I

Using MATSim

CHAPTER 1

Introducing MATSim

Andreas Horni, Kai Nagel and Kay W. Axhausen

1.1 The Beginnings

The MATSim project (MATSim, 2016) started with Kai Nagel, then at ETH Zürich, and his interest in improving his work with, and for, the TRANSIMS (TRansportation ANalysis and SIMulation System) project (Smith et al., 1995; FHWA, 2013); he also wanted to make the resulting code open-source.[1] After Kai Nagel's departure to Berlin in 2004, Kay W. Axhausen joined the team, bringing a different approach and experience. A collaboration, successful and productive for more than 10 years, was thus established, combining a physicist's and a civil engineer's perspective, as well as bringing together expertise in traffic flow, large-scale computation, choice modeling and CAS (Complex Adaptive Systems):

- **Microscopic modeling of traffic:** MATSim performs integral microscopic *simulation of resulting traffic flows* and the congestion they produce (see Section 1.3).
- **Microscopic behavioral modeling of demand/agent-based modeling:** MATSim uses a microscopic description of demand by *tracing the daily schedule* and the synthetic travelers' decisions. In retrospect, this can be called "agent-based".
- **Computational physics:** MATSim performs fast microscopic simulations with 10^7 or more "particles".
- **Complex adaptive systems/co-evolutionary algorithms:** MATSim *optimizes the experienced utilities* of the whole schedule through the co-evolutionary search for the resulting equilibrium or steady state (see Section 1.4).

[1] TRANSIMS has, since then, also become open-source (TRANSIMS Open Source, 2013); but in 2000, it was difficult to procure in Europe.

How to cite this book chapter:
Horni, A, Nagel, K and Axhausen, K W. 2016. Introducing MATSim. In: Horni, A, Nagel, K and Axhausen, K W. (eds.) *The Multi-Agent Transport Simulation MATSim*, Pp. 3–8. London: Ubiquity Press. DOI: http://dx.doi.org/10.5334/baw.1.

At the end of the 1990s, the scene was set for these research streams' mergence into a computationally efficient, modular, open-source software enabling further development on travel behavior, network response and efficient computation: MATSim.

1.2 In Brief

MATSim is an activity-based, extendable, multi-agent simulation framework implemented in Java. It is open-source and can be downloaded from the Internet (MATSim, 2016; GitHub, 2015). The framework is designed for large-scale scenarios, meaning that all models' features are stripped down to efficiently handle the targeted functionality; parallelization has also been very important (e.g., Dobler and Axhausen, 2011; Charypar, 2008). For the network loading simulation, for example, a queue-based model is implemented, omitting very complex and computationally expensive car-following behavior (see Section 1.3).

At this time, MATSim is designed to model a *single day*, the common unit of analysis for activity-based models (see, for example, the review by Bowman, 2009a). Nevertheless, in principle, a multi-day model could be implemented (Horni and Axhausen, 2012b).

As shown in Section 1.4, MATSim is based on the co-evolutionary principle. Every agent repeatedly optimizes its daily activity schedule while in competition for space-time slots with all other agents on the transportation infrastructure. This is somewhat similar to the route assignment iterative cycle, but goes beyond route assignment by incorporating other choice dimensions like time choice (Balmer et al., 2005b), mode choice (Grether et al., 2009), or destination choice (Horni et al., 2012b) into the iterative loop.

A MATSim run contains a configurable number of iterations, represented by the loop of Figure 1.1 and detailed below. It starts with an initial demand arising from the study area population's daily activity chains. The modeled persons are called agents in MATSim. Activity chains are usually derived from empirical data through sampling or discrete choice modeling. A variety of approaches is suitable, as evidenced in the scenarios' chapters (cf. Chapter 52). During iterations, this initial demand is optimized individually by each agent. Every agent possesses a memory containing a fixed number of day plans, where each plan is composed of a daily activity chain and an associated score. The score can be interpreted as an econometric utility (cf. Chapter 51).

In every iteration, prior to the simulation of the network loading with the MATSim *mobsim (mobility simulation)* (e.g., Cetin, 2005), each agent selects a plan from its memory. This selection is dependent on the plan *scores*, which are computed after each mobsim run, based on the executed plans' performances. A certain share of the agents (often 10 %) are allowed to clone the selected plan and modify this clone (*replanning*). For the network loading step, multiple mobsims are available and configurable (see Horni et al., 2011b, and Section 4.3 of this book).

Plan modification is performed by the *replanning* modules. Four dimensions are usually considered for MATSim at this time: departure time (and, implicitly, activity duration) (Balmer et al.,

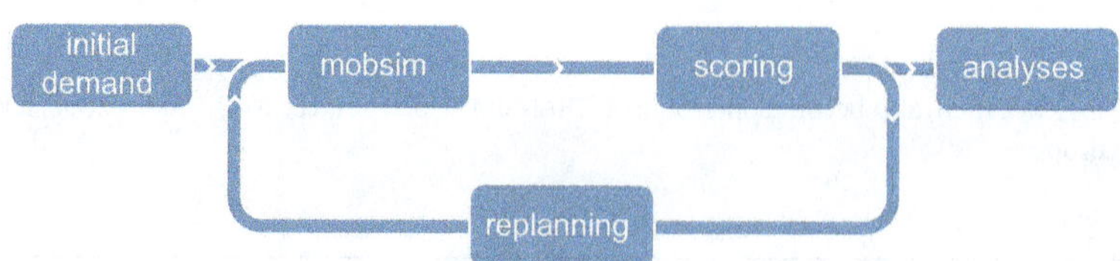

Figure 1.1: MATSim loop, sometimes called the MATSim cycle.

2005b), route (Lefebvre and Balmer, 2007), mode (Grether et al., 2009) and destination (Horni et al., 2009, 2012b). Further dimensions, such as activity adding or dropping, or parking and group choices are currently under development and only available experimentally. MATSim replanning offers different strategies to adapt plans, ranging from random mutation to approximate suggestions, to best-response answers where, in every iteration, the currently optimal choice is searched. For example, routing often is a best-response modification, while time and mode replanning are random mutations.

Initial day chains do not have to be very carefully defined for the replanning dimensions included in the optimization process. Plausible values just speed up the optimization process.

If an agent ends up with too many plans (configurable), the plan with the lowest score (configurable) is removed from the agent's memory. Agents that have not undergone replanning select between existing plans. The selection model is configurable; in many MATSim investigations, a model generating a logit distribution for plan selection is used.

An iteration is completed by evaluating the agents' experiences with the selected day plans (*scoring*). The applied scoring function is described in detail in Chapter 3.

The iterative process is repeated until the average population score stabilizes. The typical score development curve (Figure 1.2, taken from Horni et al., 2009) takes the form of an evolutionary optimization progress (Eiben and Smith, 2003, Figure 2.5). Since the simulations are stochastic, one cannot use convergence criteria appropriate for deterministic algorithms; for a discussion of possible approaches for the MATSim situation, see Sections 47.3.2.2 and 48.2 as well as Meister (2011).

MATSim offers considerable customizability through its modular design. Although implementing alternative core modules, such as an alternative network loading simulation, may entail substantial effort, in principle, every module of the framework can be exchanged. MATSim modules are described in Chapter 5 and following.

MATSim is strongly based on events stemming from the mobsim. Every action in the simulation generates an event, which is recorded for analysis. These event records can be aggregated to evaluate any measure at the desired resolution. The event architecture is detailed in Section 45.2.5.

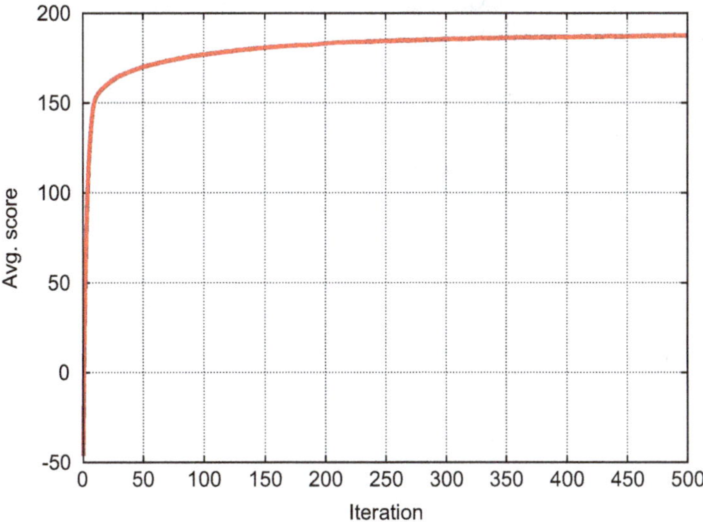

Figure 1.2: Typical score progress.

1.3 MATSim's Traffic Flow Model

MATSim provides two internal mobsims: QSim and JDEQSim (Java Discrete Event Queue Simulation); in addition, external mobility simulations can be plugged in. Some years ago, the DEQSim written in C++ and described by Charypar (2008); Charypar et al. (2007b,a, 2009) was plugged into MATSim and frequently used. The multi-threaded QSim is currently the default mobsim.

Charypar et al. (2009) distinguishes between

- physical simulations, featuring detailed car following models,
- cellular automata, in which roads are discretized into cells,
- queue-based simulations, where traffic dynamics are modeled with waiting queues,
- mesoscopic models, using aggregates to determine travel speeds, and
- macroscopic models, based on flows rather than single traveler units (e.g., cars).

As MATSim is designed for large-scale scenarios, it adopts the computationally efficient queue-based approach (see Figure 1.3). A car entering a network link (i.e., a road segment) from an intersection is added to the tail of the waiting queue. It remains there until the time for traveling the link with free flow has passed and until he or she is at the head of the waiting queue and until the next link allows entering. The approach is very efficient, but clearly it comes at the price of reduced resolution, i.e., car following effects are not captured. In JDEQSim, for computational reasons, the waiting-queue approach is combined with an event-based update step (Charypar et al., 2009). In other words, there is no time-step-based updating process of any agent in the scenario. Instead agents are only touched if they actually require an action. For example, links do not have to be processed while agents traverse them. Update events triggering is managed by a global scheduler. QSim, however, is time-step based. The MATSim traffic flow model is strongly based on the two link attributes: storage capacity and flow capacity. Storage capacity defines the number of cars fitting onto a network link.

Flow capacity specifies the outflow capacity of a link, i.e., how many travelers can leave the respective link per time step. It is an individual attribute of the link. The current implementation of QSim has no *maximum* inflow capacity specified. In contrast, in the earlier DEQSim and current JDEQSim, an inflow capacity can also be specified, which may move jams at merges from the end of the first common link, where the QSim generates them, upstream to where the links merge and where they plausibly should be (Charypar, 2008, p. 99). However, additional data is needed for this, which is often not available.

This basic traffic flow model has been extended with various modules: Signals and multiple lane modeling have been added (Chapter 12), backward-moving gaps, as investigated by Charypar (2008), are included in JDEQSim, but only available on an *experimental basis* for QSim (Section 97.5). Interactions between different modes are described in Section 4.6 and Chapter 21.

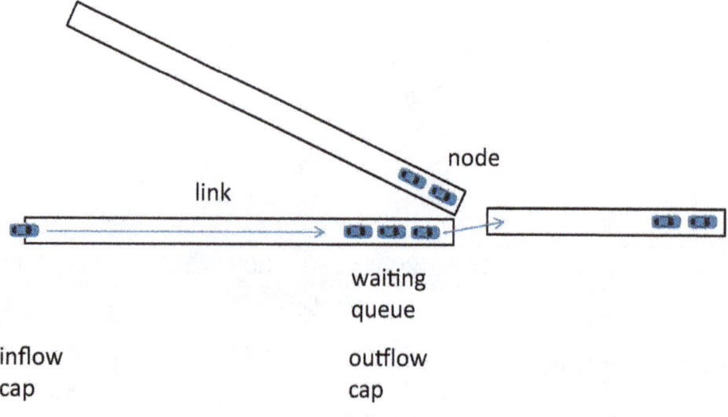

Figure 1.3: Traffic flow model.

1.4 MATSim's Co-Evolutionary Algorithm

As illustrated in Figure 1.4, the MATSim equilibrium is searched for by a *co-evolutionary algorithm* (see, e.g., Popovici et al., 2012). These algorithms co-evolve different species subject to interaction (e.g., competition). In MATSim, individuals are represented by their plans, where a person represents a species. With the co-evolutionary algorithm, optimization is performed in terms of agents' plans, i.e., across the whole daily plan of activities and travel. It achieves more than the standard traffic flow equilibria, which ignores activities. Eventually, an equilibrium is reached, subject to constraints, where the agents cannot further improve their plans unilaterally.

Note that there is a difference between the application of an evolutionary algorithm and a *co*-evolutionary algorithm. An evolutionary algorithm would lead to a system optimum, as optimization is applied with a global (or population) fitness function. Instead, the co-evolutionary algorithm leads to a (stochastic) user equilibrium, as optimization is performed in terms of *individual* scoring functions and within an agent's set of plans.

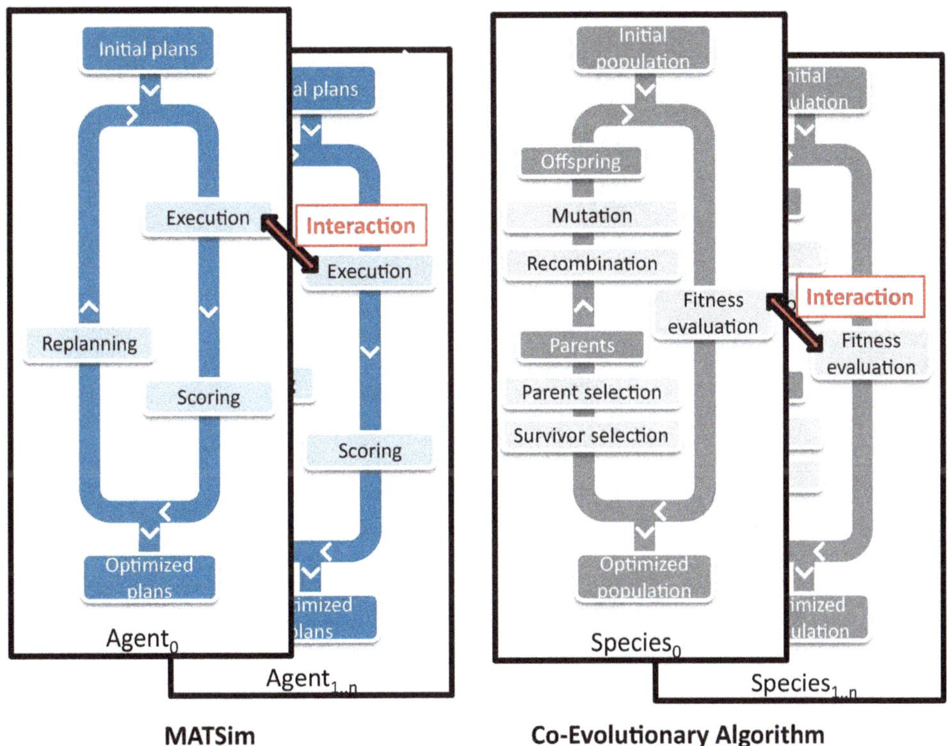

Figure 1.4: The co-evolutionary algorithm in MATSim.

CHAPTER 2

Let's Get Started

Marcel Rieser, Andreas Horni and Kai Nagel

This chapter explains how to set up and run MATSim and describes the requirements for building a basic scenario. Updated information may be available from `http://matsim.org`, in particular from `http://matsim.org/docs`.

Getting the source code into different computing environments and extending MATSim through the API is described in Part II, Chapter 45.

2.1 Running MATSim

2.1.1 Setting Up MATSim

To run MATSim, you must install the Java SE (Java Standard Edition) that complies with the appropriate MATSim version. At this time, this is Java SE 7.

Download of the release You also need the official *MATSim release*, a zip file (usually designated with the version number `matsim-yy.yy.yy.zip`), that includes everything required to run it. It can be downloaded following the "release" link under `http://matsim.org/downloads`. Unzip results in the **MATSim directory tree**. Continue with Section 2.1.2.

The MATSim directory tree on the web If you want to look at the development version, or look at things without downloading and installing a zip file: On GitHub, the root of the **MATSim directory tree** (i.e., excluding so-called contribs and playgrounds) is at `https://github.com/matsim-org/matsim/tree/master/matsim`.

How to cite this book chapter:
Rieser, M, Horni, A and Nagel, K. 2016. Let's Get Started. In: Horni, A, Nagel, K and Axhausen, K W. (eds.) *The Multi-Agent Transport Simulation MATSim*, Pp. 9–22. London: Ubiquity Press. DOI: http://dx.doi.org/10.5334/baw.2.

Download of nightly builds If you prefer to use the more up-to-date, but less stable, *nightly builds*, you should download, via the same URL (Uniform Resource Locator) `http://matsim.org/downloads`,

- the MATSim JAR (Java ARchive) file (usually tagged with the revision number `MATSim_ryyyy.jar`), and
- the required external libraries (`MATSim_libs.zip`). Unzipping this collection of 3rd-party libraries, you should then get a directory `libs`, with several JAR files inside. If the directory `libs` is in the same directory as the MATSim JAR file, the libraries are found automatically and do not have to be added manually to the `classpath`.

Maven A relatively new feature is that one can use MATSim as an Apache Maven plugin; both release versions and snapshots are available. See again `http://matsim.org/downloads` for more information. For someone who has used Apache Maven before, this is probably the best option. In this case, one may use the simple Java programming approach of Section 5.1.1.4 to get started.

2.1.2 *Running MATSim*

When this book was written, only the nightly built MATSim JAR file could be started by double-clicking. A minimal GUI (Graphical User Interface), as shown in Figure 2.1, opens and the MATSim run can be configured and started. This feature will appear in the releases, starting with version 0.8.

For the release 0.7, MATSim does not provide a GUI; thus, you must be able to handle and access a command line tool. In Linux or Mac OS X, this is typically a Terminal application; in Microsoft Windows, the Power Shell or Command Prompt. At the command prompt type the following command in one line, but substitute the correct paths:

On Linux or Mac OS X, something like:

```
java -Xmx512m -cp /path/to/matsim.jar org.matsim.run.Controler /path/
   to/config.xml
```

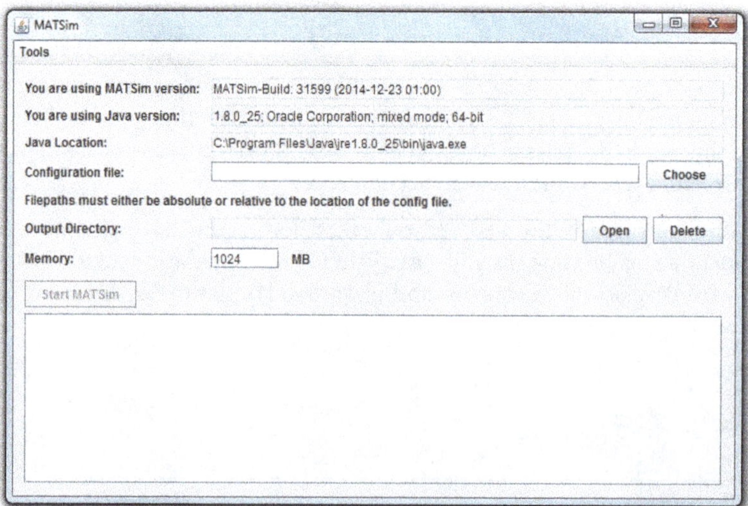

Figure 2.1: Minimal MATSim GUI.

On Windows, an example command could be:

```
java -Xmx512m -cp C:\MATSim\matsim.jar org.matsim.run.Controler
   C:\MATSim\input\config.xml
```

Such a command consists of multiple parts:

- `java` tells the system that you want to run Java.
- `-Xmx512m` tells Java that it should use up to 512 MB (Megabyte) of memory. This is typically enough to run the small examples. For larger scenarios, you might need more memory, e.g., `-Xmx3g` would allow Java to use up to 3 GB (Gigabyte) of RAM (Random Access Memory).
- `-cp /path/to/matsim.jar` tells Java where to find the MATSim code.
- `org.matsim.run.Controler` specifies which class (think of an "entry point") should be run. In most cases, the default MATSim `Controler` is the class you will need to run simulations.
- `/path/to/config.xml` tells MATSim which config file is to be used.

2.1.3 Configuring MATSim

MATSim is configured in the config file, building the connection between the user and MATSim and containing a settings list that influences how the simulation behaves.

All configuration parameters are simple pairs of a parameter `name` and a parameter `value`. The parameters are grouped into logical groups; one group has settings related to the `Controler`, like the number of iterations, or another group has settings for the mobsim, e.g., end time of the mobsim. As shown in Chapter 5, numerous MATSim modules can be added to MATSim and configured by specifying the respective configuration file section.

The list of available parameters and valid parameter values may vary from release to release. Although we try to keep this stable, software changes, mainly new features, may cause settings to change. For a list of all available settings available with the version you are working with, run the following command:

```
java -cp /path/to/matsim.jar org.matsim.run.CreateFullConfig fullConfig.xml
```

This command will create a new config file `fullConfig.xml`, containing all available parameters, along with their default values and often an explanatory comment, making it easy to see what settings are available. To use and modify specific settings, lines with their corresponding parameters can be copied to the config file, specific to the scenario to be simulated, and the parameter values can be modified in that file. See `http://matsim.org/javadoc` → main distribution → `CreateFullConfig` for more information.

A fairly minimal config file contains the following information:

```
<module name="network">
   <param name="inputNetworkFile" value="<path-to-network-file>" />
</module>

<module name="plans">
   <param name="inputPlansFile" value="<path-to-plans-file" />
</module>

<module name="controler">
   <param name="firstIteration" value="0" />
   <param name="lastIteration" value="0" />
</module>

<module name="planCalcScore" >
   <parameterset type="activityParams" >
```

```
        <param name="activityType" value="h" />
        <param name="typicalDuration" value="12:00:00" />
    </parameterset>
    <parameterset type="activityParams" >
        <param name="activityType" value="w" />
        <param name="typicalDuration" value="08:00:00" />
    </parameterset>
</module>
```

For a working example, see the MATSim directory tree (cf. 2.1.1) under `examples/tutorial/config/example1-config.xml`.

In the example, supply is provided by the network and demand by the plans file. Typical input data is described in Section 2.2.2. The specification that the first and last iteration are the same, means that no replanning of the demand is performed. What *is* executed is the mobsim (Figure 1.1), followed by each executed plan's performance scoring. To function, the scoring needs to know, from the config file, all activity types used in the plans and the typical duration for each activity type.

Further configuration possibilities are described in Chapter 4.

2.2 Building and Running a Basic Scenario

This section provides information on typical input data files used for a MATSim experiment, as well as the standard output files generated. It presents a minimal example scenario and briefly explains units, conventions and coordinate systems used in MATSim. Then, hints on practical data requirements are provided.

2.2.1 Units, Conventions, and Coordinate Systems

2.2.1.1 Units

MATSim tries to make few assumptions about actual units, but it is sometimes necessary for certain estimates. In general, MATSim expects similar types of variables (e.g., all distances) to be in the same unit wherever they are used. In the following short overview, the most important (expected) units are listed.

Distance Distance units are for example used in links' length. They should be specified in the same unit the coordinate system uses, allowing MATSim to calculate beeline distances. As the much used UTM (Universal Transverse Mercator) projected coordinate systems (see Section 2.2.1.3) use meters as the unit of distance, this is the most commonly used distance unit in MATSim.

Time MATSim supports an hour:minute:second notation in several places, but internally, it uses seconds as the default time unit. This implies, for example, that link speeds must be specified in distance per second, typically meters per second. One notable exception to this rule are scoring parameters, where MATSim expects values per hour.

Money Money is unit-free. Units are implicitly given by the marginal utility of money (cf. Equation (3.4) below). Thus, when one moves from Germany to Switzerland, the parameter β_c must be changed from "utility per Euro" to "utility per Swiss Franc".

2.2.1.2 Conventions

MATSim uses IDs intensely. These identifiers can be arbitrary strings, with the following exceptions: IDs should not contain any whitespace characters (incl. tabs, new lines, etc.) or commas, semicolons, etc., because those characters are typically used for separating different IDs from each other on IDs lists.

2.2.1.3 *Coordinate Systems*

Preparing Your Data in the Appropriate Coordinate System In several input files, you need to specify coordinates, e.g., for network nodes. We strongly advise not to use WGS84 coordinates (i.e., GPS (Global Positioning System) coordinates), or any other spherical coordinates (coordinates ranging from −180 to +180 in west-east direction and from −90 to +90 in south-north direction). MATSim has to calculate distances between two points in several sections of the code. Calculation of distances between spherical coordinates is very complex and potentially slow. Instead, MATSim uses the simple Pythagoras theorem, but this requires Cartesian coordinate system coordinates. Thus, we emphatically recommend using a Cartesian coordinate system along with MATSim, preferably one where the distance unit corresponds to one meter.

Many countries and regions have custom coordinate systems defined, optimized for local usage. It might be best to ask GIS (Geographic Information System) specialists in your region of interest for the most commonly used coordinate system there and use that for your data.

If you have no information about what coordinate system is used in your region, it might be best to use the UTM coordinate system. This system divides the world into multiple bands, each six degrees wide, and separated into a northern and southern part, which it calls UTM zones. For each zone, an optimized coordinate system is defined. Choose the UTM zone for your region (Wikipedia has a good map showing the zones) and use its coordinate system.

Telling MATSim About Your Coordinate System For some operations, MATSim must know the coordinate system where your data is located. For example, some analyses may create output to be visualized in Google Earth or by QGIS (Quantum GIS). The coordinate system used by your data can be specified in the config file:

```
<module name="global">
   <param name="coordinateSystem" value="EPSG:32608" />
</module>
```

This allows MATSim to work with your coordinates and convert them whenever needed.

You have multiple ways to specify the coordinate system you use. The easiest one is to use the so-called "EPSG (European Petroleum Survey Group) codes". Most of the commonly used coordinate systems have been standardized and numbered. The EPSG code identifies a coordinate system and can be directly used by MATSim. To find the correct EPSG code for your coordinate system (e.g., for one of the UTM zones), the website `http://www.spatialreference.org` is extremely useful. Search on this website for your coordinate system, e.g., for "WGS 84 / UTM Zone 8N" (for the northern-hemisphere UTM Zone 8), to find a list of matching coordinate systems along with their EPSG codes (in this case `EPSG:32608`).

As an alternative, MATSim can also parse the description of a coordinate system in the WKT (Well-Known Text) format.

2.2.2 *Typical Input Data*

Minimally, MATSim needs the files

- `config.xml`, containing the configuration options for MATSim and presented above in Section 2.1.3,
- `network.xml`, with the description of the (road) network, and
- `population.xml`, providing information about travel demand, i.e., list of agents and their day plans.

Thus, `population.xml` and `network.xml` might get quite large. To save space, MATSim supports reading and writing data in a compressed format. MATSim uses GZIP-compression for this. Thus, many file names have the additional suffix `.gz`, as in `population.xml.gz`. MATSim acknowledges whether files are compressed, or should be written compressed, based on file name.

2.2.2.1 *An Outlook on Extending MATSim in Part II of this Book*

Chapter 7 provides some information about MATSim's technical tools for initial input generation. With the basic setting, MATSim agents perform their activities on a specific link. If further information about activity locations needs to be specified, this can be carried out with facilities described in Section 6.4. Further, for the *simulation* of public transport, the base scenario must be extended by additional files as shown in Section 16.4.1 and Chapter 16. Count data are a common evaluation measure in transport planning. In MATSim, count data can be provided for the simulation, as shown in Section 6.3.

In more detail, the network and population files resemble the following; for the config file, see Section 2.1.3 above.

2.2.2.2 `network.xml`

Network is the infrastructure on which agents (or vehicles) can move around. The network consists of nodes and links (in graph theory, typically called vertices and edges). A simple network description in MATSim's XML (Extensible Markup Language) data format could contain approximately the following information:

```
<network name="example network">
    <nodes>
        <node id="1" x="0.0" y="0.0"/>
        <node id="2" x="1000.0" y="0.0"/>
        <node id="3" x="1000.0" y="1000.0"/>
    </nodes>
    <links>
        <link id="1" from="1" to="2" length="3000.00" capacity="3600"
                freespeed="27.78" permlanes="2" modes="car" />
        <link id="2" from="2" to="3" length="4000.00" capacity="1800"
                freespeed="27.78" permlanes="1" modes="car" />
        <link id="3" from="3" to="2" length="4000.00" capacity="1800"
                freespeed="27.78" permlanes="1" modes="car" />
        <link id="4" from="3" to="1" length="6000.00" capacity="3600"
                freespeed="27.78" permlanes="2" modes="car" />
    </links>
</network>
```

For a working example, check the `examples/equil` directory in the MATSim directory tree (cf. Section 2.1.1).

Each element has an identifier `id`. Nodes are described by an `x` and a `y` coordinate value (also see Sections 2.2.1.3 and 7.1). Links have more features; the `from` and `to` attributes reference nodes and describe network geometry. Additional attributes describe traffic-related link aspects:

- The `length` of the link, typically in meters (see Section 2.2.1).
- The flow `capacity` of the link, i.e., number of vehicles that traverse the link, typically in vehicles per hour.
- The `freespeed` is the maximum speed that vehicles are allowed to travel along the link, typically in meters per second.
- The number of lanes (`permlanes`) available in the direction specified by the 'from' and 'to' nodes.
- The list of `modes` allowed on the link. This is a comma-separated list, e.g., `modes="car, bike, taxi"`.

All links are uni-directional. If a road can be traveled in both directions, two links must be defined with alternating `to` and `from` attributes (see links with id 2 and 3 in the listing above).

2.2.2.3 population.xml

File Format MATSim travel demand is described by the agents' day plans. The full set of agents is also called the population, hence the file name `population.xml`. Alternatively, `plans.xml` is also commonly used in MATSim, as the population file essentially contains a list of day plans.

The population contains the data in a hierarchical structure, as shown in the following example. This example illustrates the data structure; minimal input files need less information, as illustrated later.

```
<population>
    <person id="1">
        <plan selected="yes" score="93.2987721">
            <act type="home" link="1" end_time="07:16:23" />
            <leg mode="car">
                <route type="links">1 2 3</route>
            </leg>
            <act type="work" link="3" end_time="17:38:34" />
            <leg mode="car">
                <route type="links">3 1</route>
            </leg>
            <act type="home" link="1" />
        </plan>
    </person>
    <person id="2">
        <plan selected="yes" score="144.39002">
            ...
        </plan>
    </person>
</population>
```

For a working example, check the `examples/equil` directory in the MATSim directory tree (cf. Section 2.1.1).

The population contains a list of persons, each person contains a list of plans, and each plan contains a list of activities and legs.

Exactly one plan per person is marked as selected. Each agent's selected plan is executed by the mobility simulation. During the replanning stage, a different plan might become selected. A plan can contain a score as attribute. The score is calculated and stored in the plan after its execution by the mobility simulation during the scoring stage.

The list of activities and legs in each plan describe each agent's planned actions. Activities are assigned a type and typically have—except for the last activity in a day plan—a defined end time. There are some exceptions where activities have a duration instead of an end time. Such activities are often automatically generated by routing algorithms and are not described in this book. To describe the location where an activity takes place, the activity is either assigned a coordinate by giving it an `x` and `y` attribute value, or it has a link assigned, describing from which link the activity can be reached. Because the simulation requires a link attribute, `Controler` calculates the nearest link for a given coordinate when the link attribute is missing.

A leg describes how an agent plans to travel from one location to the next; each leg must have a transport mode assigned. Optionally, legs may have an attribute, `trav_time`, describing the expected travel time for the leg. For a leg to be simulated, it must contain a route. The format of a route depends on the mode of a leg. For car legs, the route lists the links the agent has to traverse in the given order, while for transit legs, information about stop locations and expected transit services are stored. MATSim automatically computes initial routes for initial plans that do not contain them.

An agent starts a leg directly after the previous activity (or leg) has ended. The handling of the agent in the mobsim depends on the mode. By default, car and transit legs are well-supported by the mobsim. If the mobsim encounters a mode it does not know, it defaults to teleportation. In this case, an agent is removed from the simulated reality and re-inserted at its target location after the leg's expected travel time has passed.

A Minimal Population File The population data format is one of the most central data structures in MATSim and might appear a bit overwhelming at first. Luckily, to get started, it is only necessary to know a small subset. A population file needs, approximately, only the following information:

```
<population>
   <person id="1">
      <plan>
         <act type="home" x="5.0" y="8.0" end_time="08:00:00" />
         <leg mode="car" />
         <act type="work" x="1500.0" y="890.0" end_time="17:30:00" />
         <leg mode="car" />
         <act type="home" x="5.0" y="8.0" />
      </plan>
   </person>
   <person id="2">
      ...
   </person>
</population>
```

For a working example, check the `examples/equil` directory in the MATSim directory tree (cf. Section 2.1.1).

The following items can be used for simplification:

- Each person needs exactly one plan.
- The plan does not have to be selected or have a score.
- Activities can be located just by their coordinates.
- Activities should have a somewhat reasonable end-time.
- Legs need only a mode, no routes.

When a simulation is started, MATSim's `Controler` will load such a file and then automatically assign the link nearest to each activity and calculate a suitable route for each leg. This makes it easy to get started quickly.

2.2.3 *Typical Output Data*

MATSim creates output data that can be used to analyze results as well as to monitor the current simulation setup progress. Some of the files summarize a complete MATSim run, while others are created for a specific iteration only. The first type of files goes directly to the output folder's top level, which can be specified in the `controler` section of the config file. The other files are stored in iteration-specific folders `ITERS/it.{iteration number}`, which are continuously created in the output folder. For some files (typically for large ones, such as population), the output frequency can be specified in the config file. They then go only to the respective iteration folders. The files summarizing the complete MATSim run are built 'on the fly', i.e., after every iteration, currently computed iteration values are stored, allowing continuous monitoring of the run. Some files are created by default (such as the score statistics files); others need to be triggered by a respective configuration file section (such as count data files).

The following output files are continuously built up to summarize the complete run.

Log File: During a MATSim run, a log file is printed containing information you might need later for your analyses, or in case a run has crashed.

Warnings and Errors Log File: Sometimes, MATSim identifies problems in the simulation or its configuration; it will then write warning and error messages to the log file. Because the log file contains so much information, these warnings can be overlooked. For this reason, a separate log file is generated in the run output directory, containing only warnings and error messages. It is important to check this file during/after a run for possible problems.

Score Statistics: Score statistics are available as a picture (`scorestats.png`), as well as a text file (`scorestats.txt`). They show the average best, worst, executed and overall average of all agents' plans for every iteration. An example score plot is shown in Figure 1.2.

Leg Travel Distance Statistics: Leg travel distance statistics (files `traveldistancestats.png` and `traveldistancestats.txt`) are comparable to score statistics, but instead, they plot travel distance.

Stopwatch: The stopwatch file (`stopwatch.txt`) contains the computer time (so-called wall clock time) of actions like replanning or the execution of the mobsim for every iteration. This data is helpful for computational performance analyses, e.g., how long does replanning take compared to the mobility simulation?

The following output files are created for specific iterations:

Events: Every action in the simulation is recorded as a MATSim event, be it an activity start or change of network link; see Fig. 2.2. Each event possesses one or multiple attributes. By default, the time when the event occurred is included. Additionally, information like the ID of the agent triggering the event, or the link ID where the event occurred, could be included. The events file is an important base for post-analyses, like the visualizers. Events are discussed in detail in Section 45.2.5.

Plans: At configurable iterations, the current state of the population, with the agents' plans, is printed. The final iteration's plans are also generated on the top level of the output folder.

Leg Histogram: In every iteration, a leg histogram is plotted. A leg histogram depicts the number of agents arriving, departing or en route, per time unit. Histograms are created for each transport mode and for the sum of all transport modes. Each file starts with the iteration number and ends with the transport mode (e.g., `1.legHistogram_car.png` or `1.legHistogram_all.png`). A text file is also created (e.g., `1.legHistogram.txt`), containing the data for all transport modes.

Trip Durations: For each iteration, a trip durations text file (e.g., `1.tripdurations.txt`), listing number of trips and their durations, on a time bin level for each activity pair (e.g., from work to home or from home to shopping), is produced.

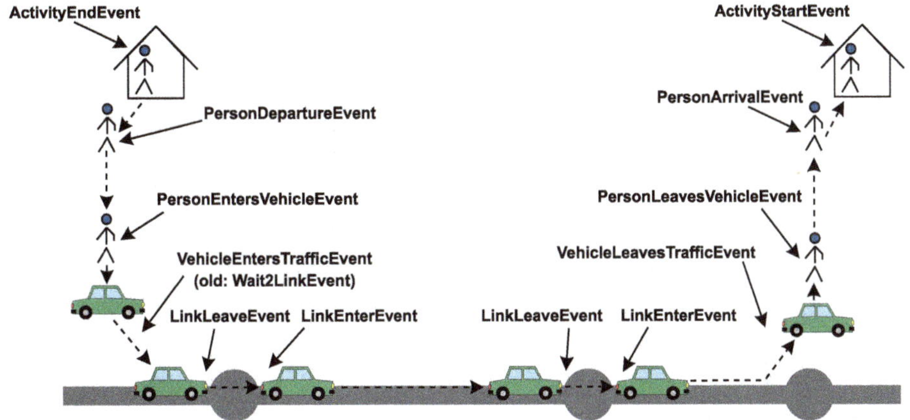

Figure 2.2: Mobsim events.

Link Stats: In each iteration, a link stats file containing hourly count values and travel times on every network link is printed. Link stats are particularly important for comparison with real-world count data, as introduced in Section 6.3.

2.2.4 An Example Scenario

The MATSim release is shipped with an example scenario named `equil` in the folder `examples/equil`, containing these files: `config.xml`, `network.xml`, `plans100.xml`, and `plans2000.xml.gz`, containing, respectively, 100 and 2000 persons with their day plans, using car mode only. A tiny population containing only 2 persons (`plans2.xml`), one using public transport, the other using car mode, is also provided. An example for count data is also found in the folder (`counts100.xml`).

In addition, there is also a file with 100 *trips* (`plans100trips.xml`), i.e., demand going only from one location to another, using a `dummy` activity type at each end. This is provided to show that MATSim can also be run as a fully trip-based approach, without considering any activities. Clearly, it loses some of its expressiveness, but the basic concepts, including route and even departure time adaptation, still work in exactly the same way.

The scenario network is shown in Figure 2.3.

The following lines explain the scenario by discussing the most important sections from the config file `config.xml`.

"strategy" section of the config file As shown in the config file excerpt below, this scenario uses replanning. 10 % of the agents reroute their current route (module `ReRoute`). The remaining 90 % select their highest score plan for re-execution in the current iteration (module `BestScore`). Plans are deleted from the agent's memory if it is full, defined by `maxAgentPlanMemorySize`. By default, the plan with the lowest score is removed; this is configurable and currently being researched (see Section 97.3).

```
<module name="strategy">
   <param name="maxAgentPlanMemorySize" value="5" />
      <!-- 0 means unlimited -->

   <parameterset type="strategysettings" >
      <param name="strategyName" value="ReRoute" />
      <param name="weight" value="0.1" />
   </parameterset>
```

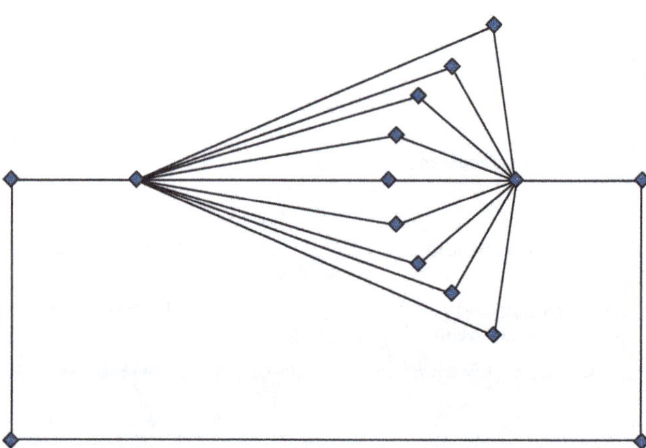

Figure 2.3: Equil scenario network.

```
    <parameterset type="strategysettings" >
      <param name="strategyName" value="BestScore" />
      <param name="weight" value="0.9" />
    </parameterset>

</module>
```

"planCalcScore" section of the config file The section `planCalcScore` defines parameters used for scoring, explained in Chapter 3. As seen in the example, two activity types, `h` (home) and `w` (work), are specified. All activity types contained in the population file (cf. Section 2.2.2.3) must be defined in the `planCalcScore` section of the config file.

```
<module name="planCalcScore" >
    <parameterset type="activityParams" >
      <param name="activityType" value="h" />
      <param name="typicalDuration" value="12:00:00" />
    </parameterset>
    <parameterset type="activityParams" >
      <param name="activityType" value="w" />
      <param name="typicalDuration" value="08:00:00" />
    </parameterset>
</module>
```

"controler" section of the config file The scenario is run for 10 iterations, writes the output files to `./output/equil` (Section 2.2.3) and uses QSim as the mobsim (more on mobsims in Section 1.3, 4.3 and 11).

```
<module name="controler">
    <param name="outputDirectory" value="./output/equil" />
    <param name="lastIteration" value="10" />
    <param name="mobsim" value="qsim" />
</module>
```

Visualization Simulation results can be visualized with Via (Chapter 33) or OTFVis (On The Fly Visualizer) (Chapter 34).

2.2.5 Data Requirements

2.2.5.1 Population and Activity Schedules

Demand estimation is an important component of MATSim. That means that, in theory, only demand components that do *not* change from one simulated average working day to the next need to be provided to MATSim. Examples are: population and its residential and working locations. In practice, however, MATSim is not yet prepared to endogenously model complete travel demand. Sequence and preferred durations of activities, for example, must be provided as input. As a result, all travel demand choices not covered by the MATSim loop have to be exogenously estimated.

For population generation, two possibilities exist: the comfortable way is to translate a full population census and the slightly more demanding way is to generate a synthetic population (e.g., Guo and Bhat, 2007), based on sample or structure surveys. For MATSim, both methods have been used based on e.g., Swiss Federal Statistical Office (BFS) (2000) and Müller (2011a).

Travel demand is usually derived from surveys: for Switzerland, from the microcensus (Swiss Federal Statistical Office (BFS), 2006). Newer data sources, such as GPS or smartphone travel diaries, are currently being investigated (e.g., Zilske and Nagel, 2015).

A critical topic in demand and population generation is workplace assignment, as commuting traffic is still a major issue, particularly during peak hours. Switzerland's full census work location was surveyed at municipality level. Such comfortable data bases are rare, however.

Having generated the residential population of the study area, additional demand components might be necessary, for example, cross-border and freight traffic. As these components often cannot be endogenously modeled, MATSim offers the feature to handle different subpopulations differently (Section 4.5). One can specify that border-crossing agents, for example, are not allowed to make destination choices within the study area, or that freight agents are not allowed to change their delivery activity to a leisure activity.

2.2.5.2 Network

In simulation practice, two different network types are used: planning networks and navigation networks (compare Swiss examples in Figure 2.4(a) and Figure 2.4(b) for the Zürich region). The former are leaner and often serve as initial explorative simulation runs, while the latter are often used for policy runs, usually offering far more details, such as bike and even pedestrian links. Data are available from official sources like federal offices, free sources, such as OSM (OpenStreetMap), and commercial sources, including navigation network providers.

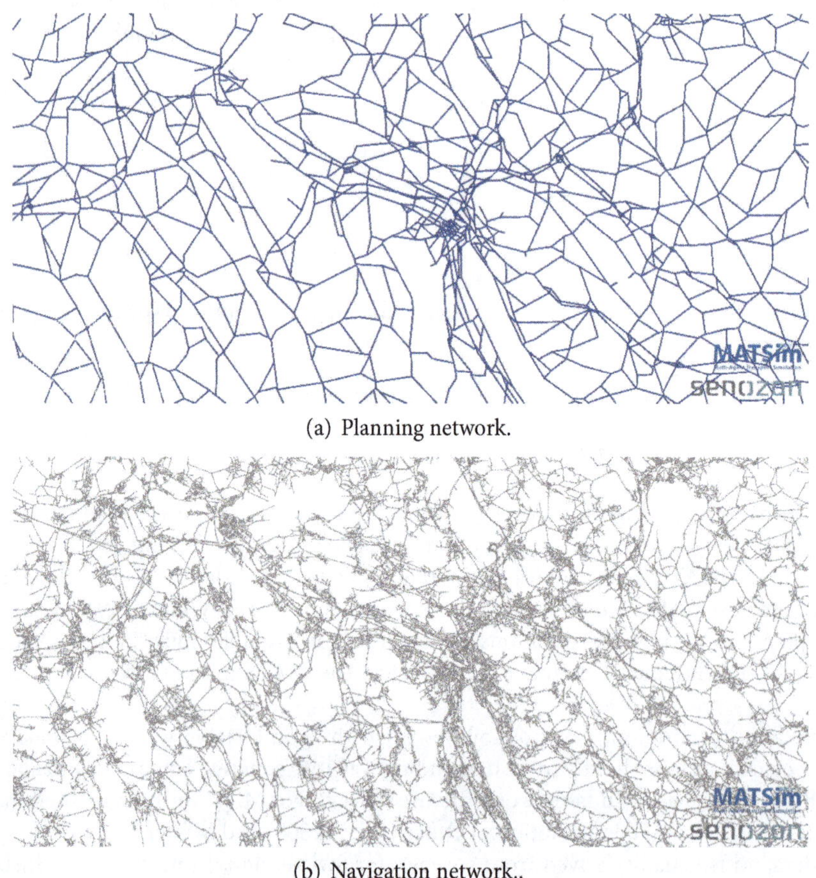

(a) Planning network.

(b) Navigation network..

Figure 2.4: Zürich networks

2.2.6 Example Scenario Input Data

Some example scenarios are included in the MATSim main distribution, in the directory "examples".

More pre-packaged scenarios can be found under `http://www.matsim.org/datasets`.

2.3 MATSim Survival Guide

There are many options and possibilities available with MATSim, and finding them can be a daunting exercise. Here are a couple of recommendations, derived from our own frequent use of the system.

1. *Always start with and test a small example.*
2. *Always test large scenarios with one percent runs first (e.g., a randomly drawn subsample of your initial demand).* The MATSim GUI (Figure 2.1) allows creating sample populations with the command `Tools...Create Sample Population`.
   As described in Section 4.3, this requires adaptation of parameters, in particular, the mobsim's `flowCapacityFactor` and `storageCapacityFactor` factors. As shown in Part II, Section 6.3, sample scenarios also require parameter adaption for count data comparisons.
3. *If your set-up does not work any more,* immediately *go back to a working version and proceed from there in small steps.*
4. *Check* `logfileWarningErrors.log`.
5. *Check the comments that are attached to the config file options.*
   One finds them in the file `output_config.xml.gz`, or near the beginning of `logfile.log`.
6. *Try setting as few config file options as possible.*
   This has two advantages: (i) Except for the deliberately set options, your simulation will move along with changed MATSim defaults, and thus with what the community currently considers the best configuration. (ii) You will not be affected by changes in the config file syntax as long as they are different from your own settings.
7. *Search for documentation via* `http://matsim.org/javadoc`.
8. *Search for the latest tutorial via* `http://matsim.org/docs`.

CHAPTER 3

A Closer Look at Scoring

Kai Nagel, Benjamin Kickhöfer, Andreas Horni and David Charypar

3.1 Good Plans and Bad Plans, Score and Utility

As outlined in Section 1.4 and by Figures 1.1 and 1.4, MATSim is based on a co-evolutionary algorithm: Each individual agent learns by maintaining multiple plans, which are scored by executing them in the mobsim, selected according to the score and sometimes modified. In somewhat more detail, the iterative process contains the following elements:

mobsim The mobility simulation takes one "selected" plan per agent and executes it in a synthetic reality. This may also be called network loading.

scoring The actual performance of the plan in the synthetic reality is taken to compute each executed plan's score.

replanning consists of several steps:

1. If an agent has more plans than the maximum number of plans (a configuration parameter), then plans are removed according to a (configurable) plan selector (choice set reduction, plans removal).
2. For some agents, a plan is copied, modified and then selected for the next iteration (choice set extension, innovation).
3. All other agents choose between their plans (choice).

An agent's plans in a given iteration may be considered the agent's **choice set** in that iteration. As a result, steps 1 and 2 of replanning modify the choice set, while step 3 implements the actual **choice** between options. Choice is typically based on the score; higher score plans are more likely to be selected. This is discussed in more detail in Chapters 47 and 49. For the time being, note that the three steps of replanning must cooperate for the approach to work: the plans removal step should remove "bad" plans, the innovation step should generate "good" plans, and the choice should, ingeneral,

How to cite this book chapter:

Nagel, K, Kickhöfer, B, Horni, A and Charypar, D. 2016. A Closer Look at Scoring. In: Horni, A, Nagel, K and Axhausen, K W. (eds.) *The Multi-Agent Transport Simulation MATSim*, Pp. 23–34. London: Ubiquity Press. DOI: http://dx.doi.org/10.5334/baw.3.

select good plans. Here, "good" means "able to obtain a high score in the mobsim/scoring". Fortunately, due to its evolutionary concept, the approach is fairly robust: the innovation step does not always have to generate good solutions; it is sufficient if *some* of the solutions are good and lead to a high score.

With this, it is clear that scoring is a central element of MATSim. Only solutions obtaining a high score will be selected by the agent and survive the plans removal step. Thus, the scoring function needs to be "correct" for a given scenario, meaning, more or less, that plans "performing well" obtain a higher score than plans that "do not perform well". Whether a performance is good or not, is decided, in the end, by travelers living in a region: some may prefer a congested car trip, others may prefer a crowded, but affordable, trip by public transit, while others may prefer using the bicycle, even in bad weather.

The typical way to bridge this gap is to use econometric **utility** functions, for example, from random utility models (e.g., Ben-Akiva and Lerman, 1985; Train, 2003) for the score. However, in AI (Artificial Intelligence), utility functions may also be used in a more general way: for example, the score that each individual agent (or the system as a whole) wants to, or should, optimize (Russel and Norvig, 2010). For these reasons, the terms "score" and "utility" are normally interchangeable in the MATSim context. Since we will need the concept of a marginal utility, this chapter will mostly speak of 'utility', since it is a bit unusual to talk about 'marginal score'.

The user can configure numerous parameters to specify the scoring function. When users are ready to extend MATSim in the next part of the book, they will also learn how to plug in their own customized scoring function.

However, because MATSim is based on complete day plans, the application of choice models for parts of day plans only (for example, mode choice) is not straightforward, as detailed in Section 97.4.4. Because of the absence of complete-day utility functions in the literature, MATSim has started with the so-called Charypar-Nagel scoring or utility function (Section 3.2). This scoring function was, at times, modified, extended, or replaced for specific investigations (Section 3.5). Readily applicable estimates for a full-day utility function are not yet available, as discussed in Section 97.4.4.

3.2 The Current Charypar-Nagel Utility Function

3.2.1 *Mathematical Form*

The first, and still basic, MATSim scoring function was formulated by Charypar and Nagel (2005), loosely based on the *Vickrey* model for road congestion, as described by Vickrey (1969) and Arnott et al. (1993). Originally, this formulation was established for departure time choice. However, all studies performed so far indicate that the MATSim function is also appropriate for modeling further choice dimensions. It is, however, almost certainly not appropriate for activity dropping and activity addition (see Section 3.3).

Basic Function For the basic function, utility of a plan S_{plan} is computed as the sum of all activity utilities $S_{act,q}$ plus the sum of all travel (dis)utilities $S_{trav,mode(q)}$:

$$S_{plan} = \sum_{q=0}^{N-1} S_{act,q} + \sum_{q=0}^{N-1} S_{trav,mode(q)} \tag{3.1}$$

with N as the number of activities. Trip q is the trip that follows activity q. For scoring, the last activity is merged with the first activity to produce an equal number of trips and activities.

Activities The utility of an activity q is calculated as follows (see also Charypar and Nagel, 2005, p.377ff):

$$S_{act,q} = S_{dur,q} + S_{wait,q} + S_{late.ar,q} + S_{early.dp,q} + S_{short.dur,q} \,. \tag{3.2}$$

The individual contributions are defined as follows:

- The expression
$$S_{dur,q} = \beta_{dur} \cdot t_{typ,q} \cdot \ln(t_{dur,q}/t_{0,q}) \tag{3.3}$$
is the utility of performing activity q, where opening times of activity locations are taken into account. $t_{dur,q}$ is the performed activity duration, β_{dur} is related to the marginal utility of activity duration (or marginal utility of time as a resource, the same for all activities; see Section 3.2.4), and $t_{0,q}$ is the duration when utility starts to be positive.
- The expression
$$S_{wait,q} = \beta_{wait} \cdot t_{wait,q}$$
denotes waiting time spent, for example, in front of a still-closed store; β_{wait} is the so-called *direct* (see Section 3.2.4) marginal utility of time spent waiting; and $t_{wait,q}$ is the waiting time. We recommend leaving β_{wait} at zero; also see Section 3.2.5.
- The expression
$$late.ar,q = \begin{cases} \beta_{late.ar} \cdot (t_{start,q} - t_{latest.ar,q}) & \text{if } t_{start,q} > t_{latest.ar,q} \\ 0 & \text{else} \end{cases}$$
specifies the late arrival penalty, where $t_{start,q}$ is the activity starting time q and $t_{latest.ar}$ is the latest possible penalty-free activity starting time (for example, the starting time of the office core hours, or the starting time of an opera or theater performance).
- The expression
$$S_{early.dp} = \begin{cases} \beta_{early.dp} \cdot (t_{end,q} - t_{earliest.dp,q}) & \text{if } t_{end,q} > t_{earliest.dp,q} \\ 0 & \text{else} \end{cases}$$
defines the penalty for not staying long enough, where $t_{end,q}$ is the activity ending time and $t_{earliest.dp,q}$ is the earliest possible activity end time q. We normally recommend leaving $\beta_{early.dp}$ at zero, except if really good data about this effect is available.
- The expression
$$S_{short.dur,q} = \begin{cases} \beta_{short.dur} \cdot (t_{short.dur,q} - t_{dur,q}) & \text{if } t_{dur,q} < t_{short.dur,q} \\ 0 & \text{else} \end{cases}$$
is the penalty for a 'too short' activity, where $t_{short.dur}$ is the shortest possible activity duration. We normally recommend leaving $\beta_{short.dur}$ at zero, except if really good data about this effect is available.

The config syntax (config version v2) is approximately

```
<module name="planCalcScore" >
   <param name="performing" value="6.0" />
   <param name="waiting" value="-0.0" />
   <param name="lateArrival" value="-18.0" />
   <param name="earlyDeparture" value="-0.0" />
   <parameterset type="activityParams" >
      <param name="activityType" value="work" />
      <param name="typicalDuration" value="08:00:00" />
```

```
      <param name="openingTime" value="07:00:00" />
      <param name="latestStartTime" value="09:00:00" />
      <param name="closingTime" value="19:00:00" />
      ...
   </parameterset>
   ...
</module>
```

Travel Travel disutility for a leg q is given as

$$\begin{aligned} S_{trav,q} \quad = \quad & C_{mode(q)} + \beta_{trav,mode(q)} \cdot t_{trav,q} + \beta_m \cdot \Delta m_q \\ & +(\beta_{d,mode(q)} + \beta_m \cdot \gamma_{d,mode(q)}) \cdot d_{trav,q} + \beta_{transfer} \cdot x_{transfer,q} \end{aligned} \tag{3.4}$$

where:

- $C_{mode(q)}$ is a mode-specific constant.
- $\beta_{trav,mode(q)}$ is the *direct* (see Section 3.2.4) marginal utility of time spent traveling by mode. Since MATSim uses and scores 24-hour episodes, this is in addition to the marginal utility of time as a resource (again, see Section 3.2.4).
- $t_{trav,q}$ is the travel time between activity locations q and $q+1$.
- β_m is the marginal utility of money (normally positive).
- Δm_q is the change in monetary budget caused by fares, or tolls for the complete leg (normally negative or zero).
- $\beta_{d,mode(q)}$ is the marginal utility of distance (normally negative or zero).
- $\gamma_{d,mode(q)}$ is the mode-specific monetary distance rate (normally negative or zero).
- $d_{trav,q}$ is the distance traveled between activity locations q and $q+1$.
- $\beta_{transfer}$ are public transport transfer penalties (normally negative).
- $x_{transfer,q}$ is a 0/1 variable signaling whether a transfer occurred between the previous and current leg.

The config syntax (config version v2) is approximately

```
<module name="planCalcScore" >
   <param name="marginalUtilityOfMoney" value="1.0" />
   <param name="utilityOfLineSwitch" value="-1.0" />
   <parameterset type="modeParams" >
      <param name="mode" value="car" />
      <param name="constant" value="0.0" />
      <param name="marginalUtilityOfDistance_util_m" value="0.0" />
      <param name="marginalUtilityOfTraveling_util_hr" value="-6.0" />
      <param name="monetaryDistanceRate" value="-0.0002" />
   </parameterset>
   ...
</module>
```

Equation (3.4) is the direct utility contribution of travel; see Section 3.2.4 for the the full indirect utility as well as the relation to the VTTS (Value of Travel Time Savings), and Chapter 51 for a more general discussion.

Note that distance contributes to disutility in two ways. First, it is included in a direct manner via $\beta_{d,mode(q)}$, which is normal for modes involving physical effort, like walking or cycling. Second, distance is also included monetarily via $\beta_m \cdot \gamma_{d,mode(q)}$, which is normal for car or pt mode, where monetary costs increase depending on distance.

3.2.2 *Illustration*

Figure 3.1 illustrates the scoring function. Time runs from left to right. The example shows part of an executed schedule, with home, work, and lunch activities, connected by a car and walk leg.

Activities are scored with concave functions, modeling decreasing returns to spending more time at the same activity. Travel, in contrast, is modeled with downward sloping straight lines, where the slope may differ for different modes of transport and there may be an initial offset (alternative-specific constant). Note the delay between arrival at the workplace and workplace opening time, reflected in no score accumulation during that period. Agents accumulate those scores over a day, reflected in the bottom graph.

When one assumes all other things (particularly travel times) are equal, then agents maximize their score when activity durations are such that all activities have the same slope (= the same marginal utility; red lines). This follows from basic economic theory (cf. Section 51.2), but can also be seen intuitively; if red lines did not all have the same slope, the agent could gain by extending those activities with steeper slope at the expense of others. Clearly, this holds only when all other things remain constant, particularly travel times.

3.2.3 *The "Wrapping Around" of the Utility Function*

The MATSim mobsim typically starts at midnight and runs until all plans have reached their final activity. By itself, the mobsim, is not limited to a day. However, as already stated in Section 3.2.1, the standard scoring function assumes that plans "wrap around" to 24-hour days. Thus, the last activity is merged with the first into one activity. For example, if the first activity ends at 7 am and the last activity starts at 11 pm, then it is assumed that this is the *same* activity, with a duration of eight hours.

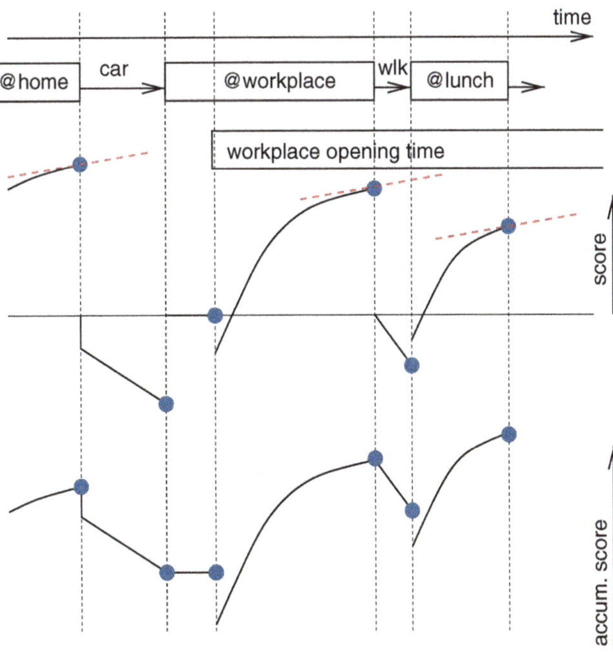

Figure 3.1: Illustration of the scoring function. TOP: Individual contributions of activities and legs. BOTTOM: Score accumulation over a day.

Note that scoring the two activities separately would lead to a different result, because of the nonlinear (logarithmic) form of the utility of performing. For example, $\ln(1)+\ln(7)=\ln(7)\neq\ln(1+7)=\ln(8)$.

3.2.4 *MATSim Scoring, Opportunity Cost of Time, and the VTTS*

As a result of the wrap-around concept, travel receives, beyond the typically negative direct marginal utility $\beta_{trav,mode}$, an additional implicit penalty from the **marginal utility of time as a resource**: If travel time could be reduced by Δt_{trav}, the person would not only gain from avoiding $\beta_{trav}\cdot\Delta t_{trav}$, but also from additional time for activities (effect of the opportunity cost of time). The **(total) marginal utility of travel time savings** is thus:

$$mUTTS = -\frac{\partial}{\partial t_{trav}}S_{trav} + \frac{\partial}{\partial t_{dur}}S_{dur}.$$

which is

$$mUTTS = -\beta_{trav} + \beta_{dur}\cdot\frac{t_{typ,q}}{t_{dur,q}} \tag{3.5}$$

and at the typical duration of an activity

$$mUTTS\Big|_{t_{dur,q}=t_{typ,q}} = -\beta_{trav} + \beta_{dur},$$

where it can be imagined q is the activity immediately following the trip (cf. Section 51.2). The marginal utility of travel time savings, $mUTTS$, can thus be defined as the indirect effect on the overall time budget, corrected by an offset β_{trav} that denotes how much better, or worse, it is to spend that time traveling, rather than "doing nothing".[1] To differentiate β_{trav} from the indirect effect, it is sometimes called **direct marginal utility** of time spent traveling.

The marginal utility of travel time savings can be transformed to the more common **VTTS (Value of Travel Time Savings)** by dividing it by the marginal utility of money, β_m:

$$VTTS = \frac{mUTTS}{\beta_m} = \frac{-\beta_{trav}+\beta_{dur}\cdot\frac{t_{typ,q}}{t_{dur,q}}}{\beta_m},$$

and at the typical duration of an activity

$$VTTS\Big|_{t_{dur,q}=t_{typ,q}} = \frac{mUTTS}{\beta_m}\Big|_{t_{dur,q}=t_{typ,q}} = \frac{-\beta_{trav}+\beta_{dur}}{\beta_m}$$

This is important for calibration of the utility function.

3.2.5 *The Resulting Modeling of Schedule Delay Costs*

Arriving Early In the same way as the marginal utility of travel time savings is not only given by $-\beta_{trav}$, but instead by $-\beta_{trav}+\beta_{dur}\cdot\frac{t_{typ,q}}{t_{dur,q}}$, the marginal utility of waiting time savings is given

[1] This is an approximate statement; in the full theory, the reference marginal utility is not given by "doing nothing", but by a Lagrange multiplier related to the constraint that a day has 24 hours; again, cf. Section 51.2.

by $mUWTS = -\beta_{wait} + \beta_{dur} \cdot \frac{t_{typ,q}}{t_{dur,q}}$: Even when the direct marginal utility of waiting, β_{wait}, equals zero, then "doing nothing" still eats into the overall time budget and thus incurs the same opportunity cost of time as traveling does. Intuitively, one can imagine that one must leave the previous activity earlier to have a longer waiting time, thus reducing the score of the previous activity.

Thus, as long as one cannot estimate β_{wait} separately from β_{dur}, we recommend leaving β_{wait} at zero.

Arriving Late Arriving late incurs a marginal utility of β_{late}, typically negative. Here, no additional opportunity cost of time is involved. Intuitively, arriving later implies having left the previous activity later. That is: the current activity is shortened by the same amount that the previous activity was extended, leaving the overall score unaffected (cf. Section 51.2).

Vickrey Parameters As a result, the Vickrey parameters of α (marginal penalty for arriving early), β (marginal penalty for traveling) and γ (marginal penalty for arriving late) (as defined by Arnott et al., 1990) are consistent with the following equations:

$$\begin{aligned} -\beta_{wait} + \beta_{dur} \cdot \frac{t_{typ,q}}{t_{dur,q}} &= \alpha \\ -\beta_{trav} + \beta_{dur} \cdot \frac{t_{typ,q}}{t_{dur,q}} &= \beta \\ -\beta_{late} &= \gamma \,. \end{aligned} \tag{3.6}$$

3.3 Implementation Details

This section summarizes the current implementation of the default MATSim scoring function. The section can be skipped if the reader understands that what has been summarized up to this point is not the full story.

3.3.1 *Zero Utility Duration*

The duration when an activity's utility is exactly zero is computed by the somewhat cryptic expression

$$t_{0,q} := t_{typ,q} \cdot \exp\left(-\frac{10h}{t_{typ,q} \cdot prio}\right), \tag{3.7}$$

where *prio* is a configurable parameter. This is designed so that all activities with the same value of *prio* obtain, at their typical duration, i.e., when $t_{dur,q} = t_{typ,q}$, the same utility value of $10 \cdot \beta_{dur}$, with the idea that this makes them equally likely to be dropped in a time shortage situation (Charypar and Nagel, 2005).[2] However, this does not work as intended, since activities receiving this utility value from a short duration have a larger utility accumulation per time unit than others and are thus dropped later. In consequence, without additional constraints, the "home" activity gets dropped

[2] Starting from Equation (3.3) and inserting Equation (3.7), one obtains

$$\begin{aligned} S_{dur,q}\Big|_{t_{dur,q}=t_{typ,q}} &= \beta_{dur} \cdot t_{typ,q} \cdot \ln\left(\frac{t_{typ,q}}{t_{typ,q} \cdot \exp\left(-10h/(t_{typ,q} \cdot prio)\right)}\right) \\ &= \beta_{dur} \cdot t_{typ,q} \cdot \ln\left(\exp\left(10h/(t_{typ,q} \cdot prio)\right)\right) = 10h \cdot \beta_{dur}/prio \,, \end{aligned}$$

which is indeed the same for all activities with the same value of *prio*.

first, which is clearly not plausible. See Section 97.4 for a discussion of alternatives. In the meantime, the recommendations are:

- Do not set the `priority` value in the config away from its default value.
- Recognize that the current MATSim default scoring/utility function is not suitable for activity dropping.

3.3.2 *Negative Durations*

In MATSim, somewhat oddly, it is possible to have activities with negative durations. This can happen because of the "wrap-around" mechanism, where the last activity of a plan is stitched together with the first activity of the plan, and only that merged activity is scored (cf. Section 3.2.3). In this situation, it can happen that an agent arrives at the last activity of the plan at a later 24-hour-time than when the first activity ended. For example, an agent could stay at home until 3 am (end of first activity), then go through her daily plan including a very late party, and return home at 6 am the next morning (Figure 3.2). In this case, the duration of the wrap-around home activity would be *minus* three hours. Originally, a score of zero was assigned to these negative duration activities. However, the adaptive agents quickly found out that they could use this to their advantage, expanding this negative duration without a penalty would lead to more time elsewhere, which the agent could use to accumulate score. For an adaptive algorithm, a penalty like this needs to be defined so that it guides the adaptation back into the feasible region. The penalty must increase with increasing negative duration. It also needs to be more strongly negative than any score value for a positive activity duration. The latter is, however, impossible to achieve with a logarithmic form, which tends to $-\infty$ as $t_{dur,q}$ approaches zero from above. The current approach is to take the slope of the expression $\beta_{dur} \cdot t_{typ,q} \cdot \ln(t_{dur,q}/t_{0,q})$ when it crosses zero, and extend this towards minus infinity (Figure 3.3).

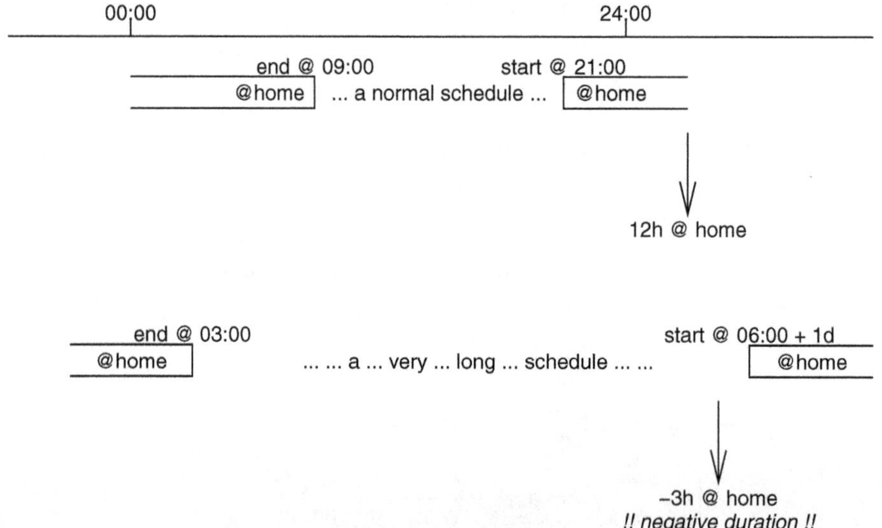

Figure 3.2: Illustration of wrap-around scoring. TOP: Normal situation. BOTTOM: Situation where final activity starts at a later time of day then when the first activity ended, resulting in negative duration.

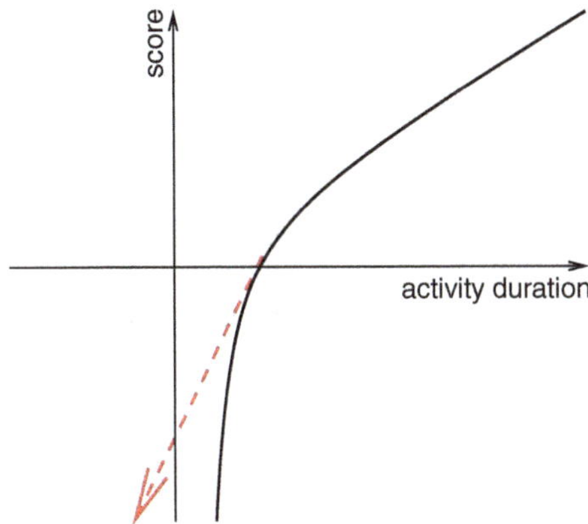

Figure 3.3: Extending the slope when the utility function crosses the zero line to negative durations.

First and Last Activity not the Same Clearly, the wrap-around approach fails if the first and last activity are not the same. The present code does not look at locations, but gives a warning and problematic results if they are of different types.

3.3.3 *Score Averaging*

The score S that is computed according to the rules given in this chapter is not assigned directly to the plan, rather, it is exponentially smoothed according to

$$S^k = \alpha\, S + (1-\alpha)\, S^{k-1}\,, \tag{3.8}$$

where S^k is the newly memorized score, S^{k-1} is the previously memorized score, S is the score obtained from the plan's execution in the mobsim, and α is a "learning" or "blending" parameter. The default value of α is one; it can be configured by the line

```
<param name="learningRate" value="..." />
```

in the config file.

Non-executed plans just keep their score.

3.3.4 *Forcing Scores to Converge*

For many situations, both practical and theoretical (see Section 47.3.2.2), it is desirable that each plan's score converges to its expectation value. Equation (3.8) will not achieve that; it just dampens the fluctuations. A well-known approach to force convergence to the expectation value is MSA (Method of Successive Averages):

$$S^m = \frac{1}{m}\, S + \frac{m-1}{m}\, S^{m-1}\,. \tag{3.9}$$

This resembles Equation (3.8), with two important differences: (1) The fixed blending parameter α is now replaced by a variable $1/m$, and (2) m is not the iteration number but counts how often a plan was executed and thus scored. This is necessary in MATSim since a plan is not executed and scored in every iteration.

This behavior can be switched on by the following config option:

```
<param name="fractionOfIterationsToStartScoreMSA" value="..." />
```

This is plausibly used together with innovation switch off (Section 4.5.3), meaning that MSA operates on a fixed set of plans.

3.4 Typical Scoring Function Parameters and their Calibration

The current MATSim default values are

$$\begin{array}{lcrl} \beta_m & = & 1 & utils/monetaryunit \\ \beta_{dur} & = & 6 & utils/h \\ \beta_{trav,mode(q)} & = & -6 & utils/h \\ \beta_{wait} & = & 0 & utils/h \\ \beta_{short.dur} & = & 0 & utils/h \\ \beta_{late.ar} & = & -18 & utils/h \\ \beta_{early.dp} & = & 0 & utils/h. \end{array} \tag{3.10}$$

They are very loosely based on the Vickrey bottleneck model (e.g., Arnott et al., 1990).

An additional insight is that, in many of the systems that we model, traveling does not seem to be less convenient than "doing nothing". Thus, the *direct* marginal utility of traveling, β_{trav}, is close to zero and sometimes even positive (see, e.g., Redmond and Mokhtarian, 2001; Pawlak et al., 2011). Based on this, a possible approach to calibration is as follows:[3]

1. Set $\beta_m \equiv$ `marginalUtilityOfMoney` to whatever is the prefactor of your monetary term in your mode choice logit model.

   If you do not have a mode choice logit model, set to 1.0. (This is the default.)

   This is normally a positive value (since having more money normally increases utility).

2. Set $\beta_{dur} \equiv$ `performing` to whatever the prefactor of car travel time is in your mode choice mode, while changing that parameter's sign from its typical $-$ to a $+$.

   If you do not have a mode choice logit model, set to $+6.0$. (This is the default.)

   This is normally a positive value (since performing an activity for more time normally increases utility).

3. Set $\beta_{tt,car} \equiv$ `marginalUtilityOfTraveling...` to 0.0.

   It is important to understand this: Even if this value is set to zero, traveling by car will be implicitly punished by the opportunity cost of time: If you are traveling by car, you cannot perform an activity; thus, you are (marginally and approximately) losing β_{dur}. See Section 3.2.4.

4. Set all other marginal utilities of travel time by mode *relative to the car value.*

   For example, if your logit model says something like

   $$\ldots - 6/h \cdot tt_{car} - 7/h \cdot tt_{pt} \ldots,$$

   then

   $$\beta_{dur} = 6\ ,\ \ \beta_{tt,car} = 0\ ,\ \text{and}\ \beta_{tt,pt} = -1\ .$$

   If you do not have a mode choice logit model, set all $\beta_{tt,mode} \equiv$ `marginalUtilityOf Traveling...` values to zero (i.e., same as car).

[3] Different groups have different systems; this one is typical for VSP, although it uses ideas from Michael Balmer.

5. Set distance cost rates `monetaryDistanceRate...` to plausible values, if you have them.

   Note that this needs to be negative: distance consumes money at a certain rate.

6. Use the alternative-specific constants $C_{mode} \equiv$ `constant` to calibrate your modal split.

   (This is, however, not completely simple; one must run iterations and look at the result; especially for modes with small shares, one needs to have innovation switched off early enough near the end of the iterations.)

If you end up having your modal split right, but its distance distribution wrong, you probably need to look at different mode speeds. In our experience, this works better for this than using the $\beta_{tt,mode}$.

Calibrating schedule-based public transport (see Chapter 16) goes beyond what can be provided here.

3.5 Applications and Extensions

The default scoring function has been applied and extended for various purposes. Thus, the historical development is accompanied by various conceptual and technical modifications leading to the current utility function described above. This also means that the reported parameter settings in the literature are an indication, not a direct recommendation.

Important applications for large scenarios are described in Chapter 52.

Special utility functions have been developed for car sharing (see Chapter 22), social contacts and joint trips (see Chapter 28), parking (see Chapter 13), road pricing (see Chapter 15) and destination innovation (see Chapter 27), also describing facility loading scoring and inclusion of random error terms.

Future topics, available on an experimental basis, are: a full-blown utility function estimation (Section 97.4.4), inclusion of agent-specific preferences (Section 97.4.5) and application of alternative utility function forms (Section 97.4).

CHAPTER 4

More About Configuring MATSim

Andreas Horni and Kai Nagel

This chapter describes configuration options that can be used together with the three basic elements: config file, population and network. Part II discusses various options to extend MATSim beyond these three elements, sometimes using only additional files, or using additional JAR files beyond the MATSim core JAR file, by writing "scripts in Java" or by adding or replacing functionality.

MATSim writes configuration files in several locations; for example, in the logfile, in the iteration output directory, or with the `CreateFullConfig` functionality described in Section 2.1.3. As explained in Section 2.3, these files come with comments explaining configuration options. This is often the best source for configuration options.

4.1 MATSim Data Containers

4.1.1 Network

The config file section `network` specifies which network file will be used in the simulation (Section 2.1.3 and 2.2.2.2). Further configuration options, e.g., specification of time-variant networks, are presented in Section 6.1.

4.1.2 Population

The config file section `plans` specifies which population file with its day plans will be used (Section 2.1.3 and 2.2.2.3). Further configuration options, e.g., specification of arbitrary agent attributes or subpopulations, are presented in Section 6.2.

Further MATSim containers are described in Chapter 6.

How to cite this book chapter:

Horni, A and Nagel, K. 2016. More About Configuring MATSim. In: Horni, A, Nagel, K and Axhausen, K W. (eds.) *The Multi-Agent Transport Simulation MATSim*, Pp. 35–44. London: Ubiquity Press. DOI: http://dx.doi.org/10.5334/baw.4. License: CC-BY 4.0

4.2 Global Modules and Global Aspects

4.2.1 *Controler*

The controler is an indispensable module for running MATSim; its parameters are set in the `controler` config file section. The MATSim run's output directory, its number of iterations and the plans and events output interval can be specified here. The expected mobsim can be defined (Section 4.3). The routing algorithm is defined here by using

```
<module name="controler" >
   <param name="routingAlgorithmType" value="{Dijkstra
   | FastDijkstra | AStarLandmarks | FastAStarLandmarks}" />
   ...
</module>
```

Possibilities for extending the `Controler` functionality are given in Chapter 45.

4.2.2 *Events*

Events are continuously generated, reporting on all activities in the mobsim, as discussed in more detail in Section 45.2.5.

Please note that, besides these mobsim events, there is a less prominent type of events, namely `ControlerEvents`, which are created by the `Controler` to report on its current state. `ControlerEvents` are also further explained in Section 45.2.5.

4.2.3 *Parallel Computing*

MATSim uses multi-threading to accelerate computing speeds. Related configuration parameters can be found in several config modules; they are combined into one section here.

Global Setting The `global` section contains

```
<module name="global" >
   <param name="numberOfThreads" value="2" />
   ...
</module>
```

This number is used in several places; most importantly, innovative strategies, where multiple routing requests are distributed to multiple threads.
A good starting point is using the number of available cores.

Parallel Event Handling The config file section `parallelEventHandling` is used to define the number of threads used for event handling. As described in Waraich et al. (2009), the simulation can be substantially accelerated when using multiple threads for the events handling, which can be a bottleneck in MATSim simulation runs.

Parallel QSim The number of threads for the parallel QSim (cf. Dobler (2013)) can be configured by

```
<module name="qsim" >
   <param name="numberOfThreads" value="10" />
   ...
</module>
```

General Recommendations Generally, computations using threads are not necessarily faster with more threads, which is also true for MATSim. Some experimentation is necessary for each combination of scenario and hardware. Here are some recommendations:

- For the "global" number of threads, a good starting point is the number of available cores.
- It is no longer possible to switch off parallel event handling completely; setting it to '0' or 'null' or '1' eventually achieves the same result. Setting it to values larger than one sometimes leads to performance gains, but they are rarely significant.
- The most sensitive parameter is that for the QSim. For somewhat older hardware (e.g., Apple Macbook Pro from 2010), using all three remaining cores—in addition to the parallel event handling—led to negligible performance gains but left the machine useless for interactive tasks such as normal office work. For new hardware (e.g., Apple Macbook Pro from 2014), using six of the available eight cores for the QSim can make the mobsim more than a factor of two faster and the machine can still be used for office tasks. Experiences with older servers show that one must carefully investigate the number of threads for the mobsim, since using more threads often slows it down (Dobler, 2013). No experiences with new servers are currently available.
- HPCC (High-Performance Computing Clusters) are often available to researchers, allowing access to high-quality machines with reduced management overhead. Typically, one pays for computation time, either directly, or by a loss of priority, with an amount proportional to the reserved resources, that is, the time the job took to finish, multiplied by the number of reserved cores. In this kind of situation, the number of cores used throughout the whole process should be stable to avoid paying for unused resources. A recommendation in this case is thus to set the number of threads for the QSim to the best value (see above), say n, parallel events handling to 1, the "global" number of threads to $n + 1$, and submit the job requesting $n + 1$ cores. Also note that fewer threads are almost always better in terms of throughput. In addition, for both calibration and "what-if" scenario exploration, one typically needs to run a large number of simulations with different parameters or input data. As total RAM memory is usually not an issue on a cluster, it is often more efficient to run a large number of simulations simultaneously with a low number of threads, rather than a low number of simulations with lots of threads.

4.2.4 *Global*

In the config file section `global`, the simulation's random seed, the "global" number of Java threads (see Section 4.2.3) and the coordinate system (cf. Section 2.2.1) can be defined. Note that no matter if you explicitly define the random seed or not, MATSim always starts from a fixed random seed, which is either the one you define, or an internal constant. That is, if you start the same version of MATSim twice from the same config file, you will get the same sequence of random numbers, and thus exactly the same simulation. If you want to change this behavior, you need to change the random seed explicitly.

4.3 Mobility Simulations

An overview of MATSim mobility simulations is given by Dobler and Axhausen (2011).

4.3.1 *QSim*

The queue-based and time-step based QSim (Gawron, 1998; Simon et al., 1999; Cetin et al., 2003; Dobler and Axhausen, 2011; Dobler, 2010) is MATSim's default mobsim. Its parameters are set in the `qsim` config file section. Important parameters are: By specifying

```
<param name="numberOfThreads" value="..."/>
```

QSim can be run in parallel, see Section 4.2.3. Importantly, the `qsim` parameters

```
<param name="flowCapacityFactor" value="..." />
<param name="storageCapacityFactor" value="..." />
```

need to be set accordingly when running sample scenarios. For example, for a 10 % sample, these factors need to be 0.1.

Currently, QSim is implemented as a single-queue model (see Chapter 50). Back-propagating gaps as discussed in Section 1.3 are available experimentally (see Section 97.5) and configurable with the parameter

```
<param name="trafficDynamics" value="..." />
```

As shown in Section 4.6.1, QSim can handle multimodal scenarios.

A somewhat ancient configuration parameter is the stuck time. It determines after how many seconds of non-movement a vehicle is moved across an intersection despite violating the storage constraint of the destination link. This parameter was introduced to resolve grid-locks, i.e., geometrical arrangements where no vehicle can move any more. With the QSim model, it is possible to add vehicles beyond the storage constraint to an overcrowded link. This corresponds to maintaining a minimal flow even under very congested conditions. The default value of this parameter is set to 10, i.e., non-moving vehicles are moved forward after 10 simulation time steps of non-movement. This may seem a rather short time, but systematic investigations (unfortunately never published) have shown that the simulations become, in comparison to traffic counts data, less realistic when this parameter is increased.

4.3.2 JDEQSim

JDEQSim (Waraich et al., 2009) was used for project *KTI Frequencies* (Balmer et al., 2010). It is is a Java reimplementation of DEQSim (Waraich et al., 2009; Charypar et al., 2007b, 2009) and provides parallel event handling, but no parallel simulation (Balmer et al., 2010, p.11). Back-propagating gaps (Section 1.3) are supported, but traffic lights, public transport and within-day replanning are not.

To run JDEQSim, the parameter `mobsim` of `controler` config file section must be set to `JDEQSim` and a `jdeqsim` config file section must be provided.

4.4 Scoring

The config file section `planCalcScore` specifies the parameters used for scoring agents' plans (Section 2.1.3); parameters are explained in Chapter 3.

4.5 Replanning Strategies

Replanning strategies are the basic innovation modules available in MATSim. We do not call them *choice* modules, although they are involved in people's choice making. The choice process is performed over the iterations with an *implicit* choice set and is not based on explicit probability function drawing. One can differentiate between modules that affect the set of plans that each agent holds, and others that only select between these plans. For a detailed discussion of MATSim in choice modeling context, see Chapter 49.

All strategy modules are called by configuring the strategy module in the configuration file as shown in the following example.

```
<module name="strategy" >
    <parameterset type="strategysettings" >
        <param name="strategyName" value="ChangeLegMode" />
        <param name="weight" value="0.1" />
    </parameterset>
    <parameterset type="strategysettings" >
        <param name="strategyName" value="TimeAllocationMutator"/>
        <param name="weight" value="0.2" />
    </parameterset>
    <parameterset type="strategysettings" >
        <param name="strategyName" value="SelectExpBeta" />
        <param name="weight" value="0.7" />
    </parameterset>
</module>
```

Each module is given a weight determining the probability, by which the course of action represented by the module is taken. The strategy modules' weights are normalized, in case they do not sum to one. In this example, each agent changes her leg mode with probability 0.1 and her plan timing with probability 0.2. Otherwise, the agent chooses a plan from her set of plans according to a logit model.

By specifying the parameter `subpopulation`, replanning strategies can be applied to distinct subpopulations: e.g.,

```
<parameterset type="strategysettings" >
    <param name="strategyName" value="ChangeLegMode" />
    <param name="weight" value="0.1" />
    <param name="subpopulation" value="urbanTravelers"/>
</parameterset>
```

In older versions of the config file, you will find a deprecated configuration syntax using numbered strategy modules.

Please note that combining strategy modules that are extensions (see Section 5.1.1), like destination innovation together with public transport, may not always work as expected. Combine them with care and contact the mailing list if you are unsure.

4.5.1 *Plans Generation and Removal (Choice Set Generation)*

4.5.1.1 *Time Innovation*

Time innovation is applied by defining its parameters in the config file section `TimeAllocationMutator` and by adding

```
<param name="strategyName" value="TimeAllocationMutator" />
```

plus its weight to the strategy modules.

The module shifts activity end times randomly within a configurable range as described by Balmer et al. (2005b); Raney (2005).

4.5.1.2 *Route Innovation*

Route innovation is applied by adding

```
<param name="strategyName" value="ReRoute" />
```

plus its weight to the strategy modules, and by specifying the routing algorithm in the `controler` config file section (Section 4.2.1). MATSim routing is described by Lefebvre and Balmer (2007).

4.5.1.3 Mode Innovation

Mode innovation is applied by adding[1]

```
<param name="strategyName"
   value="{ChangeLegMode | ChangeSingleLegMode |
   SubtourModeChoice}" />
```

plus its weight to the strategy modules. In the config file, a section with one of the mode innovation strategies needs to be added, i.e.,

```
<module name="{changeLegMode | changeSingleLegMode |
subtourModeChoice}" >
   ...
</module>
```

`ChangeLegMode` randomly picks one of a person's plans and changes the mode of transport. By default, the supported modes are: driving a car and using public transport. Only one mode of transport per plan is supported. When using different modes for sub-tours on a single day, the `SubtourModeChoice` module is required. Optionally, car availability is respected. `ChangeSingleLegMode` randomly picks one of a person's plans and changes one single leg's (picked randomly) mode of transport. In contrast to `ChangeLegMode`, it allows for multiple modes in one plan. By default, supported modes are: driving a car and using public transport. Also, this module can (optionally) respect car availability.

Mode innovation is described by Rieser et al. (2009); Meister et al. (2010); Ciari et al. (2008, 2007).

4.5.1.4 Plans Removal

The maximum number of plans per agent is configured by the setting

```
<module name="strategy" >
   <param name="maxAgentPlanMemorySize" value="5" />
   ...
</module>
```

If an agent ends up having more plans, MATSim will start removing plans, one by one, until the maximum number of plans is reached. Plans to be removed are selected by the setting configured by

```
<module name="strategy" >
   <param name="planSelectorForRemoval" value="..." />
   ...
</module>
```

Starting with release 0.8.x, the config file comments give possible options.

This option is not yet well investigated, cf. Section 97.3. Per default, the plan with the lowest score is removed if the agent's memory is full.

4.5.2 Plan Selection (Choice)

Selectors and their weight are also added to the strategy modules

```
<param name="strategyName" value="KeepLastSelected | BestScore |
SelectExpBeta ChangeExpBeta | SelectRandom | SelectPathSizeLogit" />
```

[1] The names may be changed into `ChangeTripMode` and `ChangeSingleTripMode`, please keep your eyes open.

Selectors work as follows:

- `KeepLastSelected` keeps the plan selected in the previous iteration.
- `BestScore` selects the plan with the highest score from the previous iteration.
- `SelectExpBeta` performs MNL (Multinomial Logit Model) selection between plans. It can be configured by the `BrainExpBeta` parameter from the scoring config group[2] being the scale parameter in discrete choice models, as shown in Equation 49.2. We recommend keeping this parameter at its default value of 1.0.
- `ChangeExpBeta` changes to a different plan, with probability dependent on $e^{\Delta_{score}}$, where Δ_{score} is the score difference between the two plans. This will also sample from an MNL (see Sec. 47.3.2.1).
- `SelectRandom` performs random selection between the plans.
- `SelectPathSizeLogit` selects an existing plan according to the path size logit described by Frejinger and Bierlaire (2007). It can be configured by the `PathSizeLogitBeta` parameter from the scoring config group.[3] This selector has never been investigated systematically.

Note that the `BestScore` should be used with care; it tends to get stuck with sub-optimal plans. Plans badly rated due to a random fluctuation in one single iteration, e.g., a rare traffic jam, will never be tested again. Thus, we recommend using this only in conjunction with `SelectRandom`.

4.5.3 Innovation Switch-Off

For theoretical (Section 47.3.2.3) reasons, it makes sense to eventually switch off the innovative modules, thus keeping the set of plans for each agent fixed from then on. This behavior can be configured by

```
<param name="fractionOfIterationsToDisableInnovation" value="..."/>
```

It makes sense to use this together with MSA averaging of the scores (Section 3.3.4).

4.6 Other Modes than Car

The MATSim software began with the car mode of transport, since it was then the main mode in many regions. The idea of integrating other modes has always been a theme.

The following sections describe current MATSim multi-modal capabilities. The material covers not only options that can be enabled with just config options, but also gives an overview of multi-modal extensions, described in Part II of the book.

4.6.1 QSim Side

4.6.1.1 Multiple Vehicular Modes on the Same Network

The approach described so far fails as soon as more than one vehicle type is involved. Therefore, recently the ability to allow multiple modes on the same network was introduced. It is defined by the `qsim` config option of type

```
<module name="qsim">
   <param name="mainMode" value="car,truck,bicycle" />
   ...
</module>
```

[2] This is in the scoring config group for historical reasons.

[3] Also in the scoring config group for historical reasons.

This examines the plan leg mode; if that leg mode corresponds to one of the listed main modes, it will generate a vehicle for that leg and make it enter the network.

It is currently not possible to generate different vehicle types from the config alone; one either needs to provide a vehicles file (see Section 6.6 and Section 11.1), or write a script-in-Java to generate the vehicle fleet (again see Section 11.1).

4.6.1.2 *So-Called Teleportation*

All modes *not* registered with the QSim as "main modes" will be teleported. That is, the QSim will, without problems, process legs such as

```
<leg mode="pedelec" >
   <route type="generic" trav_time="00:14:44" distance="2374" />
</leg>
```

The QSim will generate a departure event (for events, see Section 2.2.3) after the end of the previous activity and an arrival event 14 minutes and 44 seconds later. The leg will be recorded with a distance of 2 374 meters. If distance is not used for scoring (cf. Chapter 3), it can also be left out of the route (the situation in most set-ups).

4.6.1.3 *Explicitly Simulated Passenger Modes*

With "driver" modes, such as car, bicycle, or also walk, travelers are also drivers, i.e., the entities making decisions about turns at intersections, as well as arrivals (or not) on links. With "passenger" modes, such as public transit or taxi, this changes; for example, the traveler boards a bus, the bus moves around in the network; the only decision the traveler has to make if she or he wants to get off or not at the current stop. The bus, in turn, is a normal participant in the corresponding traffic system, i.e., buses and taxis operate on the normal road network and can be caught in the same congestion as cars and trucks. This is exactly how it works in the MATSim QSim; taxis typically operate on the same network as cars; pt vehicles may operate on the same network if their routes are defined so that they use the same links as regular cars. In these cases, their interactions are captured by the simulation.

4.6.1.4 *Departure Handlers*

It is possible to register a separate departure handler for each mode; see Section 45.2.8 for the syntax. There are also pre-configured extensions using this approach:

- The "multimodal" contribution moves all registered modes on separate, congestion-free networks. This is better than teleportation, since the vehicles (or pedestrians) have defined positions at each point in time, meaning that they can also re-plan, e.g., re-route (see Chapter 21).
- The public transport extension moves all registered modes with specific public transit vehicles (see Chapter 16).
- The dynamic transport systems contribution will eventually be able to move a `taxicab` mode with taxis (see Chapter 23).

4.6.2 *Routing Side*

The previous Section 4.6.1 has described how the QSim handles various modes when they are requested by the plans. Correspondingly, it now needs to be considered how non-car plans, or more specifically non-car routes inside non-car legs, are generated.

4.6.2.1 *Network Modes*

The following syntax defines modes for which the router should generate network routes, i.e., routes that contain a sequence of links to follow:

```
<module name="planscalcroute" >
   <param name="networkModes" value="car, truck" />
   ...
</module>
```

The above configuration specifies that plans containing

```
<leg mode="car"...>
```

as well as

```
<leg mode="truck"...>
```

will be treated by the network router.

As of the writing of this text, the router will route all these modes on the "car" links of the network. This means that, say, denominating some links as "car only" or "truck only" will not be picked up by the current router.[4]

Note that, per the network file DTD (Document Type Description), "car" is the default mode of each link as long as long as the link's mode field is not explicitly filled.

4.6.2.2 *Teleportation ...*

... with Teleported Mode Free Speed Factor A config entry such as

```
<module name="planscalcroute" >
   <parameterset type="teleportedModeParameters" >
      <param name="mode" value="pt" />
      <param name="teleportedModeFreespeedFactor" value="2.0" />
      <param name="teleportedModeSpeed" value="null" />
      <param name="beelineDistanceFactor" value="null" />
   </parameterset>
   ...
</module>
```

means that if the router encounters a leg with mode pt, it generates a "teleportation" route whose travel distance is the same as, and travel time is twice that of, a freespeed car route.

This models public transit, assuming it travels along roughly the same routes as a car trip, but takes twice as long (cf. Reinhold, 2006).

... with Teleported Mode Speed Setting, in the above, something like

```
   <param name="teleportedModeFreespeedFactor" value="null" />
   <param name="teleportedModeSpeed" value="4.167" />
   <param name="beelineDistanceFactor" value="1.3" />
```

will, instead, generate a teleportation route whose travel distance is 1.3 times the beeline distance, and whose travel time is that distance divided by 4.167 meters per second.

This is useful when teleported mode travel times should not change in tandem with car freespeed travel times, perhaps as a policy change result, or when teleported mode travel times are unrelated

---

[4] Check https://matsim.atlassian.net/browse/MATSIM-330 for developments.

to car travel times. One disadvantage: this approach does not take obstacles like water or mountain areas, into account for the teleported modes.

4.6.2.3 Other Routing Options

It is possible to register separate routers for specific modes. This syntax is discussed in Section 45.2.7. The pre-configured extensions and contributions discussed in Section 4.6.1.4, "multimodal", public transport, taxis, come with corresponding routers.

In addition, the so-called "matrix based pt router" (Chapter 20) uses a list of transit stops and a matrix of stop-to-stop travel times and travel distances; based on this information, it computes a teleported walk leg to the next stop, another to the destination stop, and a last teleported walk leg to the final destination.

The matrix-based pt router also illustrates that, given the QSim teleportation capability, it is possible to come up with arbitrary algorithms for arbitrary modes, as long as they generate (expected) travel times and (expected) travel distances. As said earlier, the teleportation facility of the QSim will just use these two attributes at face value. Although with such an approach neither congestion nor en-route replanning are or can be included, it is flexible and allows a fully modular addition of arbitrary modes without having to interact with the QSim.

4.6.3 Scoring Side

For all modes mentioned in the plans, a corresponding scoring section must exist. See Section 3.2.1 for an example.

4.7 Observational Modules

4.7.1 Travel Time Calculator

The routing module, for example, needs travel time estimations for all network links. To keep computational effort feasible, travel time estimations need to be aggregated to time bins. Parameters of this aggregation, such as bin size, can be specified in the configuration file section `travelTimeCalculator`.

4.7.2 Link Stats

The `linkStats` config file section can specify the output interval of individual links' simulation statistics. It is configurable if the simulated volumes are written per iteration or averaged over multiple iterations. As one of their many functions, link stats are used for comparison with count values, as introduced in Section 6.3.

PART II

Extending MATSim

CHAPTER 5

Available Functionality and How to Use It

Andreas Horni and Kai Nagel

In this chapter you will learn about possibilities to extend and customize MATSim (Multi-Agent Transport Simulation) through provided functionality. In Chapter 45, you will see how you can hook your own extensions into MATSim.

5.1 MATSim Modularity

MATSim follows a modular concept, but a "module" is not a very specific term;[1] thus, modules can exist at many levels in a software framework. Also in MATSim, a range of different functionality types, such as config functions, replanning components, contributions, or even external tools,[2] are sometimes described as modules. Metaphorically speaking, a module can thus be seen as the greatest common divisor (gcd) of different functionality provided in MATSim. Much more important is understanding the different levels of access stemming from the generally modular architecture.

5.1.1 Levels of Access

MATSim currently provides five levels of access:

1. using the MATSim core only,
2. using the MATSim main distribution,

[1] According to the Merriam-Webster (`http://www.merriam-webster.com`), a module is "one of a set of parts that can be connected or combined to build or complete something" or more specifically "a part of a computer or computer program that does a particular job".

[2] Standalone tools referencing MATSim as a library, such as the network editor, or the visualizer Via.

How to cite this book chapter:
Horni, A and Nagel, K. 2016. Available Functionality and How to Use It. In: Horni, A, Nagel, K and Axhausen, K W. (eds.) *The Multi-Agent Transport Simulation MATSim*, Pp. 47–52. London: Ubiquity Press. DOI: http://dx.doi.org/10.5334/baw.5.

3. using MATSim main distribution, contributions and possibly extensions,
4. writing "scripts in Java" and finally
5. writing your own extensions.

5.1.1.1 *Using the Core Only*

To use only the core, one needs to do the following (see Section 2.1):

- Download a MATSim release or a nightly build, by following the respective links at `http://matsim.org/downloads`.
- Obtain a network file and an initial plans file. Small versions can be typed by hand; larger versions should be generated automatically by some computational method.
- Write or edit a config file.
- Click on the MATSim jar file[3] and follow the instructions.

We think that the MATSim core is already quite powerful; for example, synthetic persons already follow full daily plans with a full daily scoring function; thus, opening times for activity types, departure time choice and schedule delay can be investigated.

5.1.1.2 *Using MATSim Main Distribution*

The extensions in the MATSim main distribution are, by design, very close to the MATSim core, thus requiring even less configuration than for contributions, as shown below. Often, providing additional files together with a respective config file entry is sufficient to use them; required steps are described below, case by case. Extensions contained in the main distribution are listed in a separate section at `http://matsim.org/extensions`.

5.1.1.3 *Using One or More Contribs or Other Extensions*

Contributions are in a separate part of the repository, separate from the MATSim main distribution. The documentation is not yet fully organized; information about contributions and other extensions can be found at `http://matsim.org/extensions`. For the contributions, there are also release versions and nightly builds, which can be found by following the links at `http://matsim.org/downloads`.

In general, contributions should provide main methods for use. We may eventually provide clickable jar files here as well, but for the time being, contributions need to be bundled with core MATSim (and potentially other contributions). As shown at `http://www.matsim.org/docs/extensions`, the syntax is roughly

```
java -Xmx2000m -cp MATSim.jar:contrib/contrib.jar org.matsim.contrib.run.RunXxx
   config.xml
```

where

- `-Xmx2000m` increases the Java heap space, so that most MATSim runs fit in,
- `MATSim.jar` needs to be replaced by a relative or absolute path to the MATSim jar to be used,
- `contrib/contrib.jar` needs to be replaced by a relative or absolute path to the contribution jar to be used,

[3] This has worked since winter 2014/15 and should be in the 0.8.x release.

- `org.matsim.contrib.run.RunXxx` needs to be replaced by the full Java class name containing the desired main method (given by the contribution documentation), and
- `config.xml` needs to be replaced by a relative or absolute path to a config file, which may contain additional sections specific to the contribution.

It is possible to combine several contributions in this way, provided someone has made a corresponding main method available. This can, in principle, be done relatively quickly, so those wishing to run studies with combinations of existing contributions, but without programming skills, can ask someone with those skills and with access to the repository for help.

5.1.1.4 *Writing "Scripts in Java"*

The contributions are written so that they can be plugged into MATSim via extension points (see Chapter 45). If a specific combination or configuration of modules is not (yet) available, one can write it. The syntax is roughly:

```
... main( ... ) {
    // construct the config object:
    Config config = ConfigUtils.xxx(...) ;
    config.xxx().setYyy(...) ;
    ...

    // load and adapt the scenario object:
    Scenario scenario = ScenarioUtils.loadScenario( config ) ;
    scenario.getXxx().doYyy(...) ; // (*)
    ...

    // load and adapt the controler object:
    Controler controler = new Controler( scenario ) ;
    controler.doZzz(...) ; // (**)
    ...

    // run the iterations:
    controler.run() ;
}
```

Extension points, especially at `(*)` and `(**)`, are described in more detail in Chapter 45.

5.1.1.5 *Writing Your Own Extensions*

If the existing extensions are not sufficient to plug your own study together, the next option is to write your own extension. Again, when writing an extension, one should use the extension points described in Chapter 45, since this is the only way an extension can later become a contribution.

5.1.2 *The Ideas Behind this Setup*

The setup, as described above, arose from the observation that an-ever growing monolithic MATSim would eventually overwhelm the MATSim team and its core developers group. Therefore, a set-up was sought allowing them to concentrate on central infrastructure, while specific functionality like road pricing, multimodal simulations, signals, additional choice dimensions, or analysis modules could be written and contributed by the community. Clearly, a plug-in architecture had to be the solution, but it took (and still takes) time and effort to make the extension points sufficiently capable and robust.

At the same time, MATSim is a research platform; research investigates innovative questions, which often means that the questions were not foreseen when the code was designed. Quite often, scripting languages are the solution to such problems; for example, python is allowed in QGIS,[4] VISUM (Verkehr In Städten – UMlegung),[5] EMME (Equilibre Multimodal Multimodal Equilibrium), or SUMO (Simulation of Urban Mobility) (via the TraCI interface)[6] for plug-ins. Scala (SCAlable LAnguage) was discussed for MATSim, but ultimately, it was decided to just use Java itself as the scripting language, with the advantage that users between development and MATSim application do not need to learn two languages. In addition, the TU (Technische Universität) Berlin team can continue to teach Java both as an entry point to MATSim and as a general professional skill.

5.2 An Overview of Existing MATSim Functionality

Figure 5.1 shows where common MATSim modules are coupled with the MATSim loop. Some modules have a single connection point (shown around the loop, connected to the respective loop

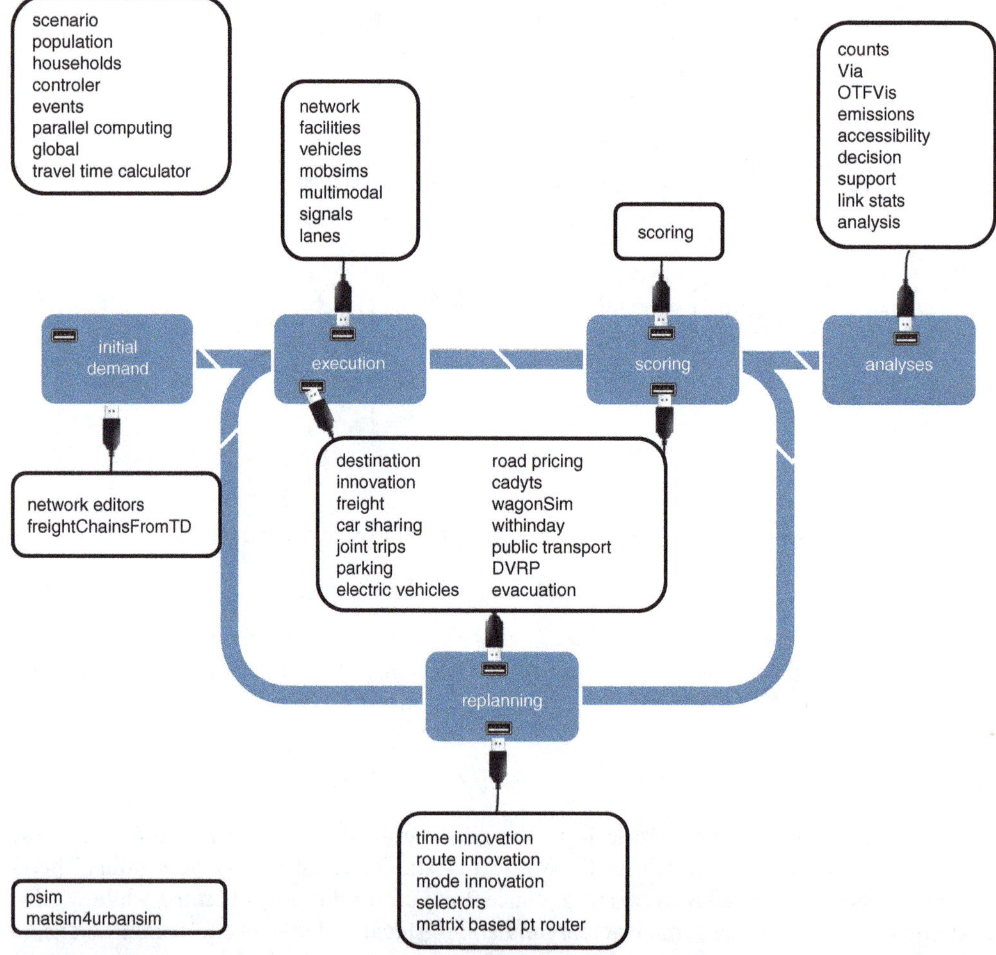

Figure 5.1: MATSim functionality.

4 http://docs.qgis.org/testing/en/docs/pyqgis_developer_cookbook/
5 PTV (2011)
6 http://sumo.dlr.de/wiki/TraCI

element), while others have multiple connection points (shown in the middle of the circle) and yet others work on a global range (shown on the left upper and lower corners).

The technical details for module usage, in particular, the parameter sets, are described at `http://matsim.org`, especially `http://matsim.org/javadoc` and `http://matsim.org/extensions`.

As a result of the distributed and project- and dissertation-driven MATSim contribution process (see Chapter 44), modules are often implemented for a specific practical purpose, leading to limitations of the respective module. For example, modules might only work for a specific mode, or for a defined calling order. Normally, additional effort is needed to generalize the module; in consequence, the combination of a specific module with other functionality is often not a straightforward task. This means that a user will have to systematically test any specific combination of modules before productively applying it.

The description of the modules in Chapter 4, and the following chapters, is based on the categorization shown in Table 5.1.

Global Modules and Global Aspects	Section 4.2
Controler	Section 4.2.1
Events	Section 4.2.2
Parallel Computing	Section 4.2.3
Global	Section 4.2.4
MATSim Data Containers	Section 4.1 and Chapter 6
Network	Section 4.1.1 and 6.1
Population	Section 4.1.2 and 6.2
Counts	Section 6.3
Facilities	Section 6.4
Households	Section 6.5
Vehicles	Section 6.6
Scenario	Section 6.7
Network Editors	
MATSim JOSM Network Editor	Chapter 8
Map-to-Map Matching Editors in Singapore	Chapter 9
The "Network Editor" Contribution	Chapter 10
Observational Modules	Section 4.7
Travel Time Calculator	Section 4.7.1
Link Stats	Section 4.7.2
Scoring	Section 4.4
Basic Strategy Modules	Section 4.5
Time Innovation	Section 4.5.1.1
Route Innovation	Section 4.5.1.2
Mode Innovation	Section 4.5.1.3
Selectors	Section 4.5.2
Mobsims	
QSim	Section 4.3.1 and Chapter 11
JDEQSim	Section 4.3.2
Individual Car Traffic	
Signals and Lanes	Chapter 12
Parking	Chapter 13

Continued on next page

Electric Vehicles	Chapter 14
Roadpricing	Chapter 15
Other Modes Besides Individual Car	
Public Transport	Chapter 16
The "Minibus" Contribution	Chapter 17
Semi-Automatic Tool for Bus Route Map Matching	Chapter 18
Events-Based Public Transport Router	Chapter 19
matrix-based pt router	Chapter 20
Multi-Modal Contribution	Chapter 21
Car Sharing	Chapter 22
Dynamic Transport Systems	Chapter 23
Commercial Traffic	
Freight Traffic	Chapter 24
wagonSim	Chapter 25
freightChainsFromTravelDiaries	Chapter 26
Additional Choice Dimensions	
Destination Innovation	Chapter 27
Joint Trips and Social Networks	Chapter 28
Socnetgen	Chapter 29
Within-Day Replanning	
Within-day Replanning	Chapter 30
Belief Desire Intention (BDI) Framework	Chapter 31
Automatic Calibration	
Cadyts	Chapter 32
Visualizers	
Via Visualizer	Chapter 33
OTFVis Visualizer	Chapter 34
Analysis	
Accessibility	Chapter 35
Emissions	Chapter 36
Interactive Analysis and Decision Support	Chapter 37
The "analysis" contrib	Chapter 38
Computational Performance Improvements	
PSim	Chapter 39
Other Modules	
Evacuation	Chapter 41
MATSim4UrbanSim	Chapter 42

Table 5.1: MATSim functionality overview.

SUBPART ONE

Input Data Preparation

CHAPTER 6

MATSim Data Containers

Marcel Rieser, Kai Nagel and Andreas Horni

6.1 Time-Dependent Network

The network container was already described in Section 4.1.1. An important additional feature of the network module is using time-dependent network attributes. Network state changes can thus be considered, as e.g., implied by accidents, or adaptive traffic control, with varying speed limits or driving directions of lanes on multi-lane roads with heavily unbalanced loads over the course of a day. Attributes that can be adapted are "free speed", "number of lanes" and "flow capacity".

The adaptation can be specified by adding the following two lines to the network config file section:

```
<param name="timeVariantNetwork" value="true" />
<param name="inputChangeEventsFile"
    value="path_to_change_events_file" />
```

An example snippet setting the free speed of three network links to zero looks something like this:

```
<networkChangeEvent startTime="03:06:00">
   <link refId="12487"/>
   <link refId="12489"/>
   <link refId="12491"/>
   <freespeed type="absolute" value="0.0"/>
</networkChangeEvent>
```

For a working example, see the file networkChangeEvents.xml in the examples/equil-extended directory in the MATSim directory tree.

Alternatively, network change events can be added directly to the code. An example can be found in the RunTimeDependentNetworkExample class under http://matsim.org/javadoc → main distribution.

How to cite this book chapter:

Rieser, M, Nagel, K and Horni, A. 2016. MATSim Data Containers. In: Horni, A, Nagel, K and Axhausen, K W. (eds.) *The Multi-Agent Transport Simulation MATSim*, Pp. 55–60. London: Ubiquity Press. DOI: http://dx.doi.org/10.5334/baw.6. License: CC-BY 4.0

Note that change values of type absolute need to be given in SI units, which means speeds in meters per second and flow capacities in vehicles per second.

6.2 Person Attributes and Subpopulations

The population container was also already discussed earlier, in Section 4.1.2. A powerful extension of a standard population can be achieved by specifying further agent attributes in an `ObjectAttributes` file input to MATSim by the parameter `inputPersonAttributesFile`.

See `http://matsim.org/javadoc` → main distribution → `RunSubpopulationsExample` class for an example. That example looks as if coding in Java is necessary, but this is really not the case; Java is just used to generate the subpopulations, which could also be done by other means.

6.3 Counts

By providing a counts input file and configuring the `counts` config file section, MATSim plots link volume comparisons between hourly simulated and counted values for motorized individual traffic (Horni and Grether, 2007).

Simulating sample populations requires scaling simulated volumes by the `countsScaleFactor` parameter, e.g., for a 10 % population this parameter needs to be set to 10.

Input The following listing shows an example of a `counts.xml` input file required for traffic count comparisons.

```
<?xml version="1.0" encoding="UTF-8"?>
<counts name="example" desc="example counting stations" year="2015">
   <count loc_id="2" cs_id="005">
      <volume h="1" val="10.0"></volume>
      <volume h="2" val="1.0"></volume>
      <volume h="3" val="2.0"></volume>
      <volume h="4" val="3.0"></volume>
      <volume h="5" val="4.0"></volume>
      <volume h="6" val="5.0"></volume>
      <volume h="7" val="6.0"></volume>
      <volume h="8" val="7.0"></volume>
      <volume h="9" val="8.0"></volume>
      <volume h="10" val="9.0"></volume>
      <volume h="11" val="10.0"></volume>
      <volume h="12" val="11.0"></volume>
      <volume h="13" val="12.0"></volume>
      <volume h="14" val="13.0"></volume>
      <volume h="15" val="14.0"></volume>
      <volume h="16" val="15.0"></volume>
      <volume h="17" val="16.0"></volume>
      <volume h="18" val="17.0"></volume>
      <volume h="19" val="18.0"></volume>
      <volume h="20" val="19.0"></volume>
      <volume h="21" val="20.0"></volume>
      <volume h="22" val="21.0"></volume>
      <volume h="23" val="22.0"></volume>
      <volume h="24" val="23.0"></volume>
   </count>
</counts>
```

For a working example, check the `examples/equil` directory in the MATSim directory tree (cf. Section 2.1.1).

It starts with a header containing general descriptive information about the counts, including a year to describe how current the data is. Next, for each link having real world counts data, hourly volumes can be specified. The network-link is referenced by the `loc_id` attribute; in the example, it is link 2. The attribute `cs_id` (counting station identifier) can be used to store an arbitrary description of the counting station. Most often, it is used to note the original real world counting station to simplify future data comparison. The hourly volumes, specified by the hour of the day and its value, are optional: That is, a value does not have to be given for every hour. If, for a counting station, data is only available for certain hours of the day (e.g., only during peak hours), it is possible to omit the other hours from the XML listing. Note that the first hour of the day, from 0:00 am to 1:00 am, is numbered as "1", and *not* by "0" as is often the case in computer science.

Output The counts module prints overview summaries for the whole network, but also analyzes for individual links. Also, a google maps-based visualization is available, showing each station with a its load curve (see the example in Figure 6.1) in a pop-up window.

Balmer et al. (2009a) have performed link volume comparisons for the Zürich scenario, with data based on city level, cantonal level and national level (ASTRA, 2006). Usually, it is helpful to exclude a substantial part of the outer range of the modeled study region in order to remove boundary effects.

6.4 Facilities

Facilities are an optional element of MATSim; some modules, such as the destination innovation module (Chapter 27), depend on it. If MATSim facilities are used, agents perform their activities in a specific facility attached to a network link.

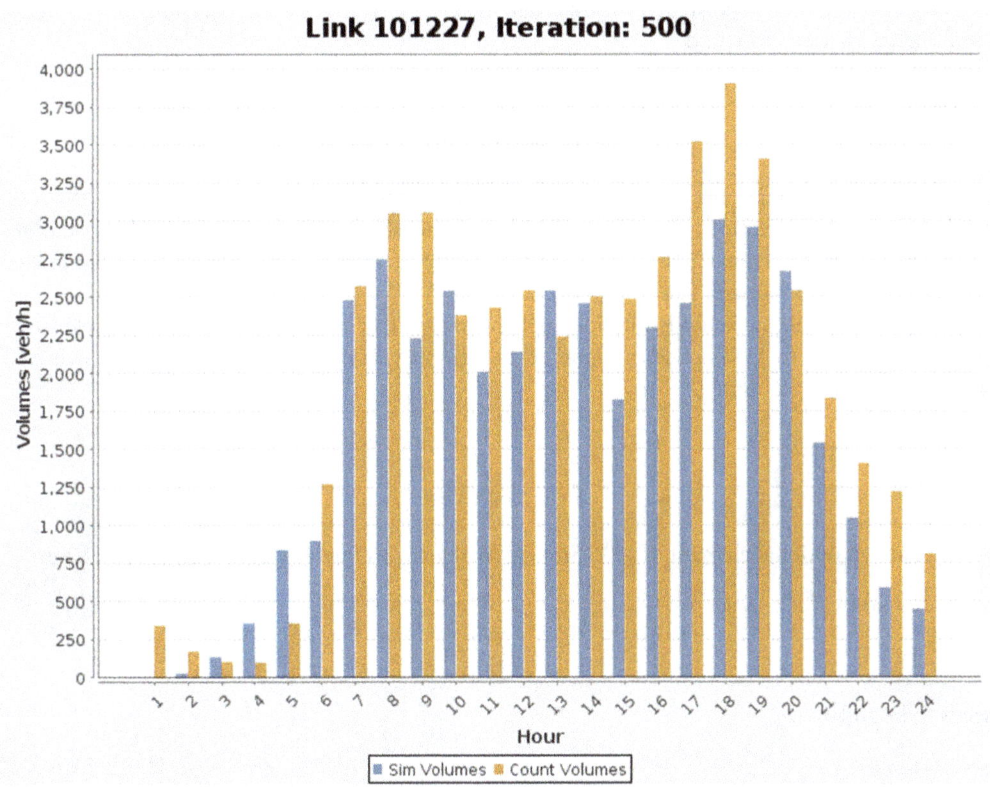

Figure 6.1: Example for a link volumes comparison between simulation and road count values.

Facilities are included in the scenario by defining the `facilities` config file section and providing a facilities file, approximately as follows.

```
...
<facilities name="test facilities for triangle network">
   <facility id="1" x="60.0" y="110.0">
      <activity type="home" />
   </facility>
   <facility id="10" x="110.0" y="270.0">
      <activity type="education" />
   </facility>
</facilities>
```

An example is given in `http://matsim.org/javadoc` → main distribution → `RunWithFacilitiesExample` class.

In addition to activities that can be done in the facility, further location attributes, such as opening times, can be specified. A working facilities example file can be found in the MATSim directory tree in the `examples/siouxfalls-2014` directory.

Facilities are mostly used by the MATSim Zürich group, in particular in the Zürich scenario, where they are derived from the Federal Enterprise Census 2001 (Swiss Federal Statistical Office (BFS), 2001) providing hectare level information. Detailed technical description of facilities generation is given by Meister (2008). Comparable data is available in most countries from official sources, such as censuses, and commercial sources, such as navigation network providers, yellow pages publishers or business directories, and last but not least google and OSM (OpenStreetMap, 2015).

Note that loading a facilities file into MATSim by itself does not mean they will be used; the functionality needs to be switched on by other means. Currently, this is only possible by using some class with a `main` method.

6.5 Households

Households are another optional element of MATSim. To load households into a scenario, the config file must contain a section `households`. This section should specify the paths to a file containing households (parameter `inputFile`) and a file containing further household attributes (parameter `inputHouseholdAttributesFile`).[1]

Again, loading the households file does not mean that it is used anywhere in the code; such functionality needs to be switched on separately. Currently, no such functionality can be switched on from the config file alone.

6.6 Vehicles

Vehicles are an optional element of MATSim. To load vehicles into a scenario, a config section

```
<module name="vehicles" >
   <param name="vehiclesFile" value="/path/to/vehicles.xml.gz" />
</module>
```

needs to be added.[2]

[1] There used to be an additional "useHouseholds" config switch. In release 0.8.x, that switch will be gone.
[2] There used to be an additional "useVehicles" config switch. In release 0.8.x, that switch will be gone.

Once more, just loading the vehicles does not use them; that needs to be configured separately. See Section 11.1 for details.

6.7 Scenario

`Scenario` is a super-container containing all the other data containers, accessible, for example, as `scenario.getNetwork()`. It used to have configuration options, but these are all gone now, so `Scenario` is only visible once you are programming in Java.

CHAPTER 7

Generation of the Initial MATSim Input

Marcel Rieser, Kai Nagel and Andreas Horni

As explained in Section 2.2, the minimal MATSim input, besides the configuration, consists of the network and population with initial plans. For illustrative scenarios, all three can be generated with a text editor. For more complicated and/or realistic scenarios, they need to be generated by other methods. People with knowledge in a scripting language may use that scripting language to generate the necessary XML files, possibly honoring the MATSim DTDs. We ourselves use Java as our scripting language for these purposes. Java is not necessarily the best choice here; this may be discussed elsewhere. We do use it, for the following reasons:

- Most of us also program MATSim extensions and these currently have to be in Java. Thus, using Java as a scripting language for initial input generation saves us the effort of becoming proficient in another programming language.
- The MATSim software, by necessity, already contains all file readers and writers for MATSim input, saving the effort of re-implementing them and one automatically moves forward with file version updates. Additionally, one can directly use the MATSim data containers.
- Once one starts writing MATSim scripts-in-Java (Section 5.1.1.4), in many situations, it makes sense to modify the input data after reading the files. The programming techniques for this are the same as for other initial input generation.

Part IV will show how initial input was generated on a practical level—discussing, e.g., the different types of original input data—for different scenarios. This section presents MATSim's technical tools for initial input generation.

How to cite this book chapter:
Rieser, M, Nagel, K and Horni, A. 2016. Generation of the Initial MATSim Input. In: Horni, A, Nagel, K and Axhausen, K W. (eds.) *The Multi-Agent Transport Simulation MATSim*, Pp. 61–64. London: Ubiquity Press. DOI: http://dx.doi.org/10.5334/baw.7.

7.1 Coordinate Transformations in Java

Section 2.2.1.3 has given information about coordinate systems. When programming in Java and MATSim for input data generation, coordinate transformations derived from geotools (Geotools, accessed 2015) can be used. For example,

```
CoordinateTransformation ct =
  TransformationFactory.getCoordinateTransformation("WGS84", "WGS84_UTM33N");
```

would transform data given in WGS84 coordinates to data in UTM coordinates.

7.2 Network Generation

7.2.1 *From OpenStreetMap*

A fairly standardized way to generate a MATSim network is from OSM data. The process (roughly) goes as follows:

1. Download the necessary `xxx.osm.pbf` file from `http://download.geofabrik.de/osm`.
2. Download a recent Osmosis build from `http://wiki.openstreetmap.org/wiki/Osmosis`.
3. The necessary command to extract the road network (approximately) is:

   ```
   java -cp osmosis.jar --rb file=xxx.osm.pbf \
      --bounding-box top=47.701 left=8.346 bottom=47.146 right=9.019 \
      completeWays=true --used-node --wb allroads.osm.pbf
   ```

   The bounding box can, e.g., be obtained from `http://www.osm.org`; it is in WGS84 coordinates.
4. It makes good sense to add the large roads of a much larger region. The necessary command (approximately) is

   ```
   java -cp osmosis.jar --rb file=xxx.osm.pbf --tf accept-ways \
      highway=motorway,motorway_link,trunk,trunk_link,primary,primary_link \
      --used-node --wb bigroads.osm.pbf
   ```

5. The two files are merged with (approximately) the following command:

   ```
   java -cp osmosis.jar --rb file=bigroads.osm.pbf --rb allroads.osm.pbf \
      --merge --wx merged-network.osm
   ```

An example script of how to convert the resulting `merged-network.osm` file into a MATSim network file can be found under `http://matsim.org/javadoc` → main distribution → `RunPNetworkGenerator` class.

7.2.2 *From Other Sources*

Networks can also be obtained from other sources. An example script of how to convert an EMME network to MATSim can be found under `http://matsim.org/javadoc` → main distribution → `RunNetworkEmme2MatsimExample` class. A problem with EMME network files is that they use user-defined variables in non-standardized ways, meaning that each converter has to be adapted to the specific situation.

Material to read VISUM files can be found by searching for the string "visum" in the code base, but is currently not systematically maintained.

7.3 Initial Demand Generation

7.3.1 Simple Initial Demand

A simple script to generate a population with a single synthetic person with one initial plan can be found under `http://matsim.org/javadoc` → main distribution → `RunPOnePersonPopulationGenerator`. A somewhat larger synthetic population is generated by `RunPPopulationGenerator`.

Note that coordinates in the population need to be consistent with coordinates in the network. Roughly speaking, coordinates mentioned in the population file need to be in the same range as coordinates mentioned in the network. Note that, in the examples presented here, coordinates of the network generated in Section 7.2.1 are *not* consistent with the demand generated by the `RunP*` -scripts; these need to be adapted accordingly.

7.3.2 Realistic Initial Demand

A script to illustrate the generation of a more realistic population and initial demand can be found under `http://matsim.org/javadoc` → main distribution → `RunZPopulationGenerator`, generating a sample population from a census file and writing it to a file.

Here, network coordinates generated in Section 7.2.1 *are* consistent with demand generated by the `RunZ*`-script.

CHAPTER 8

MATSim JOSM Network Editor

Andreas Neumann and Michael Zilske

8.1 Basic Information

Entry point to documentation:

`http://matsim.org/extensions` → josm-plugin

Invoking the module:

Can be loaded as a plug-in from the JOSM editor.

Selected publications:

Kühnel (2014) (in German)

8.2 Introduction

A plugin for the JOSM (Java Open Street Map Editor) (JOSM, 2014), is available, simplifying the process of creating and editing MATSim networks. This plugin fully integrates with JOSM, benefiting from its built-in functionality.

8.2.1 Features

The MATSim JOSM network editor lets a reader preview, edit and save a MATSim network directly from the map. Basic support for converting and editing public transport networks is implemented. The plug-in allows automatic post-processing of a network by removing unnecessary intermediate nodes and links.

Convert MATSim networks from OSM. Load map data for a selected area directly from the Internet or load it from a local OSM file. Specify conversion parameters and save a MATSim network.

How to cite this book chapter:

Neumann, A and Zilske, M. 2016. MATSim JOSM Network Editor. In: Horni, A, Nagel, K and Axhausen, K W. (eds.) *The Multi-Agent Transport Simulation MATSim*, Pp. 65–66. London: Ubiquity Press. DOI: http://dx.doi.org/10.5334/baw.8. License: CC-BY 4.0

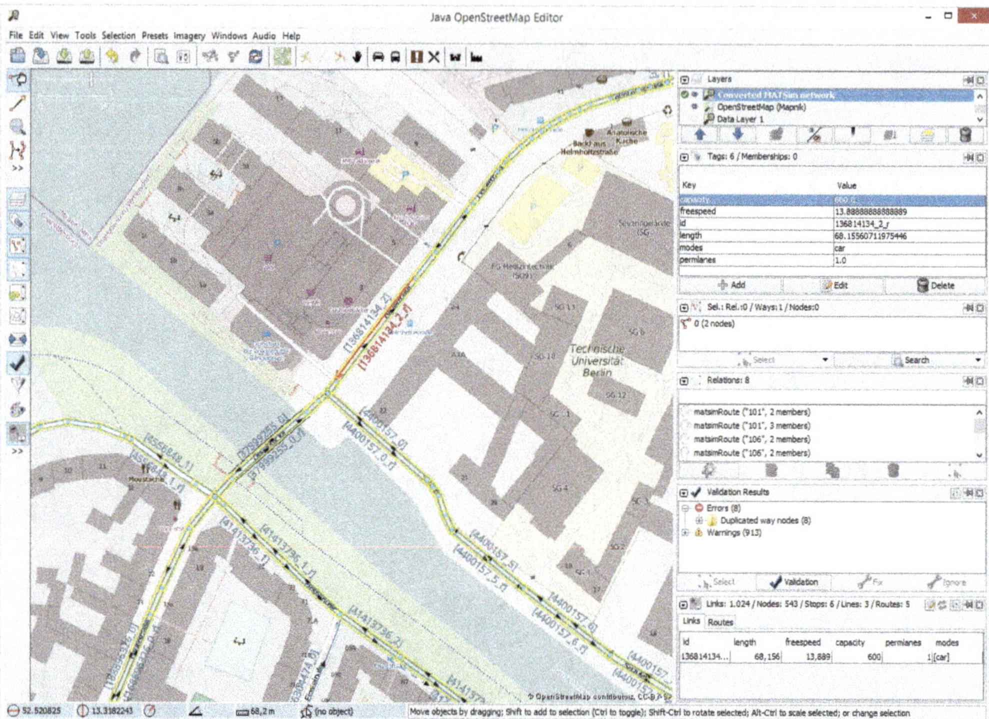

Figure 8.1: JOSM with converted MATSim network and OSM background imagery. Map data taken from OpenStreetMap (2014).

Visualize an existing or newly converted MATSim network along with other data like satellite imagery or other JOSM-supported layers.

Edit an existing or newly converted MATSim network with the available JOSM tools you know. Use the build-in undo and search functions of JOSM. Changes to the underlying OSM data are immediately reflected by the converted MATSim network. Use MATSim-specific presets to minimize errors.

Validate an existing or newly converted MATSim network to comply with requirements of the MATSim network file description. Visualize errors and fix them (automatically).

The next version will support public transport networks.

8.2.2 *Installing the Plug-In*

You do not need to download the source; it is in the JOSM plug-in repository. Just start JOSM and look for the MATSim plug-in under `Edit...Preferences...Plugins`. Download the list of available plug-ins and search for "matsim". Tick the box, press `ok` and restart JOSM.

8.2.3 *Getting the Code*

The source code is hosted on github (`https://github.com/matsim-org/josm-matsim-plugin`). Unlike MATSim, the build is not based on Apache Maven, but on Gradle. Editing the `Manifest`, downloading JOSM for compilation and building a flat JAR are easier in Gradle. Use your favorite IDE (Integrated Development Environment) to import the Gradle project and/or see the comments in `build.gradle` for details. You can run JOSM and the plug-in in the debugger.

CHAPTER 9

Map-to-Map Matching Editors in Singapore

Sergio Arturo Ordóñez

9.1 Basic Information

Entry point to documentation:

`http://matsim.org/extensions` → networkEditorSingapore

Invoking the module:

See `http://matsim.org/extensions` → networkEditorSingapore for more information

Selected publications:

Ordóñez Medina (2011a)

For the Singapore scenario and supply data, a high resolution network was obtained from the NAVTEQ company. This network consists of a graph representing every road in the island: very convenient for a high resolution model like MATSim. However, the information on travel capacities and network link free speeds is not accurate. To offset, local authorities provided the network model used for planning, which includes only major roads and simplified intersections, but capacities and free speed are accurately estimated. Figure 9.1 shows lower travel capacities of many primary roads in the navigation model (right), than in the planning model (left).

This section describes a semi-automatic tool developed to match these two network models (Ordóñez Medina, 2011a), allowing updating of navigation network (high-res network) main links/capacities and free speeds with those of the planning network (low-res network).

How to cite this book chapter:

Ordóñez, S A. 2016. Map-to-Map Matching Editors in Singapore. In: Horni, A, Nagel, K and Axhausen, K W. (eds.) *The Multi-Agent Transport Simulation MATSim*, Pp. 67–72. London: Ubiquity Press. DOI: http://dx.doi.org/10.5334/baw.9.

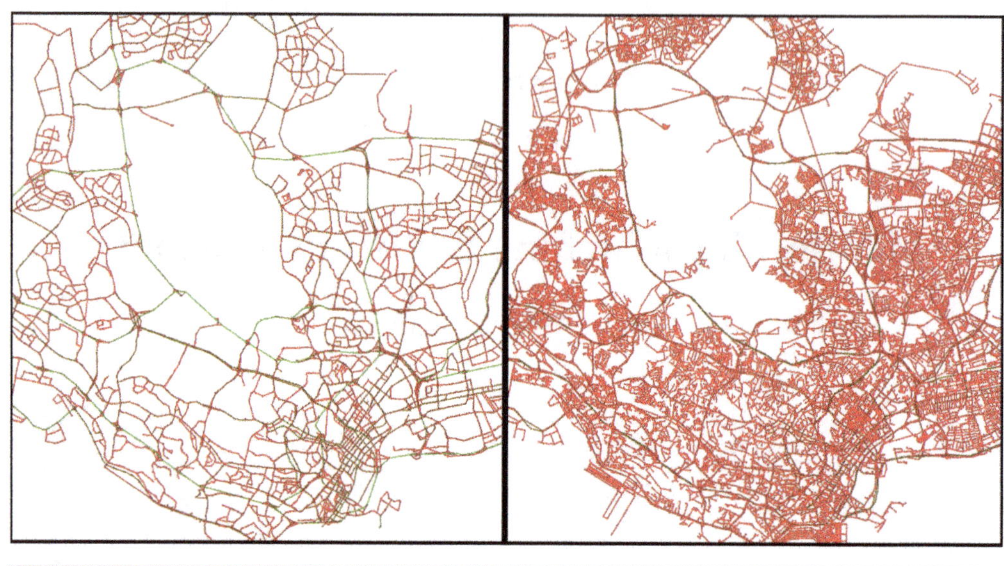

Figure 9.1: Difference in the travel capacities between the Singapore planning network model (left) and a navigation network model (right).

9.1.1 General Procedure

Although many authors try to solve matching problems for two networks in a formal way, this work follows a semi-automatic approach. This means that automatic algorithms will be used to try and solve the problem, but the user knows the solution will not be perfect; some manual work must be done. Hence, interactive tools are also provided to manually improve solutions.

The map-to-map procedure is based on the algorithm developed by Balmer et al. (2005a). It consists of the following steps:

1. Classify nodes according to their topology (e.g., source, sink, one way start, crossing) in both networks.
2. Reduce networks according to previous classification, and save relations to the original nodes.
3. Find crossings (set of close nodes) in both networks and relate them.
4. **Assuming not all crossings were found in the previous step, use the interactive tool shown in the Figure 9.2 to find all crossings in both networks and relate them.**
5. Recognize links or sequences of links joining crossings found in (3) and (4).
6. **Assuming not all links or paths found in the previous step are correct, use the link-link matching interactive tool shown in the Figure 9.3, to find or modify links or sequences of links joining the crossings**
7. Update capacities and free speeds of matched links found in (5) and (6).

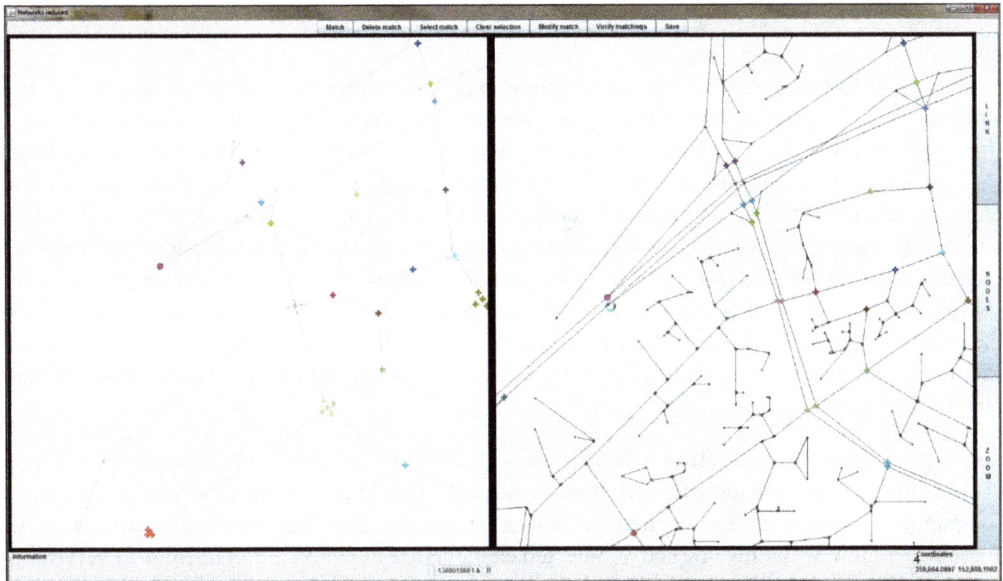

Figure 9.2: Crossing-crossing matching application. A second node, matching the pink node on the (left) low-res network, is selected from the high-res network on the right.

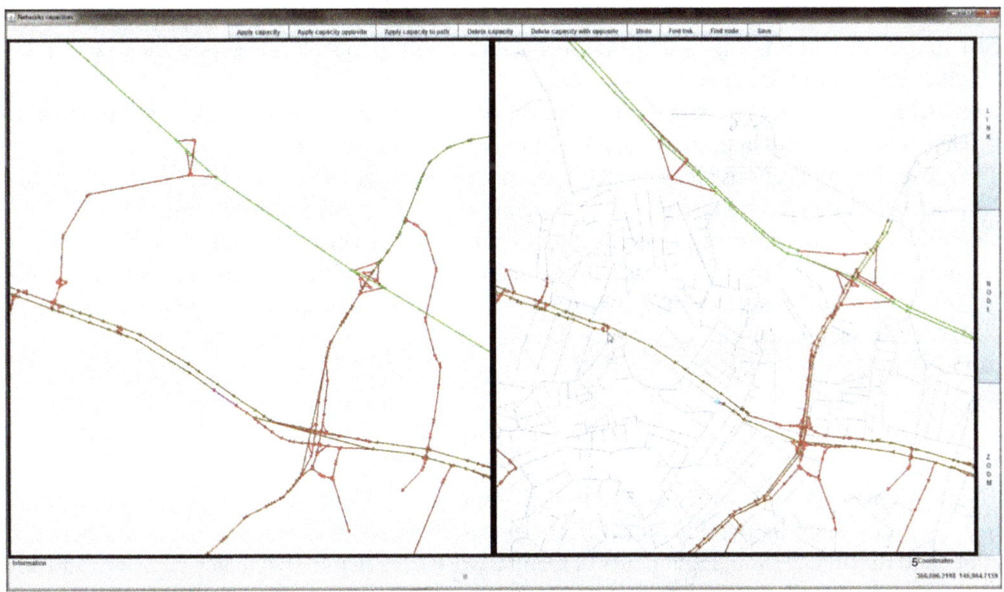

Figure 9.3: Link-link matching application. A shortest path algorithm to select a sequence of right-hand network links will be executed when clicking the destination node.

9.1.2 *Interactive Tools Characteristics*

As shown in Figure 9.2, the application allows interactive modifying of crossing-crossing relationships. A very similar interactive tool was also developed to modify link-link relationships between the high and low resolution networks. They can be found at the package `playground.sergioo.networksMatcher2012`, in the playgrounds project of MATSim. To run the crossings-crossing application graphic interface, use the class `gui.DoubleNetworkMatchingWindow`, and use the class `gui.DoubleNetworkCapacitiesWindow` for the link-link application. These applications write simple text files of the relationships located. The program found at the class `ApplyCapacities` overwrites capacities and/or free speeds, according to simple text files and writes the new resulting network XML file. This multiple-steps design enables running interactive applications several times, or in parallel. The interactive tools' developed functional requirements and quality attributes are:

- Visualization: Two navigation networks are displayed in two modes. The first mode splits the window in two, showing each network on one side and maintaining them at the same geographical position and zoom when navigating. The second superimposes both networks in the same window, with only one active. Selected elements are drawn in different colors. Everything is displayed in a bi-dimensional interactive way, showing the cursor location in the working coordinates and including panning, zoom and view-all options. The crossing-crossing application displays matched sets of nodes (crossings), with the same color in both networks. The link-link application tool also allows visualization of the capacity (or free speed) property value of both networks' links, using a color scale, as shown in Figure 9.3.
- Selection: The applications enable selection of links and nodes from both networks. The crossing-crossing option allows only selection of node sets. The link-link application allows selection of links' sequence. This can be done directly, or by selecting an origin node, a destination node and running a "select shortest path algorithm tool". It is also possible to select the other link instead of the first one chosen.
- Matching and Deletion: The applications allow creation of a similarity relationship between elements selected in both networks, sets of nodes, or sequences of links.
- Saving: The applications allow located relationships to be saved.
- Loading: The applications allow the loading of previously located relationships.
- Others: The crossing-crossing application executes and automatically verifies currently found matching, to avoid repeated nodes. It also enables clearing of the current selection. The link-link application allows automatic navigation to a link, or node, specified by the user, using its ID. It also enables the undoing of previous matching.

9.1.3 *Results*

All low-res network links were matched to high-res links, updating the corresponding link properties. Figure 9.4 shows the differences in travel capacities between original navigation network values and the final version. Eight hours of manual work were required to match crossings and ten hours of manual work to match links. Obviously, improvements in accuracy and completeness of the automatic matching algorithms reduce the manual work time.

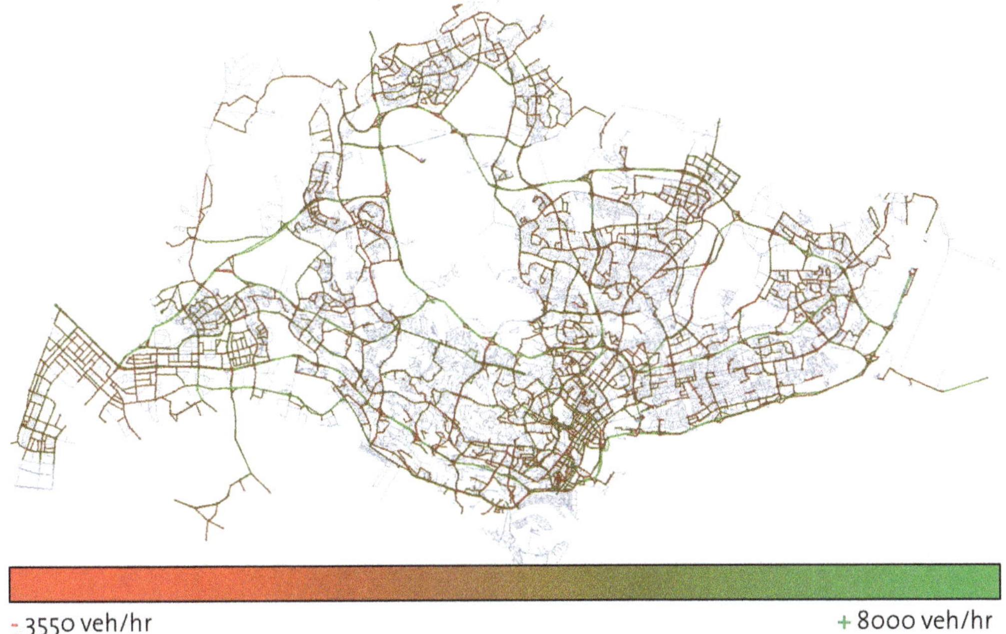

Figure 9.4: Resulting changes in navigation network travel capacity property.

CHAPTER 10

The "Network Editor" Contribution

Kai Nagel

10.1 Basic Information

Entry point to documentation:

`http://matsim.org/extensions` → networkEditor

Invoking the module:

`http://matsim.org/javadoc` → networkEditor → `RunNetworkEditor` class

Selected publications:

none

10.2 Short Description

This is, beyond the two network editors described in Chapters 8 and 9, a third network editor. It is older than the other two and has not been systematically maintained, but it still seems to be working and so it is still there.

How to cite this book chapter:

Nagel, K. 2016. The "Network Editor" Contribution. In: Horni, A, Nagel, K and Axhausen, K W. (eds.) *The Multi-Agent Transport Simulation MATSim*, Pp. 73–74. London: Ubiquity Press. DOI: http://dx.doi.org/10.5334/baw.10.

SUBPART TWO

Mobsim

CHAPTER 11

QSim

Marcel Rieser, Kai Nagel and Andreas Horni

11.1 Vehicle Types and Vehicles

For a variety of reasons—e.g., vehicle-specific emissions calculations (Chapter 36), or vehicle-specific maximum speeds (see below)—it may become necessary to assign different vehicle types to different persons, modes, or trips. The (arguably) most "honest" approach to vehicles, in terms of micro-simulation, is to generate a synthetic vehicle fleet. Each leg would then have to know which vehicle it wants to use. This is indeed possible with the planned vehicle ID that MATSim route objects can store. This functionality is switched on by first loading an additional vehicles file (see Section 6.6) and then configuring the QSim as

```
<module name="qsim">
   <param name="usePersonIdForMissingVehicleId" value="false" />
   <param name="vehiclesSource" value="fromVehiclesData" />
   ...
</module>
```

(available with release 0.8.x). This states that every time the QSim needs a vehicle with a specific ID, it will search for it in the vehicles data container, throwing an exception if it is not found there.

A Fallback for Routes that do not Contain Vehicle IDs In the above approach, vehicular routes need to provide vehicle IDs, otherwise the QSim will throw an exception. Since algorithms to compute and maintain vehicle IDs during replanning are currently not well developed within MATSim, an alternative is to assume that persons use a vehicle with the same ID as the person. This is switched on by

How to cite this book chapter:
Rieser, M, Nagel, K and Horni, A. 2016. QSim. In: Horni, A, Nagel, K and Axhausen, K W. (eds.) *The Multi-Agent Transport Simulation MATSim*, Pp. 77–80. London: Ubiquity Press. DOI: http://dx.doi.org/10.5334/baw.11.

```
<module name="qsim">
   <param name="usePersonIdForMissingVehicleId" value="true" />
   ...
</module>
```

which is also the current default. With this configuration, it will still search for the vehicle ID in the route, but if this is unavailable, it will instead use the person ID as vehicle ID. Without additional configuration, it will then still search for the vehicle under that ID in the vehicles file.

Alternative Vehicles Sources A default vehicles source is defined by

```
<module name="qsim">
   <param name="vehiclesSource" value="defaultVehicle" />
   ...
</module>
```

This generates a default vehicle (typically a medium-sized 4-seater) every time a vehicle is needed and is currently the default configuration.

At the moment, alternative approaches to vehicle generation need to be programmed as script-in-Java. See, e.g., `RunMobsimWithMultipleModeVehiclesExample` under `http://matsim.org/javadoc` → main distribution for a reference to a script that generates mode-specific typical vehicles for each mode. Simulation experiments using this feature have been performed for the Patna scenario as reported in Chapter 77.

Vehicle Behavior Vehicles need to be available where they are needed. It is, for example, impossible to perform a trip by car, then another (non-circular) trip by public transit and then make another trip with the same car as before, since the car will not be available at that location. The QSim is able to enforce such behavior, with the setting

```
<module name="qsim">
   <param name="vehicleBehavior" value="exception" />
   ...
</module>
```

This means that if a necessary vehicle is not available at the location where it is needed by the traveler, the QSim throws an exception and aborts. The idea here is that such synthetic travelers should have within-day replanning strategies (see Chapter 30) to cope with unexpectedly unavailable vehicles; any attempt to use an unavailable vehicle points to an error in the driver's behavioral logic.

In many standard situations, the above behavior will be too strict. For example, a vehicle may be shared between family members, but one member will be late in returning a vehicle. For such situations,

```
<module name="qsim">
   <param name="vehicleBehavior" value="wait" />
   ...
</module>
```

may be an option. Here, a driver will wait if a vehicle is not available. Errors in the coordination logic, i.e., very long waits, will be punished via the MATSim scoring logic, thus leading to more robust coordinations.

A final alternative is

```
<module name="qsim">
   <param name="vehicleBehavior" value="teleport" />
   ...
</module>
```

With this setting, vehicles will be teleported to locations where they are needed.

Initial Vehicle Placement For vehicle behavior of type `exception` and type `wait`, vehicles need to be at the correct location when the QSim starts. Here, the default simulation currently places all vehicles at the home location—for other variants, some additional code needs to be written or used, such as the car sharing extension (Chapter 22).

PassingQ Once various vehicles have different maximum speeds, the standard QSim, even with multiple main modes, is no longer sufficient, since it uses FIFO (First In, First Out) as the queuing discipline, meaning that fast vehicles cannot pass slower vehicles. Here, the so-called `Passing(Vehicle)Q` can be used instead. It replaces the FIFO sorting criterion—where vehicles are sorted by the sequence in which they arrive on the link—by a sorting employing the so-called earliest link exit time, computed from link enter time and freespeed travel time. Now, using the minimum of vehicle and link maximum speeds, the freespeed travel time can be differentiated between vehicles, allowing fast vehicles to obtain an earlier link exit time, even if they enter later than slow vehicles. Details and resulting fundamental diagrams are given by Agarwal et al. (2015b).

This option can be enabled by using

```
<module name="qsim">
   <param name="linkDynamics" value="passingQ" />
   ...
</module>
```

in the `qsim` section of the config file.

11.2 Other

The simulation is able to handle time-variant networks (Lämmel et al., 2010), within-day replanning (Dobler, 2009, see Chapter 30) and traffic lights (Neumann, 2008; Grether et al., 2011b, 2012, see Chapter 12). An earlier multimodal approach, targeted at overcoming the teleportation estimates of non-motorized modes, and particularly focused on pedestrians, is presented in Chapter 21.

SUBPART THREE

Individual Car Traffic

CHAPTER 12

Traffic Signals and Lanes

Dominik Grether and Theresa Thunig

12.1 Basic Information

Entry point to documentation:

`http://matsim.org/extensions` → signals

Invoking the module:

`http://matsim.org/javadoc` → signals → `RunSignalSystemsExample` class

Selected publications:

Grether et al. (2011a); Grether (2014)

12.2 Motivation

Traffic signals ensure security of travelers at junctions and regulate right of way. Furthermore, by assigning green times to the different approaches of a junction, they determine and evaluate junctions' performance. There are different strategies for traffic signal control: fixed-time traffic signal control, for example, periodically repeats the same schedule for signalization, while traffic-responsive signal control reacts dynamically to the prevailing traffic patterns to improve the junction or system performance. Traffic signal control can improve the traffic conditions at a single junction, but the whole system can be worse if a single junction is improved. Hu and Mahmassani (1997) argue that second order or network effects should be taken into account when effects of signal control strategies are tested. Network effects include drivers' reactions: not only route choice, but also scheduling. Thus, traffic control, especially traffic-responsive signals, need certain constraints. Otherwise, traffic may become unstable: rapidly at two nearby junctions, or at the network level (Lämmer and Helbing, 2010). MATSim can capture most of these effects. This chapter reviews

How to cite this book chapter:
Grether, D and Thunig, T. 2016. Traffic Signals and Lanes. In: Horni, A, Nagel, K and Axhausen, K W. (eds.) *The Multi-Agent Transport Simulation MATSim*, Pp. 83–88. London: Ubiquity Press. DOI: http://dx.doi.org/10.5334/baw.12.

concepts, usage and restrictions of the traffic signal control extension for MATSim. The chapter is particularly interesting for MATSim users, who plan to simulate traffic signals microscopically. If one wishes to capture signalization effects on a rather coarse level, consider the approach presented in Charypar (2008, pp. 139), that can be realized with the time variant network feature of MATSim (Lämmel et al., 2010). Before we go into detail on motivating traffic signals with MATSim, a case study is reviewed.

12.2.1 Case Study

The Cottbus scenario presented in Chapter 66 is applied to illustrate the influence of traffic signal control. This section summarizes results published in Grether et al. (2011a); Grether (2014). Readers interested in details are referred to these publications.

The runs sequence is performed with three different signal control strategies: In a first simulation sequence, all traffic signals are switched off. This can be used as a lower bound for results of signal control, since it assumes that vehicles are able to traverse a crossing without an accident, i.e., they are able to drive "through each other". The next sequence uses the fixed-time setup. In the third and final, sequence, all traffic signals are controlled by a traffic-actuated stage length control. The control is based on pre-timed, fixed-time schedules. The green times of the fixed-time schedules are reduced to a minimal green time of 5/10 seconds. If vehicles are still approaching at the end of this reduced green time, it is extended up to a predefined maximum.

Simulation results for iteration 1 000 of the Cottbus commuter scenario are depicted in Figure 12.1(a). The number of vehicles simultaneously on the road is plotted over the time-of-day. The results are quite similar for all signal control strategies; differences are small because of the lack of heavy congestion in the Cottbus scenario.

A change of signal control has more effect if unexpected traffic occurs in the network. It is assumed that the local soccer club, "FC Energie Cottbus", has a tournament taking place on a normal weekday, interfering with regular commuter traffic. In iteration 1 000 of the commuter scenario, in addition to the commuters 0 to 2 000 vehicles drive to the Cottbus soccer stadium during the evening peak. It is assumed that 25 % of these fans come from Cottbus, while the other 75 %

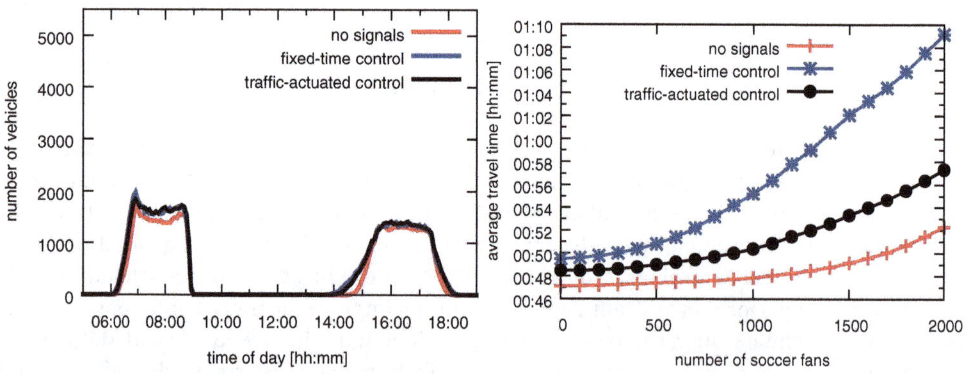

(a) No vs. fixed-time vs. traffic-actuated signal control, commuter traffic, iteration 1 000.

(b) Average travel time for unexpected event traffic, iteration 1 000.

Figure 12.1: Simulation results for the Cottbus traffic signal scenario: The simulated change of traffic signal control results in small travel pattern changes in the relatively quiet commuter scenario (left). If unexpected traffic occurs on the network, the traffic-actuated signal control enables travel time savings (right).
Source: Grether (2014)

come from the "Spree-Neiße" area around Cottbus, and that all fans start their trips between 5 pm and 6 pm.

Figure 12.1(b) plots the number of soccer fans on the x-axis, and the average travel time of all travelers on the y-axis. Without any additional vehicles, the traffic-actuated signal control leads to a gain of approximately 1 minute per traveler. The more additional traffic approaches the stadium, the more the traffic-actuated control saves travel time. When 2 000 additional vehicles are on the road, travel time savings reach approximately 15 minutes per traveler.

Summarizing: Slightly jammed commuter scenarios, where a change in traffic signal control leads to noticeably decreased overall travel time, have not yet been simulated with MATSim. Looking at different objectives with more fine-grained analysis tools can reveal network wide effects (e.g., see the analysis using macroscopic fundamental diagrams Grether, 2014, pp.114), but this is work in progress. More heavily jammed scenarios can increase the overall traffic impact of a change in traffic signal control. Nevertheless, the case study shows significant effects of traffic-responsive signal control when something unexpected happens and travelers do not react.

12.2.2 Overview MATSim & Traffic Signals

This case study highlights some previously researched MATSim traffic signals simulations aspects. MATSim is not always the traffic signal control "tool of choice" for all questions. The code base, however, can help simulate other use cases, e.g., evacuation or air transport scenarios; MATSim's open source nature provides hooks and interfaces for extension. But one must consider the amount of work required, the current state of development and specific project planning. The rest of the chapter goes into more detail. Section 12.3 provides some traffic signal control background, vocabulary, and options for modeling traffic signals with MATSim. Technical details can be found in the traffic signals user guide. Section 12.4 goes into details on network and traffic flow modeling. Iterations and learning are discussed in Section 12.5. When it comes to agent based learning, MATSim is very fast—the presented case study requires, on average, 17 seconds computation time per iteration—for scoring, replanning, and output. One complete run sequence: (1 000 iterations, single core mobility simulation, multi-core replanning) was simulated in 9 hours and 12 minutes. The simulation speed allows exploration of network-wide behavioral reactions to traffic signal control changes and the resource efficient simulation enables the joint simulation of several policies. Before publishing results, one should consider several specific aspects of evaluation and simulation results interpretation. Hints are provided in the conclusion, Section 12.6.

12.3 Traffic Signal Control

On a coarse level, control strategies for traffic signals can be classified in fixed-time and traffic-responsive strategies.

Fixed-time traffic signal control periodically assigns green times for each junction approach. Cycle time and green split are not modified within short time periods. To establish green waves between adjacent junctions, the green light start for approaches within the cycle can be adjusted by a global timer; these shifts are referred to as (coordination) offsets. For optimization of fixed-time signals, different equilibrium traffic flow regimes are determined for several periods of time, e.g., weekday morning, midday, evening and night plus a separate estimate for weekends. Optimization may target all signalized junction parameters—green split, cycle, offsets, and phase composition, but it is not possible to react to current changes in equilibrium traffic flows.

Traffic-responsive control reacts to current traffic patterns, adjusting traffic signal control parameters on the fly. In principle, all available information on prevailing traffic patterns can be used. The diversity of traffic-responsive control algorithms is wide; for a review, see Grether (2014).

MATSim's traffic signal module is designed to simulate every traffic signal control strategy. The module provides a default implementation for fixed-time control. Traffic-responsive strategies require custom implementation of the control algorithm, but can use existing data formats and fixed-time control infrastructure. Data is divided into five different types of input:

Signals & Systems: The location of the traffic signal hardware on the network is usually independent from the control strategy. Signals can be located at the end of a link or a lane (see the next section for further discussion of lanes). Signals are attached to a system that reflects, e.g., all signals of a junction or even larger units. Each signal system is controlled by exactly one control algorithm at a time.

Signal Groups: Traffic signals must be attached to a group. A group of signals shows the same color at the same time. Each time a signal group changes its state, a MATSim event is triggered. There is no explicit phase representation; if required, this can be realized over signal groups.

Signal Control: Specifies the control algorithm for each signal system. Data comprises information for fixed-time control and can be extended to capture custom control algorithms' parameters.

Amber: Specifies the amber phase at the beginning and end of green time. Currently, driving is not permitted if a traffic signal group shows amber light and this information is used only for visualization purposes.

Intergreens: The inter-green time specifies minimal time period between the ending of one and beginning of another signal group's green time. This information is important because MATSim's traffic flow model does not contain any collision detection. A validation module reads the event stream and triggers a warning, or an error, if security constraints are violated. Further, customized control strategies can access this information to ensure security aspects' control validity.

For detailed information on file structures and how to link them in the MATSim config file, we refer to the user guide in the contribution "signals".

The next section explains network representation and traffic signal location in more detail.

12.4 Network Representation & Traffic Flow

This section explains transport network representation with microscopically modeled traffic signals. In MATSim, transport network representation is a static, directed graph, consisting of nodes and links. Links depict road segments, while nodes can be interpreted as decision points in space with a coordinate as attribute, but no spatial dimension.

Figure 12.2(a) illustrates a typical layout of a real-world road segment, with several turn pockets at its end. If the whole road segment is modeled as a single link with MATSim's queue model, the first vehicle stopping at a red traffic signal at the end of this link will block all other vehicles approaching upstream, see Figure 12.2(b). In respect to the road layout shown in Figure 12.2(a), this is unrealistic. Figure 12.3(a) sketches the network layout for a more realistic modeling. Vehicles with distinct turn intentions do not block each other until the available space for queuing on the turn pocket is used completely, see Figure 12.3(b).

In principle, one can model each turn pocket as a link and put traffic signals at its end; but considering overall project constraints, this has implications for network modeling and routing.

In MATSim, all domain-relevant attributes differing from geospatial location, e.g., traffic count data, transit stops, transit lines, or speed limits, are attached to links. If one of this attributes changes, one must model several links. Frequently, geospatial location of such attributes is insufficient for a fully automatic matching of attributes to links; some data requires manual

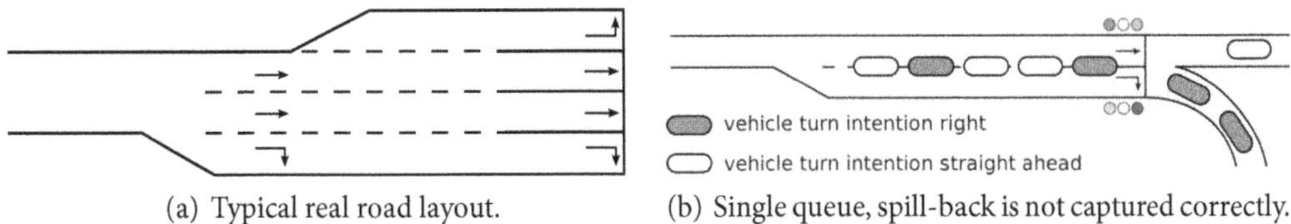

(a) Typical real road layout. (b) Single queue, spill-back is not captured correctly.

Figure 12.2: Transition from a real road segment to a graph layout with a single queue: the missing turn pockets representation prevents vehicles passing each other and cannot capture the traffic signal control for different turning moves.
Source: Grether et al. (2012)

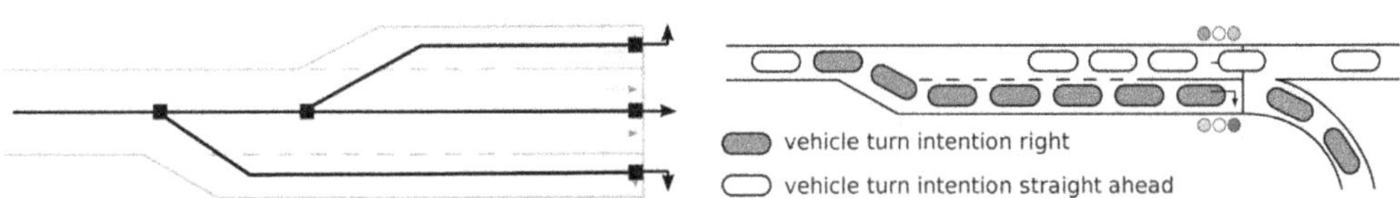

(a) Part of the graph required to model the road layout. (b) Multiple queues, spill-back is captured correctly.

Figure 12.3: Transition from a real road segment to a graph layout with multiple queues: each turn pocket is represented by its own queue. Traffic signal control for different turning moves is captured; vehicles can pass each other, unless the queue spills over.
Source: Grether et al. (2012)

post-processing. To simulate traffic signals and turn pockets with an already existing scenario, carefully consider the matching process before changing the network.

Travelers' routes are specified by link sequences within MATSim and routes are generated by a shortest path algorithm requiring a cost function for links. In standard MATSim, link travel time is part of a link's cost. When modeling turn pockets as links, the shortest path algorithm is responsible for selecting the appropriate turn pocket on a route. If modeling includes turn restrictions, ensure that they are captured by the shortest path algorithm and note that the required number of iterations increases if many turn pockets lead to the same downstream link. It is important to understand route generation and network modeling interaction when modeling turn pockets as links.

If network modeling or routing issues clash with other project goals, there is an alternative. MATSim allows the modeling of a subgraph on top of each link to reflect the structure shown in Figure 12.3(a). The links of the subgraph are then called *lanes*. At the beginning of a link, only one lane can be modeled; at the end of a link, different lanes can exist to model turn pockets. A vehicle must be in the correct turning lane for the next downstream link of its route. If there is only one lane towards the downstream link, the vehicle uses this lane. If there is more than one lane leading to the next downstream link, the vehicle is placed on the lane currently containing the fewest other vehicles. Using lanes, specific turning moves can be forbidden because the shortest path algorithm underlying network graph is modified; thus, turn restrictions are considered when the network graph is created. The shortest path calculation captures the effects of lanes without further modification (see Grether, 2014, pp. 21).

As well the differences mentioned above, lanes exhibit behavior similar or equal to links. Vehicles entering or leaving lanes trigger events with the same structure and information as link enter and leave events. Traffic signals can be placed at the end of links and lanes. Traffic on each lane is simulated the same way as for links. Traffic flow increase is linear in a signal's green time for both links and lanes.

The decision to use or not use lanes is arbitrary. Most MATSim scenarios with signals are set up using lanes; the code base is well debugged. Without lanes, the code for traffic signals is also tested; one should check carefully for artifacts and understand influences on route generation.

12.5 Iterations & Learning

This section discusses interaction between traffic signals and travelers within the MATSim iteration cycle.

Meneguzzer (1997) defines the combined traffic assignment and control problem as finding a tuple (f^*, g^*) of traffic flows f and signal settings g under policy P that fulfills

$$f^* = f^e[g^P(f^*)] \text{ or equivalently } g^* = g^P[f^e(g^*)]$$

where f^e is a function mapping signal settings to equilibrium traffic flows and g^P a function mapping traffic flows to signal settings under policy P. The formulation neatly shows the mutual interaction of traffic patterns and signal settings. The formulations do not capture the time horizon where these interactions take place.

Traffic signal interpretation within the MATSim iteration cycle depends strongly on signal control type and learning mechanism interpretation. For fixed-time control, the fixed-point interpretation can be valid, at least if one does not anticipate unexpected events on the demand side. For traffic-actuated signal control strategies, no standard interpretation can be provided. Readers seeking more detail are referred to Grether (2014, pp. 75). We conclude with this advice; clearly document what and how was simulated and provide an interpretation that makes sense for each individual project.

12.6 Conclusion

MATSim can simulate traffic signal control microscopically. However, certain traffic signal effects are not represented by MATSim without further customization and implementation, e.g., microscopic deceleration and acceleration as a reaction to traffic control. Evaluations must be checked and interpreted against the simulation setup to ensure that everything derived from simulation results is also appropriately simulated. This chapter provides an overview of traffic signals in MATSim, detailing what to consider before taking first steps in larger scenarios. Details for implementation can be found in the javadoc documentation referenced above. For the detailed scientific discussion of modeling aspects the reader is referred to Grether (2014).

We think that MATSim is a superior tool for microscopic simulated traffic-responsive signal control that should be analyzed network-wide, assuming heterogeneous user reactions.

CHAPTER 13

Parking

Rashid A. Waraich

13.1 Basic Information

Entry point to documentation:

`http://matsim.org/extensions` → parking

Invoking the module:

`http://matsim.org/javadoc` → parking → `RunParkingExample` class

Selected publications:

Waraich and Axhausen (2012); Waraich et al. (2013a); Waraich (2014); Waraich et al. (2014b)

13.2 Introduction

The MATSim simulation, by default, does not consider parking infrastructure or supply constraints. However, this can lead to artificially high car traffic to city centers in the model, often not the case in the real world, due to limited parking. The modeling of parking is also important because traffic-related policies can be designed around parking; e.g., raising prices for parking at certain times of the day, or reducing parking supply in an area, can impact travel demand.

This chapter describes work done to bridge this gap via parking models for MATSim .

13.3 Models

For technical reasons, parking modeling efforts in MATSim were divided in two parts: parking choice and parking search, described in the following two subsections.

How to cite this book chapter:

Waraich, R A. 2016. Parking. In: Horni, A, Nagel, K and Axhausen, K W. (eds.) *The Multi-Agent Transport Simulation MATSim*, Pp. 89–92. London: Ubiquity Press. DOI: http://dx.doi.org/10.5334/baw.13. License: CC-BY 4.0

13.3.1 Parking Choice Model

The first approach for modeling did not change the MATSim traffic simulation; it extended it to capture parking supply through `controler listeners` and event handling. This means that no rerouting due to parking took place during the simulation. However, changed routes could be incorporated in a post-processing step, as described in Waraich and Axhausen (2012).

In the most general case, a parking choice model performed the following simulation steps; when a vehicle arrived at a destination in MATSim, the parking choice model assigned a parking spot in the agent's area, according to a customizable algorithm (e.g., utility maximization). The assigned parking place was marked as occupied on arrival and became unoccupied again when the agent departed, allowing the model to simulate supply side constraints with the same temporal resolution as the basic MATSim model.

A simple parking choice model version was able to consider only walk distance minimization, ignoring other user preferences and park at the closest available public parking. A simple model like this was able to partially solve one of the main problems of the un-constrained parking model in MATSim; it made an area with little parking less attractive as a car destination due to longer walk distances. Parking model integration with MATSim was achieved by adding a term for the parking operation to the agent's overall plan scoring function, as follows:

$$S_{parking} = S_{walking} + S_{parking\ costs} + S_{parking\ search\ time} \tag{13.1}$$

Beyond walking distance disutility, this scoring function could also include additional features like cost, or even estimated parking search times, using models like Horni et al. (2013a).

A Zürich city study, which implemented a parking choice model and included trade-off between walk distance and parking cost, was presented in Waraich and Axhausen (2012). This study also distinguished between public, private and reserved parking, where only certain people (e.g., disabled) or certain vehicles could park (e.g., electric vehicles). Figure 13.1 shows parking choice models employed in this study, where a distinction between public, private and reserved parking was made. In Waraich et al. (2013c), another study for modeling parking in MATSim was reviewed, exploring individual gender and age parking preferences. Utility function parameters used in this study were based on a stated preference survey in Switzerland.

13.3.2 Parking Search

The parking choice model presented in the previous section could capture many relevant aspects of parking. However, it did not model parking search behavior; studies conducted around the world suggest that, on average, around 30 % of city centers traffic could be due to parking search traffic Shoup (2004). Thus, it seems extremely important to capture parking search related traffic in transportation models.

A first idea about model parking search traffic in MATSim was presented in Waraich et al. (2012). The basic idea came from surveys suggesting that people select certain strategies they think will be beneficial for them when starting the parking search process (Axhausen and Polak, 1989). Proof of this concept for development was attempted, using within-day replanning (see Chapter 30 and Dobler et al. (2012)). However, this path was aborted after development of several initial strategies, where performance and integration issues led to dead ends (Waraich et al., 2013c); performance after optimization was around 24 times slower than the original runs without parking operations.

An alternative path closer to the idea presented in Waraich et al. (2012) was successfully attempted, using a JDEQSim based model (see Section 4.3.2) with within-day support and travel time approximation, as seen in PSim (see Chapter 39, Fourie et al. (2013)). This removed overhead,

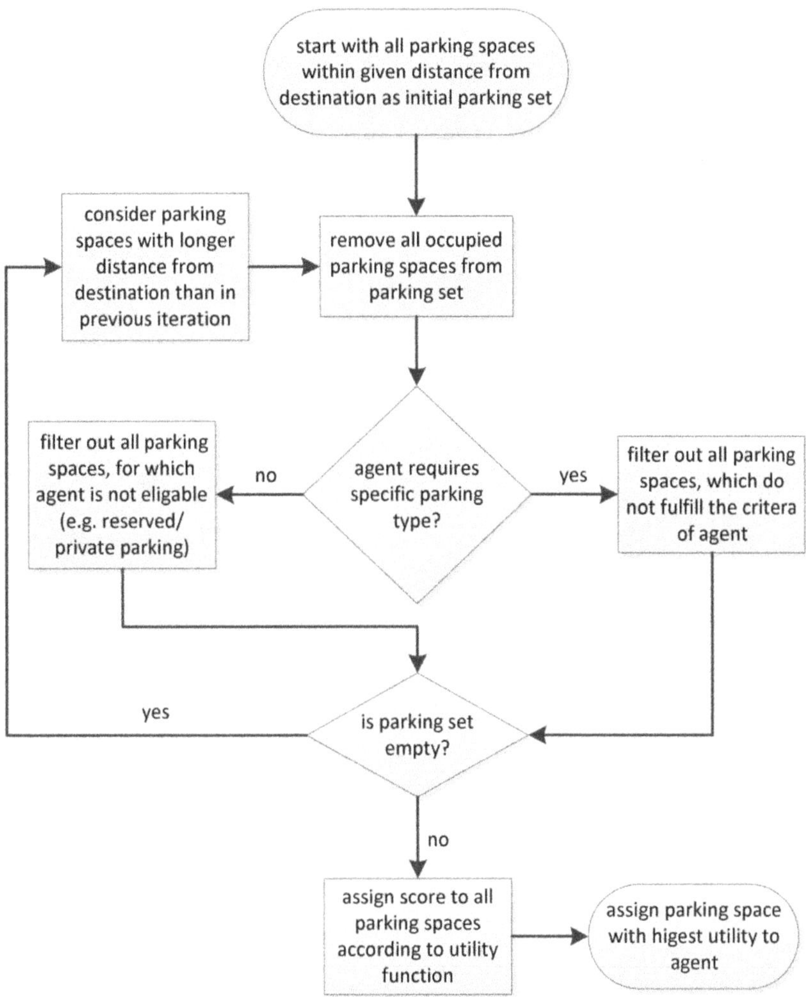

Figure 13.1: Parking choice algorithm.

Source: Waraich and Axhausen (2012)

present in the previous approach, enabling flexibility to implement many of the parking strategies presented in Axhausen and Polak (1989) and beyond. Publication of this approach's first results are expected in 2015.

Unfortunately, the approach is not available in packaged form to other users of MATSim.

13.4 Applications

Clearly, the parking model applications presented were important, diverse and especially well-suited for policy design; one example of traffic policy design by means of targeted reduction of parking supply was presented in Waraich and Axhausen (2012). Waraich et al. (2013c) explained an application of performance-based pricing for parking in MATSim, where iteratively parking prices were adapted to match demand. An integration of parking choice and electric vehicle charging was presented in Waraich et al. (2014a) for a Zürich case study and Bemetz and Hohenfellner (2014) described an even more sophisticated test model for parking and EV (Electric Vehicle) charging, with various types of charging speed and prices.

13.5 Usage

A general parking choice model was included in the parking contribution of MATSim, which provided various extension interfaces; examples were included in the parking contribution to provide help with extension.

CHAPTER 14

Electric Vehicles

Rashid A. Waraich and Joschka Bischoff

Entry point to documentation:

`http://matsim.org/extensions` → transEnergySim

Invoking the module:

No predefined invocation. Starting point(s) under `http://matsim.org/javadoc` → transEnergySim → `RunTransEnergySimExample` class.

Selected publications:

Waraich et al. (2013d); Galus et al. (2009, 2012b); Waraich (2013); Galus et al. (2012a); Waraich (2012a,b); Waraich et al. (2014a); Waraich and Axhausen (2013)

14.1 Introduction

Research related to EV modeling in MATSim started in 2008/2009, with an electricity grids project (Waraich et al., 2013d); it's goal was to uncover potential bottlenecks and/or constraint violations in Zürich city's lower voltage grid due to future EV charging. A framework emerged from the research for EV modeling, called TESF (Transportation Energy Simulation Framework) (Waraich et al., 2014a). This resulted in various framework extensions and enabled simulation of various scenarios (Waraich et al., 2014a; Waraich, 2013; Abedin and Waraich, 2014; Schieffer, 2011; Galus and Andersson, 2011; Galus et al., 2012a; Bischoff, 2013; Bischoff and Maciejewski, 2014). This chapter provides advice on these research directions and serves as a starting point for modeling EVs in MATSim.

14.2 Models

The main reason for modeling EVs in TESF was simple: it was essential to keep track of the battery charging state in the EVs. This meant that, as the EV was driving, depletion of the

How to cite this book chapter:

Waraich, R A and Bischoff, J. 2016. Electric Vehicles. In: Horni, A, Nagel, K and Axhausen, K W. (eds.) *The Multi-Agent Transport Simulation MATSim*, Pp. 93–96. London: Ubiquity Press. DOI: http://dx.doi.org/10.5334/baw.14. License: CC-BY 4.0

batteries was simulated. It was also important to consider the charging process of EVs at charging infrastructures.

While the basic EV modeling mechanisms were simple, there were many details to ponder when modeling scenarios. The TESF framework provided both interfaces and implementations to cope with more complex cases, e.g., defining a vehicle that can charge without contact while driving, for example, by using dynamic inductive charging. Furthermore, charging mechanisms themselves could also be quite complex. The following sections provided some details on this, as well as different models involved.

14.2.1 Energy Consumption Model

When a vehicle was defined in TESF, it could be assigned an energy consumption model, defining how much energy the vehicle used while driving. For conventional vehicles, just energy consumption could be logged using such a model; however, for electric vehicles, the energy consumption model was used to update the on-board battery system state of charge. PHEVs (Plugin Hybrid Electric Vehicles) can use both electricity and gasoline for driving and therefore had two different energy consumption models assigned to them for modeling these two modes. When this was written, a series hybrid model were implemented in TESF (Chan, 2007), which used electricity as long as the battery charge state was above a certain threshold value, then switched to gasoline. This type of vehicle could also be charged using a plug, like a battery electric vehicle. For PHEVs, car manufacturers often defined rules governing when a vehicle should switch between battery and gasoline use. The TESF framework provided interfaces and examples of how more advanced control strategies for PHEVs could be implemented.

14.2.2 Charging Infrastructure

In addition to plug-based charging, inductive charging infrastructure was also modeled in TESF, with two types: dynamic and stationary. The dynamic inductive charging infrastructure was often embedded in roads; vehicles able to use such infrastructure could charge while they drove. Stationary inductive charging was, more or less, modeled like plug charging; however, charging interfaces between vehicle and the charging infrastructure had to match for the charging process to function.

Another fast route to a full battery was to replace/swap the used battery for a new one at a specialized infrastructure, sometimes referred to as a swapping station (Li et al., 2011). A basic modeling of this approach was provided in TESF, which could be extended and detailed further according to specific scenario needs.

14.2.3 Charging Schemes

When an EV connected to any infrastructure for charging, a scheme was needed to define how the vehicle charging would operate; should the vehicle start charging immediately, or would charging depend on an agent's pricing preferences, which could vary with time and location? Negotiations between the vehicle computer and grid operator were also possible, which perhaps allowed for some electricity grid temporal flexibility, while fully charging a vehicle's battery before departure (sometimes referred to as "smart charging"). Various charging schemes were part of the TESF and were be used to model other more complex charging schemes; TESF-simulated examples of various charging schemes can be found in Waraich et al. (2013d).

14.2.4 Vehicle-to-Grid

When studying electric vehicles, charging is not the only topic of interest; V2G (Vehicle-to-Grid) applications where electric vehicle batteries supply power and energy back to the grid (Kempton and Tomic, 2005) were analyzed. While the integration of V2G models for MATSim was limited at any given time, an application related to V2G and intermittent energy generation at wind parks using MATSim can be found in Galus and Andersson (2011) and a preliminary attempt to integrate V2G in TESF was described in Waraich et al. (2014a); Schieffer (2011).

14.2.5 Vehicle Choice

When conducting electric vehicle studies, each vehicle owner usually has to be assigned a specific type: e.g., electric vehicle, conventional vehicle, plug-in hybrid, etc. Sometimes, these assignments were random, while ensuring vehicle type share constraints for the scenario (e.g., Waraich et al., 2014a). Often, however, possible financial or infrastructural incentive implications, e.g., different toll prices, parking fees or fuel prices for different vehicle types, had to be evaluated. A replanning module for vehicle choice, also covering EVs, was recently implemented; first results should published soon and can also be integrated in TESF.

This section provided an overview of the various TESF framework parts and the following section an application of a TESF contribution, that modeled electric taxis.

14.3 Application: Electric Taxis

Combination and extension of both the TESF and VRP (Vehicle Routing Problem) contribution (see Chapter 23) allows simulation of BEVs (Battery Electric Vehicles) taxi fleets. For electric vehicles, vehicle charging process was adapted; for taxis, the concept of taxi ranks and a modified optimizer sending idling taxis to the rank and only dispatching vehicles with sufficient battery charge were introduced.

14.3.1 Taxi Ranks

After dropping off passengers, taxis proceeded to the nearest rank location, unless there was an immediate follow-up request. Queuing took place at the rank location; the taxi that arrived first would leave the rank first. Other types of queuing were also tested, e.g., a dispatch by battery SOC (State of Charge). Ranks were not mandatory; however, driving there between trips would be typical German taxi driver behavior.

14.3.2 Charging Process

Chargers could be located at taxi ranks or any other link. Following any given BEV `AgentArrivalEvent` at a charger location link, charging would begin if

- there was a free charging spot,
- the vehicle's SOC was under a certain threshold,
- at least two minutes of time passed required for parking the car and plugging it in.

Electric taxi simulation has been used in Mielec, Poland (Bischoff, 2013; Bischoff and Maciejewski, 2014). When this was written, an application for Berlin was in progress.

14.4 Usage

The TESF contribution contained many features described above and interfaces were provided for framework extension. Examples were also given for the setup of different scenarios: e.g., energy consumption model, vehicle types, charging schemes, etc.

CHAPTER 15

Road Pricing

Kai Nagel

15.1 Basic Information

Entry point to documentation:

`http://matsim.org/extensions` → roadpricing

Invoking the module:

`http://matsim.org/javadoc` → roadpricing → `RunRoadPricingExample` class

Selected publications:

Rieser et al. (2007a, 2008); Grether et al. (2008)

15.2 Introduction

Roadpricing is a controversial policy measure (e.g., Button and Verhoef, 1998). Its implementation in MATSim is conceptually straightforward (Rieser et al., 2007a, 2008; Grether et al., 2008): Essentially, for each vehicle entering a link at a given time, the appropriate toll is computed and charged to the vehicle's driver. The scoring function will pick this up by the term (see Equation (3.4))

$$S_{trav,\,car,\,q} = \ldots + \beta_m \cdot \tau + \ldots\,,$$

where τ is change in the monetary budget invoked by all toll payments (usually negative) and β_m is the marginal utility of money (also see Chapter 3 and Chapter 51). The driver then takes this into account making decisions, e.g., for route choice, departure time choice, mode choice, destination choice, etc., and then trades off toll payments with other elements of his or her scoring function.

How to cite this book chapter:

Nagel. K. 2016. Road Pricing. In: Horni, A, Nagel, K and Axhausen, K W. (eds.) *The Multi-Agent Transport Simulation MATSim*, Pp. 97–102. London: Ubiquity Press. DOI: http://dx.doi.org/10.5334/baw.15.

It should be clear that this automatically picks up all kinds of heterogeneities, for example:

- Traveling at a different time may lead to a different toll, but possibly also to different schedule delay costs (Section 3.2.5).
- Different vehicle types may be charged different tolls (Kickhöfer and Nagel, 2013).
- Different travelers may have different time values (Nagel et al., 2014), which may even vary according to the time of day.

However, one challenge is that the innovative modules (Section 4.5) must be consistent with the scoring now modified by road pricing. The approach just described will not work if, for example, the router consistently generates toll-avoiding routes for a synthetic person with a high time value, who would normally wish to pay for a faster option. In a case like this, if a suitable route is never generated, the scoring cannot identify it, giving the choice process no chance to select it in subsequent iterations.

However, processing every detail for each individual, i.e., not only the marginal utility of money, but also specific time pressure at the route search time, is quite complex.

An alternative approach is to make the router *randomizing*, i.e., to run it with a randomly generated time value every time necessary for a given person. Computational experiments with this approach produce solutions for synthetic travelers approximately as good, or even better, than an "engineered" router (Nagel et al., 2014). At the same time, the software consistency burden is significantly reduced, noticeable in the smaller amount of information to be extracted from the agent during each router call.

15.3 Some Results

15.3.1 Effect of an Afternoon Toll on Morning Traffic

In a first demonstration of capabilities, an afternoon toll for the Zürich area was simulated. While this is an unlikely policy scheme, it still clearly demonstrated the advantage of the integrated approach over other approaches. Not only did the synthetic travelers switch to public transit, but they also did so for the morning rush hour, where no toll was charged (Figure 15.1). Thus, the MATSim approach proved its ability to affect the whole daily plan, not just the trip. For more information, see Rieser et al. (2008).

15.3.2 Income-Dependent Values of Time

Similar to Rieser et al. (2008), Kickhöfer et al. (2010); Kickhöfer (2014) introduced a distance-based morning peak toll on the same links between 6:30 am and 9 am. Toll levels were incrementally increased from 0.28 CHF/km up to an almost prohibitive price of 44.80 CHF/km. The studies assume income-dependent utility functions with a decreasing marginal utility of money. The goal was to (i) identify the welfare-maximizing (see e.g., Tirachini et al., 2014, Section 2.5) toll level, which is potentially dependent on the aggregation rule of user benefits (see Chapter 51), and (ii) to investigate distributional aspects of such pricing schemes. The studies showed that changes in travel patterns resulting from the morning peak toll impacted the whole day, affecting traffic patterns in the afternoon. Furthermore, the study showed that such a parametric approach is capable of identifying the welfare-maximizing toll level. However, results also indicated that the overall welfare effect level depends strongly on the aggregation rule for user benefits, i.e., if one first monetizes

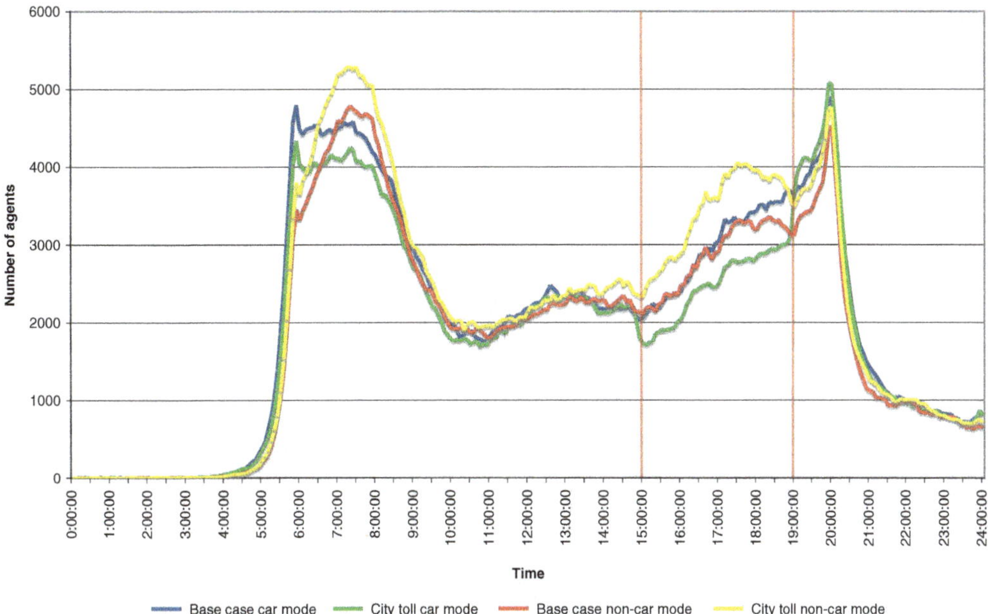

Figure 15.1: An afternoon city toll (between 3 pm and 7 pm) affects mode choice not just during the toll time, but also in the morning.

Source: Rieser et al. (2008)

individual utilities and then adds up, or first adds up utilities and then monetizes. Even the sign of that effect might not be stable depending on that choice. For more information, please refer to the two studies above.

15.3.3 Integrated Passenger and Freight Toll Simulation for the Gauteng Province in South Africa

A large scale application was undertaken for the Gauteng province in South Africa (Chapter 69). It is based on the so-called e-toll, which was switched on in December 2013. The e-toll should, logically, charge different rates for different vehicle types, with higher rates for heavy trucks. Again, logically, this should go along with higher time values of the driver-vehicle-units. Somewhat surprisingly, this turned out to be difficult to do with the MATSim software structure in place when the project was started in 2008. While it was easy to charge the freight vehicles a higher toll, it was difficult to give different replanning methods and different scoring function to the freight population; it was essentially impossible to feed the router with different time values for the freight population. This was an important driver for much development in recent years, including making the scoring function more accessible (Section 45.2.10), allowing different replanning strategies for different sub-populations (Section 4.5), and reducing consistency requirements between the router, the vehicle-based toll and the driver-based scoring function (Nagel et al., 2014).

The simulation, as expected, predicts reduced traffic volumes on the tolled roads and increased volumes elsewhere (Figure 15.2).

Figure 15.2: Predicted differences in link volumes after introduction of the toll (red: higher volumes, green: lower volumes).

15.4 Invocation

15.4.1 Minimal

A minimum amount of infrastructure is necessary when running roadpricing from the command line. For this, the MATSim JAR, its libraries, *and* the roadpricing JAR need to be downloaded, either from a release or from the nightly builds (Section 44.3.6). After unzipping all zip files, the necessary command is (may need slight refactoring with new formats):

```
java -Xmx2000m -cp MATSim.jar:roadpricing-.../roadpricing-
   ...jar org.matsim.roadpricing.run.RunRoadPricingExample config.xml
```

where `config.xml` needs to contain a section

```
<module name="roadpricing" >
...<param name="tollLinksFile" value="<path>/<tollfilename>" />
</module>
```

The toll file looks like this:

```
<roadpricing type="link" name="abc">
   <links>
      <link id="11">
         <cost start_time="05:00" end_time="10:00" amount="1." />
```

```
        <cost start_time="17:00" end_time="20:00" amount="1." />
      </link>
      <link id="12" />
    </links>

    <!--this is for all links with no cost entry above:-->
    <cost start_time="05:00" end_time="10:00" amount="2.00"/>

</roadpricing>
```

As one can see, there is a section where each link can be entered separately. A separate cost structure for each link is also possible. All links that are listed without a cost structure employ the general cost structure listed at the end. Links not listed are without toll.

15.4.2 Toll Schemes

Link toll The example refers to the "link" toll scheme, indicated by `type="link"`. It charges the amount specified on the link.

Distance toll Another useful scheme is "distance", indicated by `type="distance"`. Here, the `amount` is interpreted as amount per length unit (see Section 2.2.1). This is most useful, with only a list of tolled links and a uniform distance cost for all these links noted at the end of the file.

Area toll The simulation of an *area toll*—i.e., a toll where one has to pay a flat fee for a given time period, often a day, once one drives anywhere inside the area—suffers from a combinatorial challenge: driving through the tolled area early in the day may only pay off if one can re-use the permit later in the day. The code, in principle, addresses that by routing the agent twice: once under the assumption of a zero toll and once under the assumption of a very large toll. Afterward, the toll is added to the generalized cost of the first option, then both options are compared. In the end, the approach suffered from the same consistency burden as the general approach (see end of Section 15.2): the router made the decision about the better variant, rather than leaving the decision to the agent. It should be re-implemented using the same principles as Nagel et al. (2014).

Cordon toll The cordon toll scheme was derived from the area scheme; one could use the same file, listing all area links, for the cordon toll as well. The code ensured that toll was only charged when a vehicle moved from an untolled link to a tolled link—thereby effectively crossing the cordon. One difficulty with this approach: confusion ensues if there is no connected area and several links in sequence are tolled instead. Then, if these links are connected, the toll is only charged on the first of them; if there is a small section missing, perhaps overlooked, the toll is charged again.

SUBPART FOUR

Other Modes Besides Individual Car

CHAPTER 16

Modeling Public Transport with MATSim

Marcel Rieser

16.1 Basic Information

Entry point to documentation:

`http://matsim.org/extensions` → pt

Invoking the module:

The module is invoked by enabling it in the configuration.

Selected publications:

Rieser (2010)

16.2 Introduction

Public transport—or *transit* as it is sometimes called—plays an important role in many transport planning measures, even those initially targeting only non-transit modes. By making other modes more or less attractive (e.g., by providing higher capacity with additional lanes, allowing higher speeds, or charging money by setting up area road pricing), travelers might reconsider their mode choice and switch to public transport (*pt*) from other modes, or vice versa. Such changes can also occur when transit infrastructure is changed; additional bus lines, changed tram routes with different stops served, or altered headways—all are important for travelers on specific lines, or public transport in general. Around 2007, interest grew in extending MATSim to support detailed simulation of modes other than private car traffic, particularly public transport.

In a first step, MATSim was extended so that modes other than *car* would be teleported; agents would be removed from one location and placed at a later point of time—corresponding to estimated travel time—at their destination location, where they could commence their next activity. Together with a simple mode-choice module, randomly replacing all transport modes in all plan

How to cite this book chapter:

Rieser, M. 2016. Modeling Public Transport with MATSim. In: Horni, A, Nagel, K and Axhausen, K W. (eds.) *The Multi-Agent Transport Simulation MATSim*, Pp. 105–110. London: Ubiquity Press. DOI: http://dx.doi.org/10.5334/baw.16. License: CC-BY 4.0

legs and a simple travel time estimation for modes different than *car*, first case studies resulting in modal share changes were performed using MATSim (Rieser et al., 2009; Grether et al., 2009). This teleportation mode is now available, by default, in MATSim and still a very good fallback option to get a multimodal scenario up and running with as little data as possible.

In a second step, QSim was extended to support detailed simulation of public transport vehicles serving stops along fixed routes with a given schedule (Rieser, 2010). The next section describes, in more detail, data required and resulting features for this detailed public transport simulation.

16.3 Data Model and Simulation Features

MATSim supports very detailed modeling public transport; transit vehicles run along the defined transit line routes, picking up and dropping off passengers at stop locations, while monitoring transit vehicles' capacities and maximum speeds. Data used to simulate public transport in MATSim can be split in three parts:

- stop locations,
- schedule, defining lines, routes and departures, and
- vehicles.

This data is stored in two files; vehicles are defined in one file, stop locations and schedule in another. Examples of such files can be seen in Section 16.4.1 and Section 16.4.2, respectively.

The data model is comparable to other public transport planning software, but simplified in several respects. A line typically has two or more routes; one for each direction and additional routes when vehicles start (or end) their service at some point on the full route (coming from, or going to, a depot). Each transit route contains a network route, specifying on which network links the transit vehicle drives, as well as a list of departures, providing information about what time a vehicle starts at the first route stop. A route also includes an ordered list of stops served, along with timing information specifying when a vehicle arrives or leaves a stop. This timing information is given as offsets only, to be added to departure time at the first stop. Each departure contains the time when a vehicle starts the route and a reference to the vehicle running this service. Because timing information is part of the route, routes with the same stops sequence may exist, differing only in time offsets. This is often the case with bus lines, that take traffic congestion and longer rush hour waiting times at stops into account in the schedule.

Stop locations are described by their coordinates and an optional name; they must be assigned to exactly one line of the network for the simulation. Thus, they can be best compared to "stop points" in VISUM. There is, currently, no logical grouping of stop locations to build a "stop area"; this is a cluster of stops often sharing the same name, but located on different intersection arms, served by different lines, many with transfer corridors for passengers.

Each vehicle belongs to one vehicle 'type', which describes various characteristics, like seating and standing capacity (number of passengers), its maximum speed and how many passengers can board or depart a vehicle per second.

This data model already supports several advanced public transport modeling aspects: varying travel speeds along routes during different times of day (important for improved simulation realism), using diverse vehicle types on routes at different times of day (interesting for schedule economic analysis) and re-using transit vehicles for multiple headways along one or different routes (allows vehicle deployment planning optimization, or research on delay-propagation effects).

With these data sets, the QSim will simulate all transit vehicle movements. The vehicles will start with their first route stop at the given departure time, allow passengers to enter and then drive along their route, serving stops. At each stop, passengers can enter or leave the vehicle. The simulation generates additional, transit-related events whenever a transit vehicle arrives or departs at a stop, when passengers enter or leave a vehicle, but also when a passenger cannot board a vehicle because its capacity limit is already reached. This allows for detailed analyses of MATSim's public transport simulations.

For passengers to use public transport in MATSim, they must be able to calculate a route using transit services. For this, MATSim includes a public transport router that calculates the best route to the desired destination with minimal cost, given a departure time. Costs are typically defined only as travel time and a small penalty for changing lines, but other, more complex cost functions could be used.

The routing algorithm is based on Dijkstra's shortest path algorithm (Dijkstra, 1959), but modified to take multiple possible transit stops, around the start and end coordinates, into account to find a route. Multiple start and end stops must be considered to generate more realistic transit routes; otherwise, agents could be forced to travel first in the wrong direction, or wait at an infrequently served bus stop, instead of going a bit further to a busy subway stop location. By modifying the shortest path algorithm to work with multiple start and end locations, a considerable performance gain was achieved when compared to the basic (and somewhat naive) implementation that calculated a route for each combination of start/end location and then chose the best outcome.

16.4 File formats

16.4.1 `transitVehicles.xml`

To simulate public transport in MATSim, two additional input files are necessary. One is `transitVehicles.xml`, which describes vehicles serving the lines: big buses, small buses, trains or light rail vehicles and description of each vehicle's passenger transport capacity.

Public transport vehicle description can be split into two parts; first, vehicle types must be described, specifying how many passengers a vehicle can transport (Note that the term "vehicle" can refer to multiple vehicles in reality, e.g., a train with several wagons should be specified as one long vehicle with many seats). Second, actual vehicles must be listed. Each vehicle has an identifier and is a previously specified vehicle type. The following shows an example of a such a file, describing one vehicle and two vehicles of the same type.

```
<?xml version="1.0" encoding="UTF-8"?>
<vehicleDefinitions xmlns="http://www.matsim.org/files/dtd"
       xmlns:xsi="http://www.w3.org/2001/XMLSchema-instance"
       xsi:schemaLocation="http://www.matsim.org/files/dtd
                           http://www.matsim.org/files/dtd/
                           vehicleDefinitions_v1.0.xsd">
        <vehicleType id="1">
      <description>Small Train</description>
      <capacity>
         <seats persons="50"/>
         <standingRoom persons="30"/>
      </capacity>
      <length meter="50.0"/>
   </vehicleType>
   <vehicle id="tr_1" type="1"/>
   <vehicle id="tr_2" type="1"/>
</vehicleDefinitions>
```

16.4.2 `transitSchedule.xml`

The second, rather complex, file necessary to simulate public transport is `transitSchedule.xml`, containing information about stop facilities (bus stops, train stations, or other stop locations) and transit services.

In the first part, stop facilities must be defined; each one is given a coordinate, an identifier and a reference to a network link. The stop can only be served by vehicles driving on that specified link. It is also possible to specify both a name for the stop and whether other vehicles are blocked when a transit vehicle halts at a stop. This last attribute is useful when modeling e.g., different bus stops, where one has a bay, while at another, the bus must stop on the road.

After stop facilities, transit lines, their routes and schedules are described. This is a hierarchical data structure; each line can have one or more routes, each with a route profile, network route and list of departures. The following listing is an example of a basic, but complete transit schedule.

```
<?xml version="1.0" encoding="UTF-8"?>
<!DOCTYPE transitSchedule SYSTEM "http://www.matsim.org/files/dtd/
                                  transitSchedule_v1.dtd">
<transitSchedule>
  <transitStops>
    <stopFacility id="1" x="990.0"  y="0.0"   name="Adorf"
        linkRefId="1" isBlocking="false"/>
    <stopFacility id="2" x="1100.0" y="980.0" name="Beweiler"
        linkRefId="2" isBlocking="true"/>
    <stopFacility id="3" x="0.0"    y="10.0"  name="Cestadt"
        linkRefId="3" isBlocking="false"/>
  </transitStops>
  <transitLine id="Blue Line">
    <transitRoute id="1">
      <description>Just a comment.</description>
      <transportMode>bus</transportMode>
      <routeProfile>
        <stop refId="1" departureOffset="00:00:00"/>
        <stop refId="2" arrivalOffset="00:02:30"
                        departureOffset="00:03:00"
                        awaitDeparture="true"/>
        <stop refId="3" arrivalOffset="00:05:00"
                        awaitDeparture="true"/>
      </routeProfile>
      <route>
        <link refId="1"/>
        <link refId="2"/>
        <link refId="3"/>
      </route>
      <departures>
        <departure id="1" departureTime="07:00:00"
                          vehicleRefId="12"/>
        <departure id="2" departureTime="07:05:00"
                          vehicleRefId="23"/>
        <departure id="3" departureTime="07:10:00"
                          vehicleRefId="34"/>
      </departures>
    </transitRoute>
  </transitLine>
</transitSchedule>
```

Each transit line must have a unique ID and each transit route has an ID, which must be unique within that one line, allowing the same route ID to be used with different lines. The `transportMode` describes network links where the line runs. (Actually, this is not yet in force, although it might be in the future. It would be possible to let a bus run on train links in the simulation.)

The `routeProfile` describes the stops this route serves; the route itself describes the series of network links the transit vehicle's driver must navigate, often referred to as network route. Note that the complete route, i.e., all links the vehicle traverses, must be listed in the route, not only those with stops. All specified stops should occur along this route in correct order. Time offsets given for each stop in the `routeProfile` describe relative time offsets to an actual departure time. If a bus departs at 7 am, and stop 2 has a `departureOffset` of 3 minutes, this must be read that the bus is expected to depart at 7:03 am from the specific stop. All stops in the route profile must have a departure offset defined, except the last one. All stops, except the first one, can, optionally, have an arrival offset defined. This is useful for large trains that stop for several minutes at a station; helping the routing algorithm find connecting services at the correct time, namely the expected train arrival time.

As the last part of a transit route description, a departures list should be given. Each departure has an ID, which must be unique within the route, giving the departure time at the first stop of the specified route profile. The departure also specifies the vehicle (which must be defined in the previous transit vehicle list) with which the service should be run.

Because of its complexity, transit schedules often contain small mistakes that will return in an error when the simulation runs. Typical examples include: missing links in the network route, or incorrect defined stop order on the network route. To ensure a schedule avoids such issues before the simulation starts, a special validation routine is available:

```
java -Xmx512m -cp /path/to/matsim.jar
     org.matsim.pt.utils.TransitScheduleValidator
     /path/to/transitSchedule.xml /path/to/network.xml
```

If run, this validator will print out a list of errors or warnings, if any are found, or show a message that the schedule appears to be valid.

16.5 Possible Improvements

While the ability to simulate public transport was a big advance for MATSim, several shortcomings still require attention:

- The data model (and thus, the simulation) does not yet fully support some real world transit lines: for example, circular lines with no defined start and end cannot yet be easily modeled. Some bus or train lines also have stops where only boarding or alighting the vehicle is allowed, but not both (e.g., overnight trains with sleeper cabins). At the moment, MATSim always allows boarding and alighting at stops, leading to agents e.g., using a train with sleeper cabins for a short trip; in reality, they would be denied boarding without a reservation for a longer trip.
- A stop location, as seen by passengers in the real world, is typically modeled as a number of stop facilities in MATSim, detailing different locations where transit vehicles stop (depending on their route and direction). For analysis, one is often interested in aggregated values for such logical stop locations, not for individual stop facilities. Such a logical grouping is still missing in MATSim data format.
- Running simulations with a reduced population sample leads to artifacts when public transport is used. In a simulation with a sampled demand, network capacity is reduced accordingly, to accommodate the fact that fewer private cars are on the road. But because 100 % of public transport vehicles must run (albeit with reduced passenger capacity), calibration becomes difficult. This should be solved, in the future, not by reducing network capacity, but by giving each vehicle and agent a weighting, specifying how much each should count.
- The public transport router available and used by MATSim by default is strictly schedule-based. It assumes vehicles can keep up with the schedule and that enough passenger capacity is provided. In some regions, where transit is chronically delayed and overcrowded, MATSim's router will consistently advise agents to use routes that will perform badly in the simulation. Additional feedback from the simulation back to the router, as already done in the MATSim private car router, will be needed.

- Last, but not least, the current router, based on a modified shortest path algorithm of Dijkstra, can become rather slow and memory-intensive for larger areas with extensive transit offerings. Improved algorithms to generate the routing graph, or different routing algorithms altogether (like the non-graph based Connection Scan Algorithm (Dibbelt et al., 2013)) must be explored in the future.

16.6 Applications

Public transport simulation has been used in myriad applications of MATSim world-wide. The following list highlights some of these applications, pinpointing their special public transport simulation features.

- Berlin: the Berlin scenario (see Chapter 53) was one of the first real applications using public transport simulation in MATSim. The road and rail network, as well as the full transit schedule, was converted from a VISUM model. It is still one of the few known models where bus and tram lines share a common network with private car traffic, enabling full interaction between private and public vehicles (like transit vehicles) getting stuck and delayed in traffic jams.
- Switzerland: Senozon AG maintains a model of Switzerland containing the full timetable of all buses, trams, trains, ships, and even cable cars, in the Swiss alps. The schedule data is retrieved from the official timetable, available in a machine-readable format called "HAFAS (HaCon Fahrplan-Auskunfts-System) raw data format".
- Singapore: The model of Singapore (see Chapter 57) makes heavy use of public transport, and continually pushes the boundaries of what is currently possible to simulate. Due to the very large number of buses on Singapore's roads and strong demand for public transport, many extensions had to be implemented to realistically model pt in this context.
- Minibus: The minibus contribution (see Chapter 17) added an optimization layer to public transport functionality in MATSim, allowing automatic generation of an optimized transit schedule for a specific region.
- WagonSim: In the WagonSim contribution (see Chapter 25) public transport simulation was used to simulate rail-bound freight traffic. While the simulation was still moving around transit vehicles and letting passengers enter and leave these vehicles, the scenario had been customized so that vehicles corresponded to freight trains and passengers corresponded to actual goods being transported. Custom implementations of transit driver logic replaced vehicle capacity definition by an alternative definition, ensuring that the trains vehicles represent did not get too long or heavy. The network was constructed so that changing vehicles at stops took minimum time, corresponding to the time needed for switching wagons at freight terminals.

In addition to applications mentioned in the list above, many additional scenarios now use public transport simulation in MATSim. Importantly, the list also shows, that with some custom extensions and imagination, public transport functionality can be used for far more than "just simulating public transport"; it can be employed to solve complex problems previously handled by operations research groups.

CHAPTER 17

The "Minibus" Contribution

Andreas Neumann and Johan W. Joubert

17.1 Basic Information

Entry point to documentation:

`http://matsim.org/extensions` → minibus

Invoking the module:

`http://matsim.org/javadoc` → minibus → `RunMinibus` class

Selected publications:

Neumann (2014)

17.2 Paratransit

Paratransit is an informal, market-oriented, self-organizing public transport system. Despite the significance of this transport mode, it is mainly unsubsidized, relying on collected fares. Paratransit systems can be categorized by route pattern and function, by driver organization, type of stops and fare type. Most case studies covered by the Neumann (2014) thesis indicate that paratransit services are mainly organized as route associations operating 8-15 seater vans on fixed routes. Most of the services run in direct competition to a public transport system belonging to a public transit authority. Such a service—minibuses with fixed routes, but without fixed schedule—is often called a jitney service. The minibus module of MATSim is based on the most common characteristics, with the understanding that the jitney/minibus service is only one of many possible paratransit services.

How to cite this book chapter:

Neumann, A and Joubert, J W. 2016. The "Minibus" Contribution. In: Horni, A, Nagel, K and Axhausen, K W. (eds.) *The Multi-Agent Transport Simulation MATSim*, Pp. 111–114. London: Ubiquity Press. DOI: http://dx.doi.org/10.5334/baw.17.

The minibus model is integrated in the multimodal multi-agent simulation of MATSim. In the model, competing minibus operators begin to explore the public transport market, offering their services. With more successful operators expanding and less successful operators going bankrupt, a sustainable network of minibus services evolves. In Neumann (2014), the model is verified through multiple illustrative scenarios, analyzing the model's sensitivity to different demand patterns, transfers, and interactions of minibuses and a formal operator's fixed train line.

The minibus model can be applied to two different transport planning fields. First: in the simulation of real paratransit targeting the inner workings of different paratransit stakeholders' relationships, the model can create "close-to-reality" minibus networks in a South African context. Neumann et al. (2015) gives an in-depth presentation of the module application and South African paratransit in general. Given the informal and emergent nature of minibus paratransit in developing countries, routes, schedules and fares are usually not published; they can only be captured in the tacit knowledge of operators and frequent users. Applying the minibus model has proven valuable in gaining a better understanding of how routes evolve. Instead of imposing routes and schedules *on* the MATSim model, as is usually the case for formal transit, the modeler observes and gets the paratransit routes as an output *from* the model. As each operator aims to maximize their profit, the resulting network often favors the operators' business objectives, instead of the connectivity and mobility of the mode's users. This model feature accurately captures route-forming behavior in the South African case, where commuters are often required to take multiple, longer trips instead of direct trips.

Second, the same model provides a demand-driven approach to solving a formal transit authority's network design problem; it can be used as a planning tool for the optimization of single transit lines or networks. For more details on the second form of application, see Section 17.3.

For further reading: Neumann (2014) provides an understanding of the underlying principles of paratransit services, namely minibus services, its stakeholders, fares, route functions, and patterns. Furthermore, it contains an in-depth description of the minibus model, its theoretical background, and its application to illustrative scenarios, as well as real world examples. The website of MATSim also hosts latest implementation documentation at `http://matsim.org/doxygen`.

17.3 Network Planning or Solving the Transit Network Design Problem with MATSim

A public transport system's success depends primarily on its network design. When transport companies try to optimize a line using running costs as the main criteria, they quickly find that demand must be taken into consideration. The best cost structure is unsustainable if potential customers leave the system and opt for alternatives, like private cars. The basic problem to solve: find sustainable transit lines offering the best possible service for the customer.

More specifically,

- the customer's demand side asks for direct, uncomplicated connections, and
- the operator's supply side asks for profitable lines to operate.

Informal public transit systems around the world, often referred to as paratransit, are examples of market-oriented, self-organizing public transport systems. For an in-depth coverage of paratransit, see Section 17.2, with references. Despite the significant and increasing importance of this transport mode, it is mainly unsubsidized and relies only on collected fares. Thus, the knowledge of paratransit—and its ability to identify and fill market niches with self-supporting transit services—provides an interesting approach to solving a formal public transit company's network design problem.

The minibus module of MATSim provides a demand-driven approach to solving a formal transit authority's network design problem; it can be used as a planning tool for the optimization of single transit lines or networks. In the Neumann (2014) thesis, the model was applied to two different planning problems of the Berlin public transit authority BVG (Berliner Verkehrsbetriebe). In the first scenario, the model constructed a transit system, from scratch, for the district of Steglitz-Zehlendorf. The second scenario analyzed the Tegel airport closure impact on BVG's bus network. Apart from Tegel itself, the rest of the bus network was unaffected by the airport closure. The resulting minibus model transit system resembled the changes BVG had scheduled for Tegel's closure.

In conclusion, the minibus model developed in the thesis automatically adapted supply to demand. The model not only grew networks from scratch, but also tested an existing transit line's sustainability and further optimized the line's frequency, time of operation, length, and route. Again, the optimization process was fully integrated into the behavior-rich, multi-agent simulation of MATSim, reflecting passenger reactions, as well as those from competing transit services and other road users. Thus, the minibus model can be used, along with more complex scenarios, like city-wide tolls or pollution analyses.

CHAPTER 18

Semi-Automatic Tool for Bus Route Map Matching

Sergio Arturo Ordóñez

18.1 Basic Information

Entry point to documentation:

`http://matsim.org/extensions` → gtfs2matsimtransitschedule

Invoking the module:

`http://matsim.org/javadoc` → GTFS2TransitSchedule → `GTFS2MATSimTransitSchedule` class

Selected publications:

Ordóñez Medina and Erath (2011)

Current public transport assignment models adapt network assignment models to work with public transport traffic. Many commercial software products like EMME/2 (Version 2 of EMME), VISUM and OmniTRANS offer sophisticated procedures that include timetable-based route search. However, these models do not include interaction between public transport services and private transport. As mentioned above, the MATSim implementation handles private car traffic and public transport traffic in an integrated way, but it needs accurate public transport line routing on the transport network. While this is usually straightforward for rail-based public transport modes, the routing problem for buses requires more attention; experience shows that assumption of a shortest-path between two consecutive stops leads to unsatisfactory results. To overcome this shortcoming, one can either draw the routes manually or employ map-matching algorithms dependent on tracking data. Due to the burden of manual procedures, and the increasing availability of GPS tracking data, map-matching is becoming increasingly relevant. However, common map matching algorithms are usually not designed to account for the peculiarities of public transport routing; the procedure is very sensitive to errors in network coding, inaccurate bus stop locations and the simplified link shapes in the model.

How to cite this book chapter:

Ordóñez, S A. 2016. Semi-Automatic Tool for Bus Route Map Matching. In: Horni, A, Nagel, K and Axhausen, K W. (eds.) *The Multi-Agent Transport Simulation MATSim*, Pp. 115–122. London: Ubiquity Press. DOI: http://dx.doi.org/10.5334/baw.18.

This section presents a semi-automatic procedure combining public bus routes information (sequences of consecutive stop locations and sequences of geo-referenced points) with a highresolution network (Ordóñez Medina and Erath, 2011). The objective is to obtain a sequence of links for every route of every line and to associate each bus stop with one single link in the network. The procedure was designed to prepare the Singapore scenario public transport extension, but the tools developed can be used to set up any other scenario with similar initial data (timetable and high resolution network).

18.2 Problem Definition

Generally, the problem can be defined as follows. Given:

- a set of stop locations (two-dimensional point coordinates),
- a set of route profiles (sequence of consecutive stops),
- a set of GPS points sequences (sequence of two-dimensional point coordinates), and
- a high resolution navigation network (two-dimensional directed graph with attributes),

the task is to associate each stop with a network link, and translate each route to a network path (connected sequence of links). Figure 18.1 illustrates the problem by providing an example of the available input information and correct output.

Input Information The GTFS (General Transit Feed Specification) is a recent, but already widely-used format for specifying public transport systems, created by Google for feeding its geographic information applications. As of April 2011, the Singapore public transport system featured

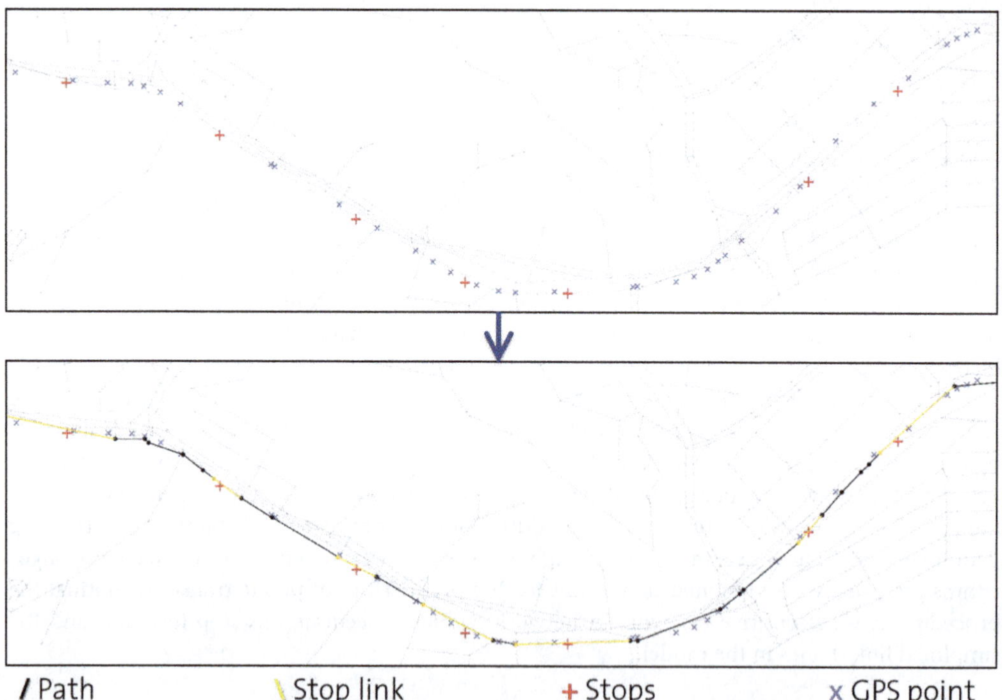

Figure 18.1: Input data and expected solution of the map-matching problem.

Source: Reprinted from Ordóñez Medina and Erath (2011, p.753), Copyright (2011), with permission from Hong Kong Society for Transportation Studies

4 584 bus stops serviced by 355 bus lines, all recorded on GTFS. Each line had several routes, i.e., different outward and return routes (due to one-way streets), as well as different coverage of serviced bus stops on weekdays and weekends. GTFS records the name and location of each bus stop; for bus lines, it records constituent bus routes as a sequence of stops, along with their shape (a sequence of GPS points) as additional information.

The GTFS data must be mapped to a high resolution network; for Singapore, this is a navigation network developed by NAVTEQ. The network is a directed graph where streets and intersections are represented as links and nodes. The links between nodes record attributes like street name, number of lanes, length, flow, free speed and capacity. Nodes are simply recorded as two-dimensional point coordinates. This network has a total of 79 835 links and 43 118 nodes.

Special Restrictions There are some intrinsic characteristics of the public transport system that should be considered serious restrictions. First, when a certain stop is assigned to a network link, this link should be a part of all paths belonging to this stop's routes. In other words: once established, stop-link relationships are fixed for resolving the missing routes. If the GPS points from a route including a specific stop suggest it should be associated with a different nearby link, then all other routes including that stop must be resolved again. Hence, the order in which the routes are resolved is important; it is preferable to resolve those routes first, when we completely trust supporting information quality (e.g., GPS trails).

Second, while many lines run in two directions, with most bus stops having a corresponding stop in the opposite direction (stop located on the other side of the street), this cannot be used to our advantage, because links defined by each return route are different, locations of stops are not necessarily exactly opposite to those in the opposite direction and return routes do not always use the same street.

However, some routes on the same line have an inclusion relationship; in peak hours, segments of bus routes with high demand are served by additional buses running on partial routes to meet demand. In these cases, if a full route is resolved, its partial routes solutions are included.

18.3 Solution Approach

It is not possible to automatically map-match the given GPS position with the network, as standard methods usually require at least 10 points for each link (Schüssler and Axhausen, 2009). In the Singapore GTFS, distance between consecutive points averages about 65 meters, and average link length is about 91 meters; thus, we have fewer than two points per link, on average. Furthermore, not all the routes have GPS points, which inhibits using a full automatic solution; in the Singapore GTFS, there are 38 bus routes without GPS points.

Consequently, the strategy for resolving each route consists of a semi-automatic procedure. Figure 18.2 illustrates the process. First, a simple map-matching algorithm is applied if the route is not part of a bigger route already solved (inclusion relationship described above). In this case, only a previous solution's partition is needed to obtain a first solution. Then, an automatic verification (described below) is performed. If the verification ends with a positive outcome, one can decide to finish the route and save the solution, or to continue editing. If one decides, or is forced, to modify the solution, there are two ways to proceed: changing parameters and running the automatic algorithm again, or editing the solution interactively with a graphical interface editing tool. In both cases, automatic verification must be executed again. If previously saved stop-link relationships are modified, prior routing solutions containing one of the involved stops are erased.

As long as more solutions are obtained, it becomes easier and faster to solve further routes, similar to a machine learning process. This happens for two reasons; first, because of the inclusion relationships that omit the algorithm and second because the increasing number of fixed stop-link relationships relaxes the algorithm (functioning explained in the following section).

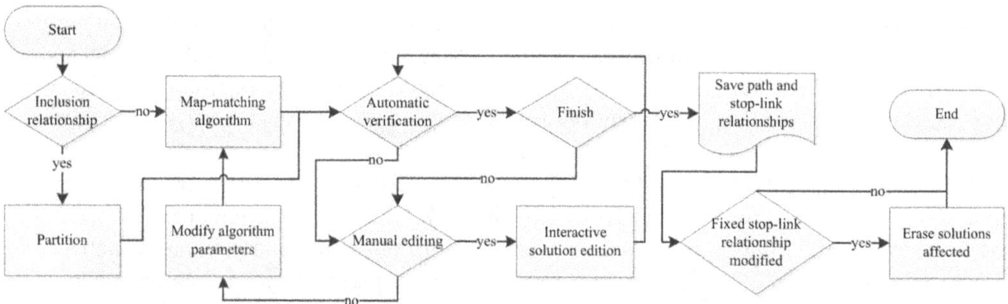

Figure 18.2: Semi-automatic process for one route.

Source: Reprinted from Ordóñez Medina and Erath (2011, p.754), Copyright (2011), with permission from Hong Kong Society for Transportation Studies

18.4 Map-Matching Automatic Algorithm

This algorithm's objective is to generate a solution (path or sequence of connected network links and a set of stop-link relationships) for one route, knowledge of its profile, a sequence of GPS points and a set of stop-link relationships. The algorithm is designed to deal with:

- low GPS point resolution,
- sporadic low network spatial resolution,
- long distances between two express routes stops, and
- understanding that the nearest link to a stop point is not always the correct one.

The route map-matching process is illustrated in Figure 18.3. Except for the first stop, the algorithm solves for each stop in the route profile, a portion of the links sequence (from previous to current) and, if this stop has no fixed link, a set of link candidates pooled from the one link selected.

Link candidates are defined as follows: the *NL* closest links to the stop point, within a distance D_{max}, define a set of candidates. Each set's element could be subjected to more restrictions; the closest point, between the stop point and the infinite line defined by the link, must be inside its line segment and the angle between the link direction and the nearest GPS points sequence direction must be lower than α_{max}.

The link's selection is performed as follows; from the previous stop link to each defined candidate, an A star search algorithm is applied for finding the shortest path. For running this algorithm, each link's cost depends on the link's travel time and distance to the GPS points. A product with flexible exponents was proposed as a first model:

$$C_{link} = \exp \frac{L_{link}}{S_{link}} w_1 \exp D_{GPS} w_2 \tag{18.1}$$

where L_{link} is its length, S_{link} is its free speed, D_{GPS} is its distance to the GPS points sequence and w_1 and w_2 are positive weights with a standard value of 1, but modifiable by the user, according to existence or quality of the GPS points sequence. The definition of D_{GPS} can also be modified; in the simplest approach, it is the minimum distance between the link and all GPS points (point-segment distance). From all calculated paths, the shortest is selected and added to the general route solution. The corresponding link candidate is also related to the stop.

If the current stop has a stop-link relationship, only the shortest path to this stop defines the solution. Thus, the process continues with the next stop in the route profile. If the first stop of the

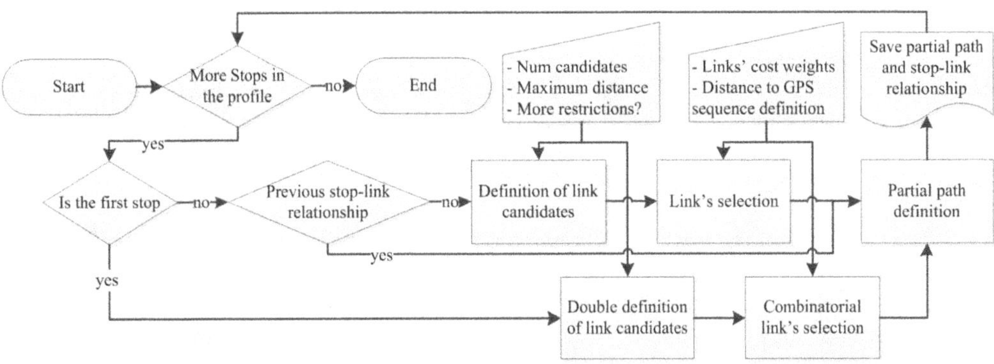

Figure 18.3: Map-matching algorithm.

Source: Reprinted from Ordóñez Medina and Erath (2011, p.755), Copyright (2011), with permission from Hong Kong Society for Transportation Studies

profile has no fixed link, a similar algorithm between the first and the second stop is performed. The definition of candidates' procedure is applied to the first and the second stops. Then, the candidates' selection procedure consists of obtaining the shortest path of all combinations between the two sets of candidates, then selecting the shortest one. This path defines links for both stops.

18.5 Automatic Verification

In this step, accuracy of the routing solution is automatically checked by performing the following ordered verification:

1. Is the path joined?
2. Is the path without U turns?
3. Is the path without repeated links?
4. Does every stop of the route have a stop-link relationship?
5. Is every link related to a stop inside the path?
6. Is the related links' order in the path the same as the corresponding stops' order in the route profile?
7. Is the nearest point between the stop point and the infinite line defined by the link inside its line segment in every stop-link relationship?
8. Are the first and last links of the path related to the first and last stops of the route profile?

Verifications (2), (3) and (7) are not mandatory and can be deactivated through the user interface. User interaction is necessary to (i) cover possible errors, and (ii) include actual route characteristics: some bus routes do include U turns, some repeat exactly the same street, in the same direction, during their travel and the geometric restriction presented in (7) is not always valid in big stop facilities, like bus interchanges.

18.6 Manual Editing Functionalities and Implemented Software

The edit functions' objective is to allow the user to modify the automatically generated routing solution. Even if the automatic algorithm generates a correct solution based on input data, problems like recent changes in routes, differences in release dates between GPS points and network data, erroneous GPS points, or lack of network element all require manual changes. Although one also could modify and correct the input data, or the generated solution, with direct data modifications, two-dimensional visualization and keyboard-mouse user interaction are two quality attributes that help reduce time and effort. Developed functional requirements and quality attributes are:

1. Visualization: A navigation network is displayed, including all relevant information for working with a single route. This includes the route's profile, given sequence of GPS points, and its current solution (path and stop-link relationships). Selected elements are drawn in a different color. Everything is displayed in a two-dimensional and interactive way, including the cursor location in working coordinates, panning, zoom and view-all options.
2. Selection: Different options for selecting solution elements, or elements from the network, are provided. It is possible to select the nearest link from the solution or from the network, the nearest node from the network, or the nearest stop from the solution, to a point indicated by the user. When a stop that already has a stop-link relationship is displayed, its corresponding link is highlighted as well. If a solution path link is selected and does not have a subsequent link connected, a new one from the network is selected with one click; the selected link is that with the angle most similar to the line defined by the end node of the initial link and a point indicated by the user.
3. Path modification: The first link of the sequence can be added by selecting any network link. If a solution path link does not have a subsequent link connected, it is possible to add one, according to the selection function described in (2). If there are two unconnected sequential links in the solution (a gap), a sub-sequence connecting these links is added, using the shortest path algorithm, with the current parameters. Further, selecting one solution path link, it is possible to delete it, or to delete all links before or after it. Finally, stop-link relationships can be modified by selecting either elements. If the modified relationship was fixed, the user is prevented from modifying the relationship, because the tool will erase the solutions of the routes to which the selected stop belongs.
4. Network modification: New nodes to the road network can be added. In addition, with any node selected, it is possible to add a new link selecting the end node.

These functions were implemented in a software package developed from scratch in Java and using the Java2D library for graphics. The package reproduces the described solution approach, looking for non-solved routes, and running the map-matching algorithm and the automatic verification for each one. Figure 18.4 shows the user interface and a demo video can be accessed at `http://www.vimeo.com/27137889`.

18.7 Conclusion and Outlook

The semi-automatic procedure designed for map-matching bus lines with a high resolution navigation in Singapore was successful, allowing the solving of all bus routes and stops in only ten days, even taking into account the quality of the input information offered, highlighting the low spatial and temporal resolution of the GPS points given for each route. Analysis indicates that reducing manual modification time is the best way to improve the procedure, which can be done by modifying the automatic algorithm to obtain more accurate results for the initial routes to be solved, or in other words, for routes not affected by the learning process.

As GTFS is becoming so popular for defining public transport systems and the code in which this process is implemented is open source, it can be used for matching routes with high

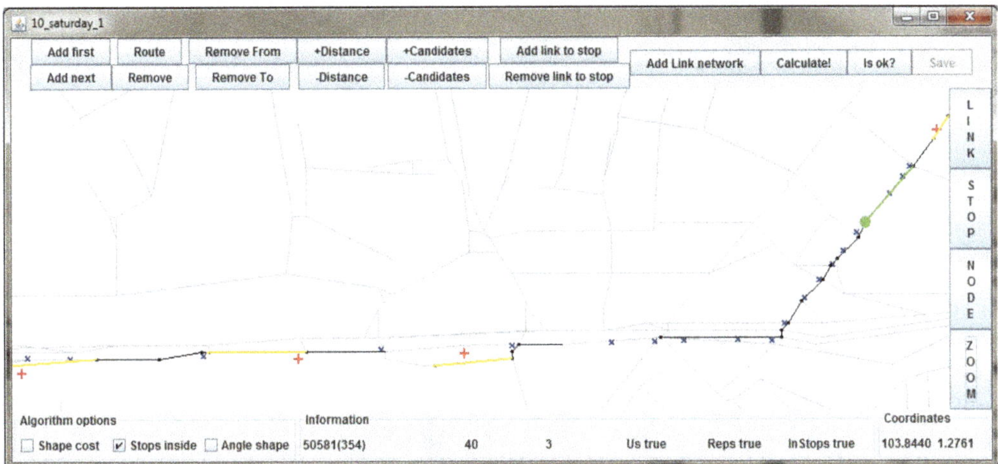

Figure 18.4: User interface of the application to edit automatic solutions.

Source: Reprinted from Ordóñez Medina and Erath (2011, p.757), Copyright (2011), with permission from Hong Kong Society for Transportation Studies

resolution networks of any GTFS-specified place. The tools are available as a MATSim contribution (`GTFS2TransitSchedule`). For generating MATSim simulation scenarios, the procedures have been used by research teams in the province of Gauteng, South Africa, on the Toronto scenario and on a different public transport simulation model developed by SMART-MIT in Singapore.

CHAPTER 19

New Dynamic Events-Based Public Transport Router

Sergio Arturo Ordóñez

19.1 Basic Information

Entry point to documentation:

http://matsim.org/extensions → eventsBasedPTRouter

Invoking the module:

http://matsim.org/javadoc → eventsBasedPTRouter → RunControlerWS, RunControlerWSV, RunControlerWW classes

Selected publications:

Ordóñez Medina and Erath (2013b)

In public transport route choice, decisions and actions of a particular user depend not only on his/her own preferences, like value of time, crowd avoidance or willingness to pay. They also depend on the decisions and actions of many other public transport users, operators and authorities. Even private transport users' decisions are also involved, as everybody shares the same infrastructure.

This implementation of MATSim used a SBPTR, as mentioned above, meaning that when an agent needed a route for a given start time, origin and destination, the SBPTR found the shortest path in a schedule-based network (assuming public transport vehicles are always on time and always have space). Within the mobility simulation, a vehicle could arrive early or late and/or it could be full, thus not allowing additional passengers to board. With a negative result, the agent obtained a bad score and this plan would have probably been replaced with a more favorable one during the iterative learning process. This scenario's problem occurred when the agent tried to

How to cite this book chapter:

Ordóñez, S A. 2016. New Dynamic Events-Based Public Transport Router. In: Horni, A, Nagel, K and Axhausen, K W. (eds.) *The Multi-Agent Transport Simulation MATSim*, Pp. 123–132. London: Ubiquity Press. DOI: http://dx.doi.org/10.5334/baw.19. License: CC-BY 4.0

find a new route for the same start time, origin and destination, the public transport scheduled network shortest path remained the same; agents could not improve their experiences by changing the route.

To address this shortcoming, a new EBPTR (Events-Based Public Transport Router) was proposed (Ordóñez Medina and Erath, 2013b), modeled, implemented and tested. It took the given schedule as a base for the first iteration, but updated information on travel times, occupancy of the public transport vehicles, and waiting times was propagated between subsequent iterations. Thus, when same day executions were performed, new routes could be generated for the same start time, origin and destination, because the system is remembered delayed bus services (longer travel times), or train services where the vehicle arrived full (longer waiting times). However, the network used to route agents required a new topology to account for such variables. This approach allowed then to account for emergent phenomena; in situations where overcrowded vehicles prohibited boarding, it made sense for some agents to travel a few stops in the outbound direction. They could then transfer to an inbound vehicle with sufficient capacity and board. Although more memory was needed, similar or even better computation times were achieved when shortest path calculations awe performed, due to the simpler network topology. Furthermore, to achieve user equilibrium required a significantly smaller number of iterations.

19.2 Events-Based Public Transport Router

A new EBPTR was developed for MATSim to more realistically model public transport route choice, where agents learn, over time, that transit vehicles are not always on time, do not always have sufficient space to allow boarding and trips with more comfort are often preferable.

Network Topology Figure 19.1(b) shows the structure of the proposed public transport network, compared with the original structure (Figure 19.1(a)). Inspired by the network designed by Spiess and Florian (1989) this implementation had two types of nodes. The first type represented a stop facility (green-black squares) as point in space, while the second type (yellow-red dots) represented a stop-route relation which could be seen as a physical or virtual platform for each line passing a particular stop facility. For example, different platforms in a metro system needed to be modeled as different stop facilities, because different services arrived at each platform and walking paths were needed to change from one platform to another. For bus stop facilities, they represented virtual platforms; in reality, buses from different lines serving the same bus stop would normally use the same physical infrastructure e.g., a bus bay. To connect those nodes, there were four types of links. The in-vehicle links joined two consecutive stop-route nodes in the direction of the correspondent route. The boarding links connected a stop node with each corresponding stop-route node. The alighting links were opposite, connecting stop-route nodes with their corresponding stop node. Finally, walking links connected a stop node with all other stop nodes located within walking distance.

Link Costs Each link in this network had a related time-dependent disutility function. Different costs were saved for different times in the day for a given time bin (at this time, 15 minutes). In-vehicle link disutilities depend on vehicle travel time, travel distance, level of occupancy and a fare rate, if this system is distance-based. Boarding link disutilities depended on waiting times, a transfer cost, and a fixed fare if this system was entry-based; thus it was possible to relate specific stop-route waiting times to these links. As the first waiting link was not a transfer, this cost had to be subtracted from the whole path cost, but this detail did not affect the shortest path calculation.

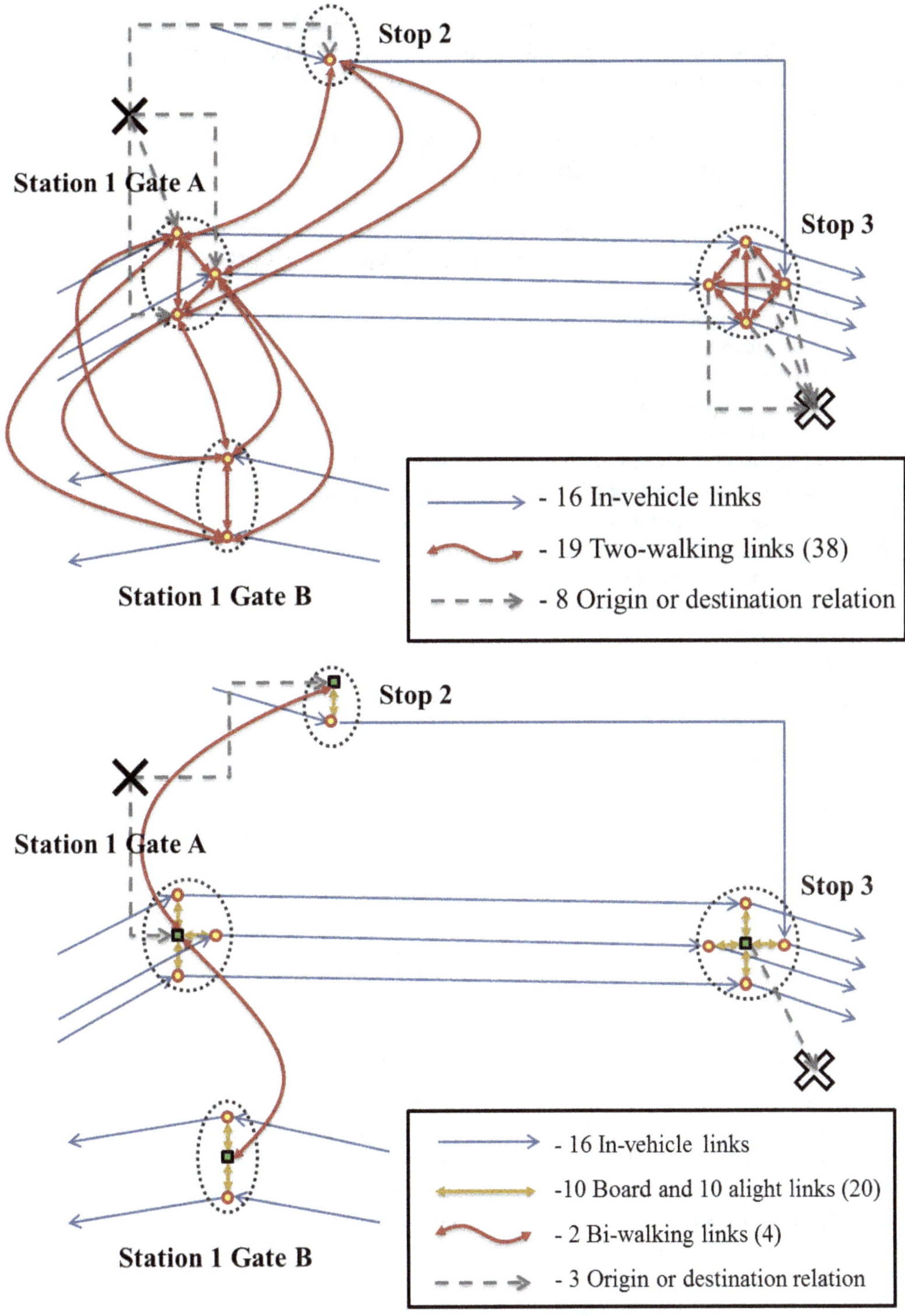

Figure 19.1: Comparison of the network topologies of the schedule-based transit router (a) and the new events-based transit router (b).

Alighting links had no associated cost, but a fare could be related to them. Finally, walking links depended on the walking travel time and distance. Equation (19.1) shows linear versions of these functions used in this model assuming a distance-based fare system.

$$\begin{aligned}
&C_{iv}(t) = (\beta_{iv} * t_{iv}(t))(1 + g(p_{oc}(t))) + \beta_{vd} * l_{iv} + f_{iv} * l_{iv} \\
&C_{bo}(t) = \beta_{wt} * t_{wt}(t)) + c_{tr} \\
&C_{al}(t) = 0 \\
&C_{wk}(t) = \beta_{wk} * t_{wk} + \beta_{wd} * l_{wk} \\
&C_{path}(t) = \sum C_{iv}(t') + \sum C_{bo}(t') + \sum C_{al}(t') + \sum C_{tr}(t') - c_{tr}
\end{aligned} \tag{19.1}$$

C_{path}: Total cost of the path.
C_{iv}: Cost of one in-vehicle link.
C_{bo}: Cost of one boarding link.
C_{al}: Cost of one alighting link.
C_{wk}: Cost of one walking link.
β_{iv}: Personalized cost per unit of time traveling in a vehicle.
β_{vd}: Personalized cost per unit of distance traveling in a vehicle.
β_{wt}: Personalized cost per unit of time waiting in a stop.
β_{wk}: Personalized cost per unit of time walking.
β_{wd}: Personalized cost per unit of distance walking.
c_{tr}: Personalized cost for making a transfer.
f_{iv}: Vehicle dependent fare rate by distance traveled.
$t_{iv}(t)$: In-vehicle travel time (from Stop-stop travel times structure).
$t_{wt}(t)$: Waiting time (from Stop-route waiting times structure).
t_{wk}: Walking time.
l_{iv}: In-vehicle distance.
l_{wk}: Walking distance.
$p_{oc}(t)$: Occupancy level in the in-vehicle link (from Route-stop occupancy structure).
$g(p)$: Simplified function of how occupancy level increases the cost (Equation (19.2)).

$$g(p) = \begin{cases} 0 & \text{if} \quad p \leq p_{sit} \\ r_{sta} * p + b_{sta} & \text{if} \quad p_{sit} < p < 1 \\ b_{full} & \text{if} \quad p = 1 \end{cases} \tag{19.2}$$

p_{sit}: Occupancy level when no more seats are available.
r_{sta}, b_{sta}: Parameters of percentage increase in discomfort from standing in the vehicle.
b_{full}: Maximum percentage increase when the vehicle is full.

Shortest Path Algorithm To find a public transport route between an origin and a destination, for a given time of day, the applied method was the same as currently implemented in MATSim; first, the algorithm looked for the stop-nodes within walking distance from both origin and destination. An initial cost was associated with each of these stop-nodes, according to access and egress walking times. Then, starting from all the origin-stop-nodes with a given access cost, a multi-node time dependent Dijkstra algorithm found the shortest path, to the destination-stop-nodes with related egress costs. Thus, the path determined the best O-D (Origin-Destination) combination as well. The algorithm was time-dependent because it recognized that while it proceeded through the path, time advanced; thus, different costs are obtained from the links while time advanced. The total disutility of this path was compared with the cost of a full walking trip. If the cost is less, the path is converted to a sequence of stages: in-vehicle stages for each in-vehicle link in the path and walking stages for each walking link. Boarding and alighting links were ignored for this conversion.

Structures to Save Travel Times, Waiting Times and Vehicle Occupancy As mentioned earlier, the mobsim of MATSim generated atomic units of information called events, which described changes for each person, e.g., boarding or alighting: each vehicle, e.g., entering and leaving a link during the simulation. The goal was to save information on public transport experience in one simulation and find better public transport routes for agents in the next iteration. This feedback mechanism was already implemented in MATSim for private transport; the car router used each saved link's time-dependent travel times from a previous iteration to calculate better routes in the road network, by changing the costs of the links. To allow the EBPTR to learn from the previous iteration, information about (a) stop-stop travel times, (b) stop-route waiting times and (c) route-stop-stop vehicle occupancy, was required.

- Stop-stop travel times: To account for public transport vehicle delays, travel time between consecutive stops had be saved. Two stops are consecutive if they were consecutive for at least one public transport route. A first option was using the previously discussed travel times structure that saved time-dependent travel times for each road network link. Because a vehicle had to follow known road links between two consecutive stops, these travel times could be summed. One problem: this structure accounted for all the vehicles in the network, but travel times of cars and buses were very different, particularly in links with public transport stops. Thus, a special structure was implemented to save these stop-stop travel times. The structure averaged all the public transport vehicle times from one stop to the next during a certain time bin. More specifically, each value comprised the time from when the vehicle arrived at a certain stop until it arrived at the next stop, denoted in the simulation by consecutive `VehicleArrivesAtFacility` events. This meant that the first stop waiting time and all queue times (if the vehicle had to queue before the bay or platform was available) were included. In other words, when an agent routed the first in-vehicle link of each trip, the full dwell time would be included. Hence, this agent assumed it was the first passenger entering the vehicle. For all the other in-vehicle links the in-vehicle waiting was included. These stop-stop times were the main component of the in-vehicle link disutilities.
- Stop-route waiting times: Waiting times are a fundamental aspect of public transport route choice and can be long due to vehicle delays (i.e., due to the stop location), or full public transport vehicles of one or several consecutive services (i.e., due to the route demand and stop position within the route). For that reason, waiting times were saved for each stop-route relation. Similarly, the structure averaged all agent waiting times in a certain stop, for a certain route, during a certain time bin in the day. More specifically, each value comprised the time from when the agent arrived at the public transport stop until it entered the vehicle, denoted in the simulation by consecutive `AgentArrivesToFacility` and `PersonEnterVehicle` events. These waiting times were the principal component of boarding link disutilities. If no observations were found for a certain stop-route-time, the model returned half the corresponding headway, specified by the transit schedule.
- Route-stop occupancy: By accounting for occupancy level, one can model routing decisions where people take longer/slower routes to feel more comfortable in emptier vehicles, i.e., valuing a higher chance to travel while seated. Occupancy depends on specific route demand and the stop position within the route. Here, occupancy was assumed to be constant between two consecutive stops. When a vehicle departed from a certain stop (denoted in the simulation as `VehicleDepartsFromFacility` event) this structure averaged the occupancy level with the other vehicles on the same route departing from the same stop during the same time bin. As there were only a few vehicles recorded for each time bin, it was unlikely to find observations for a specific bin. In this case, the structure returned the value of the next time bin, where at least one observation was found for the corresponding stop and route.

19.3 Functional Results

Relaxation Process The number of iterations needed by MATSim's co-evolutionary algorithm to reach a stable state was a critical variable; efforts were made to reduce it (Meister et al., 2006; Fourie et al., 2013).

The EBPTR effectively reduced the iterations public transport users needed to reach equilibrium. Using a 25 % sample of the Singapore scenario, Figure 19.2 shows average score plan evolution for 355 207 agents over 100 iterations. These 100 iterations were executed four times to use both routers for two different replanning strategies. Agents saved five plans in memory. At iteration 0, both EBPTR and SBPTR started with routes described in the schedule; however, the EBPTR returned routes that performed better in this first simulation. This occurred because, for each pair of consecutive stops, the EBPTR used the average of all scheduled route times that contained this pair as the first estimate. On the other hand, the SBPTR used the specific scheduled time of the corresponding route. Results indicated the average stop-stop time seemed to be a more reliable estimate for this first iteration.

For the rest of the iterations, the Figure 19.2 shows how the scores evolved. The first replanning strategy stipulated that 30 % of the agents were re-routed at each iteration. This evolution is shown in the first graph of the figure. Using SBPTR, agents received the same route over and over again as the start time, origin and destination did not change between iterations. Small variations in scores occurred because of the stochastic simulation nature explained above. Although scores started in the same range, using EBPTR allowed better-performing routes to be found within a very small number of iterations.

For a more realistic comparison, a second replanning strategy was tested, where just 20 % of the agents were re-routed and the activity start times were modified randomly within a half an hour for 10 % of the agents. The second graph of the figure shows how both routers managed to improve agents' plan scores. But with the EBPTR, number of iterations needed to achieve the average executed score, achieved after 100 iterations for the SBPTR (120), was only 5. The target marginal score, as a measure of change in score over iterations, was taken arbitrarily as 0.1 utilities per iteration, or the rate produced after 200 iterations with the SBPTR. In contrast, this target rate was achieved after 77 iterations with the EBPTR, a 2.6 improvement factor .

Modeling Advantages Because of the links disutility function in the proposed network account for aspects like waiting times or occupancy levels and because MATSim allows for modeling heterogeneity among agents, the router could be a very powerful tool to model observed emergent behavior in public transport route choice. In Singapore, like many other crowded cities in the world, some commuters decide to travel backwards for a few stops and then transfer to a train in the opposite direction to find a seat or space in a public transport vehicle Chakirov and Erath (2011). With the SBPTR this kind of least cost path could not be found, but with the newer proposal, this was possible. Although proportions did not match actual observations as the Singapore scenario lacked appropriate and calibrated utility parameters for traveling and waiting time under crowded conditions, Figure 19.3 shows totals of people traveling backwards from different stops in the island after 100 iterations (see Figure 19.2 (a)).

19.3.1 Comparing Quality Attributes With the Current Implementation

Computation Time The tests described next were executed using 12 computational nodes, accessing 70 GB of shared memory, using the Singapore scenario described in Chapter 57. Before the first iteration, if plans were not routed, MATSim prepared every agent with an initial route. As mentioned before, the stop-stop travel times and stop-route waiting times were initially taken from the schedule. Because of its simpler network structure the EBPTR took 01:17:35 to initially

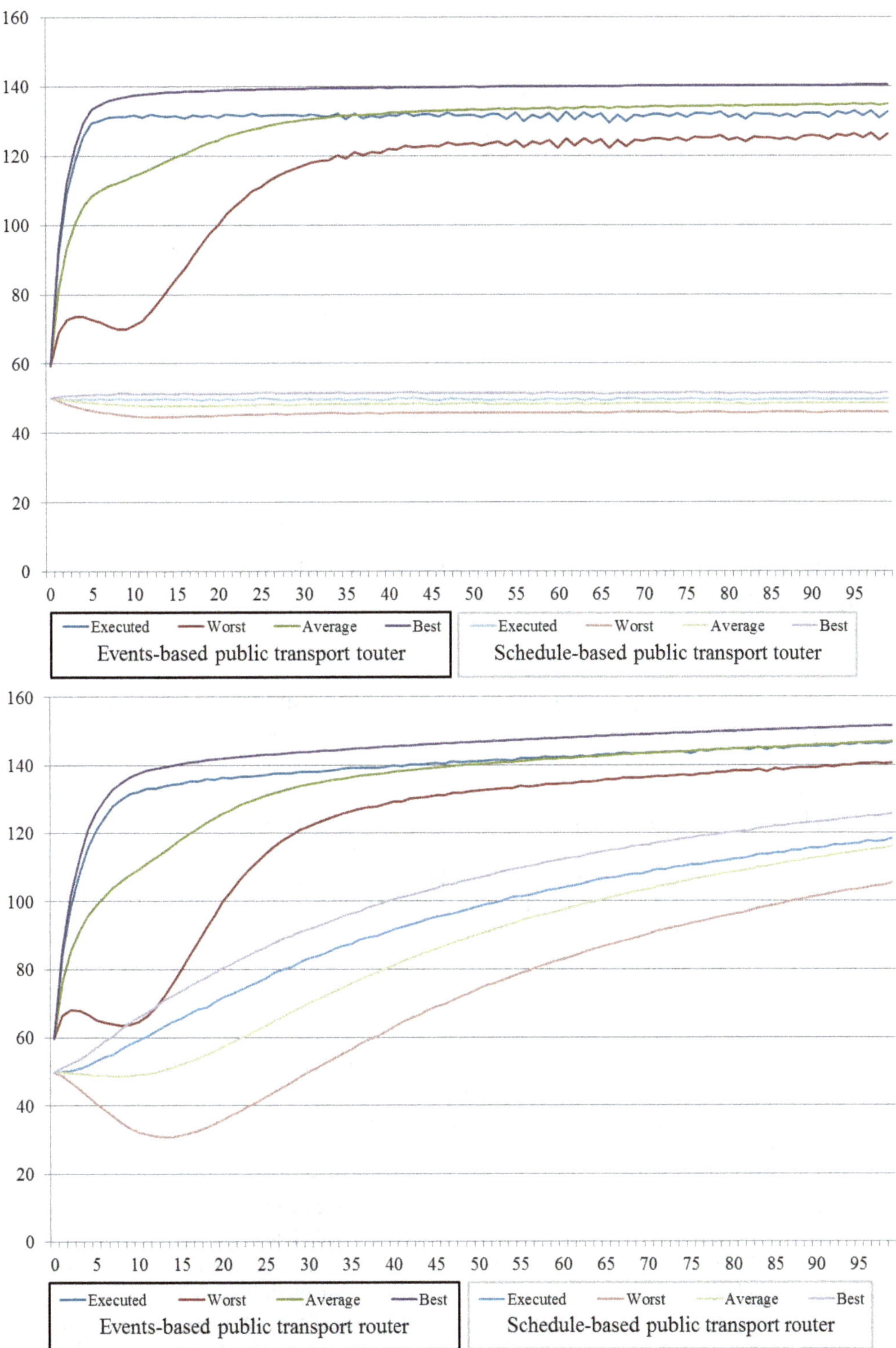

Figure 19.2: Comparison of score evolution: a) 30 % re-route, b) 20 % re-route and 10 % time allocation.

Figure 19.3: Number of agents traveling backwards at each MRT (Mass Rapid Transit, Singapore) station of the Singaporean rail system.

route the 355 207 users, compared with 01:28:55 needed by the SBPTR, producing a performance gain of about 12.7 % for this scenario. When running MATSim iterations with the EBPTR, computation times principally changed in two processes: mobility simulation (mobsim) and replanning. Figure 19.4 shows computation times measured for the first 20 iterations of the process. Although the EBPTR needed more time in mobsim, it continued to require considerably less time for rerouting during the replanning, due to a simpler network topology. The longer mobsim time was due to information saving in the new structures during the simulation. However, on average, the EBPTR outperformed SBPTR, per iteration, by about 3 minutes or 11 %. As mentioned above, 2.6 times more iterations were needed for the SBPTR to achieve a specific point in the relaxation process. For 77 iterations with the EBPTR, computation amounts 35:25:43, and for 200 iterations with the SBPTR, computation amounts 99:10:51; a 2.8 improvement factor in our experimental setting.

Memory Consumption The EBPTR needed more memory than the SBPTR, because the EBPTR managed more information. The necessary extra memory was allocated to the three structures described before. Given the Singapore scenario conditions described, the extra memory was calculated as follows. One numeric value needed eight Bytes, and with a time bin of 15 minutes, 120 bins were needed for 30 hours. The Stop-stop travel times structure saved two values (average and number of observations) for each time bin and each pair of consecutive stops. The number of pairs for the Singaporean public transport system was 6 602. Thus, this structure needed approximately 12.7 MB. Similarly, the stop-route waiting time structure saved two values (average and number of observations) for each time bin and each pair of stop/route combinations. The number of stop/route relations for the Singaporean public transport system was 27 156. Thus, this structure needed approximately 52.1 MB. Finally, the vehicle occupancy structure saved the average and number of vehicle occupancy observations for 26 353 route-stop relations for each of the 120 time bins, requiring approximately 50.7 MB. In total, less than 120 MB were needed for the three structures.

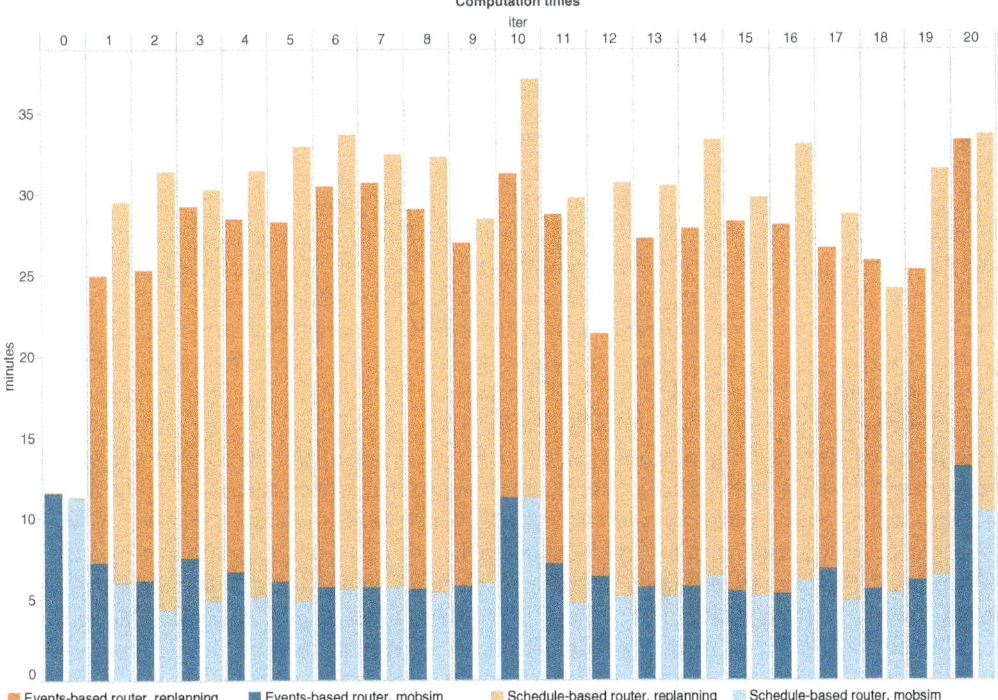

Figure 19.4: Comparison of computation times for 20 iterations.

On the other hand, the size of the network where public transport routes were calculated was smaller for the EBPTR. Although, in the case of Singapore, it created 31 939 nodes compared with 27 156 of the SBPTR (4 783 new stop-nodes), the number of links is dramatically smaller. The SBPTR created 424 070 walking links and 26 353 travel links (450 423 in total). The EBPTR created the same 26 353 travel links, plus 27 156 boarding links, plus 27 156 alighting links and just 4 390 walking links (85 055 links in total); less than a fifth in total. As a node needed 48 bytes and a link 128 bytes, the SBPTR needed roughly 46.8 MB more memory for links and just 229.6 KB less for nodes. The EBPTR saved 46.5 MB for the network, concluding that in total the SBPTR needed 70 MB less memory. This quantity was negligible compared with the total memory needed for the whole simulation (more than 40 GB).

19.4 Conclusion and Future Work

In this work, a new public transport router for MATSim was designed, implemented and tested. It produced more diverse routes in large scale scenarios, taking into account many complexities of urban public transport systems. On the supply side, the system simulated congestion, public transport vehicles occupancy levels, queues in public transport stops, bay sizes, and bus or train bunching. On the demand side, in addition to commonly used factors like in-vehicle time, number of transfers and walking time, the new router took disutility of additional waiting time due to congestion or overcrowded vehicles, comfort level inside public transport vehicles and preference heterogeneity among agents for all mentioned factors into account.

The utility of the new approach was tested in a large scale Singapore scenario. Using a simplified public transport only simulation, 100 iterations of a 25 % scenario (355 207 agents) with 30 % of the agents re-routing each iteration took just 45 hours approximately, or about 27 minutes per iteration, using 12 cores and 70 GB of memory. The computation decreased by 11 %, compared to the standard MATSim. If just 20 % of the plans were re-routed, using 35 cores accessing 85 GB of memory, the time per iteration would be reduced to less than 13 minutes, achieving 100 iterations in less than one day. But, most importantly, for computation time gains, we showed that the proposed events-based router was able to reach a steady state in a much smaller number of iterations.

If the proposed router works better than the original one, should it be changed? The current scheduled-based router of MATSim would still be relevant if the topology of its network were changed for the proposed one. It should also be applied to scenarios where the public transport system operates very reliably and punctually, with few cases of overcrowding. In that case, routing calculations would be as fast as the events-based router (with the new network structure), and the mobility simulation would be faster, as no information (in-vehicle time, wait time and occupancy) would be needed. In other words, it could be applied to city models where public transport users can reliably plan their trips using only a timetable.

Scrutinizing the resulting network loading, the biggest potential advantage of the proposed events-based router was its capacity to generate emergent behavior in congested public transport systems, in line with actual observations. Future research should aim at estimating the various route choice behavior parameters corresponding to the functionalities of the proposed system and calibrating the simulation. Although the values used came from a stated preference survey commissioned by the Land Transport Authority for the case of Singapore, advanced studies could, for example, be tailored to quantify preference heterogeneity. Furthermore, results from work in progress about the value of a seat in Singapore and discomfort disutility can improve prediction confidence. Finally, information from the Singapore smart card data could be used for revealed preference estimation of further behavioral parameters, like quality of a transfer described, e.g., by the number of escalators, to further refine the system.

CHAPTER 20

Matrix-Based pt router

Kai Nagel

20.1 Basic Information

Entry point to documentation:

`http://matsim.org/extensions` → matrixbasedptrouter

Invoking the module:

`http://matsim.org/javadoc` → matrixbasedptrouter → `RunMatrixBasedPTRouterExample` class

Selected publications:

Section 3.1 of Nicolai and Nagel (2015); Röder et al. (2013)

20.2 Summary

The matrix based PT (Public Transport) router reads a list of PT stops, and constructs "teleported" PT routes using the stops nearest to origin and destination. That is, each resulting trip will approximately look as follows:

```
<act type="previous" ... />
<!-- begin trip -->
<leg mode="walk" ... />
<act type="ptInteraction" ... />
<leg mode="pt" ... />
<act type="ptInteraction" ... />
<leg mode="walk" ... />
<!-- end trip -->
<act type="next" ... />
```

How to cite this book chapter:

Nagel, K. 2016. Matrix-Based pt router. In: Horni, A, Nagel, K and Axhausen, K W. (eds.) *The Multi-Agent Transport Simulation MATSim*, Pp. 133–134. London: Ubiquity Press. DOI: http://dx.doi.org/10.5334/baw.20. License: CC-BY 4.0

The attributes of the walk and the PT legs will be computed from the coordinates of the locations in the same way as teleportation routing (see Section 4.6.2.2), and then taken at face value in the mobsim (see Section 4.6.1.2).

Travel times and travel distances between PT stops can alternatively be given by corresponding matrices. This is particularly useful if a PT assignment exists and such information can be extracted from that. This was used by Röder et al. (2013) and by Zöllig Renner (2014).

CHAPTER 21

The "Multi-Modal" Contribution

Christoph Dobler and Gregor Lämmel

21.1 Basic Information

Entry point to documentation:

`http://matsim.org/extensions` → multimodal

Invoking the module:

`http://matsim.org/javadoc` → multimodal → `RunMultimodalExample` class

Selected publications:

Dobler and Lämmel (2014)

21.2 Introduction

MATSim's standard mobsim, QSim, has recently been enabled to model multimodal scenarios as shown in Section 4.6.

In this chapter,[1] an earlier approach to handle multimodal scenarios, the multimodal link contribution, is presented. As shown below, it is a very efficient approach, that considers persons' biking and walking speeds to improve the teleportation estimates for these modes, whereas mode interactions are not taken into account.

[1] Parts of this chapter are based on work published at the 6th International Conference on Pedestrian and Evacuation Dynamics in Zürich Dobler and Lämmel (2014).

How to cite this book chapter:

Dobler, C and Lämmel, G. 2016. The "Multi-Modal" Contribution. In: Horni, A, Nagel, K and Axhausen, K W. (eds.) *The Multi-Agent Transport Simulation MATSim*, Pp. 135–140. London: Ubiquity Press. DOI: http://dx.doi.org/10.5334/baw.21.

21.3 Modeling Approach and Implementation

21.3.1 Multi-modal Link Contribution

Figure 21.1 shows the implementation's basic concept—a multimodal contribution is added to each link object in the mobsim.

While traffic flow dynamics are simulated by MATSim's mobsim using a queue model, these flows are not taken into account in the multimodal contribution. Examining typical pedestrian and cyclist traffic flows shows that congestion is very rare compared to vehicular traffic, justifying application of this simplistic approach over a scenario. For regions with higher traffic flows, this simple model loses accuracy, but still outperforms the teleportation approach, which MATSim uses by default.

Each multimodal link contribution uses a priority queue to manage all agents traveling on that link using a non-motorized mode. The queue orders the agents based on their scheduled link leave time (see Figure 21.2). This time is calculated when an agent enters a link and is based on parameters like the agent's age and gender, as well as the links' steepness. In each time step, it is checked whether the queue contains agents who have reached their link leave time and thus must be moved to their route's next link. An agent's position on a link is not determined by the model. However, under the assumption that agents move with constant speed, their position can be interpolated. This approach is computationally very efficient, because computation effort is created only when an agent enters or leaves a link but not when it is traveling along a link. Additionally, agents can travel at different speeds, so can overtake each other.

21.3.2 Travel Times

Walk travel time calculation is based on results of a comprehensive literature review by Weidmann (1992). Starting point is a normally distributed reference speed of 1.34 meters per second with a standard deviation of 0.26 meters per second, which leads to an individual reference speed for each person. FGSV (2009) and Transportation Research Board (2010) report comparable,

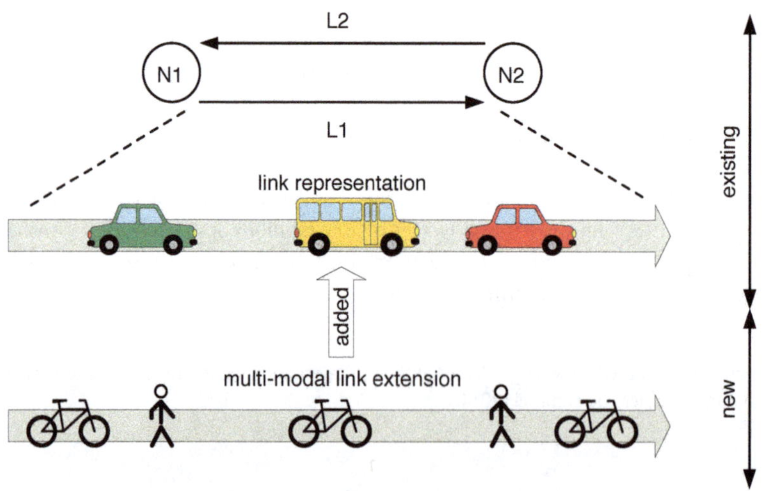

Figure 21.1: Multi-modal link contribution.

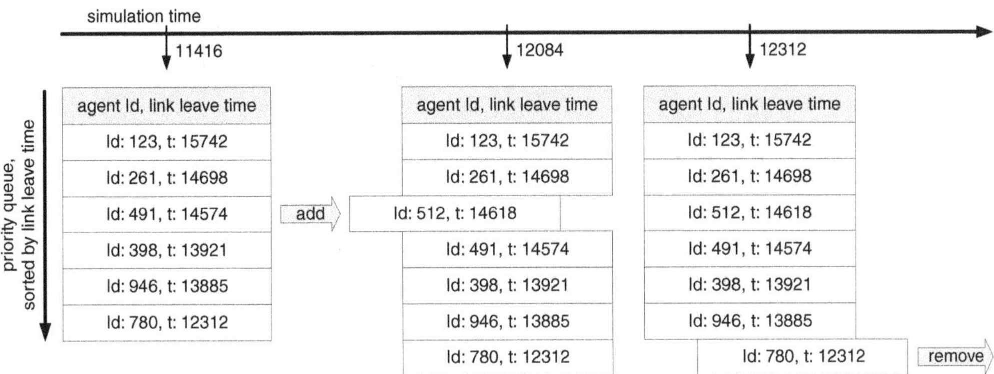

Figure 21.2: Link representation in the simple model.
At time 12 084 seconds from midnight, agent 512 enters the link and is—based on its calculated link leave time 14 618 seconds from midnight—inserted into the queue. At time 12 312 seconds from midnight, agent 780 has reached its leave time and is then removed from the queue.

but less detailed data. If a trip's purpose is known, a person's reference value can be adjusted (commuting 1.49 meters per second, shopping 1.16 meters per second, leisure 1.10 meters per second; see FGSV, 2009). Using the reference speed and referencing a person's age, gender and statistical spreading, a personalized speed is calculated (see Figure 21.3(a)). Finally, to calculate the person's travel time on a specific link, influence of the link's steepness on the person's speed is taken into account (see Figure 21.3(b)). The combination of person-specific attributes and link steepness is shown in Figure 21.3(c).

As a result, a person's speed on plain terrain is calculated as:

$$f_{\text{person}} = f_{\text{statistical spreading}} \cdot f_{\text{gender}} \cdot f_{\text{age}} \tag{21.1}$$

$$v_{\text{person, walk}} = v_{\text{reference, walk}} \cdot f_{\text{person}} \tag{21.2}$$

A link's steepness is incorporated as:

$$v_{\text{person walks on link}} = v_{\text{person, walk}} \cdot f_{\text{steepness}} \tag{21.3}$$

The speed of cyclists is determined using results from Parkin and Rotheram (2010). Starting point is, again, an individual's speed based on a normal distributed ($\mathcal{N}(6.01, 1.17)$) reference speed. Once more, a person's speed is calculated by accounting for age and gender (see Figure 21.4(a)).

When calculating the steepness factor, one must define whether a link goes uphill or downhill. When going uphill, the person's speed is reduced by a factor based on the grade and a reference factor of 0.4002 meters per second, which is scaled by the same factor as the person's reference speed. i.e., the speed drop of slow people is lower than the drop of fast people. When bike speed drops below walk speed, which happens at a grade of approximately 12 %, it is assumed that the person switches to walking (see Equation (21.5)). For downhill links, a reference factor of

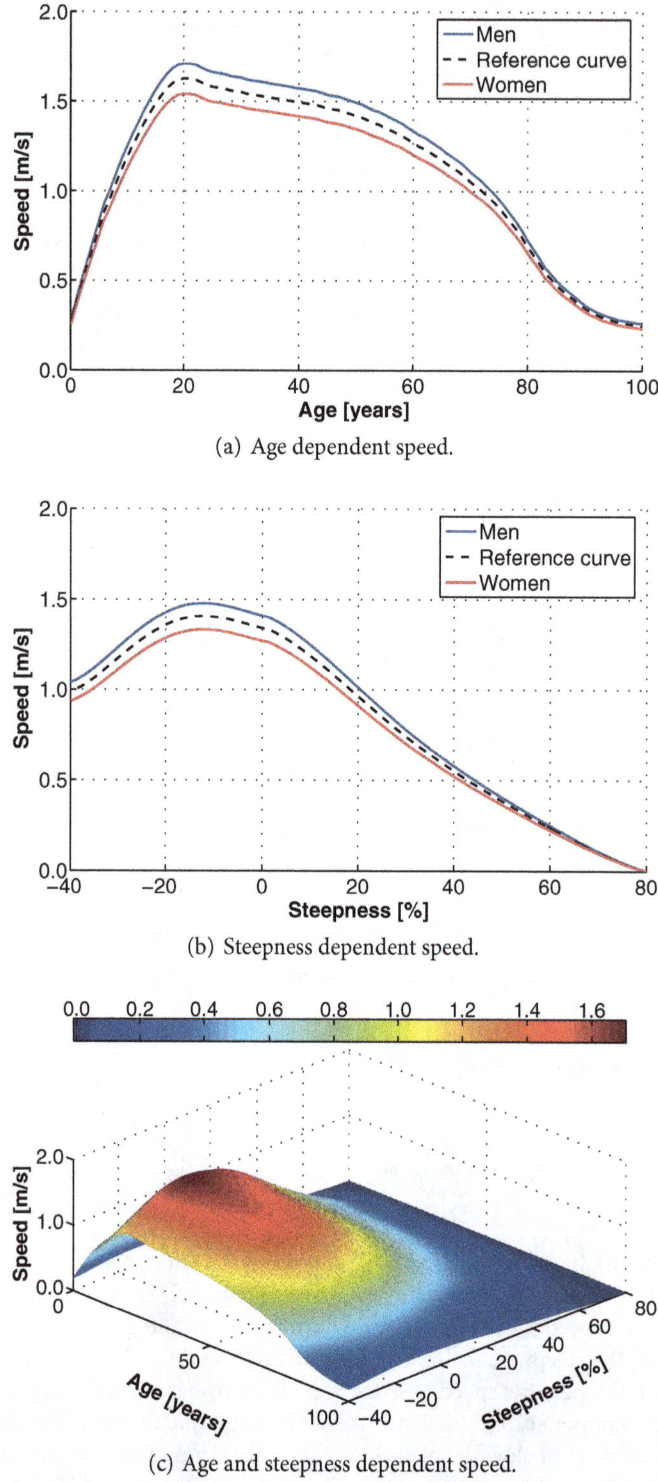

(a) Age dependent speed.

(b) Steepness dependent speed.

(c) Age and steepness dependent speed.

Figure 21.3: Age and steepness dependent speed of pedestrians.

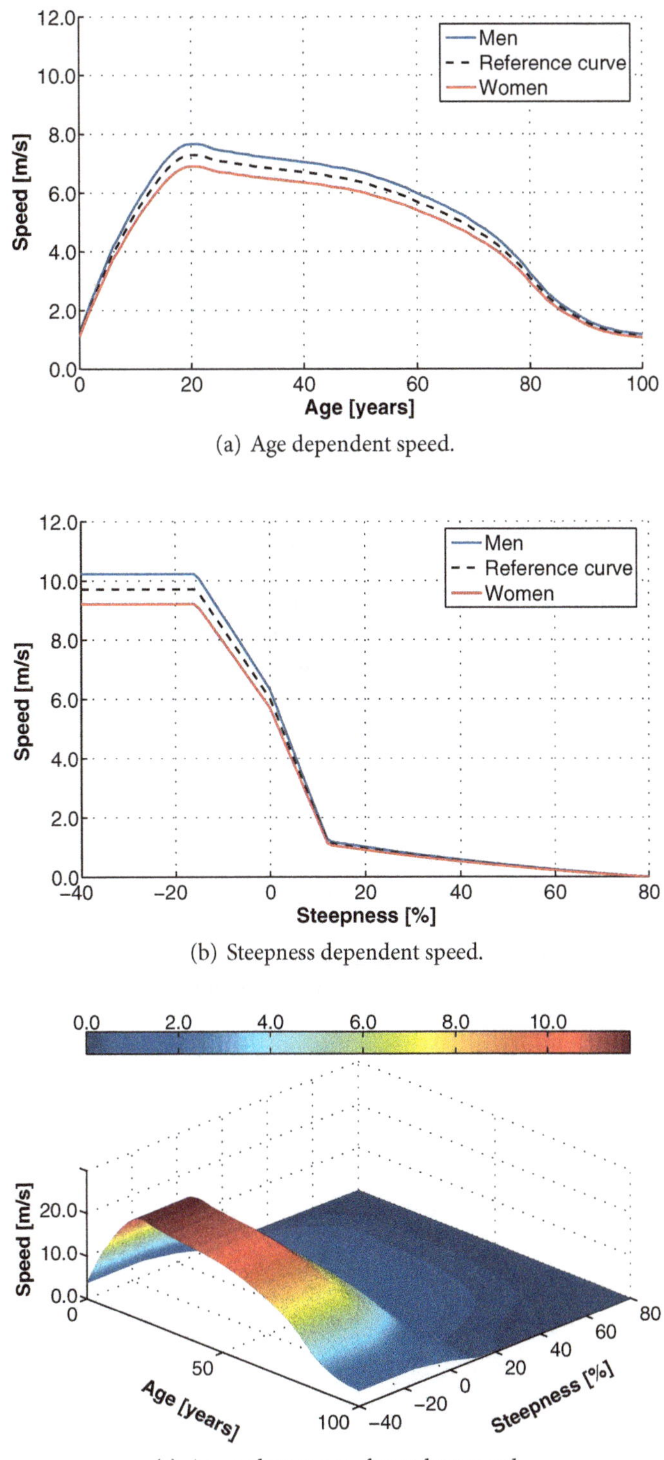

(a) Age dependent speed.

(b) Steepness dependent speed.

(c) Age and steepness dependent speed.

Figure 21.4: Age and steepness dependent speed of cyclists.

0.2379 m/s is used. Additionally, it is assumed that cyclists limit their speed to 35 kilometers per hour (9.7222 meters per second; see Equation (21.6)).

$$v_{\text{person, bike}} = v_{\text{reference, bike}} \cdot f_{\text{person}} \tag{21.4}$$

$$v_{\text{person, uphill}} = \max \begin{cases} v_{\text{person, bike, flat}} - 0.4002 \cdot |\text{grade}| \cdot f_{\text{person}} \\ v_{\text{person, walk, uphill}} \end{cases} \tag{21.5}$$

$$v_{\text{person, downhill}} = \min \begin{cases} v_{\text{person, bike, flat}} + 0.2379 \cdot |\text{grade}| \cdot f_{\text{person}} \\ 9.7222 \end{cases} \tag{21.6}$$

Another parameter affecting pedestrian and cyclist speed is the crowd density of the link where they are physically present. Data to take this effect into account is, again, presented by Weidmann (1992). However, to calculate crowd density of a link, its geometry has to be taken into account, as discussed by Lämmel (2011).

21.4 Conclusions and Future Work

The multimodal contribution allows the tracking an agent's movement in detail, essential for studies related to topics like evacuations, e-bikes, car sharing or public transport. Experiments testing the implementation and demonstrating its capabilities are described by Dobler (2013).

An application's required level of detail strongly influences the modeling approach selection. A simple model including agents' age and gender, but not incorporating agent-agent interactions, might be detailed enough for some studies (e.g., e-bikes or public transport). However, for other studies, a more detailed model, also simulating agent interactions, might be necessary.

A first implementation of a pedestrian simulation module for MATSim, which also supports agent-agent interactions, was presented by Lämmel and Plaue (2014) introducing a force-base model. The agents' high-level planning (i.e., route and destination choice) was performed on a graph representing the transport system (e.g., a MATSim network), while the low level behavior (i.e., physical interaction between the participants) was simulated with a force-based model. Due to the intense computational effort of the underlying physical model, the scenario size was limited to a few thousand agents. An attempt to bypass this limitation was presented by Dobler and Lämmel (2012). They combined the force-based pedestrian simulation module with the multimodal link contribution, creating the opportunity to simulate large-scale scenarios, by staying highly resolved where needed and being more aggregated where possible.

CHAPTER 22

Car Sharing

Francesco Ciari and Milos Balac

22.1 Basic Information

Entry point to documentation:

`http://matsim.org/extensions` → carsharing

Invoking the module:

`http://matsim.org/javadoc` → carsharing → `RunCarsharing` class

Selected publications:

Ciari (2012)

22.2 Background

The basic carsharing idea is simple; a fleet of cars can be shared by several users, who can rent a car when needed, without having to own one. The possibility of renting short-term is the main difference from traditional car rentals. This basic concept can be implemented in various ways; in the last few years, several new business models have emerged on the market. From an operational perspective, there are three main variations:

- Round-trip based: Cars are parked at dedicated stations. They can be picked up from a station and left at the same station after use.
- One-way: Cars are parked at dedicated stations. They need to be picked up from a station and left at any station after use.
- Free-floating: Cars are parked in any parking slot within a defined service area. They can be picked up and left after use anywhere within this area.

How to cite this book chapter:

Ciari, F and Balac, M. 2016. Car Sharing. In: Horni, A, Nagel, K and Axhausen, K W. (eds.) *The Multi-Agent Transport Simulation MATSim*, Pp. 141–144. London: Ubiquity Press. DOI: http://dx.doi.org/10.5334/baw.22. License: CC-BY 4.0

From a transport planning perspective, the essential element of carsharing—the importance of its availability at precise points in time and space—does not fit with traditional models, which consider vehicle-per-hour flows. It is crucial to represent availability of vehicles at the local level, thus representing individual travel with high spatial and temporal resolution. At the same time, for the choice of using carsharing it is of fundamental importance how a trip/activity is embedded in the whole activity chain. This combination of features can be found in MATSim, which is therefore a suitable framework for carsharing modeling.

22.3 Modeling of Carsharing Demand in MATSim

Carsharing as a modal option in MATSim has been introduced in a simplified manner and only in its round trip-based version, as part of a dissertation work (Ciari, 2012). Several improvements have been introduced since then and all three main types of carsharing can now be simulated.

22.3.1 *Round-Trip Based Carsharing*

The use of round-trip carsharing by an agent in the simulation is modeled in the following steps:

1. The agent finishes his/her activity, finds the closest available car and reserves it (making it unavailable for other agents),
2. walks to the station where he/she has reserved a vehicle,
3. drives the car (interaction with other vehicles is modeled),
4. parks the car at the next activity.
5. After finishing his activity the agent takes the car and drives to the next activity.
6. Before reaching the last activity in the subtour, agent ends the rental and leaves the vehicle at the starting station, making it available to other agents,
7. walks to the activity, and
8. carries out the rest of the daily plan.

22.3.2 *One-Way Carsharing*

In the case of one-way carsharing, the steps are similar, but with few significant differences:

1. The agent finishes his activity, finds the closest station with an available car and reserves the vehicle (making it unavailable for other agents),
2. walks to the station where it has reserved the car (takes the car and frees a parking spot at the station),
3. finds the closest station to his destination, with a free parking spot and reserves it (making it unavailable for others),
4. drives the car to the reserved parking spot (interacting with other vehicles in the network),
5. parks the car on the reserved parking spot and ends the rental,
6. walks to the next activity, and
7. carries out the rest of the daily plan.

22.3.3 Free-Floating Carsharing

The use of free-floating carsharing by an agent is simulated using similar steps, but the rental ends with the end of one trip:

1. rent the nearest car,
2. walk from start activity to the rented car,
3. drive to the next activity (interaction with other vehicles are modeled),
4. park the car close to the next activity, and
5. end the rental (and make the car available for other rentals).

22.3.4 Generalized Cost of Carsharing Travel

The function representing generalized cost of travel for car sharing traveling from activity $q-1$ to activity q is:

$$S_{trav,q,cs} = \alpha_{cs} + \beta_{c,cs} \cdot c_t \cdot t_r + \beta_{c,cs} \cdot c_d \cdot d + \beta_{t,walk} \cdot (t_a + t_e) + \beta_{t,cs} \cdot t \quad (22.1)$$

The same equation is used for all modeled forms of carsharing but the values of the parameters will be different. The first term α_{cs} is a constant which can be used as calibration parameter and will also be, generally, different for different types of carsharing (and for different context). The second and third terms refer to the time dependent and the distance dependent parts of the fee, respectively. t_r is the total reservation time and c_t represents the monetary cost for one hour reservation time. d is the total reservation distance and c_d is the marginal monetary cost for one kilometer travel. The parameter $\beta_{c,cs}$ represents the marginal utility of an additional unit of money spent on traveling with carsharing. The fourth term is the walk path to and from the station (access time t_a and egress time t_e) and is evaluated as a normal walk leg. The parameter $\beta_{t,cs}$ represents the marginal utility of an additional unit of time spent on traveling with carsharing, where t is the actual (in vehicle) travel time.

22.4 Carsharing Membership

Carsharing is a membership program. In order to access a specific carsharing service, individuals must become members of that carsharing program. A logit model has been estimated for Switzerland (Ciari and Weis, forthcoming) and implemented in MATSim as part of the carsharing module. The model variables are mainly individual socio-demographic characteristics. An important feature of the model, however, is that carsharing accessibility is explicitly considered, both from home and from work. Accessibility A of person p is calculated with the following formula:

$$A(n) = \ln\left(\sum_{s=1}^{m} X_s \cdot e^{-\beta \cdot d_{sh}}\right) + \ln\left(\sum_{s=1}^{m} X_s \cdot e^{-\beta \cdot d_{sw}}\right) \quad (22.2)$$

The weight parameter for distances is set to 0.2 as in Weis (2012), and more details on it are given below. Assuming m as the number of stations in the system, d_{sh} and d_{sw}, are calculated for each station as the distance between the station s and the home and work location of person n respectively; and X_s is the number of cars at station s. The model is not directly transferable to other regions but a different model can be easily implemented in the Java code created.

22.5 Validation

The simulation model has been calibrated to reproduce actual modal share for carsharing in the Zürich, Switzerland region. It was made using booking data from the Swiss operator Mobility. With the same data, the results were validated along several dimensions. Since Mobility offered only round-trip based carsharing until now, only this model could be validated. Dimensions included in the validation process were: distance from the last activity to the pick-up station, departure times, purpose of the rental (main purpose of the subtour) and temporal length of the rental.

22.6 Applications

After a long phase of creating and improving the module to simulate carsharing in all its forms, work has been recently carried out on concrete carsharing operations issues. Examples include evaluating the impact of introduction of a free-floating carsharing program in Berlin (Ciari et al., 2014) and Zürich (Ciari et al., forthcoming) on travel demand and investigating the relationship between demand and supply in both round-trip and one-way systems (Balac et al., 2015).

CHAPTER 23

Dynamic Transport Services

Michal Maciejewski

Entry point to documentation:

`http://matsim.org/extensions` → dvrp

Invoking the module:

No predefined invocation. Starting point(s) under `http://matsim.org/javadoc` → dvrp → `RunOneTaxiExample` class.

Selected publications:

Maciejewski and Nagel (2013b,c,a); Maciejewski (2014)

23.1 Introduction

The recent technological advancements in ICT (Information and Communications Technology) provide novel, on-line fleet management tools, opening up a broad range of possibilities for more intelligent transport services: flexible, demand-responsive, safe and energy/cost efficient. Significant enhancements can aid in both traditional transport operations, like regular public transport or taxis and introduction of novel solutions, such as demand-responsive transport or personal rapid transport. However, the growing complexity of modern transport systems, despite all benefits, increases the risk of poor performance, or even failure, due to lack of precise design, implementation and testing.

One solution is to use simulation tools offering a wide spectrum of possibilities for validating transport service models. Such tools have to model, in detail, not only the dynamically changing demand and supply of the relevant service, but also traffic flow and other existing transport services, including mutual interactions/relations between all these components. Although several approaches have been proposed (e.g., Regan et al., 1998; Barcelo et al., 2007; Liao et al., 2008;

How to cite this book chapter:

Maciejewski, M. 2016. Dynamic Transport Services. In: Horni, A, Nagel, K and Axhausen, K W. (eds.) *The Multi-Agent Transport Simulation MATSim*, Pp. 145–152. London: Ubiquity Press. DOI: http://dx.doi.org/10.5334/baw.23.

Certicky et al., 2014), as far the author knows, no existing solutions provide large-scale microscopic simulation that include all the components above.

23.2 DVRP Contribution

To address the problem above, MATSim's DVRP (Dynamic Vehicle Routing Problem) contribution has been developed. The contribution is designed to be highly general and customizable to model and simulate a wide range of dynamic vehicle routing and scheduling processes. Currently, the domain model is capable of representing a wide range of one-to-many and many-to-many VRPs; one can easily extend the model even further to cover other specific cases (see Section 23.3). Since online optimization is the central focus, the DVRP contribution architecture allows plugging in of various algorithms. At present, there are several different algorithms available, among them an algorithm for the *Dynamic Multi-Depot Vehicle Routing Problem with Time Windows and Time-Dependent Travel Times and Costs*, analyzed in (Maciejewski and Nagel, 2012), and a family of algorithms for online taxi dispatching, studied in (Maciejewski and Nagel, 2013b,c,a; Maciejewski, 2014).

The DVRP contribution models both supply and demand, as well as optimizing fleet operations, whereas MATSim's core is used for simulating supply and demand, both embedded into a large-scale microscopic transport simulation. In particular, the contribution is responsible for:

- modeling the DVRP domain,
- listening to simulation events,
- monitoring the simulation state (e.g., movement of vehicles),
- finding least-cost paths,
- computing schedules for drivers/vehicles,
- binding drivers' behavior to their schedules, and
- coordinating interaction/cooperation between drivers, passengers and dispatchers.

Dynamic transport services are simulated in MATSim as one component of the overall transport system. The optimizer plugged into the DVRP contribution reacts to selected events generated during simulation, which could be: request submissions, vehicle departures or arrivals, etc. Additionally, it can monitor the movement of individual vehicles, as well as query other sources of online information, e.g., current traffic conditions. In response to changes in the system, the optimizer may update drivers' schedules, either by applying smaller modifications or re-optimizing them from scratch. Drivers are notified about changes in their schedules and adjust to them as soon as possible, including immediate diversion from their current destinations. For passenger transport, such as taxi or demand-responsive transport services, interactions between drivers, passengers and the dispatcher are simulated in detail, including calling a ride or picking up and dropping off passengers.

23.3 DVRP Model

The DVRP contribution can be used for simulating *Rich VRPs*. Compared to the classic *Capacitated VRP*, the major model enhancements are:

- one-to-many (many-to-one) and many-to-many topologies,
- multiple depots,
- dynamic requests,
- request and vehicle types,
- time windows for requests and vehicles,

- time-dependent stochastic travel times and costs, and
- network-based routing (including route planning, vehicle monitoring and diversion).

Except for the travel times and costs (discussed in Section 23.3.2), which are calculated on demand, all the VRP-related data are accessible via `VrpData`.[1] In the most basic setup, there are only two types of entities, namely `Vehicles` and `Requests`. This model, however, can be easily extended as required. For instance, for an electric vehicle fleet, specialized `ElectricVrpData` also stores information about `Chargers`. This, and other examples of extending the base VRP model, such as a model of the *VRP with Pickup and Delivery*, are available in the `org.matsim.contrib.dvrp.extensions` package.

23.3.1 Schedule

Each `Vehicle` has a `Schedule`, a sequence of different `Tasks`, such as driving from one location to another (`DriveTask`), or staying at a given location (e.g., serving a customer or waiting; `StayTask`).[2] A `Schedule` is where supply and demand are coupled. All schedules are calculated by an online optimization algorithm (see Section 23.6) representing the fleet's dispatcher. Each task is in one of the following states (defined in the `Task.TaskStatus` enum): `PLANNED`, `STARTED` or `PERFORMED`; each schedule's status is one of the following:

- `UNPLANNED`—no tasks assigned
- `PLANNED`—all tasks are `PLANNED` (none of them started)
- `STARTED`—one of the tasks is `STARTED` (this is the schedule's `currentTask`; the preceding tasks are `PERFORMED` and the succeeding ones are `PLANNED`)
- `COMPLETED`—all tasks are `PERFORMED`

In general, when modifying a `Schedule`, one can freely change and rearrange the planned tasks; those performed are considered to be read-only. For the current task, one can, for instance, change its end time, although the start time must remain unchanged. Proceeding from the current task to the next one is carried out by invoking the `Schedule.nextTask()`) method.

The execution of the current task may be monitored with a `TaskTracker`.[3] In the most basic version, trackers predict only the end time of the current task. More complex trackers also provide detailed information on the current state of task execution. `OnlineDriveTaskTracker`, for example, offers functionality similar to GPS navigation, such as monitoring the movement of a vehicle, predicting its arrival time and even diverting its path.

`ScheduleImpl`, along with `DriveTaskImpl` and `StayTaskImpl`, is the default implementation of `Schedule` and offers several additional features, such as data validation or automated task handling. It also serves as the starting point when implementing domain-specific schedules or tasks (e.g., `ChargeTask` in the electric VRP model mentioned above).

23.3.2 Least-Cost Paths

MATSim's network model consists of nodes connected by one-way links. Because of the queue-based traffic flow simulation (Section 1.3), a link is the smallest traversable element (i.e., a vehicle cannot stop in the middle of a link). Besides links, the DVRP contribution also operates on a higher level of abstraction: paths. Each path is a sequence of links to be traversed to get from one location

---

1 Package `org.matsim.contrib.dvrp.data`.
2 Package `org.matsim.contrib.dvrp.schedule`.
3 Package `org.matsim.contrib.dvrp.tracker`.

to another in the network, or more precisely, from the end of one link end to the end of another link.

The functionality of finding least-cost paths is available in the `org.matsim.contrib.dvrp.router` package. `VrpPathCalculator` calculates `VrpPaths` by means of the least-cost path search algorithms available in MATSim's core (Jacob et al., 1999; Lefebvre and Balmer, 2007).[4] Because of changing traffic conditions, paths are calculated for a given departure time. Since MATSim calculates average link travel time statistics for every 15 minutes time period by default, the 15 minutes time bin is also used for computing shortest paths.

`VrpPaths` are used by `DriveTasks` to specify the link sequence to be traversed by a vehicle between two locations. It is possible to divert a vehicle from its destination by replacing the currently followed `VrpPath` with a `DivertedVrpPath`.

To reduce computational burden, the already calculated paths can be cached for future reuse (see `VrpPathCalculatorWithCache`). However, when calculating least-cost paths from one location to many potential destinations, a significant speed-up can be achieved by means of least-cost tree search (see `org.matsim.utils.LeastCostPathTree`).

23.4 DynAgent

Contrary to the standard day-to-day learning in MATSim (but see also Section 97.3.5), in the DVRP contribution, each driver behaves dynamically and follows orders coming continuously from the dispatcher. The `DynAgent` class, along with the `org.matsim.contrib.dynagent` package, provides the foundation for simulating dynamically behaving agents. Although created for DVRP contribution needs, `DynAgent` is not limited to this context and can be used in a wide range of different simulation scenarios where agent dynamism is required.

`DynAgent`'s main concept assumes an agent can actively decide what to do at each simulation step instead of using a pre-computed (and occasionally re-computed; see 30.4.2) plan. It is up to the agent whether decisions are made spontaneously or (re-)planned in advance. In some applications, a `DynAgent` may represent a fully autonomous agent acting according to his/her desires, beliefs and intentions, whereas in other cases, it may be a non-autonomous agent following orders systematically issued from the outside (e.g., a driver receiving tasks from a centralized vehicle dispatching system).

23.4.1 Main Interfaces and Classes

The `DynAgent` class is a dynamic implementation of `MobsimDriverPassengerAgent`. Instead of executing pre-planned `Activitys` and `Legs`, a `DynAgent` performs `DynActivitys` and `DynLegs`. The following assumptions underlie the agent's behavior:

- The `DynAgent` is the physical representation of the agent, responsible for the interaction with the real world (i.e., traffic simulation).
- The agent's high-level behavior is controlled by a `DynAgentLogic` that can be seen as the agent's brain; the `DynAgentLogic` is responsible for deciding on the agent's next action (leg or activity), once the current one has ended.
- Dynamic legs and activities fully define the agent's low-level behavior, down to the level of a single simulation step.

At the higher level, the `DynAgent` dynamism results from the fact that dynamic activities and legs are usually created on the fly by the agent's `DynAgentLogic`; thus, the agent does not have to plan

[4] Package `org.matsim.core.router`.

future actions in advance. When the agent has a roughly detailed legs and activities plan, he/she does not have to adhere to it and may modify his/her plan at any time (e.g., change the mode or destination of a future leg, or include or omit a future activity).

Low-level dynamism is provided by the execution of dynamic activities and legs. As for the currently executed activity, the agent can shorten or lengthen its duration at any time. Additionally, at each time step, the agent may decide what to do right now (e.g., communicate with other agents, re-plan the next activity or leg, and so on). When driving a car (`DriverDynLeg`), the agent can change the route, destination or even decide about picking up or dropping off somebody on the way. When using public transport (`PTPassengerDynLeg`), the agent chooses which bus to get on and at which stop to exit.

Incidentally, the behavior of MATSim's default plan-based agent, `PersonDriverAgentImpl`, can be simulated by `DynAgent`, combined with the `PlanToDynAgentLogicAdapter` logic. This adapter class creates a series of dynamic activities and legs that mimics a given `Plan` of static `Activity` and `Leg` instances.

23.4.2 *Configuring and Running a Dynamic Simulation*

`DynAgent` has been written for and validated against `QSim`. Dynamic leg simulation requires no additional code. However, to take advantage of dynamic activities, `DynActivityEngine` should be used, instead of `ActivityEngine`. The `doSimStep(double time)` method of `DynActivityEngine` ensures that dynamic activities are actively executed by agents and that their end times can be changed.

The easiest way to run a single iteration of `QSim` is as follows:

1. Create and initialize a `Scenario`,
2. call `DynAgentLauncherUtils`' `initQSim(Scenario scenario)` method to create and initialize a `QSim`; this includes creating a series of objects, such as an `EventsManager`, `DynActivityEngine`, or `TeleportationEngine`,
3. add `AgentSources` of `DynAgents` and other agents to the `QSim`,
4. run the `QSim` simulation, and
5. finalize processing events by the `EventsManager`.

Depending on needs, the procedure above can be extended with additional steps, such as adding non-default engines or departure handlers to the `QSim`.

23.4.3 *RandomDynAgent Example*

The `org.matsim.contrib.dynagent.examples.random` package contains a basic illustration of how to create and run a scenario with `DynAgents`. To highlight differences with plan-based agents, in this example 100 dynamic agents travel randomly (`RandomDynLeg`) and perform random duration activities (`RandomDynActivity`).

High-level random behavior is controlled by `RandomDynAgentLogic`, that operates according to the following rules:

1. Each agent starts with a `RandomDynActivity`; see the `computeInitialActivity(DynAgent agent)` method.
2. Whenever the currently performed activity or leg ends, a random choice on what to do next is made between the following options: (a) stop being simulated by starting a deterministic `StaticDynActivity` with infinite end time, (b) start a `RandomDynActivity`, or (c) start a `RandomDynLeg`; see the `computeNextAction(DynAction oldAction, double now)` method.

The lower level stochasticity results from random decisions being made at each consecutive decision point. In the case of RandomDynLeg, each time an agent enters a new link, he or she decides whether to stop at this link or to continue driving; in the latter case, the subsequent link is chosen randomly; see the RandomDynLeg(Id<Link> fromLinkId, Network network) constructor and the movedOverNode(Id<Link> newLinkId) method. As for RandomDynActivity, at each time step the doSimStep(double now) method is called and a random decision is made on the activity end time.

Following the rules specified in Section 23.4.2, setting up and running this example scenario is straightforward. RandomDynAgentLauncher reads a network, initializes a QSim, then adds a RandomDynAgnetSource to the QSim, and finally, launches visualization and starts simulation. The RandomDynAgentSource is responsible for instantiating 100 DynAgents that are randomly distributed over the network. The simulation ends when the last active agent becomes inactive.

23.5 Agents in DVRP

Realistic simulation of dynamic transport services requires a proper model of interactions and possible collaborations between the main actors: drivers, customers (often passengers) and the dispatcher. By default, drivers and passengers are simulated as agents, while the dispatcher's decisions are calculated by the optimization algorithm (see Section 23.6). This, however, is not the only possible configuration. One may simulate, for example, a decentralized system with a middleman as dispatcher rather than the fleet's manager.

23.5.1 Drivers

A driver is modeled as a DynAgent, whose behavior is controlled by a VrpAgentLogic that makes the agent follow the dynamically changing Schedule.[5] As a result, all changes made to the schedule are visible to and obeyed by the driver. Whenever a new task is started, the driver logic (using a DynActionCreator) translates it into the corresponding dynamic action. Specifically, a DriveTask is executed as a VrpLeg, whereas a StayTask is simulated as a VrpActivity. Both VrpLeg and VrpActivity are implemented so that any change to the referenced task is automatically visible to them. At the same time, any progress made while carrying them out is instantly reported to the task tracker.

23.5.2 Passengers

To simulate passenger trips microscopically, passengers are modeled as MobsimPassengerAgent instances. As part of the simulation, they can board, ride and, finally, exit vehicles. In contrast to the drivers, they may be modeled as the standard MATSim agents, each having a fixed daily plan consisting of legs and activities.

Interactions between drivers, passengers and the dispatcher, such as submitting Passenger Requests or picking up and dropping off passengers, are coordinated by a PassengerEngine[6]. Requests may be immediate (*as soon as possible*) or made in advance (*at the appointed time*). In the former case, a passenger starts waiting just after placing the order;in the latter case, the dispatched vehicle my arrive at the pickup location before or after the designated time, which means that either the driver or the customer, respectively, will wait for the other to come. To ensure proper coordination between these two agents, the pickup activity performed by the driver must implement the PassengerPickupActivity interface.

---

[5] Package org.matsim.contrib.dvrp.vrpagent.

[6] Package org.matsim.contrib.dvrp.passenger.

23.6 Optimizer

Since demand and supply are inherently stochastic, the general approach to dealing with dynamic transport services consists of updating vehicles' schedules in response to observed changes (i.e., events). This can be done by means of re-optimization procedures that consider all requests (within a given time horizon) or fast heuristics focused on small updates of the existing solution, rather than constructing a new one from scratch. Usually, re-optimization procedures give higher quality solutions compared to ocal update heuristics; however, when it comes to real-world applications, where high (often real-time) responsiveness is crucial, broad re-optimization may be prohibitively time-consuming.

In the most basic case, an optimizer implements the `VrpOptimizer` interface[7], that is, implements the following two methods:

- `requestSubmitted(Request request)`—called on submitting `request`; in response, the optimizer either adapts vehicles' schedules so that `request` can be served, or rejects it.
- `nextTask(Schedule<? extends Task> schedule)`—called whenever `schedule`'s current task has been completed and the driver switches to the next planned task; this is the last moment to make or revise the decision on what to do next.

This basic functionality can be freely extended. Besides request submission, one may, for example, consider modifying or even canceling already submitted requests. Another option is monitoring vehicles as they travel along designated routes and reacting when they are ahead of/behind their schedules. Such functionality is available by implementing `VrpOptimizerWithOnlineTracking's` `nextLinkEntered(DriveTask driveTask)` method, which is called whenever a vehicle moves from the current link to the next one on its path.

Last but not least, there are two ways of responding to the incoming events. They can be handled either *immediately* (*synchronously*) or *between time steps* (*asynchronously*). In the former case, schedules are re-calculated (updated or re-optimized) directly, in response to the calling of the optimizer's methods. This simplifies accepting/rejecting new requests, since the answer is immediately passed back to the caller. In the latter case, all events observed within a simulation step are recorded and then processed in batch mode just before the next simulation step begins.[8] By doing that, one can not only speed up computations significantly, but also avoid situations when, due to vehicles' inertia (e.g., an idle driver can stop waiting and depart only at the beginning of the simulation step), two or more mutually conflicting decisions could be made by the optimizer at distinct moments during a single simulation step, causing the latter to overwrite the former (not always intentional).

23.7 Configuring and Running a DVRP Simulation

Like in within-day replanning (see Chapter 30), dynamic transport services are typically run with the DVRP contribution as a single-iteration simulation. Setting up and running such a simulation requires carrying out the following steps:

1. Create a `Scenario` (MATSim's domain data) and `VrpData` (VRP's domain data),
2. create a `VrpOptimizer`; this includes instantiation of a least-cost path/tree calculator, e.g., `VrpPathCalculator`, and

[7] Package `org.matsim.contrib.dvrp.optimizer`.

[8] This can be achieved by using an optimizer implementing the interface `org.matsim.core.mobsim.framework.listeners.MobsimBeforeSimStepListener`.

3. call DynAgentLauncherUtils' initQSim(Scenario scenario) method to create and initialize a QSim; this includes creating a series of objects, such as an EventsManager, DynActivityEngine, or TeleportationEngine.
4. When simulating passenger services, add a PassengerEngine to the QSim; this includes instantiation of a PassengerRequestCreator that converts calls/orders into PassengerRequests; otherwise (i.e., non-passenger services), add an appropriate source of requests to the QSim, either as a MobsimEngine or MobsimListener.
5. Then, add AgentSources to the QSim; for the DynAgent-based drivers, one may use a specialized VrpAgentSource and provide a DynActionCreator.[9]
6. run the QSim simulation, and
7. finalize processing events by the EventsManager.

The org.matsim.contrib.dvrp.run package contains VrpLauncherUtils and other utility classes that simplify certain steps of the above scheme. To facilitate access to the data representing the current state of the simulated dynamic transport service, MatsimVrpContext provides the Scenario and VrpData objects and the current time (based on the timer of QSim).

The VrpOptimizer's performance may be assessed either by analyzing the resulting schedules, or by processing events collected during the simulation.

23.8 OneTaxi Example

The org.matsim.contrib.dvrp.examples.onetaxi package contains a simple example of how to simulate on-line taxi dispatching with the DVRP contribution. In this scenario, there are ten taxi customers and one taxi driver, who serves all requests in the FIFO order. Each customer dials a taxi at a given time to get from work to home. The example is made up of six classes:

- OneTaxiRequest—represents a taxi request.
- OneTaxiRequestCreator—converts taxi calls into requests prior to submitting them to the optimizer.
- OneTaxiOptimizer—creates and updates the driver's schedule.
- OneTaxiServeTask—represents StayTasks related to picking up and dropping off customers.
- OneTaxiActionCreator—translates tasks into dynamic activities and legs.
- OneTaxiLauncher—sets up and runs the scenario.

All data necessary to run the OneTaxi example is located in the /contrib/dvrp/src/main/resources/one_taxi directory.

23.9 Research with DVRP

Currently, the DVRP contribution is used in several research projects. Two of them focus on on-line dispatching of electric taxis in Berlin and Poznan (Maciejewski and Nagel, 2013b,c,a; Maciejewski, 2014). Another project deals with design of demand-responsive transport, where DVRP has been applied to the case of twin towns, Yarrawonga and Mulwala, described in Chapter 95 (Ronald et al., 2015, 2014). In a recently launched project, the DVRP contribution will be used for simulation of DRT services in three cities: Stockholm, Tel Aviv and Leuven.

The current code development focuses on increasing performance and flexibility of the implemented shortest paths search (see Section 23.3.2). An interesting future research topic, related specifically to DRT planning, is multi-modal path search, where on-demand vehicles may be combined with fixed-route buses within a single trip. Another potential research direction is adding a benchmarking functionality and standardized interfaces so that the DVRP contribution could serve as a testbed for the *Rich VRP* optimization algorithms.

---

[9] Package org.matsim.contrib.dvrp.vrpagent,

SUBPART FIVE

Commercial Traffic

CHAPTER 24

Freight Traffic

Michael Zilske and Johan W. Joubert

24.1 Basic Information

Entry point to documentation:

`http://matsim.org/extensions` → freight

Invoking the module:

`http://matsim.org/javadoc` → freight → `RunChessboard` class

Selected publications:

Schröder et al. (2012); Zilske et al. (2012)

Various MATSim freight traffic modeling approaches have been implemented in recent years.

For Zürich, available origin-destination matrices for small delivery trucks and heavy trucks have been disaggregated Shah (2010). Data was taken from the KVMZH (Kantonales Verkehrsmodell Zürich) provided by Amt für Verkehr, Volkswirtschaftsdirektion Kanton Zürich (2011) and documented in Gottardi and Bürgler (1999). This special freight sub-population is restricted to route choice.

In South Africa, freight vehicles' plans were derived from GPS records of more than 30 000 commercial vehicles tracked over a 6-month period. Activity chains' extraction from raw GPS data was documented in Joubert and Axhausen (2011); the first joint private car and freight implementation appeared in Joubert et al. (2010). In Nagel et al. (2014), we used MATSim to evaluate the impact of a complex vehicle-type specific toll structure where sub-populations, including freight, have different time values.

The most sophisticated solution, however, was the introduction of carrier agents, described in the following section.

How to cite this book chapter:

Zilske, M and Joubert, J W. 2016. Freight Traffic. In: Horni, A, Nagel, K and Axhausen, K W. (eds.) *The Multi-Agent Transport Simulation MATSim*, Pp. 155–156. London: Ubiquity Press. DOI: http://dx.doi.org/10.5334/baw.24.

24.2 Carriers

Until now, real-world scenarios set up with MATSim modeled freight traffic demand share by using plan sets with activities at the depot and pick-up and delivery locations, without variability in any dimension except route choice. We improved this situation by modeling freight vehicles as non-autonomous agents employed by, and serving the interests of, freight operators. Freight vehicle drivers' missing choice dimensions are then realized as logistics decisions made by the freight operators who employ them. In the freight transport sector, decisions are distributed among actors with different roles. Freight transport decisions include: lot-size choice, path-searches in logistical networks, vehicle choice and tour planning. A freight operator's planning problem is quite different from its passenger counterpart.

First, success of freight transport plans is not determined by the utility of time spent at activity locations, but rather by commercial success. Plans must fulfill customers' requirements, i.e., time windows and providing sufficient capacity at reasonable cost.

Second, freight operators often operate several vehicles and their options include rescheduling deliveries from one vehicle to another or even changing the number of vehicles used.

Thus, a new software layer populated by *carrier agents* was introduced into the simulation. Each carrier agent represents a firm with a vehicle fleet, depots and contracts. Contracts determine type and quantity of goods to be carried and contains the respective origin and destination as well as pick-up and delivery time windows.

The carrier agent's plan contains a tour schedule for each fleet vehicle, containing planned pick-up, delivery or arrival times at customer locations and a route through the physical network. In our basic model, all vehicle schedules of a carrier begin and end at one of its depots. When a simulation scenario is initialized, the carrier agents build a schedule for each of their vehicles, including a route through the transport network, with pick-up and delivery activities corresponding to their contracts. At the interface between the freight operator plans and the mobility simulation, the set of routed vehicles from each carrier plan is injected into the traffic demand as individual driver agents, where they move through the traffic system along with passenger vehicles. While executing their plans, the freight driver agents report their shipment-related activities back to the carrier.

When all plans have been executed, agents evaluate the success of their plan. The carrier agents use a custom utility function capturing their economic success. Their cost is calculated as a sum of vehicle-dependent distance and time costs incurred by scheduled vehicles, as well as some individual fixed costs, plus penalties incurred by missed time windows.

Finally, carrier agents create new plans to improve their performance in the next iteration. For instance, a time-dependent vehicle routing heuristic can be plugged in to replan vehicle schedules. Shipments can be switched between vehicles, or an entire vehicle added or removed. During repeated executions of their plans, passengers as well as carriers gain experience from the transport system. The carriers experience congestion and other disturbances in the traffic system when they incur a higher cost through longer vehicle usage, or by penalizing missed pick-up and delivery times.

The planning algorithms themselves are implemented in the project `jsprit`, a library separate from MATSim. In the replanning phase of each iteration, `jsprit` is called and replans the carrier plans.

The model is described in a paper by Schröder et al. (2012). For more details about the implementation, as well as more references, see the technical report by Zilske et al. (2012).

CHAPTER 25

WagonSim

Michael Balmer

25.1 Basic Information

Entry point to documentation:

`http://matsim.org/extensions` → wagonSim

Invoking the module:

`http://matsim.org/javadoc` → wagonSim → `RunWagonSim` class

Selected publications:

-

25.2 Summary

The wagonSim contribution allows use of MATSim's route-optimization process to find optimal paths for rail-based freight wagons in a given rail-based freight infrastructure.

The network links, here, define the rails, nodes define train stations and schedule transit stops define train station stopping points. Freight locomotives are driven by a strictly fixed schedule, where each locomotive is given as a single transit line with a single transit route and a single departure. Freight wagons correspond to agents with a given origin and destination (single trip agents). Routing takes various constraints into account, i.e., a minimum shunting time while switching locomotives and maximum freight train weight and length; it also differentiates between locomotive stops for shunting and stops only for waiting (without shunting possibility).

WagonSim contribution is based on specialized input data. The first step converts input data into MATSim formats (scenario data). In a second step, it allows one to manually adapt the scenario for different parametrization of train stops, shunting stations, minimum shunting times and

How to cite this book chapter:

Balmer, M. 2016. WagonSim. In: Horni, A, Nagel, K and Axhausen, K W. (eds.) *The Multi-Agent Transport Simulation MATSim*, Pp. 157–160. London: Ubiquity Press. DOI: http://dx.doi.org/10.5334/baw.25.

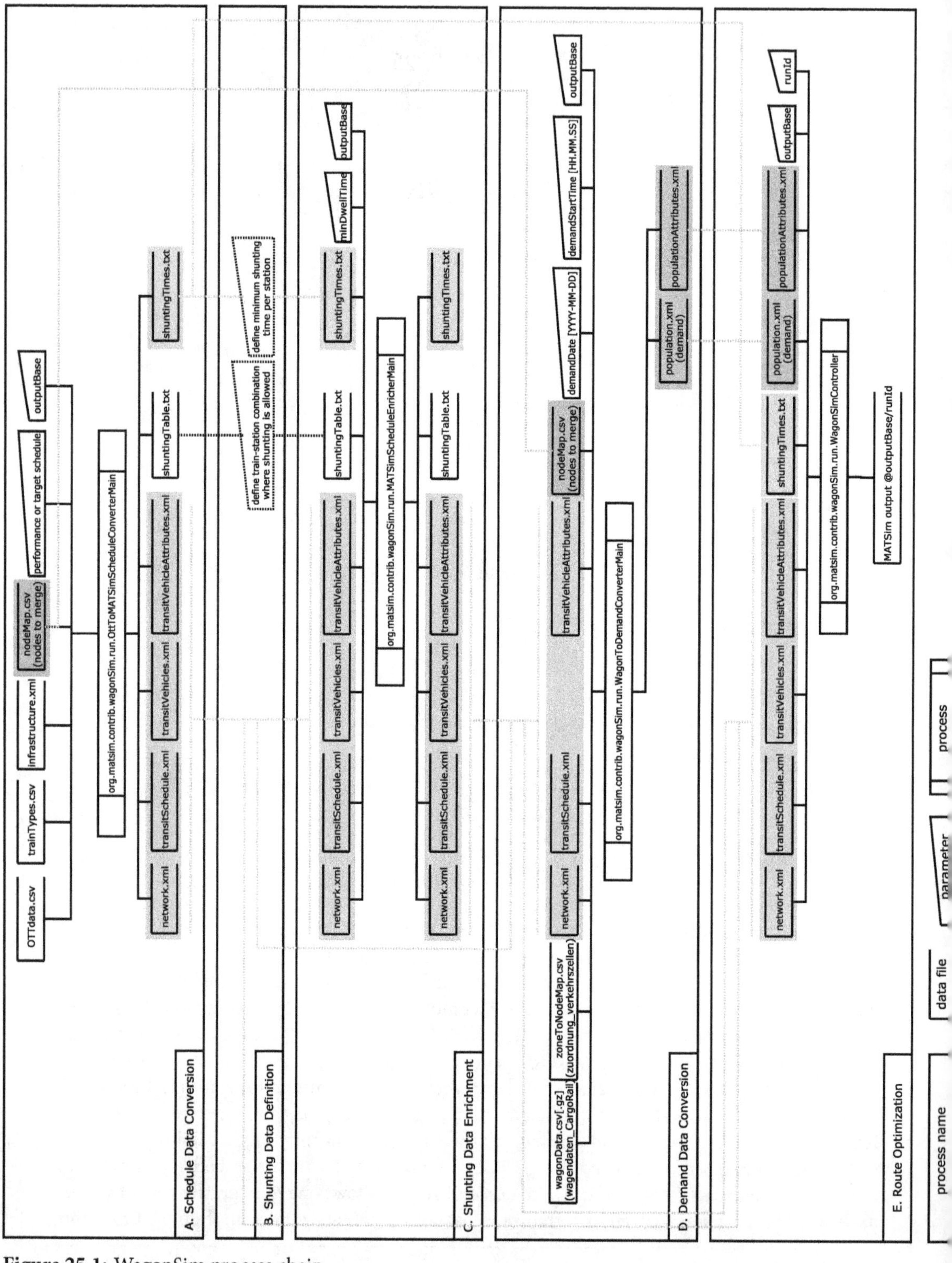

Figure 25.1: WagonSim process chain.

dwell times of trains at stops. The third step sets up route optimization configuration and runs the MATSim optimization cycle.

As shown at `http://www.matsim.org/docs/extensions/wagonSim` and in Figure 25.1, data conversion and WagonSim execution is composed of five stages, described in more detail at above referenced url:

A) schedule data conversion,

B) shunting data definition,

C) shunting data enrichment,

D) demand data conversion, and

E) route optimization.

WagonSim contribution has been applied to ETH (Eidgenössische Technische Hochschule), IVT (Institut für Verkehrsplanung und Transportsysteme – Institute for Transport Planning and Systems) Transport Systems group projects.

CHAPTER 26

freightChainsFromTravelDiaries

Kai Nagel

Entry point to documentation:

`http://matsim.org/extensions` → freightChainsFromTravelDiaries

Invoking the module:

Currently not possible.

Selected publications:

Schneider (2011)

Sebastian Schneider has done a Ph.D. dissertation about generating freight vehicle chains by essentially re-sampling the information contained in the German survey KiD (Kraftfahrzeugverkehr in Deutschland) (Steinmeyer and Wagner, 2005). Since the KiD is essentially an activity-based travel diary, the method should also be applicable to other situations. Since Sebastian has left science for the time being, he allowed us to take his code and integrate it into the repository, under the GPL (GNU General Public License). For the time being, it will just "sit" here until someone attempts to make it work.

How to cite this book chapter:

Nagel, K. 2016. freightChainsFromTravelDiaries. In: Horni, A, Nagel, K and Axhausen, K W. (eds.) *The Multi-Agent Transport Simulation MATSim*, Pp. 161–162. London: Ubiquity Press. DOI: http://dx.doi.org/10.5334/baw.26. License: CC-BY 4.0

SUBPART SIX

Additional Choice Dimensions

CHAPTER 27

Destination Innovation

Andreas Horni, Kai Nagel and Kay W. Axhausen

27.1 Basic Information

Entry point to documentation:

`http://matsim.org/extensions` → locationchoice

Invoking the module:

`http://matsim.org/javadoc` → locationchoice → `RunLocationChoiceBestResponse`, `RunLocationChoiceFrozenEpsilons` classes

Selected publications:

Horni et al. (2012b); Horni (2013)

27.2 Introduction

Generally speaking, destination choice represents an optimization problem, where every agent searches for his or her optimal destination according to an objective function, subject to various constraints such as the agent's travel time budget–as well as interactions with other agents–while competing for space-time slots in the infrastructure. The MATSim destination innovation module provides a problem-tailored heuristic algorithm to solve this problem.

MATSim's iterative base requires a mechanism (the main component of the destination innovation module), ensuring consistent probabilistic choices over the course of iterations.

Unobserved heterogeneity, usually dominant in destination choice, is captured in the adaptable objective function by random error terms (Horni et al., 2012b; Horni, 2013).

As well as considering competition for road infrastructure, the destination choice module can also be configured for activities infrastructure (for example, at shopping malls' parking lots) as shown in Section 27.3.5 and by Horni et al. (2009).

How to cite this book chapter:

Horni, A, Nagel, K and Axhausen, K W. 2016. Destination Innovation. In: Horni, A, Nagel, K and Axhausen, K W. (eds.) *The Multi-Agent Transport Simulation MATSim*, Pp. 165–174. London: Ubiquity Press. DOI: http://dx.doi.org/10.5334/baw.27. License: CC-BY 4.0

27.3 Key Issues in Developing the Module

Key issues of integrating destination innovation into MATSim include behavioral and algorithmic problems. On the behavioral side, specification of choice sets for model estimation has not yet been solved. On the algorithmic side, as mentioned above, destination innovation is, in principle, an ordinary optimization problem. However, as agents interact, and choices are embedded in a highly dynamic context, the problem becomes complex, particularly because targeted scenarios are usually large-scale. Thus, as in real-world optimization problems, solutions must be based on problem-tailored heuristics (Michalewicz and Fogel, 2004). Construction of a search space and subsequent evaluation of the search space's elements are important MATSim destination innovation components.

The main component however, is a mechanism to generate consistent random draws over iterations necessary to include the objective function's error terms (see next Section 27.3.1). This mechanism is also applicable to other choice dimensions.

27.3.1 Error Terms

As described in Chapter 49, MATSim—as a utility-maximizing model—is related to the discrete choice framework, meaning that this framework can productively guide the MATSim utility function specification. Utility in discrete choice models is composed of a deterministic part and a random error term representing the unobserved heterogeneity, i.e., it subsumes, both truly, i.e., inherently random, decisions and the modeler's missing knowledge about the choice and its context.

In MATSim, the utility function for route, mode and time innovation does not contain an explicit random error term (yet). This is at least partially compensated through replanning stochasticity, in Chapter 49 denoted by the scale parameter μ and η. An example for this might be: route and time choices are usually subject to significant competition. The co-evolutionary algorithm of MATSim, detailed below, essentially assigns the resources in a random manner to the persons. For example, two identical persons may end up with different routes, according to the order in which they undergo the replanning. Essentially, this means that an (implicit) random term is present in the choice making.

The above, however, does not add enough unobserved heterogeneity to destination choice. Further problems might, or might not, appear when trying to interpret this randomness, since it is added implicitly and somewhat unsystematically. Thus, an explicit random error term $\varepsilon_{n\ell q}$ for every person n, alternative ℓ and activity q, held stable over the iterations, is added to the scoring function during the running of the destination innovation module (Horni, 2013). Research about the necessity of error terms for the remaining choice dimensions is required, as discussed in Section 97.4.6.

27.3.2 Quenched Randomness

Due to random error terms, discrete choices are quantified by probabilities; for example, for the logit model, as $p_{n\ell q} = \exp(V_{n\ell q})/\sum_{j\in L} exp(V_{njq})$, where $V_{n\ell q}$ is person n's systematic utility of alternative ℓ for activity q. When drawing from the distribution specified by $p_{n\ell q}$ for a population, the aggregate choices are reproduced. This is basically also true when applied in iterative frameworks. However, iterative frameworks are usually associated with some kind of learning or relaxation mechanism, which is heavily distorted by repeatedly and randomly drawing from $p_{n\ell q}$ in every iteration. In this case, the $\varepsilon_{n\ell q}$ effectively fluctuate from iteration to iteration, which is disastrous for the algorithm's convergence and behaviorally implausible.

Instead, random error terms ε must remain fixed from iteration to iteration. The optimization is then performed as a deterministic search, based on the resulting utilities $U_{n\ell q}$, i.e., an alternative ℓ for person n; activity q is selected as

$$\underset{\ell \in choice\ set}{\operatorname{argmax}}\ U_{n\ell q} = V_{n\ell q} + \varepsilon_{n\ell q}\,.$$

This includes, via the systematic part $V_{n\ell q}$, the disutility of traveling to destination ℓ for activity q.

As stated above, random error terms must remain the same over the iterations (also discussed in Chapter 49). In physics, this approach would be called "quenched" (sometimes also "frozen") randomness; all randomness is computed initially and then attached to particles or destinations, rather than instantaneously generating it, which would be called "annealed" randomness. Two natural approaches for implementing quenched randomness are as follows:

(a) Freezing the applied *global* sequence of random numbers, meaning that a Monte Carlo method with the same random seed is used before and after introduction of a policy measure and over the course of iterations. Thus, error terms should come out the same way *before* and *after* the introduction of the policy measure. Differences in the outcome can thus be directly attributed to the policy measure.

(b) Computing and storing a separate $\varepsilon_{n\ell q}$ for every combination of person n, alternative ℓ and activity q.

Both strategies have flaws. Approach (a) is only an option if one is certain about every single aspect of the computational code. Literally, one additional random number, drawn in one run, but not in the other, completely destroys the "quench" for all decisions computed later in the program. Consistency is thus hard to achieve, especially in parallel or even distributed computing environments; substantial machinery is necessary to ensure consistent choices. In a modular environment, as in MATSim, designed for external plugging-in of users' own modules—possibly drawing their own random numbers—the danger of destroying the quench is prohibitively high and thus approach (a) is impractical.

Approach (b) is certainly more robust. However, for large numbers of decision makers and/or alternatives, storing error terms is difficult. For destination innovation, one quickly has 10^6 decision makers and 10^6 alternatives, resulting in $4 \cdot 10^{12}$Byte $=$ 4TB of storage space.

One may argue that this should not be a problem, since a normal person will rarely consider more than the order of a hundred alternatives in their choice set, reducing the computational problem. Aside from the necessity of storing every decision maker's choice set, this converts the computational problem into a conceptual one, since a good method to generate choice sets then needs to be found. With more conceptual progress, this may eventually be an option; at this point, a conceptually simpler approach is preferred.

The solution developed below is generally applicable in econometric microsimulators. The same *stable* error term can be *re*-calculated on the fly by using stable random seeds $s_{n\ell q} = g(k_n, k_\ell, k_q)$, containing uniformly distributed random numbers associated with k, ℓ, and q. That is, for each person n, a random number k_n is generated and stored; the same is done with each destination ℓ. Value for the activity q can be derived from its index in the plan, possibly combined with the person's value k_n. This reduces the storage space dramatically, from $N_q \cdot N_n \cdot N_\ell$ to $N_q(N_n + N_\ell)$, where N_n is the number of persons or agents and N_ℓ is the number of destinations and N_q is the average number of discretionary activities in an agent's plan. This means that storage space is reduced to approximately $2 \cdot 4 \cdot 10^6$Byte $=$ 8MB, which can be easily stored on any modern machine.

Distribution of these seeds is essentially irrelevant; any error term distribution can be generated from any basic seed distribution. In the current version, $g(k_n, k_\ell, k_q) = (k_n + k_\ell + k_q) \times v_{max}$ is used. v_{max} is the maximum (long) number that can be handled by the specific machine.

To evaluate utility for a person n visiting the destination ℓ for activity q, a sequence of Gumbel-distributed random numbers $seq_{n\ell q}$ is generated on the fly for every person-alternative-activity combination using the seed $s_{n\ell q}$. Some random number generators have problems in the initial phase of drawing, e.g., the first couple of random numbers are correlated or never cover the complete probability space. As in our procedure, the random number generator is constantly re-initialized; for these technical reasons, the error term $\varepsilon_{n\ell q}$ is not derived from the first element, but from the mth element of the sequence $seq_{n\ell q}[m]$. Here, m is set to 10. This procedure is valid, as the set of all m^{th} elements of all different sequences is also a pseudo-random sequence, following the same distribution as the sequences $seq_{n\ell q}$; clearly, *true* random number generators relying on physical phenomena, such as hardware temperature, are not applicable.

27.3.3 Search Space Construction and Evaluation

MATSim destination innovation is based on best-response, rather than random mutation; in every iteration, the best current alternative, *including* the $\varepsilon_{n\ell q}$, is chosen. This works as long as inter-iteration changes are small, which usually happens, given by the relatively small share of agents who re-plan. The best-response approach is adopted due to the usually huge number of alternatives in combination with the search space characteristics. The discrete search landscape is characterized by random noise, because error terms are not spatially correlated (see Figure 27.1(a)). For such problems—as opposed to continuous landscapes (see Figure 27.1(b))—efficient search methods, such as local search methods, generally do not work.

When searching for the best choice, the large number of alternatives—prohibiting exhaustive search—is restrained as follows (for the detailed derivation see Horni, 2013, p.51 ff.). It is assumed that travel costs are always negative and that a person drops activities with negative net utility. Then, the maximum potential travel effort a person is willing to invest is constrained by the maximum error term per person and activity. This approach is promising, as very large values for Gumbel-distributed variables are rare, meaning that a huge space must be searched for only a few persons.

This search space reduction saves a great deal of computation time; however, it is still unfeasible and further speed-ups are necessary. Most computation time is due to travel time calculation, i.e., due to routing, for evaluation of the alternatives in the search space. To reduce these huge routing costs, the Dijkstra (Dijkstra, 1959) routing algorithm is not only applied forward—providing one-to-all travel times–but also *backwards*, using an average estimated arrival time as initial time. This is an approximation; thus, a *probabilistic* best response is applied, justified by the assumption that, during the course of the iterations, the probabilistic choice will reduce the errors incurred by approximating travel times.

With this procedure, the required computational effort is dramatically reduced, allowing application of destination innovation to large-scale scenarios.

27.3.4 Destination Choice Set Specification

Choice set specification is natural for choices with few alternatives; but in contrast, for problems with a large universal choice set, specifying individual choice sets becomes a challenging computational and behavioral issue. This is particularly true for spatial choices like destination or route choice (e.g., Pagliara and Timmermans, 2009; Thill, 1992; Schüssler, 2010; Frejinger et al., 2009b). Estimates are sensitive to choice sets; at the same time, no established choice set definition procedure exists for spatial problems. This means that choice sets and, hence, estimates are dependent on the modeler.

An important extension of the standard discrete choice modeling approach to treat this problem is formed by stochastic choice set models, founded by Manski (1977); Burnett and Hanson

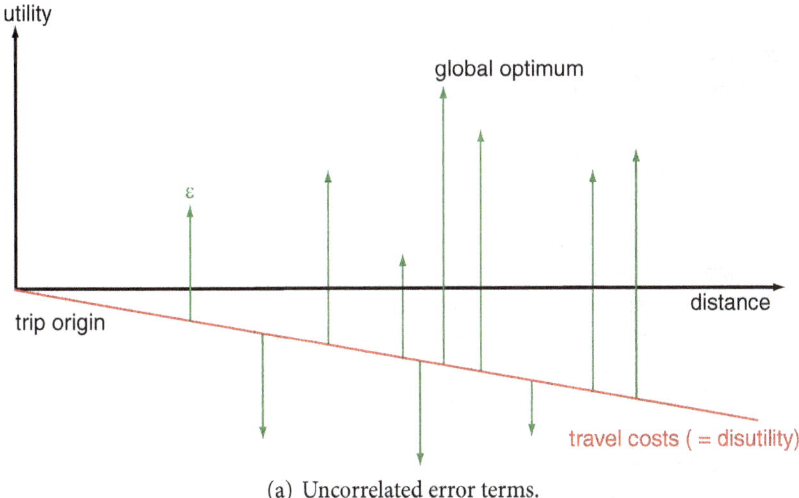

(a) Uncorrelated error terms.

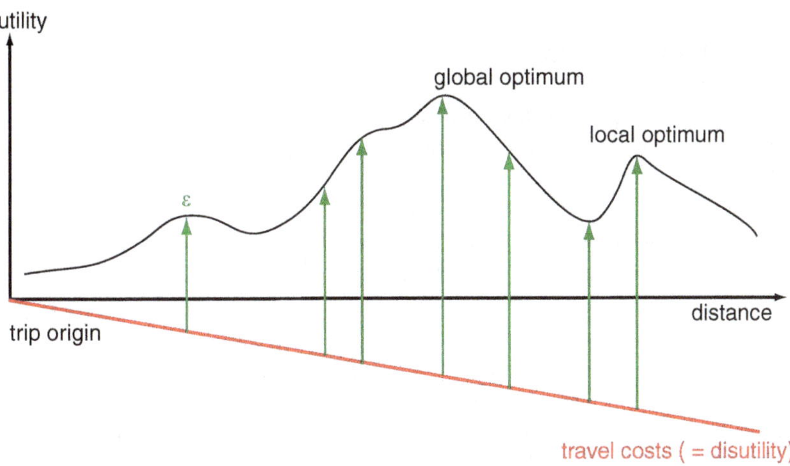

(b) Spatially correlated error terms.

Figure 27.1: Search space: The search algorithm must be able to handle correlated, as well as uncorrelated, error terms as given by the MNL model. Local search methods, such as hill-climbing algorithms are only able to handle continuous search spaces; thus, for situation (a), a best-response global search algorithm is required.

(1979); Burnett (1980); these integrate the choice set formation step into the estimation procedure by jointly estimating choice set selection and selection of a particular alternative of this choice set (Manski, 1977; Ben-Akiva and Boccara, 1995). Probabilistic choice set formation is conceptually appealing; choice sets are, in principle, not restrained a priori by exogenous criteria, as in standard choice set specification. However, the procedure is generally associated with combinatorial complexity, making it computationally intractable. As a consequence, practical approaches also require mechanisms to reduce complexity of the choice set specification problem (e.g., Ben-Akiva and Boccara, 1995, p.11). Zheng and Guo (2008), for example, make the moderate assumption of continuous store choice sets (i.e., sets without "holes") around the trip origin, while Ben-Akiva and Boccara (1995)'s random-constraints model exploits additional information on alternatives' availability for individuals.

In conclusion, the destination innovation set specification problem is still unsolved, meaning that estimated models can only be fully consistently applied for the region where the model was estimated. For MATSim, destination choice model estimation efforts are reported in Horni (2013, Chapter 5).

27.3.5 *Facility Load*

The influence of interaction in *transport* infrastructure for people's route and departure time choice was recognized almost a century ago (e.g., Pigou, 1920; Knight, 1924; Wardrop, 1952). It can also be reasonably assumed that agent interaction in *activities* infrastructure affects travel choices (Axhausen, 2006). Marketing science provides ample evidence that agent interactions influence utility (positively or negatively) of performing an activity (Baker et al., 1994, p.331), (Eroglu and Harrell, 1986; Eroglu and Machleit, 1990; Eroglu et al., 2005; Harrell et al., 1980; Hui and Bateson, 1991; Pons et al., 2006).

In Horni et al. (2009), based on the Zürich scenario, a model is presented introducing competition for activity infrastructure space-time slots. The actual load is coupled with time-dependent capacity restraints.

Activity location load, computed for 15 minute time bins, is derived from events delivered by the mobsim. The load of one particular iteration, combined with time-dependent activity location capacity restraints, is considered in the agents' choice process of the succeeding iteration. In detail, this means that the utility function term $S_{dur,q}$, described above, is multiplied by $max(0; 1 - f_{load\ penalty})$, penalizing agents dependent on the load of the location they frequented. $f_{load\ penalty}$ is a power function; this has proved to be a good choice for modeling capacity restraints (remember that the well-known cost-flow function by U.S. Bureau of Public Roads (1964) is a power function). To introduce additional activity location heterogeneity, an attractiveness factor $f_{attractiveness}$ is introduced, defined to be logarithmically dependent on the store size given by the official workplaces census.

Also for demonstration purposes, capacity restraints are exclusively applied to shopping locations; in principle, leisure activity locations could be handled similarly. However, deriving capacity restraints for leisure activity locations is expected to be much more difficult than for shopping locations, because far less data is available for leisure locations and capacity restraints vary much more between different leisure locations than between different shopping activities (hiking versus going to the movies might be a good example).

The model allows assignment of individual time-dependent capacities to the activity locations. For the sake of demonstration, the capacities of all shopping facilities can be set equal, where their values can be derived from the shopping trip information given in the Swiss microcensus (Swiss Federal Statistical Office (BFS), 2006). The total daily capacity is set so that the activity locations located in the Zürich region satisfy the total daily demand with a reserve of 50 %. In detail, the capacity restraint function for a location l is as follows:

$$f_{load\ penalty,\ell} = \alpha_l \cdot \left(\frac{load_\ell}{capacity_\ell}\right)^{\beta_\ell}$$

with $\alpha_\ell = 1/1.5^{\beta_\ell}$, $\beta_\ell = 5$. $f_{load\ penalty,\ell}$ is the penalty factor for location ℓ as described above.

Simultaneous computation of all agents' score reduction avoids the last-record problem discussed in Vovsha et al. (2002). There, a sequential choice process is proposed; alternatives are removed from later travelers' choice sets if locations are already occupied by earlier travelers. Thus, travelers' order is specified arbitrarily; the last-record problem (last travelers must go a long distance to find an available location) is significant when modeling heterogeneous travelers.

As expected, the constrained model improves result quality by reducing the number of implausibly overcrowded activity locations.

27.4 Application of the Module

The destination innovation module has been successfully applied for the Zürich scenario (Chapter 56), as reported in Horni (2013, p.99), for the Tel Aviv model (see Chapter 91) and for the MATSim 2030 project. Figure 27.2 and Figure 27.3 show that, through error term scaling, distance distributions can be nicely fitted, decreasing count data error.

27.5 The Module in the MATSim Context

The destination innovation module explicitly incorporates unobserved heterogeneity through random error terms; the standard MATSim utility function, however, does not contain error terms. Randomness measured in empirical data is included implicitly through the simulation process stochasticity, including possible randomness in the choice itself. For destination innovation, this has led to a dramatic underestimation of total travel demand, making inclusion of unobserved heterogeneity inevitable. Clearly, the problem is the impossibility of making all choices at the same level; destination choice is conditional on mode choice which, in turn, is conditional on route choice. Hierarchical choice modeling has clearly showed that randomness, expressed by the logit model scale parameter, needs to be larger in higher level decisions. This chapter addresses replacing the need for more randomness in the choice model by directly including randomness into the utility function; that randomness must be quenched, otherwise the iterative procedure will just average

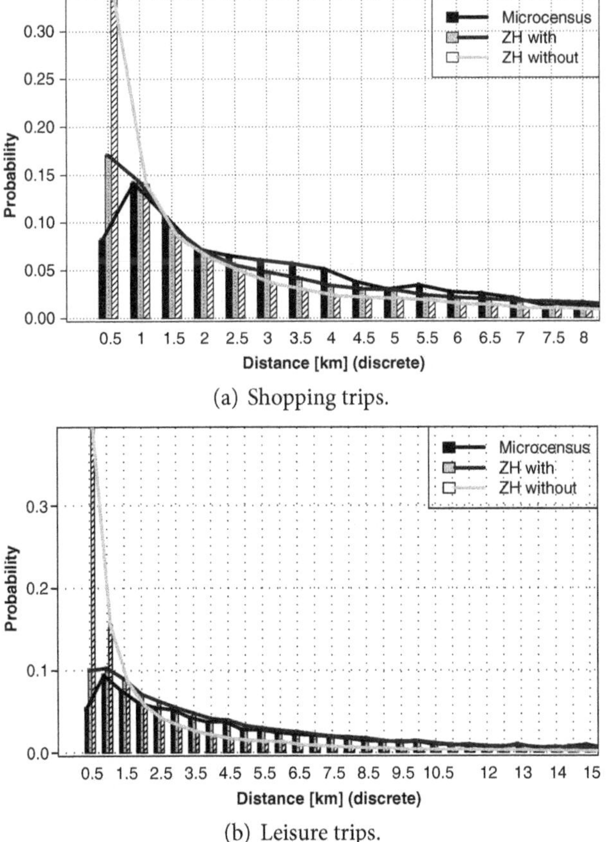

(a) Shopping trips.

(b) Leisure trips.

Figure 27.2: Error term runs for the Zürich scenario.

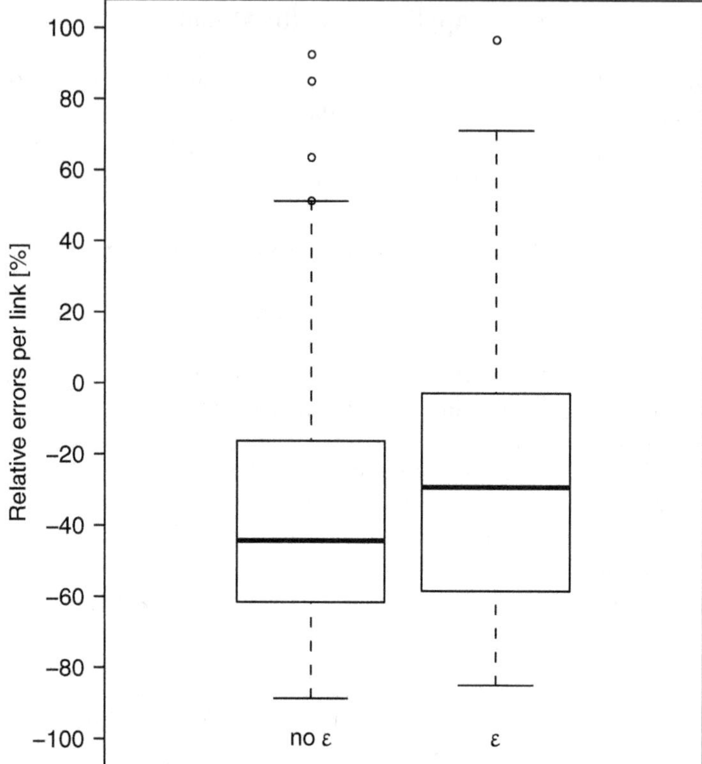

Figure 27.3: Daily traffic volumes for 123 links compared to traffic counts. Per link k the relative error is used, i.e., $(vol_{simulated,k} - vol_{counted,k})/vol_{counted,k}$.

it out. Whether the standard utility function might also profit from the innovations made for this module should be a topic for future research .

MATSim replanning offers different strategies to adapt plans, ranging from random mutation via approximate suggestions to best response answers. Destination innovation is based on best response to handle the sheer size of the alternatives set.

Although the destination innovation utility function is based on discrete choice framework, some conceptual differences about the common discrete choice models application persist. As shown above, there is no drawing from discrete choice models, but instead, maximization of an iteration-stable utility function. The set of alternatives is not necessarily limited a priori; thus, we use the notion of a search space and not of a choice set here.

27.6 Lessons Learned

Two interesting lessons were learned while developing the destination innovation module: first, a lesson on preferences and space interdependence and the necessity to evaluate them in combination. When looking at distance distributions (e.g., Figure 27.2) one might think that the functional form directly represents the preferences, but this is not necessarily the case. In our simulations, it is the result of a *linear* travel disutility, but applied in geographic space, where number of opportunities increases with the *square* of the radius, in other words, with the *square* of travel distance. A similar emergent effect appears when scaling random error terms. Although both negative and positive error terms are enlarged and the average remains stable, distribution gets more skewed toward the tail; for agents' choices, maximum values—not average values—are relevant.

The second lesson concerns simulation results' variability. Although random elements are not present only in destination choice, it was the largest contributor of endogenous variability when it was developed, necessitating the experiments presented by Horni et al. (2011a) (see also Section 48.4).

27.7 Further Reading

The main information source is Horni et al. (2012b); Horni (2013); technical details and documentation are available at Horni (2016) and in javadoc. Further reading related to destination choice is: Horni et al. (2013b),for parking, or Horni et al. (2012a), about coupling customers' and retailers' choices or, in other words, supply and demand.

CHAPTER 28

Joint Decisions

Thibaut Dubernet

28.1 Basic Information

Entry point to documentation:

`http://matsim.org/extensions` → socnetsim

Invoking the module:

`http://matsim.org/javadoc` → socnetsim → `RunExampleSocialSimulation` class

Selected publications:

Dubernet and Axhausen (2013), Dubernet and Axhausen (2014)

This chapter describes the extension of MATSim to consider what we call *joint decisions*. Section 28.2 explains what we call a joint decision, and gives an overview of why such processes are important in transportation. Section 28.3 then presents concepts to model this behavior, a generalization of the MATSim algorithm to search for solutions to the *joint planning problem*, and gives technical insights on how this implementation could be achieved, given the MATSim software architecture.

28.2 Joint Decisions and Transport Systems

28.2.1 Motivation

In recent years, there has been a growing interest in the social dimension of travel and how travel decisions are influenced, not only by the global state of the transportation system, but also by joint decisions and interactions with social contacts.

How to cite this book chapter:

Dubernet, T. 2016. Joint Decisions. In: Horni, A, Nagel, K and Axhausen, K W. (eds.) *The Multi-Agent Transport Simulation MATSim*, Pp. 175–182. London: Ubiquity Press. DOI: http://dx.doi.org/10.5334/baw.28.

A very active field of research is the study and modeling of intra-household interactions and joint decision-making, often using the classical random utility framework extended to group decision-making. Examples of household scheduling models include: Zhang et al. (2005, 2007); Kato and Matsumoto (2009); Bradley and Vovsha (2005); Gliebe and Koppelman (2005, 2002); Ho and Mulley (2013); Vovsha and Gupta (2013). Most of those models are specific to given household structures; in particular, separate models need to be estimated for different household sizes.

Another class of approaches, more oriented toward multi-agent simulation than analysis, is the use of optimization algorithms to generate households plans. These algorithms handle the household scheduling problem by transforming it into a deterministic utility maximization problem. Contrary to the previously presented approaches, those alternatives do not lead to the estimation of a model against data. Examples of approaches rooted in operations research include: Recker (1995), for which Chow and Recker (2012) designed a calibration method, or Gan and Recker (2008). Another attempt to generate plans for households uses a genetic algorithm, building on a previous genetic algorithm for individual plan generation (Charypar and Nagel, 2005; Meister et al., 2005), using a joint utility. Finally, Liao et al. (2013) formulate the problem of creating schedules for two persons traveling together as finding the shortest path in a "supernetwork", but note that their model is specific to the two person problem and that extension to larger numbers of agents may prove to be computationally expensive. All those approaches remained experimental, and were not integrated into multi-agent simulation tools.

Another class of methods aiming at multi-agent simulations is constituted rule based systems, which use heuristic rules to construct household plans, such as Miller et al. (2005); Arentze and Timmermans (2009).

Other authors have investigated the role of more general social networks on travel. One of the main incentives to conduct such studies comes from the continuous increase of the share of leisure purpose trips (Schlich et al., 2004; Axhausen, 2005). This trend represents a challenge for travel behavior modeling, as those trips are much more difficult to forecast than commuting trips; they are performed more sporadically and data from such trips is much more difficult to collect—particularly concerning location and event attributes, necessary to make models that are more than just random noise. A better understanding of how leisure trip destination choices are made is essential to improve the accuracy of those forecasts.

Various studies have been conducted to confirm that making social contacts is an important factor in leisure trip destination choice, or activity duration choice. Examples of empirical work include: Carrasco and Habib (2009); Habib and Carrasco (2011) or Moore et al. (2013). All these studies show strong influence of social contacts on the spatial and temporal distribution of activities. In a simulation experiment, Frei (2012) demonstrated that considering social interactions in leisure location choice helps increase the accuracy of predicted leisure trip distance distribution.

Another field of empirical research studies the spatial characteristics of social networks. For instance, Carrasco et al. (2008) studied the relationship between individual's socioeconomic characteristics and the spatial distribution of their social contacts. This kind of empirical work allows specification and estimation of models able to generate synthetic social networks, given sociodemographic attributes and home location. An example of such a model, based on the results of a survey in Switzerland, can be found in Arentze et al. (2012). This kind of model is essential if one wants to include social network interactions in microsimulation model.

This integration of social networks in multi-agent simulation frameworks has already been attempted by other authors. Due to their disaggregated description of the world, such models are particularly well-suited to complex social topologies representation. Han et al. (2011) present experiments using social networks to guide activity location choice set formation in the FEATHERS (Forecasting Evolutionary Activity-Travel of Households and their Environmental Repercussions) multi-agent simulation framework. Using a simple scenario with 6 agents forming a *clique*, they consider the influence of various processes like information exchange and adaptation to the behavior of social contacts to increase the probability of an encounter. They do not, however, represent *joint decisions*, such as the scheduling of a joint activity. The same kind of processes have been investigated by Hackney (2009), using more complex network topologies (within the MATSim framework) used in this paper. Ronald et al. (2012); Ma et al. (2011, 2012) present agent based systems, which integrate joint decision-making mechanisms, based on rule based simulations of a bargaining processes. They are not yet integrated into any operational mobility simulation platform.

Those remarks point the need to include explicit coordination in multi-agent simulation platforms.

28.2.2 The Joint Planning Problem

Here, we present a simulation framework able to represent *joint decisions*: that is, behavior requiring *explicit* coordination between individuals—such as shared rides, social activities or intra-household task allocation. The basic idea is that social contacts will make such a joint decision if it results in an improvement in the satisfaction of all participants. Modeling the interaction of individuals with possibly conflicting objectives has been the subject of game theory for decades, making this theoretical framework particularly well suited for the problem at hand.

Interestingly, game theoretic view of transportation systems has been popular since the seminal work of Wardrop (1952). The essential underlying concept is a view of the transportation system as a set of shared resources (road space, public transport vehicle seats...), for which individuals compete; individuals in the population try to maximize their own satisfaction, given the resources left available by others. Game theory studies *solution concepts* for such strategic interactions. A game theoretic solution concept is a definition of which states are *equilibria*: that is, *stable* under assumption of rationality—a state is considered stable if no agent/player has an incentive to change its behavior. The static, trip-based approach of Wardrop (1952) has been refined and extended with time. In particular, the equilibrium idea can be quite readily transferred to the *activity based* framework: individuals try not only to optimize their trips, but their whole day. This is, in particular, the approach of MATSim (Axhausen, 2006; Nagel and Flötteröd, 2012).

Most solution concepts in transportation are akin to the Nash equilibrium: a state where no individual can improve its satisfaction by *unilaterally* changing its behavior. This kind of solution concept does not allow to represent joint decisions. This can be illustrated by a classical game, called the *House Allocation Problem* (Schummer and Vohra, 2007). This game consists of n players and n houses. Moreover, each player has its individual ordering of the houses, from the most preferred to the least preferred, and players prefer being allocated alone to any house rather than to a house occupied by someone else. The strategy of a player centers around the house where the player chooses to live.

An interesting feature of this game is that any one-to-one allocation of players to houses is a Nash Equilibrium; no player can improve its payoff by *unilaterally* changing its strategy, as it would require choosing an occupied house. This result, however, contradicts basic intuition about the stability of such an allocation. In this particular case, a more realistic solution concept is the *Absence of Blocking Coalition*; given a one-to-one allocation of houses to players, a blocking coalition is a set of players which could all be better off by reallocating their houses among themselves. It should be noted that both solution concepts correspond to rational agents, i.e., agents having a preference ordering over outcomes. The only difference lies in the degree of communication allowed.

In the activity-based framework, this solution concept naturally becomes what we define as the *Absence of Improving Coalition* solution concept. An improving coalition for a given allocation of daily plans is a set of social contacts who can all feel themselves to be better off by *simultaneously* changing their daily plan—for instance, by switching from separate dinners at home to a joint dinner at a restaurant. The simulation of joint decision consists of searching an allocation of daily plans without such coalitions.

28.3 A Solution Algorithm for the Joint Planning Problem: A Generalization of the MATSim Process

28.3.1 Algorithm

Given this theoretical framework, one needs to design and implement an algorithm to search for allocations of daily plans to individuals that satisfy this solution concept. This implementation consists of two groups of components:

1. A `Controler` that implements the extension of the MATSim co-evolutionary algorithm, outlined hereafter. It is implemented in a modular fashion, to be easily adapted to the specific need of different simulation scenarios and
2. specific implementations of the modular components, namely replanning strategies and scoring functions, to allow explore the set of possible *joint plans* and representations of possible preferences specific to joint decisions.

Controler The MATSim framework provides a `Controler` to build and configure co-evolutionary algorithms, where agents each optimize their plan given the (evolving) state of the transport system.

Unfortunately, this approach makes choices of agents independent—which, of course, goes against the simulation of *joint decisions*. To implement an algorithm searching for states without blocking coalitions, one needs a way to represent the influence of explicit coordination on daily plan utility. This is solved by including *joint plans* constraints. A joint plan is a set of individual plans executed simultaneously. Different copies of the same individual plan can be part of different joint plans—for instance, an agent might go to a given restaurant alone, with members of its household or with a group of friends. The score of the different copies will take into account the influence of the joint plan to which it pertains. Those joint plan constraints are included using heuristic rules, applied after mutation operators are applied, and are classified as strong or weak constraints—weak constraints are considered when selecting plans for execution, but are allowed to be broken when merely selecting plans for mutation. They are then part of the evolution process. In the current application, the heuristic rules consist of joining newly created plans with joint trips (strong), or with leisure activities at the same location at the same time (weak).

To allow handling joint plans, replanning needs to be performed for groups of agents. This is straightforward for households; all agents of the same household are always handled as a single group. For more general social networks, agents are handled with all agents with whom they have a joint plan, plus some social contacts with whom new joint plans can be created.

For each group, two actions are then possible. For most groups, an allocation of existing plans—fulfilling the joint plans constraints—is selected for execution. Based on plan scores, randomized by adding an extreme value distributed error term, an algorithm inspired by the "Top Trading Cycle" algorithm used for the "House Allocation Problem" (Schummer and Vohra, 2007) searches for an allocation without improving coalitions.

For the other groups, a plan allocation is selected and copied. The copied plans then undertake mutation, to make the agents explore new alternative joint plans. Which mutation is performed determines which alternative plans will be tried out by the agent.

Agents have a limited memory size, keeping by default at most three plans per joint plan composition, and ten plans in total. If this limit is exceeded, one should keep the plans which have the highest probability of creating improving coalition: that is, preferable to the other plans in the agent's memory. To this end, a lexicographic ordering is used; the process removes the joint plan maximizing the number of individual plans which are the worst of the agents' memories. If several joint plans have the same number of worst plans, the process chooses among them to find the joint plan which maximizes the number of second worst plans, and so on, until the "worst" joint plan is unique. When the overall maximum number of plans in the memory of an agent is reached, the worst individual plan for this agent is removed along with plans of other agents of the same joint plan. Each agent keeps at least one plan that is not part of a joint plan, as there might otherwise be no state without blocking coalitions. Agents are parsed in random order, to avoid the emergence of "dictators" over iterations, whose worst plan would always be removed, even if it is the only "bad" plan of a joint plan.

Though those selection operators seem to be in accordance with the chosen solution concept, it is difficult, if not impossible, to prove that the process will actually converge towards the state searched. As noted by Ficici et al. (2005), when they perform a theoretical analysis of different selection methods in a co-evolutionary context, "co-evolutionary dynamics are *notoriously complex*. To focus on our attention on selection dynamics, we will use a simple evolutionary game-theoretic framework to eliminate *confounding factors* such as those related to genetic variation, noisy valuation, and finite population size". Those "confounding factors" can, however, not be eliminated from an actual implementation of a co-evolutionary algorithm; rigorously proving that a given algorithm actually implements a specific solution concept is very tedious, if not impossible.

With iterations, agents build a choice set of daily plans that becomes better and better, given the actions of the other agents. However, the presence of a large group of agents with plans resulting from random mutation creates noise, not only for the analyst looking at the output of the simulation, but for the agents themselves when they compute their score plans. To solve this issue, when the system reaches a stable state, agents stop performing mutation, and select plans only from their memory for a given number of iterations, using the absence of improving coalition with randomized scores. This ensures that the selected plans are the result of a behavioral model, rather than the result of random mutation operators.

28.3.2 Technical Considerations on the Implementation

As highlighted in Chapter 45, the preferred way to add new behaviors to the MATSim software is by designing *pluggable* elements, that can be added to a `Controler` from a configuring "script".

This modular approach works well in most of the cases used and makes it possible to combine different elements and design highly specific runs. There is, however, an element that one cannot modify this way: the general form of the evolutionary process. This process is exactly what has to be modified to include joint decisions—this section focuses on the challenges and solutions to undertaking such a major modification, as a reference from developers facing this exact problem.

The one important modification of the process: in the standard MATSim process, *replanning* is performed independently for each agent, whereas for joint decisions, agents must be replanned *as groups*: selection of plans needs to fulfill *joint plan constraints* and is performed using the group-level "absence of improving coalition" criterion, and mutation operators are allowed to work on several plans at the same time, for instance to insert *joint trips*, or select the location of a *joint activity*.

Doing so requires the replacement of the `ReplanningListener`, that is, the element responsible for managing the whole replanning step. This can only be done by implementing a separate `Controler`.

Modularity was kept as high as possible, in particular by providing standard ways to use the default individual-based replanning modules from this new element.

28.4 Selected Results

This section presents a few simulation results demonstrating how the approach can help improve simulation results. It uses a scenario using 2010 data, with a leisure contacts network generated using the approach of Arentze et al. (2013).

Specific replanning modules include: inclusion and removal of joint trips (by joining existing trips), and joint location choice for leisure activities. A specific scoring term is added to consider the preference for *joint activities*; individuals want to perform leisure activities with at least one social contact. Leisure time passed without any contact is penalized.

Figure 28.1 presents the repartition of "car passenger" trips by purpose, in the simulation as well as the Swiss National Travel Survey. The simulation is able to reflect the fact that most trips are

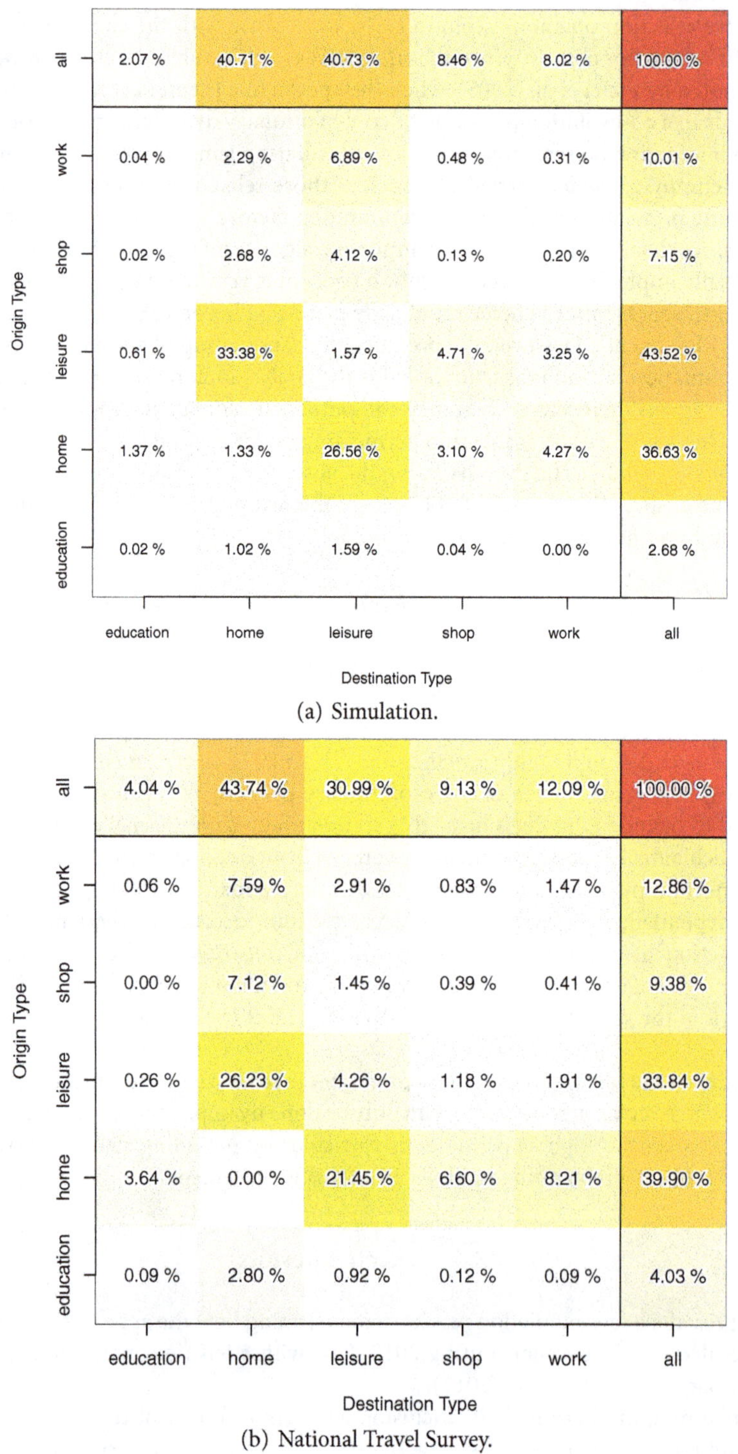

(a) Simulation.

(b) National Travel Survey.

Figure 28.1: Share of passenger trips by purpose.

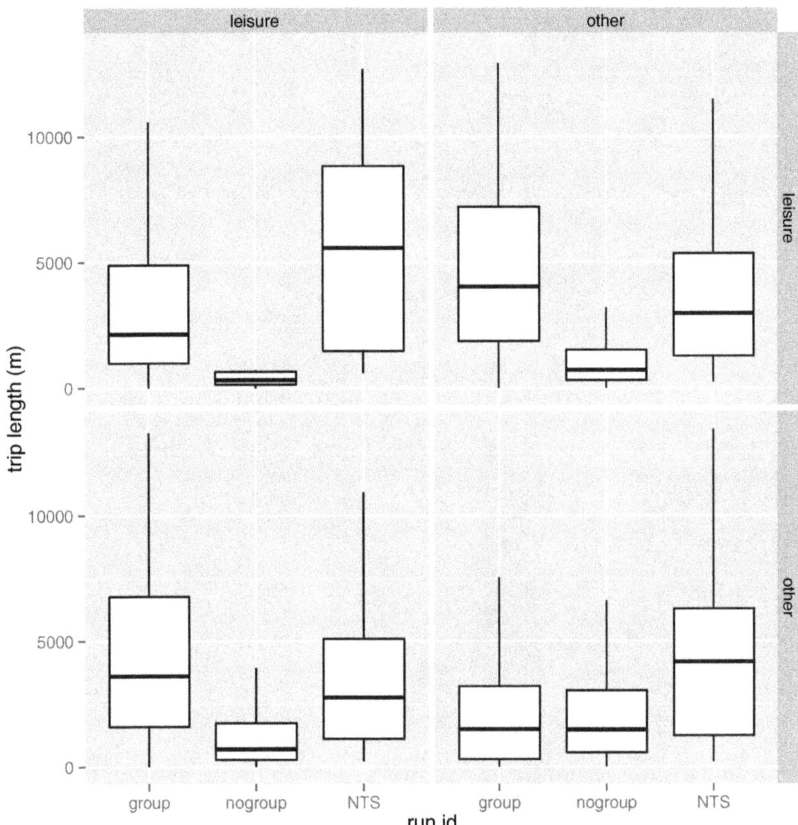

Figure 28.2: Car passenger travel distance to leisure activities.

performed *for leisure purposes*. Figure 28.2 shows the distance distribution of *car passenger* trips by purpose, with preference for group activities enabled or not, as well as in the "Swiss National Travel Survey". The preference for joint activities certainly encourages individuals to travel together for leisure, without waiting for each other, resulting in distance distributions much closer to the "Swiss National Travel Survey" data with this parameter than without.

28.5 Further Reading

The work presented in this chapter has been described in more detail in various other papers. Dubernet and Axhausen (2013) presents an early stage of the algorithm, applied to a toy scenario. Dubernet and Axhausen (2014) provides more theoretical ground, making more explicit reference to game theory and compares two *solution concepts* for solving joint planning problems in the household case: first,the absence of improving coalition presented here and second, a "joint utility" formulation, well-represented in literature. Dubernet and Axhausen (forthcoming) presents a validation of the model for the household case, using a Zürich scenario. An independent approach to model household choices developed for the Baoding scenario is presented in Chapter 61.

CHAPTER 29

Socnetgen

Kai Nagel

29.1 Basic Information

Entry point to documentation:

`http://matsim.org/extensions` → socnetgen

Invoking the module:

`http://matsim.org/javadoc` → socnetgen → `RunErgmSimulator` class

Selected publications:

Illenberger (2012)

29.2 Summary

This package contains algorithms to generate social networks that may be used on top of the MATSim population. It pre-dates the work by Dubernet presented in Chapter 28. The approach in `socnetgen` is much more lightweight than that of Chapter 28, but it also does nothing beyond just generating the social network according to given statistical criteria.

How to cite this book chapter:

Nagel, K. 2016. Socnetgen. In: Horni, A, Nagel, K and Axhausen, K W. (eds.) *The Multi-Agent Transport Simulation MATSim*, Pp. 183–184. London: Ubiquity Press. DOI: http://dx.doi.org/10.5334/baw.29.

SUBPART SEVEN

Within-Day Replanning

CHAPTER 30

Within-Day Replanning

Christoph Dobler and Kai Nagel

30.1 Basic Information

30.1.1 Implementation Alternative 1

Entry point to documentation:

`http://matsim.org/extensions` → withinday

Invoking the module:

`http://matsim.org/javadoc` → tutorial → `RunWithinDayExample` class

Selected publications:

See Section 30.4.2.

30.1.2 Implementation Alternative 2

Entry point to documentation:

`http://matsim.org/extensions` → withinday

Invoking the module:

`http://matsim.org/javadoc` → tutorial → `RunOwnMobsimAgentUsingRouter` class

Selected publications:

See Section 30.4.3.

How to cite this book chapter:

Dobler, C and Nagel, K. 2016. Within-Day Replanning. In: Horni, A, Nagel, K and Axhausen, K W. (eds.) *The Multi-Agent Transport Simulation MATSim*, Pp. 187–200. London: Ubiquity Press. DOI: http://dx.doi.org/10.5334/baw.30. License: CC-BY 4.0

30.2 Introduction

In recent years, transport planning and traffic management interest in unforeseeable, or only partially foreseeable events within scenarios has increased. Partially foreseeable events often occur with taxis and car sharing. For example, agents with a planned taxi trip cannot know in advance which taxi will be available when they need one. When using car sharing, an agent might walk to the car sharing station and check whether a car is available or not. If it is not, the agent could either decide to wait, or change its plan and switch to another transportation mode. Road accidents, terrorist attacks or disasters such as earthquakes are examples of completely unpredictable events.

As discussed earlier, traditional simulation approaches (used in default-MATSim) calculate demand-supply equilibria using an iterative process. There, it is assumed that a typical situation is simulated where agents can rely on their experience from comparable situations, like previous iterations. Applying an iterative approach to a scenario with unexpected events results in problems like illogical agent behavior, producing false results. In the next section, these problems, as well as an alternative simulation approach, are presented. On one hand, this approach—called within-day replanning—simulates only a single iteration, avoiding problems resulting from an iterative simulation process. On the other hand, this approach does require a more detailed behavioral model for the agents. Subsequently, using MATSim as a base, the iterative approach is discussed, followed by two different implementations of the within-day replanning approach into the framework, including discussions of the technical implementations.

30.3 Simulation Approaches

30.3.1 *Iterative Simulation Approaches*

An iterative day-to-day replanning approach is appropriate as long as the scenario describes a *typical* situation or day. For such scenarios, it is feasible to assume that agents are familiar with typically occurring events like traffic jams during peak hours. Therefore, they try to avoid driving during those times, or use alternative routes with less traffic. However, if the scenario contains unexpected events that the agents cannot foresee, e.g., accidents or heavy weather conditions, using an iterative approach is not an appropriate choice. First, a user equilibrium will not be reached in such a scenario because agents do not have enough information to choose optimal routes and daily activity plans. Another problem is the optimization process itself. Even if an agent chooses its routes randomly due to a lack of information, it will eventually find a good route if it tries enough different routes.

Figure 30.1 shows a simple example scenario where an iterative approach would produce illogical and faulty results. In Figure 30.1(a), an agent's planned route in a sample network is shown, including the times when the driver passes each node of the route. Clearly, those times are only valid if no exceptional event occurs. Figure 30.1(b) shows a link where an event, like an accident, blocks that link for two hours. As a result, the agent reaches its destination two hours later than expected (Figure 30.1(c)). When this scenario is iterated, the agent recognizes that its route has a much higher travel time than expected and therefore it will choose another route. The traffic jam caused by the accident will probably also increase travel times on links next to the blocked link. Therefore, the agent might find a route which is quite different than the original one (Figure 30.1(d)). A closer look at the node where the new route deviates for the first time from the original one shows that this occurs even before the accident happened, which is unfeasible and illogical.

An obvious solution to avoiding such problems is using an alternative simulation approach without an iterative optimization process. The next section discusses such an approach and the requirements that must be fulfilled.

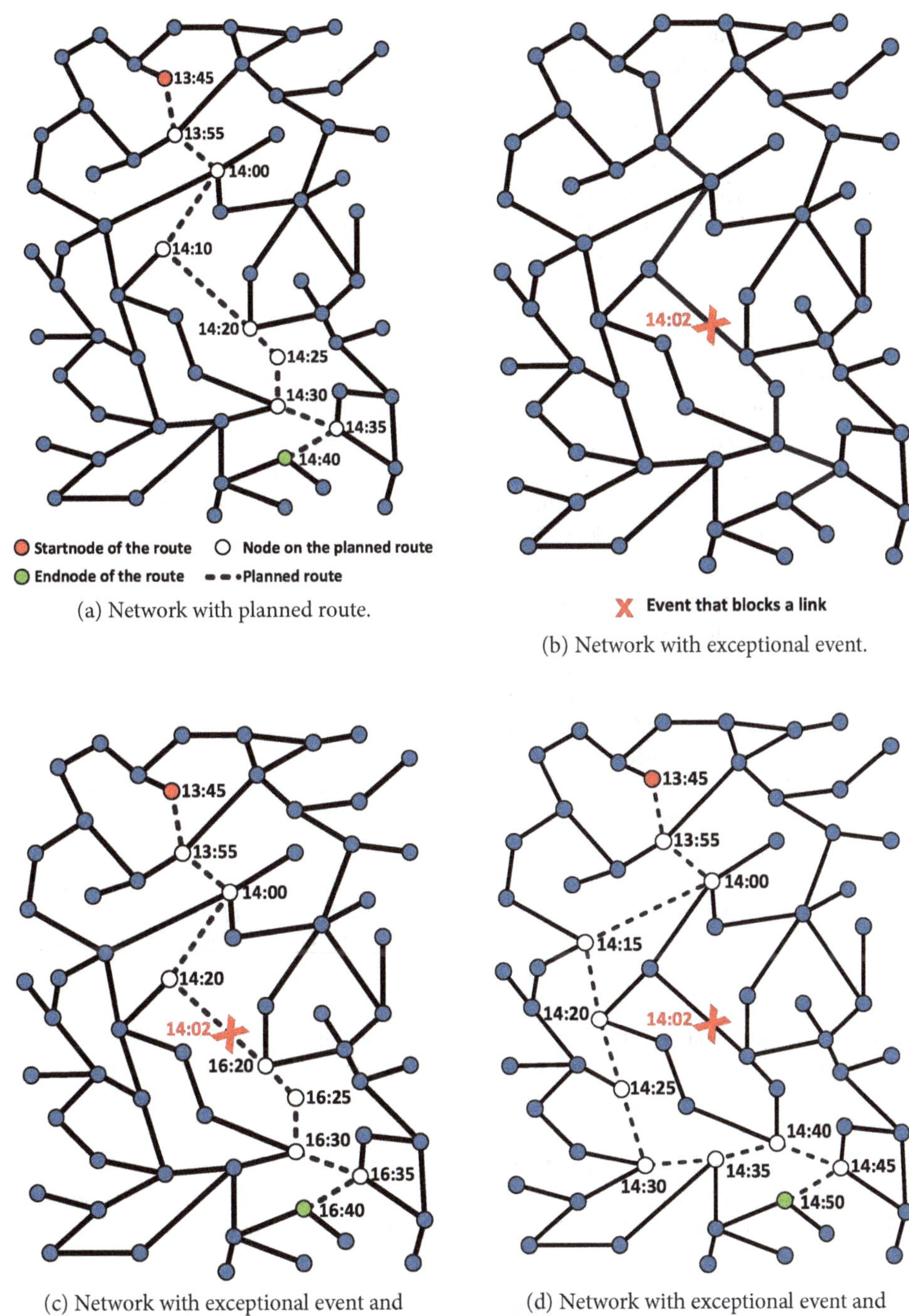

(a) Network with planned route.

(b) Network with exceptional event.

(c) Network with exceptional event and planned route.

(d) Network with exceptional event and adapted route.

Figure 30.1: Exceptional event in a network.

30.3.2 Within-Day Replanning Approach

A within-day replanning approach uses a significantly different strategy from that of an iterative approach. Instead of multiple iterations, only a single one is simulated. Thus, it is now essential that agents can adapt their plans during this iteration without having information from previous iterations available. To do so, they have to continuously collect information and take into account their desires, beliefs and intentions when they decide how to (re)act.

While iterative approaches can use best-response modules, a within-day approach has to use something that might be called a best-guess module. Travel times are an obvious example. In an iterative approach, travel times can be collected from the previous iteration or even be averaged over several past iterations. The nearer a stable system is to a relaxed state, the smaller the differences in travel times between two iterations. This is not possible in a within-day approach. Even if an agent has perfect knowledge, it can only assume how the traffic flows will evolve in the future. To do so, it can take different information into account to estimate travel times. It could, for example, take travel times from a typical day without exceptional events and combine them with information it gathers during the simulated day. Depending on the amount and the quality of this information, the agent might rely more or less on its experience.

Therefore, the decision-making process of an agent becomes an important topic. In an iterative approach, each agent has total information and can thus select the best route. Due to limited available information, this is not possible in a within-day approach. One agent could, for example, choose a route where expected travel time is very short, but also very uncertain. Another agent might not be willing to take that risk and therefore select a longer route where the assumed travel time is more reliable. Perception of information might also vary between agents; one could rely on media traffic information, another might ignore it.

Each within-day replanning action is categorized by two parameters—the replanned element of the plan (an activity or a trip) and the point in time when the replanned plan element is executed (right now or at a future point in time). If an activity is replanned, several changes are possible. Its start and end time can be adapted, its location can be changed, it can be dropped, or created new from scratch. For a trip, origin and destination, route, mode of transport and departure time can be replanned. Often replanning one single plan element results in a chain reaction that forces replanning of other plan elements. If, for example, an activity is dropped, the trips from and to this activity have to be merged.

The second parameter categorizing a replanning action depends on when the replanned plan element is executed. This could be either the currently performed plan element or one being performed in the future. Clearly, in a currently performed plan element, not all previously mentioned replanning actions could be conducted, e.g., start time of an activity or transport mode of a trip currently being performed can no longer be adapted.

Due to the limited available information, a within-day replanning approach will, in contrast to an iterative approach, not converge to a user equilibrium. Decisions made during the simulated time period may seem to be optimal when they are made. However, evaluated retrospectively, an agent might realize that they were not.

Figure 30.2 shows how within-day replanning can be integrated into MATSim's iterative optimization loop. An additional block builds another (inner) loop with the mobility simulation. Depending on the type of simulated scenario, the outer loop can be skipped.

30.3.3 Combined Approaches

An alternative to iterative, or within-day replanning only approaches, is to combine them. An obvious application is solving situations that cannot be planned exactly in advance, like parking or car sharing. An agent is, for example, able to plan a parking activity, but it cannot anticipate which parking spots will be available when the agent arrives. Thus, within-day replanning can be used when the agent starts its parking choice.

Other agents might want to share their cars, so an actual meeting must be confirmed. This can be ensured using within-day replanning. If the driver arrives too early, a *waiting* activity is added to its plan; otherwise the agent being picked up will perform a *waiting* activity until the car arrives.

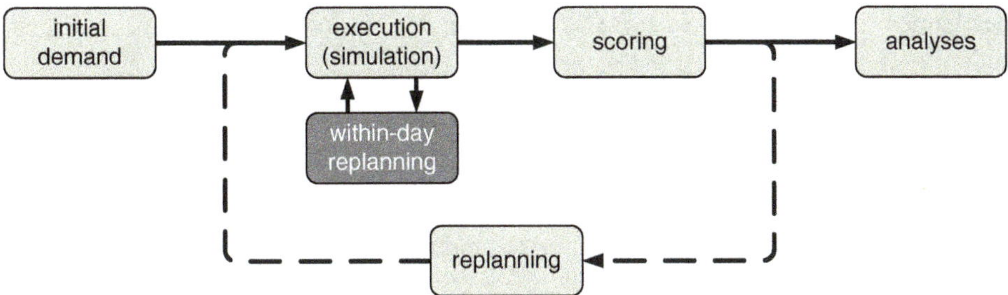

Figure 30.2: (Iterative) within-day replanning MATSim loop.

30.4 Implementation

30.4.1 General Thoughts

Within-day or en-route replanning means that travelers replan during the day or while they are on their route. This means that the simulation needs to find some way to influence the agent while the mobsim (network loading) is running. For the MATSim main network loading module, the so-called QSim, this could be achieved by inserting an agent-loop, as follows:

```
void doSimStep() {
    for ( each agent ) {    // <-- agent loop
        agent.doSimStep() ;
    }
    for ( each link ) {
        link.doSimStep() ;
    }
    for ( each node ) {
        node.doSimStep() ;
    }
}
```

In this loop, each agent has the chance to deliberate in every time step. Clearly, the agent can decide that he/she has nothing to deliberate and return immediately.

Such an approach does, however, lead to computational challenges. Going through all links and nodes in every time step is already an expensive operation and a number of efficiency improvements (such as "switching off non-active links") are contained in the code. Also, the number of links or nodes is typically an order of magnitude smaller than the number of synthetic persons in a scenario. Thus, some massive optimization would have to be undertaken in order to make the above approach computationally efficient.

An alternative approach to the above is to ask each agent only when a decision needs to be made. The most important decision for a driver is to chose the next link, i.e.,

```
class MyDriverAgent implements DriverAgent {
    ...
    @Override
    public Id<Link> chooseNextLink() {
        <algorithm to determine ID of next link>
        return nextLinkId ;
    }
}
```

Similar implementations are needed for all other queries that could be asked of the agent, for example:

- Should the trip end on the current link?
- Should the agent get off at the current stop?
- What is the ID of the vehicle to be used for a trip?

From the agent's perspective, such an approach might be called *event driven*, since the agent performs only mental activity at such events.

There is, indeed, a mechanism to program such agents and to insert them into the QSim. This is discussed in more detail in Section 30.4.3.

A challenge inherent in that approach is that the complete agent needs to be re-programmed. This agent needs to have enough capabilities to be oriented about itself; for example, it needs to be able to compute plausible routes.

On the other hand, there are situations where the capability to decide the turn at each intersection while en-route is, in fact, not needed. For example, for typical evacuation applications, it makes sense to start all agents on their normal daily plans. When an emergency warning is distributed, the simulation can go once through all agents and decide how they react. This will be done by replacing some, or all, future elements of the current plan. In some applications, this may happen more than once; for example, if recommended evacuation directions change because of a shift in the wind. In other applications, evacuating agents could become stuck in unexpected congestion which might trigger en-route re-routing. This may, however, be restricted to relatively small regions, and it may be sufficient to go through such a replanning loop, perhaps every 300 simulated seconds.

For such applications, the plan-based approach (Section 30.4.2) is more suitable. Rather than having each agent answering certain queries in every time step or at every intersection, the plan-based approach first waits for a trigger (such as an emergency warning, or unexpected congestion), then decides on the affected agents, then goes through those agents and changes the future part of their plans. This is not only conceptually easier than having every agent answer for him-/herself, but it is also computationally more efficient, since it is only called when it is triggered and impacts only the affected agents.

Overall, implementers and users will have to balance their needs. If there are relatively few times when agents should re-plan, and these times can be easily identified by, i.e., corresponding to an emergency signal, then this is an indicator for the plan-based approach. If, on the other hand, an agent goes into the simulation mostly or entirely without a plan, like an entirely reactive taxi driver, then this speaks for replacing the agent.

MATSim provides infrastructure for both approaches. The plan-based approach currently provides more support infrastructure, i.e., many important use cases can be implemented by re-using existing methods. The approach that replaces the agent, in contrast, provides more flexibility. In particular, it allows agents to make decisions at the latest possible time without additional computational overhead. While this is not entirely realistic behaviorally, such an approach is often desirable from a simulation perspective, where one does not want reproducibility of simulations depend on, e.g., random elements such as how far an agent plans ahead.

30.4.2 Implementation Alternative 1: Plan-Based Implementation

When adding within-day replanning to MATSim, its iterative loop (see Figure 1.1) has to be adapted as shown in Figure 30.2. On one hand, the additional *within-day replanning* module is added, which interacts with the mobsim. On the other hand, multiple iterations are only necessary if a combined simulation approach is used.

The implementation is realized as so-called `MobsimEngine` which can be plugged into the QSim. In every simulated time step, the QSim iterates over all registered `MobsimEngines` and allows them to simulate the current time step. Besides simulation of the traffic flows, those engines are also able to let agents start or end activities. The engine containing the within-day replanning logic (called `WithinDayEngine`) does not simulate traffic flows, but tracks agents and adapts their plans. Doing so is separated into two steps. First, agents whose plans have to be adapted in the current time step are identified. In a second step, the adaption of their plans is performed.

Figure 30.3 shows the structure of the `WithinDayEngine`. Multiple `Replanners` can be registered to the engine. Each `Replanner` represents a unique replanning strategy like re-routing or time mutation and uses a set of `AgentSelectors` that communicate with agents and select those who are given the opportunity to adapt their plans. An `AgentSelector` can be seen as an information-distributing unit, like a radio station or a policeman. Therefore, not every *AgentSelector* communicates with all agents. For example, agents at home will probably listen to the radio, but agents walking in the park will not. Each `AgentSelector` returns a list of agents to its superior `Replanner`, which then adapts those agents' plans.

Responsibilities are divided between `Replanners` and `AgentSelectors`. The first ones are responsible for adapting the agents' plans, but they should not check whether an agent should be replanned or not. If, for example, a `Replanner` updates an agent's route, it has to be ensured by the `AgentSelectors` that only agents who are currently performing a leg are replanned. In turn, `AgentSelectors` should select agents who have to be replanned but should not change their plans. As a result of this division, the often time-consuming replanning of the agents' plans can be performed using parallel threads, which leads to an almost linear speed-up. In general, simulation results do not depend on the order in which agents are replanned. *Replanners* which use random

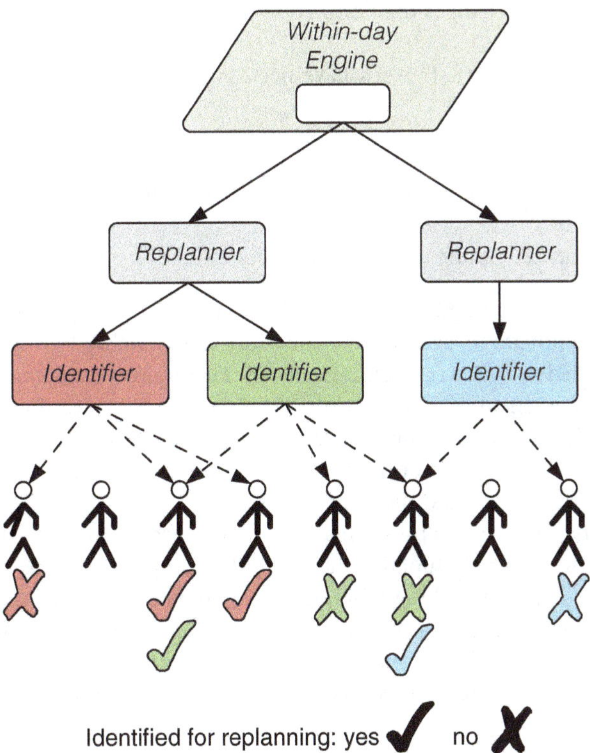

Figure 30.3: WithinDayEngine.

numbers are a special case. In the present implementation, their *random number generator* is re-initialized for every replanned agent, using a deterministic value (e.g., a combination of the agent's `ID` and the current time step). On one hand, this ensures that an agent's decisions can be reproduced even when the global sequence of random numbers changes. On the other hand, the simulation outcomes do not change if the number of threads used for the replanning is changed.

Running the `AgentSelector(s)` to select those agents who have to adapt their plans is performed sequentially. On one hand, an `AgentSelector`'s runtime is typically very short and therefore no significant performance losses are expected. On the other hand, this makes the design robust so it cannot produce race conditions which could occur if multiple instances of an `AgentSelector` run concurrently. An example would be an `AgentSelector`, which selects agents on household level, i.e., if a member of a household is identified, also all other members are added to the list of agents who have to be replanned. In an approach with parallel running instances of an `AgentSelector`, an instance could identify member "A" of a household while concurrently another instance could identify member "B" of the same household. As a result, the household's members would be duplicated in the list of agents to be replanned—once added by each `AgentSelector` instance.

`Replanner` implementations are available for any basic change of an agent's scheduled daily plan. All trips and activities can be adapted, although some replanning operations are not available when trip or activity has already been started. Possible adaptations are:

- current trip (route, destination),
- future trip (add, remove, mode, route, origin, destination),
- current activity (end time), or
- future activity (add, remove, location, type, start and end time).

For complex plan adaptations, those basic `Replanners` can be combined. If, for example, an agent currently performing a trip changes the destination of its next activity, routes of the current and next trip must be adapted.

Additionally, four basic `AgentSelectors` have been implemented so far. They identify agents, which are...

- performing an activity,
- performing an activity which will end in the current time step,
- performing a trip, or
- performing a trip and are going to move to another link.

Often, only a subset of the population, e.g., only male agents, or agents currently traveling in a car, needs to be identified. To prevent that the same functionality having to be implemented multiple times, so-called `AgentFilters` are introduced. Their task is to remove agents not meeting the filter criteria from an agent set. Using `AgentFilters` not only avoids duplicated code, but can also reduce computation effort: for example, two `AgentSelectors` which should identify only agents currently traveling in a certain part of the network. Without `AgentFilters`, each of them would have to track all traveling agents and their current positions. When this functionality is moved to an `AgentFilter`, the two `AgentSelectors` can share a single instance of that filter.

Basically, simple and re-usable functionality should be implemented as `AgentFilters`, while more complex and/or decision-making functionality should be part of an `AgentSelector`. Again, an example: e.g., a scenario modeling the search for a parking space: a filter can be utilized to take only agents currently traveling by car into account. The `AgentSelector` solves the more complex tasks, such as deciding when the agent starts its search, or selecting the searching strategy to be applied.

Three basic `AgentFilters` have been implemented so far. They filter agents which are not...

- part of a predefined agent set,
- currently using a transport mode included in a given set, or
- currently located on a link included in a predefined set.

In addition to the logic identifying agents and adapting their plans, another important within-day replanning framework component is code that continuously collects information and provides it to the `AgentSelectors`. These decide,based on that data, whether agents are replanned or not. In a time step-based approach—as realized by the QSim—collecting, analyzing and aggregating data, as well as providing it, can be easily realized. Figure 30.4 shows the structure of a QSim's time step. Each time step is separated into three phases:

Phase 1:

```
before time step
```

Phase 2:

```
do sim step
```

Phase 3:

```
after time step
```

During phase 2 all registered `MobsimEngines` simulate the current time step. Phases 1 and 3 allow code execution before or after simulation of the current time step. A class can collect data such as link travel times during phase 2. phase. Then, the collected data can be analyzed and aggregated in phase 3. In the next time step, the `WithinDayEngine`'s `AgentSelectors` can use that data for their decisions. The `WithinDayEngine` is always the first `MobsimEngine` executing its `doSimStep` method, ensuring that no agent has changed its status since phase 3 of the previous time step. As a result, the `AgentSelectors` make their decisions on current data.

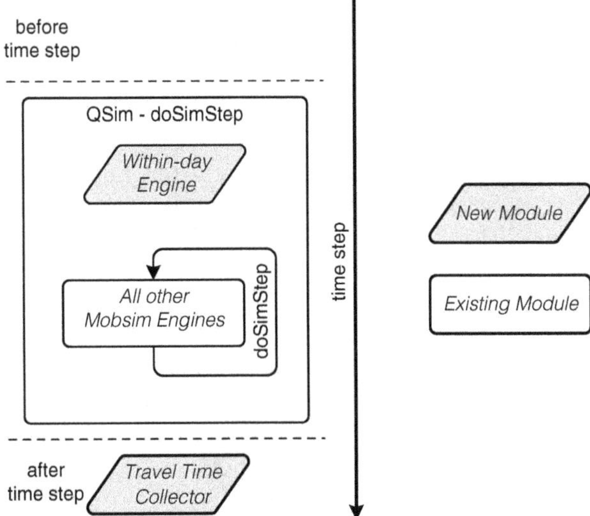

Figure 30.4: QSim time step.

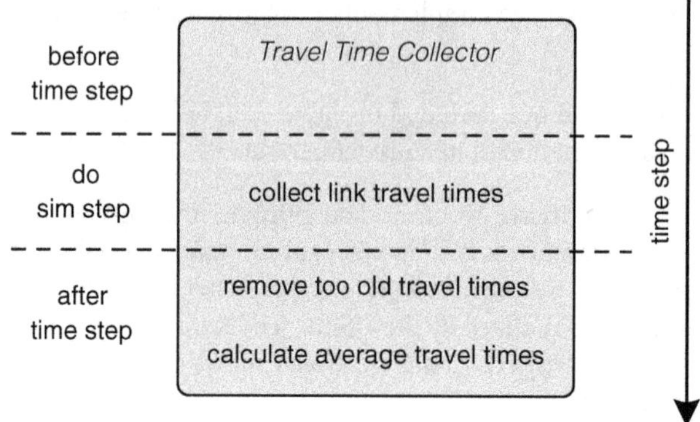

Figure 30.5: TravelTimeCollector.

An example of this type of class is the so-called `TravelTimeCollector`. It provides actual link travel times to the `Replanners` by collecting and averaging travel times of agents that have recently passed a link during a given time. A typical time span is 15 minutes; older link travel times are ignored. Specific time span duration has an important impact on travel times reported to the `Replanners`. On one hand, significant changes in link travel times will be communicated very slowly, if the time span is too long. On the other hand, a too short duration will overrate outliers.

The `TravelTimeCollector` is a simple, but efficient, implementation of a within-day travel time calculator. It does not incorporate features like traffic flow predictions or dynamic recent travel times weighting based on historic data. Because it does not factor in such features, it is very robust, even in scenarios where traffic flow conditions change dramatically.

The current MATSim code differentiates between `Person` and `MobsimAgent`. `Person` can be seen as a very simple Q-learning entity, possessing multiple `Plans` ("actions"), each with an expected score updated with every plan run. Thus, a `Person` is consistent over the iterations; in fact, the internal state of each `Person` is written to file at the end of the iterations. `MobsimAgent`, in contrast, is instantiated every time the QSim is called, and does not exist beyond the QSim running time. A `MobsimAgent` is essentially reactive, queried by the framework about decisions when approaching intersections, arrival points, or public transit stops. In the standard implementation, these queries are answered by the plan, but other implementations can be used and/or additional `MobsimAgents` can be added which do not correspond to `Persons`.

This leads to a question; should within-day adaptations to the `Plan` be passed through to the `Person`? Let us call the actual trajectory through the system the "executed plan". This can be different from the original plan, i.e., a different route, different departure times, different modes, etc. The original plan cannot just be replaced by the executed plan, since it is not clear that the executed plan, when used as input, will have itself as expected output. In consequence, it is not possible to treat the executed plan together with the just-obtained score as an action-value pair in the sense of Q-learning, since the score was obtained from the *original* plan, not from the executed plan.

As a result, the code uses a copy of the original plan and modifies the copy. The score, however, is given to the original plan. The implementation is able to *also* memorize the executed plan and add it to the set of plans. This functionality, however, is experimental.

In certain situations, setting the original to the executed plan clearly does not make sense; a parking search is one (Waraich et al., 2013c, 2012).

A person's plan contains, as destination, the location where a free parking space is expected. However, if the agent realizes in the mobility simulation that there is no free space left, it starts looking for a free parking spot. As a result, the agent's route is extended. This extension has to be local in the agent's route, since it is only necessary in the current iteration and probably not in another one, where the initially selected parking spot is available.

Capabilities of this within-day replanning implementation are shown and discussed by Dobler (2013), based on two sets of experiments. The first set is based on a model of Zürich city, where it is assumed that capacities of several city-center arterial roads are drastically reduced during the morning peak. Traveling agents are given the opportunity to bypass the resulting traffic jams by adapting their routes, using within-day replanning. As a result, average travel time of an agent affected by the incident is reduced from 42 to 23 minutes. Also interesting is that even if only 50 % of the population adapts its routes, average travel times are reduced to 25 minutes.

The second set of experiments uses within-day replanning to create agents' initial routes. The results are compared to runs where routes are created before the simulation starts, without traffic flow information. Results indicate that agents' average travel times are already very close to the values in a relaxed state. When using MATSim's traditional approach, 10 to 15 iterations must be performed before average travel times reach this level.

30.4.3 *Implementation Alternative 2: Replacing the Agent*

According to Russel and Norvig (2010), an agent is "anything that can be viewed as perceiving its environment through sensors and acting upon that environment through effectors." As stated above, MATSim has agents on two levels:

- `Person` is a Q-learning agent that is persistent over the iterations.
- `MobsimAgent` is a reactive agent that only exists during the mobsim.

For the Q-learning agent, perception works through the events; i.e., events are used to compute the score, build mental models to generate alternatives, etc. Acting on the environment works through plan selection.

For the reactive agent, perception works more directly through callback methods, such as the simulation notifying the agent it has just moved through an intersection. Acting on the environment works through making decisions at decision points, e.g., about turning directions at intersections, or whether to board a certain bus.

As discussed, the approach described in Section 30.4.2 assumes that the reactive agent still has followed (and generally follows) a plan. There may, however, be situations where this is inappropriate: for example, when the agent makes up the route as it goes, or when one wants to investigate models where each agent has its own perception and deliberation, rather than some external algorithm modifying its plan. As also mentioned earlier, there is no clear rule governing when and where an approach is better; it depends both on both project requirements and on the developer's personal preferences. Here, with this in mind, we will look at `MobsimAgents` that no longer have a pre-computed plan, but make decisions as they go. There is also a class `DynAgent`, which wraps around `MobsimAgent`, making it easier to use and providing additional infrastructure (Section 23.4).

30.4.3.1 *Agent Interface*

The `DriverAgent` interface structurally looks as follows:

- `Id chooseNextLinkId()`—agent is asked at intersections and needs to return how to proceed.
- `boolean isWantingToArriveOnCurrentLink()`—agent is asked if it wants to arrive on the current link.
- `void notifyMoveOverNode(Id newLinkId)`—agent is notified that it has traversed the intersection and entered a new link.

The rest comprises relatively simple bookkeeping methods like `getId()`—the agent needs to know its own identifier.

If it is assumed that the agent does not only replan en-route, but also while at activities, then the `MobsimAgent` interface also must be implemented. This is a bit more involved; important methods are:

- `endLegAndComputeNextState(...)`—agent is notified that the current transport leg has ended, and the agent internally needs to decide how to continue.
- `endActivityAndComputeNextState(...)`—agent is notified that current activity has ended; the agent internally needs to decide how to continue.
- `setStateToAbort(...)`—if a leg or an activity was not ended cleanly: this could happen if `chooseNextLinkId()` returns a link that is not outgoing from the current node.[1]
- `getState()`—agent needs to return its current state, which essentially either returns `ACTIVITY` or `LEG`; most important here is that the framework obtains information about whether the agent wants to start a new activity or leg.

Again, everything else concerns bookkeeping methods.

30.4.3.2 *Agent Insertion*

The code accepts several ways to insert such a self-programmed `MobsimAgent` into the code, but the preferred method is using the `AgentSource` interface, as follows:[2]

```
class MyAgentSource implements AgentSource {
    // constructor
    MyAgentSource ( Guidance guidance ) {
        ...
    }
    public void insertAgentsIntoMobsim() {
        // insert agent:
        MobsimAgent ag = new MyMobsimAgent( guidance ) ;
        qsim.insertAgentIntoMobsim(ag) ;

        // insert vehicle:
        // ...
        qsim.createAndParkVehicleOnLink(veh, linkId );
    }
}
```

`Guidance` helps the agent with making decisions, see below.

30.4.3.3 *Perception, Decision, Integration*

The agents somehow need to perceive their environment. The simulation tells the agent where it is, via `notifyMoveOverNode(Id<Link> nextLinkId)`. In general, however, this will not be sufficient. For example, the agent may want to be informed about congestion, or evacuation directions.

A general way to achieve this is to use the `Events` channel.

We would probably suggest separating observer, guidance, and the agent itself.

---

[1] Despite the name of the method, the agent can recover.

[2] See `http://matsim.org/javadoc` → main distribution → the `AgentSource` class for a pointer to a working code example.

Observer The observer would probably listen to events:

```
class MyObserver implements BasicEventHandler {
   @Override
   public void handleEvent(Event event) {
      ... // memorize information
   }
   ...
}
```

For working code, see `http://matsim.org/javadoc` → main distribution → `RunOwnMobsimAgentWithPerception` class and related.

Guidance A guidance object might give advice to agents. It could, for example, be designed as follows:

```
class MyGuidance {
   MyGuidance( MyObserver observer ) {
      ...
   }
   Id<Link> chooseNextLinkId( Id<Link> currentLinkId ) {
      ... // compute and return decision
   }
}
```

For working code, see `http://matsim.org/javadoc` → main distribution → `RunOwnMobsimAgentWithPerception` class and related.

Agent The agent needs access to the guidance object:

```
class MyAgent implements MobsimDriverAgent {
   MyGuidance guidance ;
   MyAgent( MyGuidance guidance ) {
      this.guidance = guidance ;
   }
   ...
   @Override
   Id<Link> chooseNextLinkId() {
      return this.guidance.chooseNextLinkId( this.currentLinkId ) ;
   }
   ...
}
```

For working code, see `http://matsim.org/javadoc` → main distribution → `RunOwnMobsimAgentWithPerception` class and related.

Control script This would be plugged together by a variant of the following script:

```
Controler ctrl = ... ;
...
// create observer object:
MyObserver observer = new MyObserver() ;
// add into events channel:
ctrl.addEventsHandler(observer) ;
// create guidance object:
MyGuidance guidance = new MyGuidance( observer ) ;
// create mobsim factory and set into controler:
ctrl.setMobsimFactory(new MobsimFactory(){
   public Mobsim createMobsim(Scenario sc, EventsManager ev ) {
      MobsimFactory factory = new QSimFactory() ;
```

```
        QSim qsim = (QSim) factory.createMobsim(sc, ev) ;
        // add agent source into mobsim:
        qsim.addAgentSource( new MyAgentSource( guidance ) ) ;
        return qsim ;
    }
}) ;
...
ctrl.run() ;
```

The above "script" uses an anonymous class for the `MobsimFactory`. This method of writing code is quite convenient for adapting MATSim to individual needs, also see Chapter 45.

For working code, see `http://matsim.org/javadoc` → main distribution → `RunOwnMobsimAgentUsingRouter` class and related.

30.4.3.4 `DynAgent`

As stated earlier, there is also a class `DynAgent`. It wraps around `MobsimAgent`, making it easier to use and providing additional infrastructure (Section 23.4).

CHAPTER 31

Making MATSim Agents Smarter with the Belief-Desire-Intention Framework

Lin Padgham and Dhirendra Singh

31.1 Basic Information

Entry point to documentation:

`http://matsim.org/extensions` → bdiintegration

Invoking the module:

See `http://matsim.org/extensions` → bdiintegration

Selected publications:

Padgham et al. (2014)

31.2 Introduction

In this chapter, we introduce a MATSim extension allowing a developer to program (some of) an agent's decision-making in a BDI (Belief Desire Intention) system, while actual actions and environment percepts occur within MATSim.[1] This allows sophisticated modeling of agents within a BDI framework, using the concepts of goals, hierarchical abstract plans (containing sub-goals)

[1] This work was supported by the ARC Discovery DP1093290, ARC Linkage LP130100008 and Telematics Trust grants. We would like to thank Agent Oriented Software for use of the JACK BDI platform and Kai Nagel, Todd Mason, Sewwandi Perera, Edmund Kemsley, Oscar Francis, Daniel Kidney, Andreas Suekto, Qingyu Chen, and Arie Wilsher for their contribution to the BDI platform integration framework and to these applications.

How to cite this book chapter:

Padgham, L and Singh, D. 2016. Making MATSim Agents Smarter with the Belief-Desire-Intention Framework. In: Horni, A, Nagel, K and Axhausen, K W. (eds.) *The Multi-Agent Transport Simulation MATSim*, Pp. 201–210. London: Ubiquity Press. DOI: http://dx.doi.org/10.5334/baw.31.

and percepts (information from the environment), as well as information about the current situation. For example, we used it to model residents in a bushfire[2] evacuation, as well as an incident controller in an evacuation scenario. The residents may receive information about the bushfire from the fire simulation, as well as warnings and messages from the incident controller agent. They may well have to pick up children, check on neighbors and communicate with other family members, etc. Their plans enable decision-making, which will result in actions executed within MATSim.

In standard MATSim usage, intelligence within individual agents' behavior arises from co-evolutionary algorithms in the replanning phase. This is based on agents evaluating—via a scoring function—the plan they have executed during a given day and modifying this to obtain a new plan, until all agents have acceptable plans; the system then reaches a stable state. This approach, however, only works for applications where one can assume that the agents adjust and refine their behavior over many iterations, to eventually obtain their standard modus operandi. For applications such as emergency management, agents must react immediately to the situation as it evolves, doing so in an "intelligent" manner.

The chapter on Within-Day Replanning introduces two approaches to the mobsim component which address the need to be more reactive to an evolving situation. The first allows a centralized MATSim process to identify sets of agents that should have their plans modified, then runs one or more processes to adjust agents' plans. The second rewrites the agent, so that instead of following a specified plan, the agent invokes a decision-making process at all possible decision points. By integrating a BDI agent platform with MATSim (Padgham et al., 2014), we allow autonomous individual decision making to be programmed in specialized and powerful systems developed specifically for this purpose, balancing reactive behavior and goal-based commitment. Different BDI platforms have different strengths, but are, in general, based on a simplified psychological/philosophical view of how people behave, facilitating a high level specification of complex human behavior. These systems have been demonstrated to be very efficient for building complex applications (Benfield et al., 2006). Provided the appropriate system interface support is developed, any BDI system can be coupled to MATSim, as described here. Until now, we have used three different BDI systems, for which the system level interface is available. The decisions made in the BDI system are then inserted into the relevant agents' MATSim plans, allowing the MATSim agents to operate in the same efficient manner as in standard MATSim.

31.3 Software Structure

Our framework supports independent execution of MATSim and the BDI platform, with synchronization via the infrastructure provided. They can either run within a single process (in separate synchronized threads, or sequentially in a single thread), or in two separate processes (synchronizing using inter-process communication, such as sockets). The former is, of course, considerably more efficient. Conceptually, for every MATSim agent whose decision making is to be carried out in the BDI system, a BDI agent must be created. The BDI counterpart can be regarded as "the brain" associated with the MATSim agent. It is possible to have BDI agents with no MATSim counterpart and vice versa. For example, in our bushfire application, the incident controller has no MATSim agent, as he does not move on the road network. He receives information about the fire and has some static location information; his role in the simulation is to issue warnings and evacuation advisories, which, in turn, affect the resident agents. There may also be MATSim agents that do not have a BDI counterpart. For example, in a taxi modeling application, there may be MATSim

[2] Bushfire is the Australian term for what is otherwise known as a wildfire or forest fire.

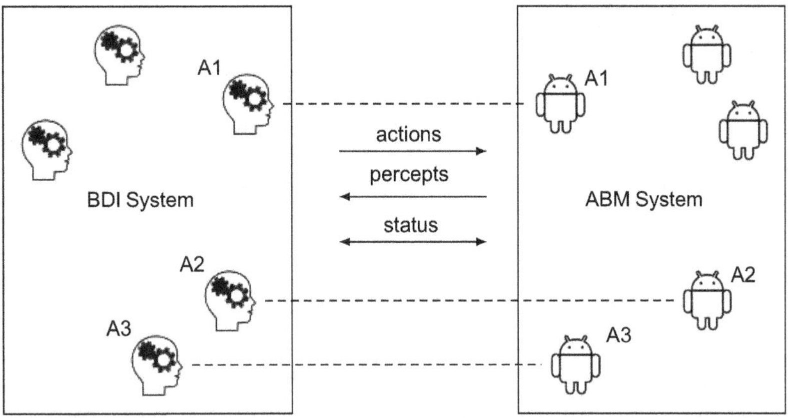

Figure 31.1: Conceptual BDI-ABMS integration architecture.

Source: Figure adapted from Padgham et al. (2014, Figure 1) distributed under the Creative Commons Attribution Non-Commercial License

agents using the road network, but with no need for complex decision-making modeling; these may exist only within MATSim.

Figure 31.1 shows the two parallel systems' basic architecture and the information passed between them at each time step.

The structure of the data components passed between the MATSim agent and its BDI counterpart is shown in Table 31.1 and consists of *BDI Actions*[3], *Percepts* and *Queries*. As indicated in Figure 31.1, BDI-actions are always initiated by the BDI system. Their status field, however, can be modified by both systems. When a BDI action such as `DriveTo(loc)` is decided by the BDI agent, the BDI system sets the status of this action as "INITIATED". MATSim will then set its status to "RUNNING", which will probably remain in this state for several steps. When the `loc` destination is reached, the MATSim routine will set the status to "PASSED" and the BDI system will continue reasoning about the next stage of agent behavior. If desired, the MATSim routine can also detect situations which should be conveyed as "FAILED" and pass this to the BDI counterpart. For example, if there is a BDI action to meet at a location and time and the MATSim agent is delayed in traffic, the BDI action implementation in MATSim can be programmed to detect the missed deadline and set the status to "FAILED", at which point the BDI agent will attempt failure recovery (as part of the BDI infrastructure). The BDI system can also set the status to "ABORTED"—for example, if information arrives requiring a different action—in which case, it is canceled within MATSim. The BDI system can also set status to "SUSPENDED", though this is not currently implemented.

To manage BDI actions, we provide a `MatsimAgentManager` class responsible for updating BDI actions status for all agents. At each step, the `MatsimAgentManager.updateActions(...)` function identifies (from the information package supplied by the BDI system) all agents initiating, aborting, or suspending actions. These are the agents which may require their MATSim plans to be modified. For each agent that has some action with s status "INITIATED", the action is passed to the agent's action handler class `MatsimActionHandler` via a call to `MatsimActionHandler.processAction(agentID, actionID, params)`. This function, based on the action, calls an appropriate helper function that performs required modifications to the MATSim plan and other relevant bookkeeping, to ensure that success and failure are observed (via

[3] We call these actions BDI Actions to distinguish them from actions in the ABMS (Agent-Based Modeling and Simulation) which may include lower level or additional actions.

Components of The Data Package Provided to Specific Agents Via The Interface:

Component Type	Component fields
BDI action	$< instance\_id, action\_type, parameters, status >$
Percept	$< percept\_type, parameters, value >$ (parameters and value may be complex objects)
Query	$< query, response >$

BDI Action Status:

State	Description
INITIATED	Initiated by BDI agent and to be executed
RUNNING	Being executed, set by the simulation agent
PASSED	Completion detected and set by the simulation agent
FAILED	Failure condition detected and set by the simulation agent
DROPPED	Aborted by the BDI agent
SUSPENDED	Temporarily suspended by the BDI agent

Table 31.1: Data Passed Between The BDI and ABMS Systems

appropriate MATSim callbacks) and that status is reported back to the BDI system. For example, for a `DriveTo` action, a `processDriveTo(agentID, loc)` function is executed to determine the leg associated with `loc`, obtain a route using the MATSim router and insert this into the MATSim agent's plan. The standard MATSim execution then follows this plan at each subsequent step. If the `processAction` function returns a success status indicating that the action was handled successfully, then `updateActions` changes the status for this action to "RUNNING"; otherwise, it sets it to "FAILED."

Sometimes, a running action can also fail in the ABMS for some reason. For instance, a `DriveTo (loc)` action could fail due to a road-closure in a bushfire evacuation simulation. While this functionality is supported by our infrastructure, it has not yet been used in the applications we have built with MATSim. Failing actions will soon be added for some applications. Aborting and suspending are also not currently implemented for MATSim. This would be accomplished by having appropriate functions declared which reset the plan contents of the agent to a 'holding state' (activity with infinite end time), maintaining the removed contents of a suspended plan in some data structure for eventual resumption.

Percepts capture information identified as necessary for the BDI agent's reasoning. Typically, this is any information leading to triggering of a BDI-goal, or causing an executing goal/plan to be re-evaluated. Approaching a destination is one example. MATSim callbacks are used to capture the relevant information within MATSim; this is then provided to the BDI counterpart via our infrastructure. The appropriate MATSim event is caught with `AgentActivityEventHandler.handleEvent(event-type)`. The `handleEvent(event-type)` function then first checks whether the agent receiving the event is one registered for a percept that triggers with this event type, and if so, calls the appropriate function to calculate the percept's value and add it to the percept container for that agent, to be sent to the BDI system. Termination conditions (PASSED and FAILED) of BDI actions are also similarly detected.

Instead of passing back the percept in these cases, the relevant action and its status is edited and passed back. For example, a BDI action `DriveTo(loc)` should succeed when the agent reaches the link closest to this location. To achieve this, we implement `handleEvent(PersonArrivalEvent)`, which will then trigger for every agent arriving anywhere. If the agent has a current (`DriveTo`) BDI action being monitored, then `arrivedAtDest(agentID,loc)` is called to ascertain whether the

PersonArrivalEvent caught does match the link closest to the coordinates of the desired destination. If it does, the action status of that DriveTo action for that agent is changed to PASSED and the action is removed from the monitoring list.

This approach conveniently uses MATSim callback infrastructure. However, we note that it will generate an event that must be processed any time any agent arrives anywhere, although most will not be an arrival at a desired destination. This is a substantial overhead; we may eventually consider collecting (some) percepts and state information for determining action status, in a separate, more efficient global processing at the end of the step.

Queries are defined for any information that the BDI system may want to request from MATSim during its reasoning process. Typically, queries are based on plans' context conditions, which must be evaluated to determine if a plan is applicable. Each query structure must be defined and the code must be supplied on the MATSim side to call the relevant functions to provide the response. Similar to the MatsimActionHandler class, we have a MATSimPerceptQueryHandler class containing a queryPercept(agent,query,response) function. This function then uses the query string received to extract the percept type and make a specific function call to obtain and provide the results. For example, if an agent agentID sends a queryPercept(agentID, ''RequestLocation agentX'', loc) query to request the location loc of some agent agentX (possibly itself), then the queryPercept function will execute the clause:

```
if percept_type = "RequestLocation"
   loc = getLocation("agentX")
```

The agentID of the requesting agent, obtained from the data package, is always provided to the query response function, in case it is required, although in this case it is not. Queries can be made at any point during the BDI execution and are answered immediately. They have no effect on the MATSim simulation.

A number of commonly used BDI actions and percepts are defined as part of our integration infrastructure. New ones can be added as part of developing a specific application, as described in Section 31.4. This structure allows all high-level decision making to be carried out by individual agents, within the BDI-system, which is designed and optimized for this purpose with regard to both representation and execution. On the MATSim side, specified functions simply modify the agents' MATSim plans (in parallel, if desired), retaining the standard MATSim simulation execution where each agent just follows its MATSim plan. This approach allows for both simplicity and efficiency at the lower level.

31.4 Building an Application Using BDI Agents

We focus here only on what must be done to integrate BDI agent reasoning into MATSim. To learn about BDI design and development, we refer the reader to Padgham and Winikoff (2004), as well as the excellent "practicals" (tutorials) available as part of the JACK platform[4]. In Figure 31.2, we show part of a taxi agent design, in an application involving taxis operating within MATSim. Here, the percept ClosetoDest (potentially) triggers a plan GrabJob. Plans have context conditions which indicate whether or not they are viable in the current situation, as a response to a percept, or a way of achieving a goal. Let us assume, in this example, that the plan GrabJob has the context condition (Location(self,loc)) $\wedge$ board.job.loc $\wedge$ (distance(board.job.loc,loc) $<$ 4km). Thus, the figure at the left of the diagram can be understood as the rule:

ClosetoDest $\wedge$ Location(self,loc) $\wedge$ board.job.loc $\wedge$ (distance(board.job.loc,loc) $<$ 4km) $\rightarrow$ GrabJob

[4] http://aosgrp.com/products/jack/

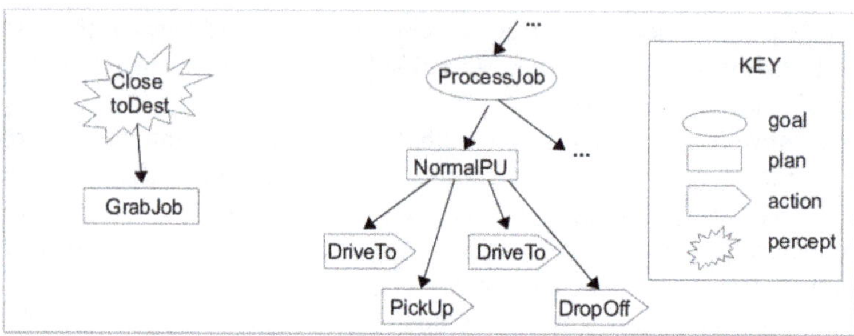

Figure 31.2: Excerpt of taxi design.

There are two pieces of information in this rule that must come from MATSim: first, the agent is close to its destination (`ClosetoDest`) and second, the agent's current location (`Location(self, loc)`). We could have MATSim send the agent location at every step. However, this is unnecessary overhead; instead, we send `ClosetoDest` as a percept. This requires the BDI agent to query its location to evaluate whether there are pending jobs whose location necessitates triggering some instance of `GrabJob`. This gives us an example of a percept and a query required in MATSim. On the right hand component in Figure 31.2, we see four different actions which will have a corresponding BDI-action on the MATSim side. We will focus here on the `DriveTo` action, but the `PickUp` and `DropOff` would be realized in a similar way, using MATSim activities rather than legs.

The following must usually be done:

- Every plan trigger which is information from MATSim must be defined as a percept.
- All information required from MATSim, that is not a trigger, must be defined either as a percept (and then stored locally), or as a query.
- All actions which should be executed in MATSim must be defined.

In the rest of this section, we describe exactly what must be provided in the MATSim application files for each of these to work as expected. Instructions and examples for the BDI application can be found in the integration repository (noted at start of chapter).

31.4.1 The `ClosetoDest` *Percept*

All functions for collecting percepts for the BDI system are defined in the `AgentActivityEventHandler` class. Perusal of existing functions can ascertain whether the desired percept is already calculated. For example, `arriveAtDest` is already defined for use as a BDI percept. If the percept collection function already exists, the developer must ensure that the appropriate agent type is registered for this percept within the relevant function. For example, in `arriveAtDest()` we have:

```
if agent.type = taxi
   AND agent.loc = dest(agent) \* obtained from infrastructure data *\
       // collect and package this percept
```

If we now want this percept provided to agents of type commuter, we must make the first line:

```
if ((agent.type = taxi) OR (agent.type = commuter))
   AND agent.loc = dest(agent) \* obtained from infrastructure data *\
       // collect and package this percept
```

The `arriveAtDest` function is triggered by the MATSim `LinkEnterEvent` event using MATSim provided callbacks. Thus, we have defined `handleEvent(LinkEnterEvent)` to call all percept collection functions triggered by this event – in this case `arriveAtDest`.

The `ClosetoDest` percept will be triggered by the same MATSim event `LinkEnterEvent`, so to add this, we must add the call to `ClosetoDest` in the `handleEvent(LinkEnterEvent)` and then define our `ClosetoDest` function within the `AgentActivityEventHandler` class. We only want to send the `ClosetoDest` percept when we first come within the defined distance of our destination, not at every step. Therefore, the `ClosetoDest` function must first check whether this percept has already been sent to this agent, for the current destination. If so, nothing more is done. If not, it is ascertained whether the link entered is within the desired "close-to" distance and, if so, the percept is registered. For efficiency, the first link "close-to" the dest can be calculated and recorded when the `DriveTo` action is initiated; in which case, one must only check whether the entered link-ID is the same as the recorded "close-to" link-ID.

In principle, percepts could also be calculated in a function executed after all agents had been stepped. The important thing is that when a percept occurs, it is recorded in the percept data package for that agent. Further work is required to ascertain which percept collection methods will be most efficient with very large numbers of agents.

31.4.2 *The* `RequestLocation` *Query*

Queries are defined in, and managed through, the `MATSimPerceptQueryHandler` class. A function `queryPercept(agent,query,response)` responds to a query by extracting the specific query and calling the relevant defined function. So, for example, to respond to the `queryPercept(ownID, ''RequestLocation agentID'',loc)` query from an agent, `queryPercept` will contain the code:

```
if percept_type = "RequestLocation"
   loc = getLocation("agentID")
```

The `getLocation` function will then ascertain the location of `agentID`, storing the value in `loc`. If the query is already defined in MATSim, nothing further is required to use it in an application.

31.4.3 *The* `DriveTo` *BDI-Action*

The `DriveTo(loc)` BDI action is, of course, the most basic and commonly used BDI action in MATSim and is already implemented in our infrastructure. As long as the appropriate BDI action and parameters are passed in the information package from the BDI system, nothing further is required within MATSim. However, for the purpose of illustration, we will assume it has not yet been implemented and we will go through the steps of defining a new BDI action with this as an example.

The `MATSimActionList` class defines mappings for all BDI actions in the system and the MATSim function calls that realize those BDI actions. Any new BDI action must first be added to this list.

The `MATSimActionHandler` defines all functions that realize BDI actions, as well as a `processAction` function which handles all BDI action strings from the BDI system, calling the appropriate helper functions. Thus any new BDI action must have its implementation defined within this class and must have the appropriate call to the function added within `processAction`. Let us call the relevant function that we will add `processDriveTo`. This function will always need the `agentID` as a parameter, as well as whatever parameters are provided in the action package.

So, in our example, we will have the function `processDriveTo(agentID, loc)` which needs to be defined. The function for the new action must perform two key tasks:

1. Obtain the MATSim plan of the relevant agent and modify it so that regular MATSim execution of the plan will have the desired effect.
   Generally, when the plan is accessed, it will have a single dummy activity with end-time infinity. The end time of this activity must be set to now and a leg must be instantiated with the link corresponding to the destination `loc` as the end point and the links to be followed, as calculated by the router. This leg must then be inserted into the plan, followed by a new dummy activity instance with end time infinity.
2. Place the action instance into the list of actions being monitored.

It is also necessary to set up recognition of when the action has finished, so that this information can be sent back to the BDI system and the agent can continue to reason about its next actions. This is done via the MATSim callbacks provided, in the same way as detecting percepts. However, the corresponding function, instead of placing information in the percept package for the agent, will modify the `status` of the relevant BDI action instance in the information package to PASSED and remove the instance from the list of actions being monitored. It is also possible to define a condition where the action should be considered FAILED and to detect this in a similar way. Alternatively, failure can be managed by sending a percept, and having the BDI agent abort the action as a result[5].

The current structure assumes that multiple actions of a single agent cannot be executed in parallel (a reasonable assumption for MATSim). It is the responsibility of the BDI system to allow only one active BDI action per agent.

Further instructions, as well as examples, can be found in our BDI-MATSim integration repository.

31.4.4 Discussion

An important aspect of a simulation design using BDI agents within MATSim is deciding on which abstraction level BDI actions should be described. So far, we have tended to have BDI actions map to a single `leg` or `activity` within a MATSim plan. However, it is certainly easy to think of BDI actions that combine several such components. Straightforward examples would be grocery shopping or taking kids to school - both involving a leg to a destination, an activity at that destination and a return leg. There are no immediately obvious advantages associated with BDI actions at higher abstraction levels (requiring coding of these actions in MATSim) vs using lower level BDI actions with the higher level coded as BDI plans/goals. Future experience and experimentation may provide insights to guide decisions.

31.5 Examples

Here, we describe two different examples of BDI agents within MATSim: a bushfire evacuation simulation, where MATSim is being used because traffic flow is a crucial component in this type of evacuation and a taxi application developed as a demonstrator for integration of a BDI system with MATSim (Padgham et al., 2014). We compare this approach to incorporating taxis with that described in Chapter 23 for incorporating dynamically scheduled vehicles and with the approaches to "within-day replanning" described in Chapter 30.

[5] The simplest way in JACK is to use a maintenance condition relying on a belief that is modified as the result of a percept.

Both our example applications use only the `Mobsim` engine (`QSim`) of MATSim and do no repeated daily cycles with plan scoring and modification. There are undoubtedly applications which could benefit from a combination of BDI agents and agents which evolve using MATSim's scoring and replanning, but we have not yet investigated them.

31.5.1 Bushfire Example

The bushfire example (currently) involves modeling of residents and their decision-making behavior about what to do regarding a nearby bushfire. Potential driving activities include picking up children from a school or other facility, checking on neighbors or friends and driving to a local or more distant destination, possibly via a specified route. Decision making may involve various factors, such as time of day, ideas about what other family members are doing, warnings and notifications from emergency services, observations of neighbors, etc. In one approach, we focused on incorporating well-developed and validated actual human decision making models in a bushfire situation, developed by a collaborator. Our contribution has been to integrate this with MATSim, using our integration framework, to provide data about any traffic-related issues, thus providing a more valuable simulation to planners. In our other approach, we model both residents and an incident controller. Here, our focus has been on technical issues that involve providing an interactive simulation suitable for use by emergency services personnel and/or communities for exploration of potential strategies.

In the interactive version, the incident controller assigns specified evacuation centers and routes to residents in certain sections of the town being evacuated. Evacuation of different areas may be started at different times. Residents follow the incident controller's instructions with some probability based on their individual situations (currently modeled very superficially). Following the suggested route is achieved by driving via suggested way points (using the `DriveTo` BDI action), with the BDI agent (potentially) re-assessing as each waypoint is reached. An alternative would be to define a new BDI action `DriveToViaWaypoints`. One issue that arose during the development of this simulation involved road congestion; MATSim routing algorithms began developing very circuitous routes, sometimes going back towards the fire threat. There were two issues illustrated here about developing a realistic simulation: one was that, realistically, people would not choose their routes based on global knowledge of current congestion; the other was that, regardless of congestion, people would not head back into the fire zone. The current solution is to use a routing algorithm not accounting for current road speeds, using only static speed limits. Going forward, one may want to assume some knowledge of congestion (based on radio broadcasts or other social media). An interesting future research question is how to best achieve responsibility sharing for realistic behavior between MATSim and the BDI decision-making program, on route selection.

31.5.2 Taxi Example

The taxi prototype application was developed purely as a 'proof of concept', allowing decisions to be made dynamically by the BDI brain on an ongoing basis, then carried out by the MATSim execution engine. There is a simple taxi administrator in the BDI system, which generates jobs, posts them to a notice board and confirms requests from taxis to take specific jobs. Taxis have plans allowing them to take jobs from the board, go to a taxi rank, or take a break. After taking a job from the board, the taxi drives to the pick-up address, picks up the passenger, then drives to the destination and drops them off. When the taxi approaches the destination, it looks on the job board for nearby jobs; if something suitable is found, it requests it from the administrator. The only BDI action implemented in this application is a simple `DriveTo`. The `ClosetoDest` percept was used as described in Section 31.4. This application was tested with the Berlin road network and the 15 963 agents in the MATSim sample files, with all agents operating as BDI taxi agents. Profiling showed that, by far, the majority of the execution time was spent in route planning, with very little in the BDI reasoning, or communication with the BDI system.

31.5.3 Discussion

Both evacuation and taxis are discussed in Chapters 30 and 23, as applications requiring a reactive approach to planning, rather than iteration over many days to find the preferred plan. Chapter 23 discusses two implementation options: one which replaces the MATSim agent with an agent that considers what to do at each relevant decision point (particularly intersections); the other leaves the agent code as is, but modifies the agent's plans when certain events occur. The BDI approach has the computational advantages of the latter, in that only a small subset of agents require changes to their plans at any simulation step and many existing MATSim routines can be used to modify the plans. However, it also has many of the advantages of the former approach; agents are still fully autonomous, with all decision making occurring within the BDI system. By registering for any percepts which could potentially cause the agent to change its mind, the agent remains fully in control at all times. However, it only needs to decide its next action when it completes the current high level action—which will almost certainly be orders of magnitude less often than at each intersection—or when a percept arrives indicating a need to reconsider. The provision of the ability to drop current BDI actions (legs or activities) provides the same level of reactive autonomy as the fully reactive within day replanning agent, but probably at a lower computational cost. Perhaps more important than the computational cost savings: agent decision making can be programmed in a framework that is at a high level of abstraction, using goals, plans and beliefs, within existing highly efficient platforms such as JACK (Winikoff, 2005), Jadex (Braubach et al., 2005) or Jason (Bordini et al., 2007). Design tools for developing such agents also already exist (Padgham and Winikoff, 2004). One study has shown that using a BDI language makes program development hugely more efficient than programming in Java (Benfield et al., 2006). The close mapping between intuitively understandable design diagrams and the program code implementing this in a BDI system is also highly advantageous for validating design of realistic agents with domain experts. We have discussed design of resident agents in a sandbagging flood scenario, with emergency services personnel extremely experienced in that domain and found the representation to be effective. We consider that this representational aspect can be a significant advantage when compared to programming the agent using the `DynAgentLogic` facility described in Chapter 23.

SUBPART EIGHT

Automatic Calibration

CHAPTER 32

CaDyTS: Calibration of Dynamic Traffic Simulations

Kai Nagel, Michael Zilske and Gunnar Flötteröd

32.1 Basic Information

Entry point to documentation:

`http://matsim.org/extensions` → cadytsIntegration

Invoking the module:

`http://matsim.org/javadoc` → cadytsIntegration → `RunCadyts4CarExample` class

Selected publications:

Flötteröd (2010); Flötteröd et al. (2011); Flötteröd et al. (2011a); Flötteröd (2008); Moyo Oliveros (2013)

32.2 Introduction

Cadyts (Calibration of Dynamic Traffic Simulations)[1]—licensed under GPLv3 (GNU General Public License version 3.0)—calibrates disaggregate travel demand models of DTA (Dynamic Traffic Assignment) simulators from traffic counts and vehicle re-identification data. Cadyts is broadly compatible with DTA microsimulators, into which it can be hooked through parsimonious interfaces.

As explained formally in Chapter 47 and 48, DTA aims at consistency between a dynamic travel demand model, defining the choice of activity-travel plans, and a dynamic network supply model, capturing spatiotemporal network flows and congestion evolution.

[1] `http://people.kth.se/~gunnarfl/cadyts.html`

How to cite this book chapter:

Nagel, K, Zilske, M and Flötteröd, G. 2016. CaDyTS: Calibration of Dynamic Traffic Simulations. In: Horni, A, Nagel, K and Axhausen, K W. (eds.) *The Multi-Agent Transport Simulation MATSim*, Pp. 213–216. London: Ubiquity Press. DOI: http://dx.doi.org/10.5334/baw.32.

Cadyts adjusts the plan choice probabilities of all agents, resulting in simulated network conditions that are consistent with measured real-world data while maintaining the behavioral plausibility of the underlying travel demand model. Within MATSim, plan choice probabilities adjustment is realized by adjusting plan scores, as explained in the next section.

32.3 Adjusting Plans Utility

When traffic counts are the empirical source, plan-specific score corrections are composed of link- and time-additive terms $\Delta S_a(k)$ for each link a and each calibration time step k (often one hour). When congestion is light and traffic counts are independently and normally distributed, these correction terms become

$$\Delta S_a(k) = \frac{y_a(k) - q_a(k)}{\sigma_a^2(k)} \tag{32.1}$$

where $y_a(k)$ is the real-world measurement on link a in time step k, $q_a(k)$ is its simulated counterpart and $\sigma_a^2(k)$ is (an estimate of) the real measurement variance (assuming its expected value coincides with the prediction $q_a(k)$ of a perfectly calibrated simulator).

The score correction of an agent's given activity-travel plan is calculated as the sum of all $\Delta S_a(k)$, given that following that plan implies entering link a within time step k. With this, the *a posteriori* choice probability of agent n's plan i given the count data $\mathbf{y} = \{y_a(k)\}$ becomes

$$P_n(i \mid \mathbf{y}) \sim \exp\left(S_n(i) + \sum_{ak \in i} \Delta S_a(k)\right) = \exp\left(S_n(i) + \sum_{ak \in i} \frac{y_a(k) - q_a(k)}{\sigma_a^2(k)}\right) \tag{32.2}$$

where $S_n(i)$ is the *a priori* score of plan i of agent n, as calculated for example with Equation (3.1) and $ak \in i$ reads: "following plan i implies entering link a in time step k".

Intuitively, if the simulated value $q_a(k)$ is smaller than the real measurement $y_a(k)$, then a score increase, and thus a choice probability increase, results. The variance $\sigma_a^2(k)$ denotes the level of trust in that specific measurement—a large $\sigma_a^2(k)$ implies a low trust level, taking effect through a large denominator in the corresponding score correction addend.

Flötteröd et al. (2011) is the key methodological reference on Cadyts. It derives the calibration approach from a Bayesian argument and provides more technical information, such as a more general correction of the utility function than in Equation (32.1) that also applies when congestion is present. A lighter presentation is Flötteröd et al. (2011a), where the formulas above are discussed in somewhat greater detail.

32.4 Hooking Cadyts into MATSim

Hooking Cadyts into MATSim is based on the following operations:

1. Initialization: When the calibration is started, it requires all available traffic counts and some further parameters. For this, the Cadyts function `void addMeasurement(...)` is called once for every measurement before the simulation starts. It registers a certain measurement type, which has been observed on a specific link.
2. Iterations: The calibration is run jointly with the simulation until (calibrated) stationary conditions are reached.
    a. Demand simulation: The calibration needs an access point in the simulation to affect the plan choice. There are various ways to realize this, depending on the simulator. Before a MATSim agent chooses a plan, it asks the calibration through the Cadyts function

```
double calcLinearPlanEffect(cadyts.demand.Plan<L> plan)
```

for all of this plans' score offsets. The agent then chooses a plan based on accordingly modified scores.
All selected plans of an iteration are registered to Cadyts by

```
void addToDemand(cadyts.demand.Plan<L> plan) .
```

Since Cadyts has its own plans format, MATSim plans need to be converted to that format beforehand.

b. Supply simulation: The calibration must observe simulated network conditions to evaluate their deviation from real traffic counts. For this, the Cadyts function

```
void afterNetworkLoading(SimResults<L> simResults)
```

is called once after each network loading. It passes a container object to the calibration that provides information about the most recent network loading results, particularly on simulated flows at measurement locations.

32.5 Applications

Cadyts has been successfully applied in studies like Ziemke et al. (2015); Zilske and Nagel (2015); Flötteröd et al. (2011a). Zürich scenario results illustrate its efficiency, as shown in Flötteröd et al. (2011b, Slide 8), reproduced in Figure 32.1.

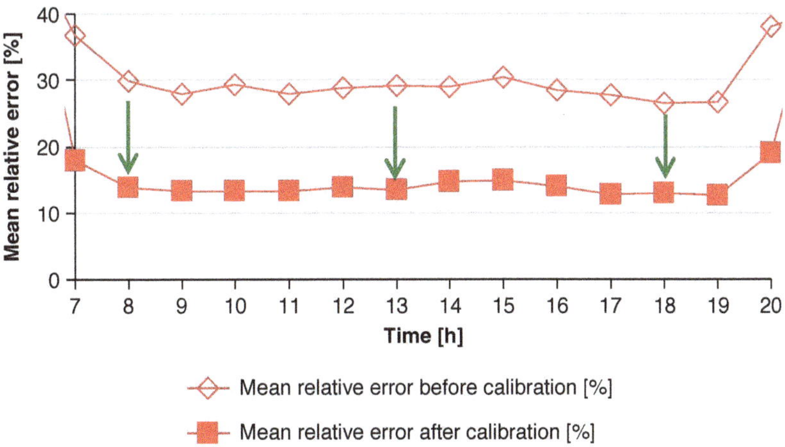

Figure 32.1: Zürich case study results: mean relative error in link volumes.
Source: Flötteröd et al. (2011b, Slide 8)

SUBPART NINE

Visualizers

CHAPTER 33

Senozon Via

Marcel Rieser

33.1 Basic Information

Entry point to documentation:

`http://matsim.org/extensions` → Via

Invoking the module:

Standalone GUI, double-clickable jar file

Selected publications:

`http://via.senozon.com` → Download → manual

33.2 Introduction

Via is an application to visualize and analyze MATSim simulation results. Unlike MATSim, *Via* is not open source; it is developed as a proprietary commercial software by Senozon AG, an ETH Spin-off company founded by two former PhD students involved in MATSim development. Shortly after the company was founded, first (potential) client presentations began; the lack of visual material was an obvious handicap. Explaining to customers that all answers to their questions were contained in a huge events file was not satisfactory; pictures or even animations made it much easier for them to understand. Thus, work on a visualization tool started as soon as the company was set up. Initially planed as a purely internal tool, it quickly became clear that a graphical visualization and analysis tool would also benefit other users of MATSim. After a beta test phase with selected MATSim users in Spring 2011, the first version of *Via* was released in July 2011. Since then, the list of features provided by the application has grown continuously.

Via is written in Java and thus works on any platform able to run MATSim. For easier deployment, the application comes as double-clickable, native executable on Windows and Mac OS X,

How to cite this book chapter:

Rieser, M. 2016. Senozon Via. In: Horni, A, Nagel, K and Axhausen, K W. (eds.) *The Multi-Agent Transport Simulation MATSim*, Pp. 219–224. London: Ubiquity Press. DOI: http://dx.doi.org/10.5334/baw.33.

partially hiding its Java nature. A limited version is available for free and can be downloaded from the product website (senozon AG, 2015). Different licenses are available for commercial usage or for research or educational purposes to serve different user group needs.

Via includes some general functionality that most people will use in the core application, like visualizing networks, facilities, vehicles and activities. Optionally available plugins provide additional features often relevant only to specialized user groups. This includes functionality related to public transport, comparison with car counts, using web maps like Google Maps or OSM as background, aggregation analyses, or movie recording.

Via allows customization of its window. The following descriptions refer to elements as they are placed in the default layout. The default configuration can be re-created by choosing `Reset Window State` from the Window menu in *Via*.

33.3 Simple Usage

Via differentiates between data sets, and how the data is visualized. It does so by managing data sources (typically MATSim files like `network.xml` or `events.xml`), and layers (e.g., displaying the network, vehicles, activity locations). A layer can use more than one data source for its visualization purposes (e.g., a network and some data from the events), and a data source can be used by multiple layers (e.g., events can be used by many different layers to visualize different things like vehicles, activities, link volumes, etc).

By default, *Via*'s window looks similar to the one shown in Figure 33.1. To add a file as a data source, the file can either be drag-and-dropped onto the layers list left of the black visualization area, or by choosing `Add Data...` from the `File` menu. To add a layer, the little plus icon in the lower left of the window can be pressed, or by choosing `Add Layer...` from the `File` menu. To get started, it's usually best to add a network and (small) events file from MATSim to Via, and create a `Network` layer and a `Vehicles` layer.

Elements shown in the visualization area like the network or vehicles can be queried. Queries are usually provided by layers, made available with buttons with question-mark icons. Clicking

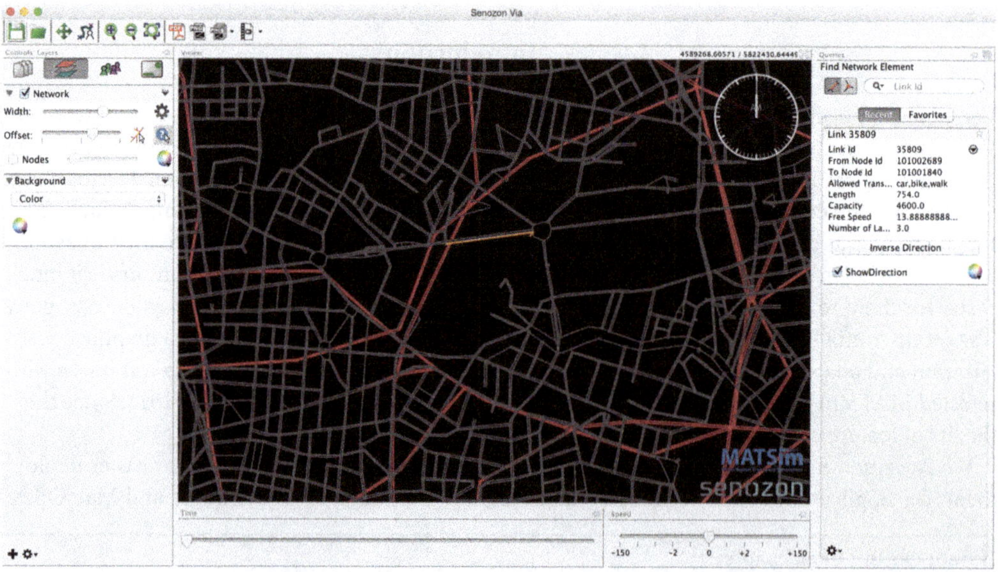

Figure 33.1: Via's window with default layout and a network query being shown.

such an icon activates the corresponding query mode, and any subsequent click on the visualization area will run the query. Query results are shown on the right side of the visualization area. Figure 33.1 shows a network query for links. One query is special, globally available, and not linked to a layer: querying an agent plan. This query is available from the toolbar, next to the icon, to shift the visualization view around.

Once a query has been made, *Via* often allows another query based on the current query results. By right-clicking in the visualization area, a pop up menu appears with more options regarding the last query, as well as additional possible queries. Examples are: `Select Link Analysis` given a link, `Select Facility Analysis` given a facility, `List Transit Lines` that use a given link, or `List Passengers` if a transit vehicles was queried in the first place.

33.4 Use Cases and Examples

33.4.1 *Agent Visualization*

The animated visualization of agents moving around in the modeled area was one of the main features in *Via*'s original development. To do this, *Via* needs only the `network.xml` and `events.xml` files from a MATSim run as data sources. For the visualization, a `Network` layer, `Vehicles` layer and activities layer must be created. With this setup, vehicles will move around in the visualization area as time progresses, and agents performing activities will be represented as colored dots.

The visualization can be further customized; with the addition of a `population.xml` file, more detailed activity coordinates can be loaded to obtain a better distribution of activity locations (MATSim's events file does not contain coordinates for activities, only the assigned link ID. So by default, all activities taking place on a link are first shown at the location of the link's to-node). Vehicles and groups of vehicles can also be styled differently; it is possible to visualize transit vehicles with a square shape with colors representing the occupancy of the vehicles, pedestrians or cyclists in a multi-modal simulation can be shown as circles and private cars can be displayed with a triangular shape with colors representing their absolute speed or their speed relative to the allowed maximum speed on their current link (see Figure 33.2). As mentioned above, arbitrary groups of vehicles can be styled differently, which is useful to highlight special agents, e.g., when simulating a fleet of electric vehicles, a car sharing fleet, or agents simulated with special routing guidance.

It is also possible to load arbitrary attributes for agents and then use those attributes for visualization purposes, e.g., having different colors for vehicles driven by agents who are employees, have a high income or are within a certain age range.

33.4.2 *Facility Analysis*

Activity facilities allow for very detailed modeling in MATSim, especially considering the functionality provided by the destination innovation module (Chapter 27). *Via* provides several unique ways to analyze the mobility effects to and from facilities.

For each facility, a detailed analysis can be performed showing the number of agents arriving at, departing from, or staying at a facility over the simulated time. The numbers can be differentiated by the type of activity the agents perform at the facility, by the transport mode they arrive or depart with, or by other arbitrary agent attributes loaded by users.

An alternative analysis is similar to the—for transport planners—well known `Select Link Analysis`, but designed for facilities: the `Select Facility Analysis`. This analysis shows the combined link loads produced by agents arriving or departing at a facility, showing the starting location for agents visiting a specific facility and what routes they use. Figure 33.3 shows such an example.

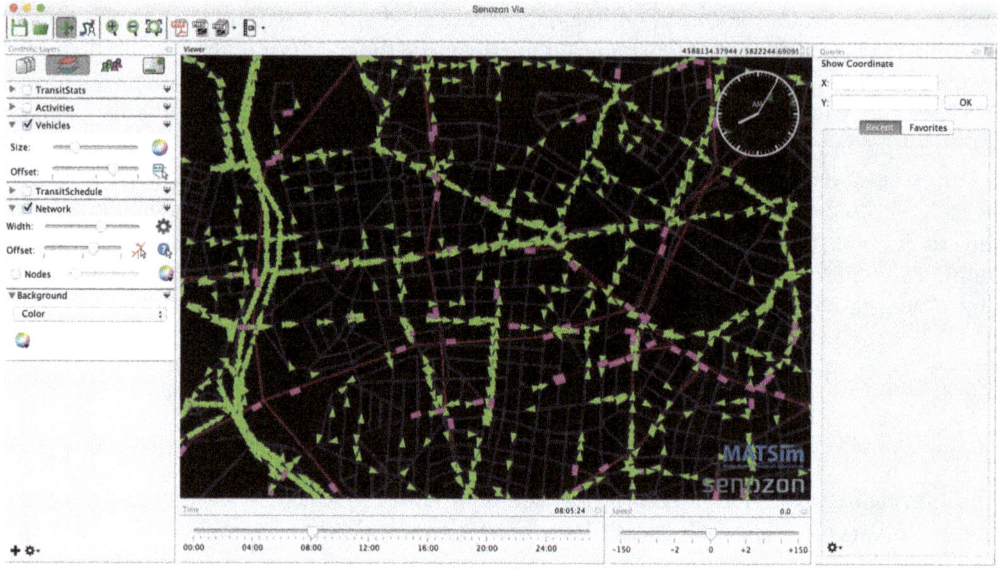

Figure 33.2: Vehicles in Via: Green triangular symbols represent private cars, pink rectangular symbols public transport vehicles.

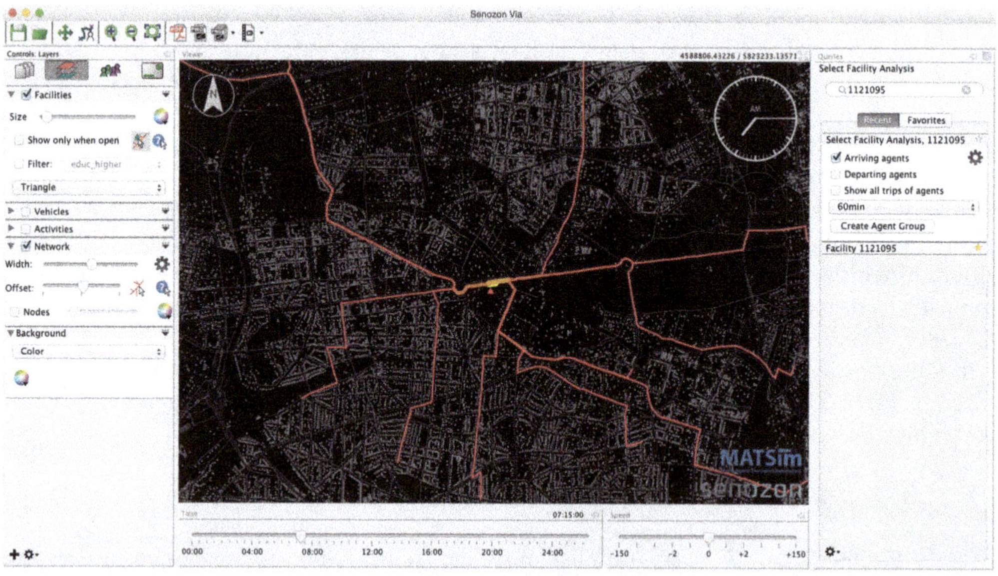

Figure 33.3: Select facility analysis: Links used to travel to and from a facility are highlighted.

33.4.3 Public Transport Analysis

The public transport plugin provides many different functions for analyzing public transport simulations. It starts with providing the specified vehicle types as agent attributes, so the vehicles can be differently visualized, based on the vehicle type they represent. Also, the absolute or relative occupancy of a transit vehicle is provided as attribute, allowing transit vehicles to be visualized accordingly. For stop locations, the number of passengers waiting for a bus or train can be plotted over the time of day, and the occupancy along a bus or train route can be visualized.

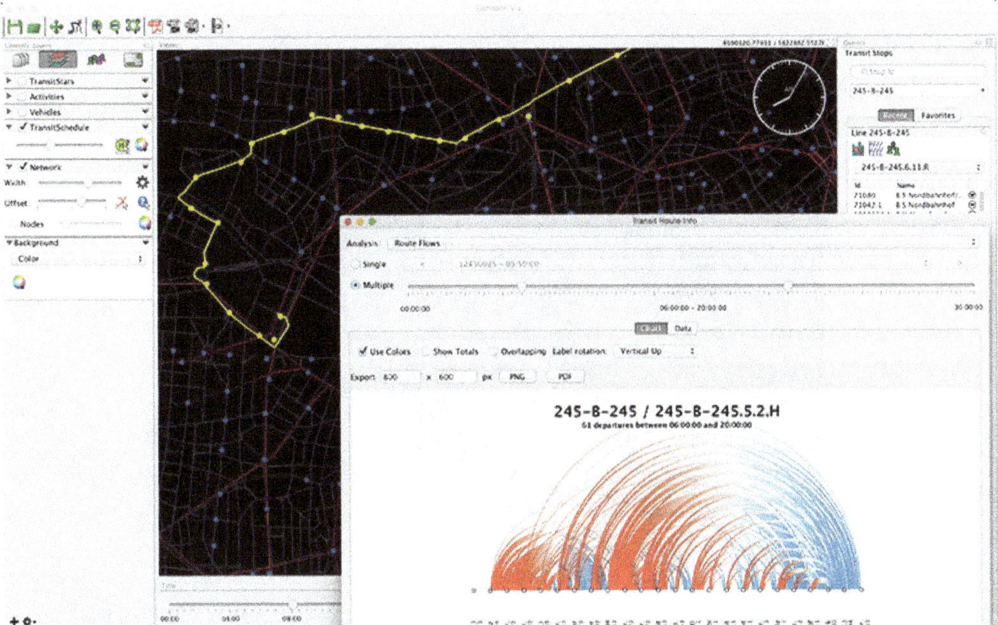

Figure 33.4: Passenger flows on a transit line.

A special, but very useful visualization is the `Route Flow` analysis. This shows, in a visually appealing way, the number of passengers traveling between two stops along a route—for all possible stop combinations. Figure 33.4 shows an example of such a route flow with the route of the transit line shown in the background. It is clear that the demand on the bus route is more or less split in two; a first travel demand up to about the first third of the route, and then it again collects passengers all wishing to go to one of the last stops along the route. This could indicate that it might make sense to split the line in two.

33.4.4 Scenario Comparisons

A typical use of MATSim is simulating a base case and then one or more case studies. Comparing scenarios then becomes an important step in the analysis of the different case studies. *Via* allows comparison of the link volumes of two scenarios visually by coloring the network with the absolute or relative difference of the link volumes between two models. In the future, other differences like average speeds will supported too. The differences are time-dependent, aggregated over time intervals as small as 15 minutes.

33.4.5 Aggregating Data

While MATSim requires and produces a lot of disaggregated data, it is still often necessary to aggregate data to make statements or predictions about a simulated scenario. *Via* provides a powerful mechanism to easily build arbitrary aggregations of available data. Such data can be either point data (like activity locations, trip start locations, GPS points or any other spatial point data) or origin-destination data (like trips with a start and end location, or the relation of an activity location to the home location of the agent performing the activity). While *Via* provides: activity locations, trip start and trip end locations, facility locations (automatically) as point data sources

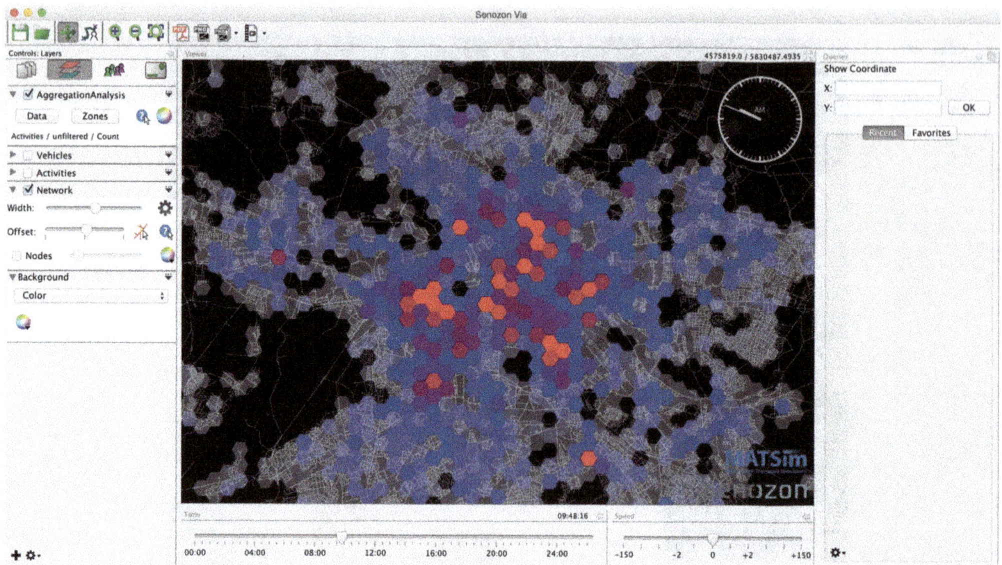

Figure 33.5: Aggregation analysis: Number of performed activities during the whole day.

for aggregation, and the trips performed by agents as O-D data sources, any tabular custom data with coordinate attributes can also be used for this.

Data can be aggregated into a rectangular or hexagonal grid, where the cell-size can be specified by the user, or into arbitrary zones provided as ESRI (Environmental Systems Research Institute) shape file by the user. The data points can be filtered by any of the available attributes, and the aggregation can either just count the data points in each region, or build the sum, the minimum or maximum or average of a data points attribute.

With the activity locations provided by an `Activities Layer`, the following (and more) aggregations are possible:

- show number of performed activities per region,
- show number of performed work activities per region,
- show number of work activities starting after 10 am per region, and
- show average duration of work activities starting after 10 am per region.

Similarly, with trip data provided by a `Vehicles` layer, the following exemplary aggregations are possible:

- show number of trip starts per region,
- show number of trip starts with mode "car" per region,
- show percentage share of trips starting with mode "car" in a region, compared to all trips starting in that region, and
- show average duration of trips starting with mode "pt" in a region after 11 am.

By using custom data tables, e.g., containing more information about trips, i.e., the 'from and to' activity types they connect, the number of line switches if it is a public transport trip (this requires the aggregation of MATSim's legs to trips for analysis purposes), many more complex analyses are just a few clicks away in *Via*, like showing the average duration of car-trips starting between 6 am and 8 am, going from "home" to "work".

CHAPTER 34

OTFVis: MATSim's Open-Source Visualizer

David Strippgen

34.1 Basic Information

Entry point to documentation:

`http://matsim.org/extensions` → otfvis

Invoking the module:

`http://matsim.org/javadoc` → otfvis → `OTFVis` class, `RunOTFVis` class

Selected publications:

Strippgen (2009)

34.2 Introduction

For most MATSim users, Via's (Chapter 33) free branch will be a good solution for their visualization needs. However, if project demand reaches beyond the given (and fixed) abilities of the Via free version, there is another—though not as stylish—option for MATSim output visualization, the OTFVis.

The short term for "On the Fly Visualizer", OTFVis was designed to support actual visualization of live simulation runs with MATSim. Therefore, one purpose of the OTFVis is the debugging of MATSim (input) data. Nonetheless, playing prerecorded movie (MVI (An OTFVis Movie File, not to be confused with the "Musical Video Interactive" file usually abbreviated mvi)) files created from MATSim events is another way to use OTFVis. Generally speaking, OTFVis serves as an open-source counterpart to the possibilities Via gives the MATSim community. The OTFVis is written in Java and available as source code to extend for different MATSim projects' special needs. Hence, it is possible and desirable to actually extend the OTFVis functionality, incorporating the user's own data sets and visualizations.

How to cite this book chapter:

Strippgen, D. 2016. OTFVis: MATSim's Open-Source Visualizer. In: Horni, A, Nagel, K and Axhausen, K W. (eds.) *The Multi-Agent Transport Simulation MATSim*, Pp. 225–234. London: Ubiquity Press. DOI: http://dx.doi.org/10.5334/baw.34.

34.3 Using OTFVis

In this chapter, we show how to achieve simple things, like creating MVI-files from MATSim run events, how to play these MVI-files and how to use a MATSim config file to view/play an actual simulation with all data (e.g., agents' plans) attached. With the latter, it is also possible to examine the data "on the fly" by sending queries into the mobsim and visualizing the results.

34.3.1 MVI Files

MVI files can be generated through the OTFVis. Under the hood, these files consist of a few binary dumps of OTFVis data packed into a zip-file. This binary data is created by Java's own serialization capabilities. Unfortunately, this setup is not very change-resistant, making it advisable to regard MVI files as temporary cached versions of your event files. These MVI files can be re-created at any time from the event files. Still, as converting one into the other is a time-prone process, the MVI files are a handy tool for temporary storage and fast loading of your visualizations.

34.3.2 Starting OTFVis

OTFVis is a MATSim contribution. There is no actual stable release of the OTFVis package; so, to acquire a working version, a "nightly build" needs to be downloaded as shown in Section 44.3.6. There, one finds the latest `otfvis-version-SNAPSHOT-build.zip` file available for download. Unzip it to the place where the `matsim.jar` already resides; do not forget to extract the `libs`-directories found in the respective zip files.

OTFVis demands substantial RAM (depending on your simulation size/MVI file); to successfully launch the visualizer, a command line like

```
java -Xmx500m -cp MATSim-XXX.jar:otfvis/otfvis-XXX.jar
   org.MATSim.contrib.otfvis.OTFVis
```

(exchange ";" with ":" depending on the used OS (Operating System)) is a good starting point. This will open the dialog window shown in Figure 34.1, asking for one choice from four possible usages of OTFVis; these will be explained in the next section.

34.3.3 Use Cases of OTFVis

With the open dialog appearing after starting the vanilla `OTVFis` class, the following options appear, as shown in Figure 34.1:

1. opening a prerecorded MVI file,
2. opening a network file (for inspection),
3. opening a live run of a MATSim config file (rather memory intensive) or
4. converting an event file (plus a given network file) to a movie (MVI) file.

Each tab stands for an individual usage. To start a visualization, one chooses the appropriate tab, fills in the necessary data and finally proceeds by pressing the `Load...` button located in the bottom left corner of the window.

The next sections provide an overview of different ways to use OTFVis.

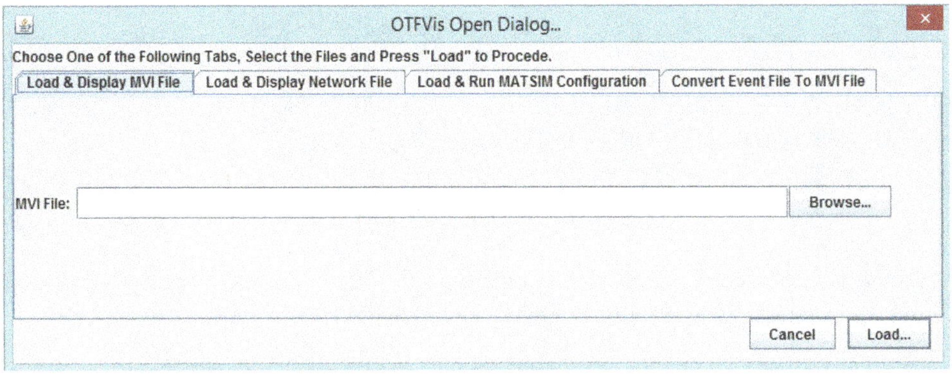

Figure 34.1: OTFVis Start Dialog.

34.3.3.1 Converting Event Files

Though the first option tab is the most used choice for OTFVis, the fourth, and last, option tab is a good starting point for exploring the visualizer; after having successfully run a MATSim simulation, there will typically be some event files at one's disposal. With any of these event files and a given (matching) network file, a MVI file can be created. Four items: event, network and movie file names, as well as a time period, must be specified for this tab to execute. The last parameter is a time period, after which a new sample of the mobsim's state is taken. This MVI-generation process might be time consuming. For smaller projects, it might be an option to display the outcome in the visualizer right away (by checking the box `Open mvi afterwards`). If the choice is to just convert the events to a MVI file, this can be opened with the first option tab of the visualizer's start dialog at any time.

From the shell, this process can be started by giving the event file, network file and, optionally, the conversion period as input parameters.

34.3.3.2 Network File Loading

The second option tab offers the opportunity to examine a network file (e.g., for errors). It will show a rendering of the given network and also, if so chosen in the preferences, the associated network link IDs for each link. This option might be helpful for debugging a freshly converted network, or inspecting specific regions and connections. Loading and interacting with a network file should be very fast.

The network file can also be given as the sole parameter to OTFVis with the shell command.

34.3.3.3 Running a MATSim Configuration

The third, and most advanced, option for running OTFVis is an actual, live running mobsim, visualized in real time (actually much faster than real time; who has all day to watch tiny cars drive around?). This option includes the possibility of exploring the data set and issuing queries into the executing mobsim. These queries can display an agent's day plan, show all links driven by agent's crossing a particular link of interest, search for a particular link or node by ID, or answer any user-defined queries. We will see later in this chapter how to program a user's own queries, but for the rest of this section we will detail OTFVis "offline" behavior.

It is also feasible to input the config file as a single parameter to OTFVis by starting it from the shell. OTFVis will make an educated guess whether the input is a config or a network file.

34.3.3.4 Loading & Displaying an MVI File

If the first and default option tab is chosen, a MVI file is selected and shown as detailed in next section 34.3.4. This is the most common use case for OTFVis; the same results can be achieved by starting OTFVis from the shell with an MVI file as an argument.

34.3.4 Viewing an MVI File

An example is illustrated in Figure 34.2. On the top left of the application, one finds buttons for controlling the file playback. A short summary of the functionality is given in Table 34.1.

This buttonbar is followed by a text field where the desired time can be written for an instant jump. In an MVI file, one can jump forward and backward in time, whereas in the live simulation case, going back in time is omitted.

Another way of iterating through the animation is to grab the time slider at the bottom of the application and drag it. Opening and closing bracket symbols are located on the left side of the slider; by clicking them, one can set the start, or end, time of a time loop to the actual time step given, making it possible to restrict playback to a certain space of time.

34.3.5 General Interaction with the Main Screen

Regardless which option for loading data was chosen, interaction with the main display area is the same.

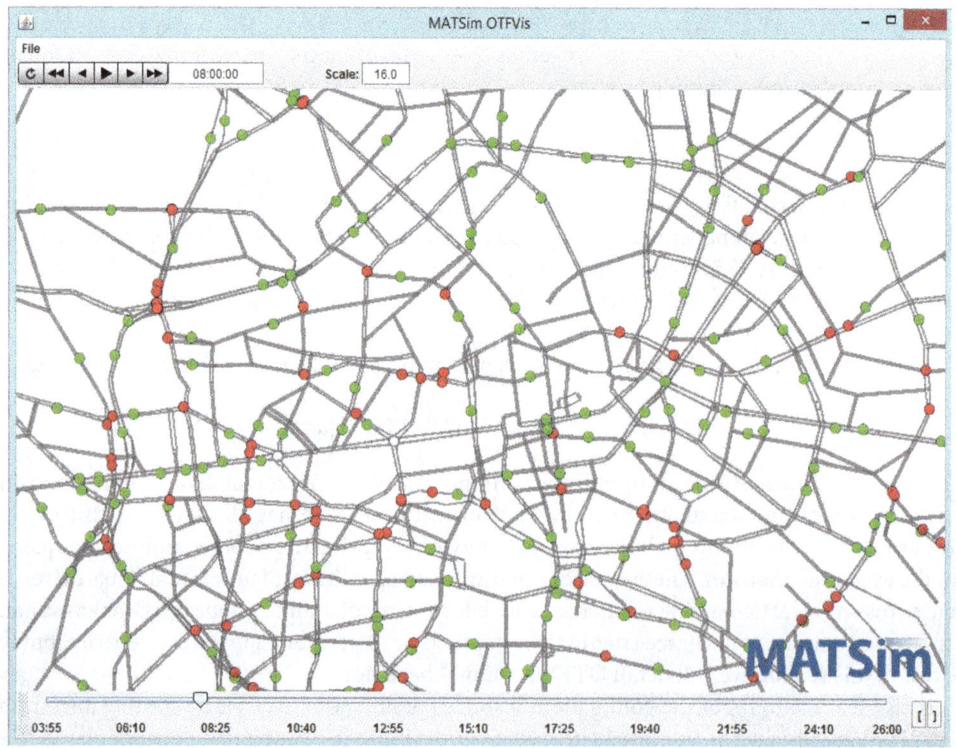

Figure 34.2: Displaying an MVI file.

Icon	Function
	Reset - set time to the start time
	Large step back
	Small step back
	Play
	Pause
	Small step forward
	Large step forward

Table 34.1: OTFVis Buttonbar.

Right button drag: Extend a rectangle for zooming into the view. Releasing the button will execute a zoom, so the chosen rectangle will best fit the screen.
Middle-Mouse-drag: Pan (translate) the screen.
Right-Mouse-Click: Show a context menu (for now only with the option to save the view settings).

34.3.6 User Interaction in the Live Mobsim

When started as a live simulation, OTFVis will look different than Figure 34.3. First, the controls of the simulation's view flow are a restricted subset of those used in MVI playback. There is no way to reset or rewind the simulation. One can still take small or large steps forward. A new option

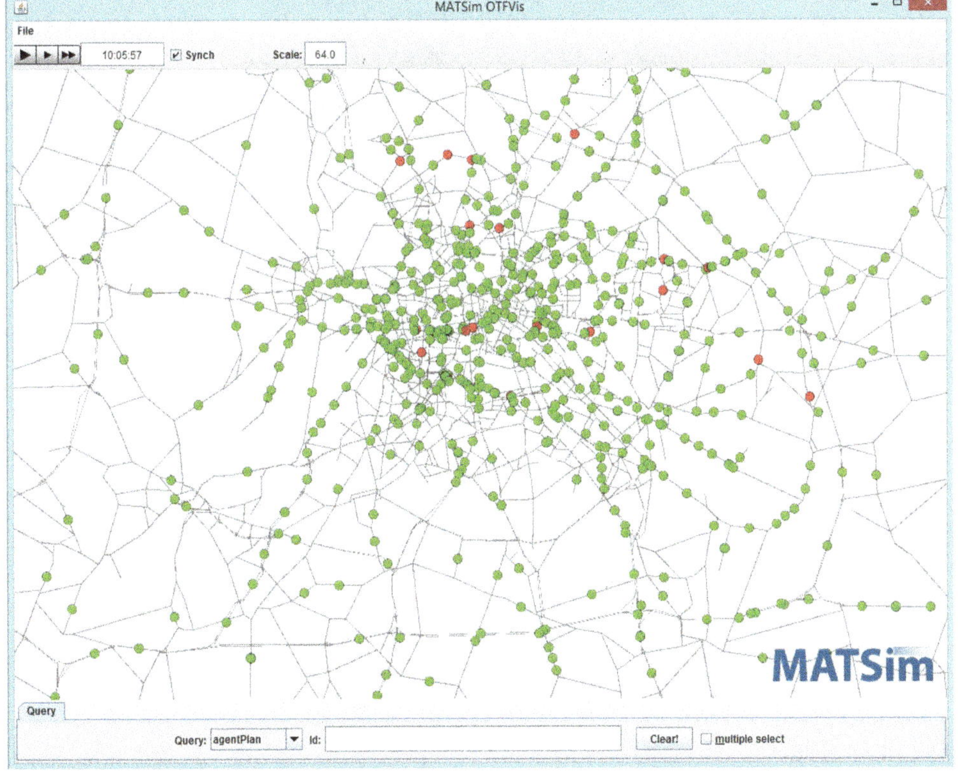

Figure 34.3: Live mode.

is given by the synch checkbox, which determines whether the mobsim will stop for each frame the OTFVis renders, or run independently. Usually the un-synched version will proceed faster, as the OTFVis output is restricted to a default of about 30 frames/updates per second and a small mobsim's simulation speed will be a magnitude higher. The time-consuming generation of visualization data will also only be necessary for a small fraction of the simulation. Length of OTFVis pauses between frames can be configured in the preferences dialog.

Apart from the reduced control set, there is another UI element new to this OTFVis option. At the bottom of the screen, the scrubbar/time line element is replaced by a "query" bar. It is possible to code "queries" into the mobsim, answering questions about its inner state. As the simulation is actually happening, all information necessary to run it is available for output. This is a clear superset of information available in the event files and in the MVI files. This rich information infrastructure can be queried and visualized in many ways. In the next session, a query example is given.

34.3.7 Running a Query in OTFVis Real Time Data

From the dropdown box, one can choose the different query types. Often, additional input is necessary, either in the text field next to it or, more often, by clicking into the network. To give an example with agent query selected, a click onto any agent's symbol will give a visualization of this particular agent's day plan. This is shown in Figure 34.4. There are other pre-defined queries. These queries are rather project-oriented, so defining own queries will probably be necessary to make best use of this option. In the second part of this chapter, we will look into defining own queries.

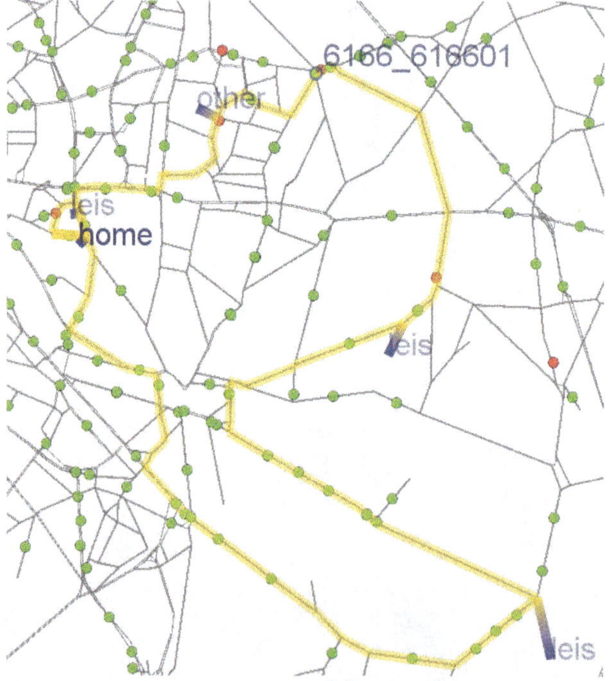

Figure 34.4: Queries.

34.4 Extending OTFVis

Because it is open source, the OTFVis is a good starting point for customizing mobsim run visualizations. OTFVis has been written in Java, but depends heavily on the JOGL (Java OpenGL) Java library. JOGL is a very thin layer within the OS hardware driver, meaning it will have OS-specific, native dependencies. These should be attended to by the maven-dependency management, but should still be kept in mind when developing for OTFVis. The displaying parts of OTFVis are based on OpenGL (Open Graphics Library). Therefore, it will be necessary to understand OpenGL to create new ways of displaying data. In the following sections, we examine how data is computed inside the OTFVis and how this can be extended.

34.4.1 Design Principles of OTFVis

The overall goal of OTFVis design was to have an easy-to-extend, fast visualizer capable of handling huge amounts of data. The specific design goals for the visualizer were:

- abstract data source (data collection) from data display (visualization),
- easy extension with own data types,
- capability for local simulation run on desktop computer,
- reduction of sent data to a minimum,
- visualization that connects to running simulation (on-the-fly),
- minimally-invasive format for existing MATSim code,
- enough speed for large scenarios,
- visualization that reads from post-mortem dump (MVI file), and
- use of hardware support for drawing.

MATSim runs can easily engage millions of agents traveling a network. To make a visualization of these large data sets feasible, two measures have been taken. A quad tree structure was implemented to ensure that only the smallest set of data necessary to display the visible sector of the network is transferred. The quad tree is a simple data structure to aggregate spatial data and retrieve parts of it efficiently for real time visualization. Apart from data structures, hardware is also used to speed up displaying the simulation. OpenGL is a platform-independent API for interfacing graphics hardware, specifically the 3D acceleration chips implemented in every contemporary computer. With the aid of 3D graphics hardware, millions of agents can be displayed in real time. Other measures were taken to segregate data extraction from data visualization, like the reader/writer pairs presented in the next section.

34.4.2 Readers and Writers

OTFVis was designed to be minimally dependent on the mobsim used. Data formats applied within the mobsim should be abstracted from data used in OTFVis, meaning that any data passed to the visualizer will have to run through some stages of abstraction.

The first stage is a writer-reader pair, responsible for transferring a certain set of data to the OTFVis. The writer will understand the data format of the hosting mobsim and convert it to simple data types, like float or string values. A set of these writers, all using a joint byte buffer to aggregate the data, will be called after each mobsim step to accumulate data. This array of bytes is then sent to the visualizer, which, in the original design, could be run anywhere in your network.

For each writer, there has to be a sibling-reader class, responsible for reading back extracted data from the byte buffer. It is crucial to ensure that these pairs work synchronously. Most

Writer/Reader-pairs are implemented in the same class, since having the source-code at the same place reduces errors in the synchronization.

Apparently, it can be necessary, or at least useful, to have different ways of visualizing data on the OTFVis front-end. Thus, actual readers are not responsible for the drawing of a certain data set. A third kind of class is responsible for that, the drawer classes.

34.4.3 Visualization of the Data

The reader objects in the quad tree will generate separate drawer objects for displaying "their" information and add these to another data structure, called `SceneGraph`, which is responsible for the actual drawing onto an OpenGL canvas. Displaying data in an interactive application will make re-draws of the display necessary for a variety of reasons: displaying menus, animations, zooming, panning and other user interactions. Not all of these changes introduce new data from the mobsim. Zooming into the network will not imply reading data from the mobsim; panning the view most certainly will. When no new data is needed, the scene graph is capable to handle all operations, no reader/writer class will be accessed and displaying is solely done with existing drawers. On the other hand, if new data is demanded, the scene graph will be "invalidated" (a term lent from the OpenGL community); thus, the graph will be dismissed and all relevant readers will be asked for new drawer objects representing the actual view. The scene graph is mainly a list of drawer objects; as an extra structuring unit, these drawers can be sent into different layers, to render them more effective.

34.4.4 Layers

To make sure that only data actually necessary for drawing the particular area visible in the viewport is sent, writers should minimize the data packets, so the quad tree can make a spatial data reduction. This seems somewhat in opposition to OpenGL or any graphics API). The API wants maximal data to be accumulated, to optimize output through the underlying hardware graphics pipeline. Think of an assembly line vs. a handcrafted item; whenever the flow of data is interrupted, the assembly line stalls and graphics performance derogates. To ease this issue, "layers" have been introduced to OTFVis. Any drawer (responsible for a bit of information) can be assigned to a layer and these layers will ultimately be summoned to draw the screen's content. It is up to the layer to optimize the execution of the drawers when necessary. For example, a network layer might store all network info from the drawer in one array, or display a list to optimize drawing of the network; (often in OpenGL, it is advisable to rather let the hardware decide what to draw. It might be faster to have all complete data residing in graphics hardware memory, rather than to transfer the reduced information set every frame). There are three layers predefined in OTFVis. The `networkLayer` contains the static street net, the `agentLayer` the actual dynamic agents and a third layer, the `miscellaneousLayer`, contains additional data.

34.4.5 Patching the Connections

In total, there are four basic elements involved in the visualization: writers, readers, drawers and layers. An additional class configures how the first two work together: `OTFConnectionManager`.

This class maps several routes for the information coming from the mobsim, building a chain of responsibility. Each data item starts at a link in our network. An `OTFDataWriter` object is responsible for extracting the desired data from the link and writing it into a `ByteBuffer`. Complementing this, an associated `OTFDataReader` is needed to retrieve data from the buffer. This item will also be responsible for adding a drawable item derived from the

class `OTFGLAbstractDrawable` to the scene graph representing the actual screen content. The connection between these items is made by adding entries into the `OTFConnectionManager`, with calls to `OTFConnectionManager.connectLinkToWriter(OTFDataWriter)` and `OTFConnectionManager.connectLinkToWriter(OTFDataWriter, OTFDataReader)`, respectively.
Example (from the `OTFClientLive.java`):

```
conMan.connectLinkToWriter(OTFLinkAgentsHandler.Writer.class);
conMan.connectWriterToReader(OTFLinkAgentsHandler.Writer.class,

  OTFLinkAgentsHandler.class);
```

34.4.6 *Sending the Data*

The class `OTFLinkAgentsHandler` should give a good example of extracting, sending, receiving and displaying data in the OTFVis context. The method `invalidate` is called whenever the actual scene graph has been dismissed and needs to be rebuilt. In this case, a valid representation of the object's state should be added to the new scene graph. This also means that for drawing the actual scene, no additional reading will take place, unless there is a change in the visible data: then, this update is triggered.

34.4.7 *Performance Considerations*

When implementing new ways to visualize data, the following guidelines should be kept in mind.

If the data is spatially distributed over the whole network and is updated frequently, an `OTFDataWriter/Reader` pair should be considered. It will reduce data updating to times when the data is actually visible, not creating, transporting or drawing the data otherwise. If a fraction of the data needs to be transferred only once—because it is static over the time of the simulation—it can be sent with the `writeConstData()` method; otherwise using `writeDynData()` is advised. If the data is sparse and little information is transmitted or it has no discernible spatial cohesion, it might be simpler to just add it to the server quad tree as additional data with a call to `OTFServerQuadTree.addAdditionalElement()`.

34.4.8 *Sending Live Data*

Flow of data within OTFVis is almost always a one way affair, except for one important issue: sending queries into the simulation. In case of a live simulation run, visualized with help from the `OTFVisLiveClient` class, queries can be sent into the simulation. Again, the methods involved in this process are threefold; queries will be realized through an object derived from the abstract class `AbstractQuery`. Such an object initiates several methods that will be used as callback over the lifetime of the query.

First, a new query is sent to the server and the method `installQuery()` is called. In this method, all relevant parts (network, population, events) of the simulation run can be accessed and data can be collected. The visualizer framework will later repeatedly call the `result()` method, to retrieve an `OTFQueryResult` object. This has to implement a `draw()` method, to visualize the results in the given screen context. If the result indicates `isAlive()`, the `query()` method of the `AbstractQuery`-derived object will be called with each frame; otherwise, only once.

SUBPART TEN

Analysis

CHAPTER 35

Accessibility

Dominik Ziemke

35.1 Basic Information

Entry point to documentation:

`http://matsim.org/extensions` → accessibility

Invoking the module:

`http://matsim.org/javadoc` → accessibility → `RunAccessibilityExample` class

Selected publications:

Nicolai and Nagel (2014); Joubert et al. (2015)

In transport science and planning, the term accessibility can refer to at least three different concepts. First, accessibility may be used to describe how well a certain transport infrastructure component can be utilized by travelers, particularly those with handicaps (Faura, 2012). In this sense, *accessibility guidelines* tell engineers and planners how to design transport infrastructure elements, such as public transport facilities, to make them accessible, i.e., useable for all travelers. Second, accessibility may be used to describe how easy/convenient the approach to a given land-use facility is. There are, for instance, studies (Fujiyama, 2004) to improve the accessibility of shopping centers by redesigning access roads and their connection to major roads. Finally, the term accessibility can be used in a more global way, to describe availability and spatial distribution of activity facilities within a given area, e.g., a metropolitan region and the ease with which these facilities can be reached from other locations in the area. MATSim's accessibility extension focuses on all these aspects; the discussion in this chapter draws on Nicolai and Nagel (2014).

How to cite this book chapter:

Ziemke, D. 2016. Accessibility. In: Horni, A, Nagel, K and Axhausen, K W. (eds.) *The Multi-Agent Transport Simulation MATSim*, Pp. 237–246. London: Ubiquity Press. DOI: http://dx.doi.org/10.5334/baw.35.

35.2 Introduction

Improvement in accessibility is often defined as a central goal of proposed transport or infrastructure schemes (Geurs et al., 2012b) and accessibility is usually a precisely-defined, quantitative measure. While Batty (2009) traces the origins of the accessibility concept back to location theory and regional economic planning in the 1920s (when transport planning began in North America; Geurs et al., 2012b), Hansen, with his widely-cited paper (Hansen, 1959), is generally credited with the first real definition of accessibility, defining it as *the potential of opportunities for interaction*. In more detail, Morris et al. (1979) define accessibility as "the ease with which activities may be reached from a given location using a particular transportation system". The concept of accessibility is a potential methodology for the assessment of transport systems, as it is a comprehensive and inclusive way to evaluate how, where and why people move, taking well-known dependencies between transport and land use into account. Hansen (1959) was probably the first to develop a procedure for quantitative consideration of accessibility, discussed in more detail in Section 35.3.

In their widely-cited review, Geurs and van Wee (2004) identify four accessibility components from existing definitions and applied measures:

1. The **land-use** component reflects the number and spatial distribution of opportunities.
2. The **transport** component describes the effort to travel from a given origin to a given destination.
3. The **temporal** component considers the availability of activities at different times of day, e.g., during morning peak hours.
4. The **individual** component addresses various socio-economic groups' different needs and opportunities, e.g., different income groups.

In this review, Geurs and van Wee (2004) list and summarize typical approaches applying the accessibility concept, focusing on the accessibility components discussed above:

1. **Infrastructure-based** measures focus on the (observed or simulated) performance or service level of transport infrastructure, e.g., represented as average travel speed. These measures are typically used in transport planning.
2. **Location-based** measures describe level of accessibility to spatially distributed activities, such as number of jobs within 30 minutes travel time from origin locations. These measures are typically used in urban planning and geographical studies.
3. **Person-based** measures analyze accessibility at the individual level, such as the activities in which an individual can participate at a given time. These measures are grounded in Hägerstrand (1970)'s space-time geography.
4. **Utility-based** measures analyze the economic benefits that people derive from access to spatially distributed activities. These measures have their origin in economic studies.

Geurs and van Wee (2004) intersects these approaches with the four accessibility components identified above, creating a matrix. This matrix illustrates how each of the four accessibility components is represented in the four different accessibility measures. There, each measure focuses on certain weaknesses in those accessibility components outside the focus of a specific measure. Accordingly, Geurs and van Wee (2004) recommend that an accessibility measure include all four discussed accessibility components. The accessibility extension of MATSim, described in the following, could be one way to achieve this goal.

In other recent research, as identified by Geurs et al. (2012b), the accessibility concept is also applied to social exclusion analysis (e.g., by examining the benefit of employment accessibility for disadvantaged populations before and after the implementation of a transport scheme), economic valuation of accessibility effects (e.g., in cost-benefit analyses and studies assessing the impact of changes in public transport accessibility on house prices) and behavior analysis vis-a-vis accessibility measures (e.g., walking behavior dependence on different residential neighborhood accessibility qualities). It has also been used to explore questions of oil dependence, climate change and other concerns (Curtis et al., 2013).

35.3 The Measure of Potential Accessibility

Today, methods to assess accessibility quality are often used in superordinate planning procedures, like regional transport planning, where a central goal is to provide citizens with a certain level of access to various services. For instance, the approach used by Germany's agency responsible for regional planning calculates travel times to major service facilities, like airports or hospitals (Bundesinstitut für Bau-, Stadt- und Raumforschung, accessed March 2015). The results, typically visualized by multi-colored maps, give useful insights into population access to certain services, thus aiding transport infrastructure planning. In this approach, travel times are calculated to a *next* airport, *next* hospital and *next* autobahn access; thus, the implicit assumption is that citizens' needs are fulfilled by one (i.e., the *next*, or closest in terms of travel times) type of facility.

An accessibility measure becomes significant, however, if not just the ability to reach *the nearest* facility serving a particular need is taken into account, but also a *set of multiple reachable* facilities serving the same need; different facilities of the same type may offer varying qualities of a given service. Services may also expand and improve when combined with complementary services provided by another facilities of the same type. For instance, a person planning to take a holiday trip by plane will probably consider several airports in his/her vicinity, instead of just looking at flights offered from the nearest airport. Thus, accessibility to airports should be made dependent on the ability to reach all local airports instead of just the nearest one. Facilities offering medical services may serve as another example. Considering the nearest hospital may be sufficient when looking at simple services like first aid, presumably available at almost *any* hospital. In other cases, however, medical services accessibility should consider several hospitals in the vicinity because they are likely to offer different specialized medical treatment. Consideration of a set of multiple facilities, potentially useful from the perspective of a person at a given location, corresponds to taking into account the **land-use component** of accessibility defined above.

Hansen (1959) considers the whole scope of potential activity facilities, where an accessibility measure *potential accessibility* is defined. Such measures of potential accessibility are specified as the (weighted) sum over the accessibilities of several specific activity facilities (e.g., shopping, leisure etc.) and take the mathematical form

$$A_\ell = g\Big(\sum_j a_j f(c_{\ell j})\Big), \tag{35.1}$$

where j are all possible destinations (opportunities), a_j describes opportunity attractiveness, $c_{\ell j}$ denotes the generalized traveling cost between origin ℓ and destination j, $f(c)$ is an impedance function which (typically) decreases with increasing distance and $g(.)$ denotes an arbitrary, but usually monotonically increasing function. The weight of each opportunity j is thus the product of the destination's attractiveness, a_j, and the ease of getting there, $f(c_{\ell j})$. As seen in its functional form, this type of accessibility measure is related to gravity models used in trip generation models, explaining why this measure is sometimes also referred to as a "gravity type" accessibility indicator (Morris et al., 1979). The (quantitative) accessibility measure used in the MATSim accessibility extension is expressed in this mathematical form and may thus be seen as a *potential accessibility* measure.

It is important to note that the above-defined measure quantifies how accessible a given location ℓ is *to* certain services j. This kind of accessibility is *outgoing accessibility*, while a measure of *ingoing accessibility* quantifies how accessible a given destination location j is *from* other locations. Nicolai and Nagel (2014) discuss circumstances under which these measures are interchangeable.

35.4 Accessibility Computation Integrated with Transport Simulation

As mentioned above, accessibility computations are often based on travel times (Bundesinstitut für Bau-, Stadt- und Raumforschung, accessed March 2015; Büttner et al., 2010), which serve as an impedance measure. Ways of calculating these travel times can, however, vary significantly. The simplest way to calculate a travel time between two locations is to measure the Euclidean distance (beeline distance) between these two locations and multiply with some average speed. According to Geurs and van Wee (2004), this is the usual approach in location-, person-, and utility-based accessibility approaches, where the focus is not specifically on the transport system.

To strengthen the **transport component** of accessibility (as introduced above) and make accessibility measure sensitive to transport infrastructure changes, a better representation of the travel impedance between origins and destinations is required. The most common approach is travel time calculation using shortest-path algorithms on a real-world transport infrastructure network representation. Many accessibility computations are embedded into GIS software, offering procedures for network-based computations (Bundesinstitut für Bau-, Stadt- und Raumforschung, accessed March 2015; Curtis et al., 2013; Büttner et al., 2010).

The accessibility extension in MATSim also offers this type of accessibility computation. To run it, an accessibility controler listener, e.g., the `GridBasedAccessibilityControlerListenerV3` must be added to the MATSim controler. An example is given in `RunAccessibilityExample` (see `http://matsim.org/javadoc` → accessibility → `RunAccessibilityExample` for details). As input, a network file and a facilities file are required (for more information on networks and facilities, refer to Section 4.1.1 and Section 6.4 of this book). This procedure is more disaggregate than many common approaches to accessibility computations, where single facilities are seldom considered; there, structural data like zone sizes, number of jobs, or total sales area are used to represent the *potential* of a given zone (Büttner et al., 2010; Gulhan et al., 2014) (also see Section 35.6).

Either way, performing an accessibility computation this way can be regarded as a *supply-based approach*, since both supply with transport infrastructure (required to reach a given location) and supply with activity opportunities at these locations are taken into account. The utilization of these two supply dimension by users, i.e., the dimension of *demand* is, however, not considered in this approach. Therefore, no *effects of competition* (Geurs and van Wee, 2004), either for transport infrastructure resources (defined by network capacities), or activity facilities capacities, are taken into account. It is obvious, however, that supply and demand interaction effects are relevant, because opportunities may disappear if they can no longer be reached within reasonable travel times, or when activity facility capacities are exceeded.

By considering demand-supply interaction effects in addition to just the supply side, the scope of the accessibility calculation can be significantly increased. Gauging these effects on *facility capacities* can be addressed by specifying facility capacities in the according value in the `facilities` input file. Observation of *network capacities* and their effects on agents' behavior is one of the core features of the MATSim transport simulation. This is also one major argument for the integration of an accessibility computation with the dynamic transport simulation system MATSim. While other accessibility tools—the majority based on GIS systems (Bundesinstitut für Bau-, Stadt- und Raumforschung, accessed March 2015; Curtis et al., 2013; Büttner et al., 2010; Liu and Zhu, 2004;

Gulhan et al., 2014)—can calculate travel times on a routed network, they do not calculate accessibilities dependent on transport infrastructure usage level. This property, is, however, essential when making accessibility measures sensitive to transport demand management policies, i.e., transport system changes that do not alter the transport infrastructure and are thus not captured by models considering only the supply side.

To take these effects into account, the MATSim accessibility extension must be run with a MATSim transport simulation. To do so, an initial plans file (as described in Chapter 2 of this book) needs to be specified in the MATSim config file. Furthermore, the value `timeOfDay` in the `accessibility` module of the MATSim config file needs to be specified. If then, as described, an accessibility controler listener is added to the MATSim controler, the best-path travel times, on which the accessibility computation will be performed, are taken from travel times observed in the MATSim transport simulation at the time specified by the value `timeOfDay`. This is useful when transport demand level varies significantly during the day; for instance, with morning and afternoon peaks; it also allows transport policy accessibility changes (and decision makers' reactions) to be better analyzed.

35.5 Econometric Interpretation

As pointed out by Morris et al. (1979), accessibility indicators provide a very useful way to summarize a large volume of information on household locations and how they relate to urban activities' distribution and connecting transport systems. They also take land use, the transport system and their inter-dependencies into account holistically. Curtis et al. (2013) explain that accessibility assessment tools overcome policy innovation restrictions associated with traditional transport planning practice, pointing out that use of such tools enables examination of a broader range of policy issues.

For effective policy decisions, accessibility assessment tools must be economically interpretable. To make an accessibility measure clearest in an econometric evaluation (e.g., cost-benefit analyses), it seems sensible to adapt equation 35.1 as follows: $g(.) = \ln(.)$, $a_j = 1, f(c_{\ell j}) = e^{-c_{\ell j}}$, and $-c_{\ell j} = V_{\ell j}$. Thus, equation 35.1 becomes

$$A_\ell := \ln \sum_k e^{V_{\ell k}} , \tag{35.2}$$

where k denotes all possible destinations and $V_{\ell k}$ equals the disutility of traveling from location ℓ to destination k. Equation (35.2) is the so-called logsum term of exponentials and can be interpreted as the expected maximum utility (e.g., Ben-Akiva and Lerman, 1985; de Jong et al., 2007). Equation 35.2 can be derived by assuming that the full utility of destination location k as perceived at origin location ℓ, is $U_{\ell k} = V_{base} + V_{\ell k} + \epsilon_{\ell k}$, where V_{base} is a base utility for performing a given activity without considering its location, $V_{\ell k}$ is the systematic or observed disutility of traveling to from origin ℓ to destination k, and $\epsilon_{\ell k}$ is a random term which absorbs the randomness of the disutility of traveling, as well as fluctuations in utility around V_{base}. Under the usual assumption that the $\epsilon_{\ell k}$ are independent and identically (iid) Gumbel-distributed random variables, the expectation value of $U_{\ell k}$ becomes

$$E(U_\ell) = E(\max_k U_{\ell k}) = \ln \sum_k e^{V_{\ell k}} + Const \equiv A_\ell + Const . \tag{35.3}$$

Const does not need to be considered, as it is invariant for all locations. As a consequence of dropping the positive *Const*, A_ℓ may take negative values.

Geurs et al. (2012a), for instance, use the logsum measure of user benefits as an alternative to the travel time savings method (i.e., rule-of-half measure) in a case study examining the effects of spatial planning on accessibility benefits and economic efficiency of public transport projects.

35.6 Spatial Resolution, Data, and Computational Aspects

In contrast to many other transport simulations, MATSim is based on coordinates (see Chapter 2 of this book), not zone-based. Therefore, accessibility computation in MATSim can also be conducted independent from any zoning system and, instead, be based on a raster with arbitrary granularity, i.e., adjustable grid size. Depending on the calculation planned (zone-based or grid-based), a `ZoneBasedAccessibilityControlerListenerV3`, or a `GridBasedAccessibilityControlerListenerV3`, respectively, need to be added to the MATSim controler. Unlike the MATSim accessibility extension, most other accessibility assessment tools rely on the zone-based approach (Curtis et al., 2013; Liu and Zhu, 2004; Büttner et al., 2010). More detail about the interpretation of cell- and zone-based accessibility measures is given by Nicolai and Nagel (2014).

Running a grid-based calculation, especially if a high spatial resolution is selected, avoids several issues that could arise (like"self-potential") if accessibility computations are based on zones (see, e.g., Nicolai and Nagel, 2014). A zone-based approach also makes the measure dependent on size and shape of the geographical units (cf. MAUP (Modifiable Areal Unit Problem)). Due to its typically lower resolution level, a zone-based approach may also not adequately represent local details (Kwan, 1998). This is especially relevant when lower-speed mode accessibilies (like walking) must be considered.

The MATSim accessibility calculation does not require typical zone-based statistical data. Instead, the calculation can be conducted on the basis of so-called VGI (Voluntary Geographic Information) like OSM, which contains activity facilities data on a coordinate-based level. Hence, no reference to any zoning system is necessary when using these data. Furthermore, data from OSM is publicly and freely available; the amount of these data are steadily increasing and quality is improving. In particular, OSM seems to have established itself as a uniform and globally-accessible standard for crowd-sourced and other geo-data, which makes the MATSim accessibility assessment highly portable.

If the coordinate-based (= grid-based = raster-based = cell-based) version of the MATSim accessibility computation is selected, its results can be interpreted as an accessibility field, i.e., as a measure that varies continuously in space. This *accessibility field*, can be visualized by calculating the values on regular grid points. Figure 35.1 gives an example of such a visualization and depicts the accessibility of work places in Nelson Mandela Bay Municipality in South Africa, as calculated by the grid-based MATSim accessibility computation with a grid size of 1 000 meters.

To calculate the accessibility A_ℓ of a given origin location ℓ to opportunity locations k, both the origin location ℓ, and opportunity locations k, are assigned to a road network. If the option to integrate the accessibility computation with the transport simulation, as described in Section 35.4, is chosen, a congested network with time-dependent travel times (as they have been simulated in MATSim) is used. For every ℓ, a so-called *least cost path tree* computation (Lefebvre and Balmer, 2007) is carried out. Accessibility of the same location at a different time of day will usually be different, since congestion patterns vary. The least cost path tree computation determines the best route and the least negative travel utility $V_{\ell k}$ from the origin location ℓ to each opportunity location k, based on Dijkstra's shortest path algorithm (Dijkstra, 1959). Once the least cost path tree has explored all nodes, the resulting disutilities $V_{\ell k}$ for all opportunities k are queried and the accessibility is calculated, as stated in Equation (35.2) (Nicolai and Nagel, 2014). A crucial question is how to choose the point, i.e., the coordinate, where the accessibility computation is anchored. Most quantitative accessibility tools use geographical centroids of given zones. This is also true when the zone-based MATSim accessibility computation is selected. Alternative ways to select a centroid (e.g., land-use-based centroids; Büttner et al., 2010) are discussed as well. If the grid-based MATSim accessibility computation is selected, the question of choosing a representative point for a spatial zone becomes less relevant, as cells are usually not selected to be as large.

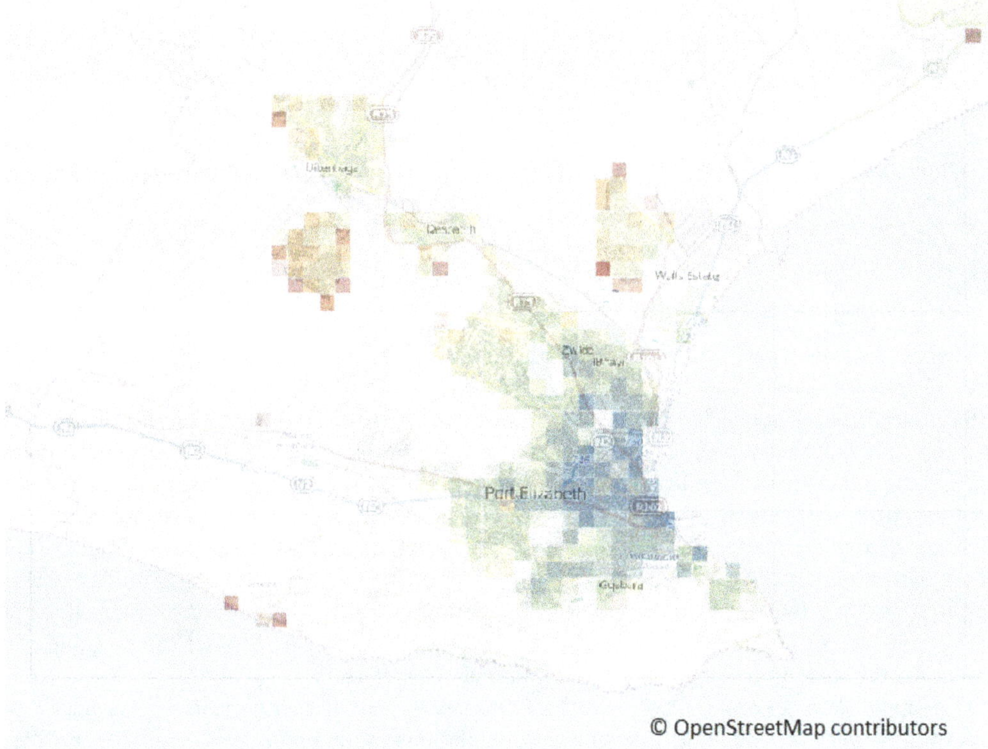

Figure 35.1: Accessibility of work places in Nelson Mandela Bay Municipality calculated by the grid-based MATSim accessibility computation

If the granularity of the grid-based MATSim accessibility computation is increased, origin locations ℓ and opportunity locations k, possibly located off the network, become increasingly important. To keep the approach consistent, the $V_{\ell k}$ calculation has to include disutility of travel to overcome the gap between locations and the road network. Therefore, the disutility of travel calculated by running the least cost path tree computation on the network has to be supplemented by the disutility to access the network from the origin ℓ (network access) and the disutility to access the destination k from the network (network egress). For origin locations ℓ, shortest distance to the network is given either by the Euclidean distance to the nearest node, or the orthogonal distance to the nearest link on the network. For destination locations k, the Euclidean distance to the nearest node is used to determine the shortest distance to the network.

This assumption (i.e., that opportunity locations are attached to the nearest network *node* rather than the nearest network *element*) is, in fact, the only approximation that the MATSim accessibility extension makes for the spatial resolution of opportunities (Nicolai and Nagel, 2014). While this assumption is unlikely to significantly alter accessibility results, it offers great potential for the optimization of computational performance, which has often been a major obstacle to higher-resolved accessibility computations (Kwan, 1998; Büttner et al., 2010). In the concrete case of the MATSim accessibility computation, exploration of the entire network by the least cost path tree is a computationally expensive task.

Thanks to the assumption, it is enough to sum over all opportunities k attached to a node j only once. The travel disutility $V_{\ell k}$ can be deconstructed as

$$V_{\ell k} = V_{\ell j} + V_{jk} \; \forall k \in j \, , \tag{35.4}$$

where $k \in j$ denotes all opportunities k attached to node j,

$$\sum_{k \in j} e^{V_{\ell k}} = \sum_{k \in j} e^{(V_{\ell j} + V_{jk})} = \sum_{k \in j} e^{V_{\ell j}} e^{V_{jk}} = e^{V_{\ell j}} \sum_{k \in j} e^{V_{jk}} =: e^{V_{\ell j}} \cdot Opp_j \,. \tag{35.5}$$

It is thus sufficient to compute Opp_j once for every network node j, and compute accessibilities as

$$A_\ell = ln \sum_k e^{V_{\ell k}} = ln \left[\sum_j e^{V_{ij}} \cdot Opp_j \right] . \tag{35.6}$$

Therefore, the loop performing the calculation does not have to run over all opportunities k, just over all network nodes j.

Similarly, for each origin location ℓ, the nearest road network node is identified. Locations ℓ that share the same nearest node have different travel disutilities to reach that node, but from then on have the same travel disutility to any other network node j. Exactly like the destinations, the least cost path tree is executed only once and calculated disutilities on the network are reused for all origins ℓ that are mapped on the same nearest network node. Therefore, only the calculation of the network access disutility needs to be performed individually for each origin ℓ. Nicolai and Nagel (2014) show that, due to this run time optimization, computation time increases sub-linearly with resolution. At the same time, they find that no significant further insights can be gained by increasing the resolution beyond a grid resolution of 100 meters.

The application example `RunAccessibilityExample` (see `http://matsim.org/javadoc` → accessibility) performs multiple accessibility computations for different types of activity facilities (e.g., accessibility of workplaces or accessibility of leisure facilities) by adding multiple instances of `GridBasedAccessibilityControlerListenerV3` to the MATSim controler. Other ways of performing distinct accessibility assessments for parts of the land-use system are just as feasible. Figure 35.1 is an example of work place accessibilities.

35.7 Conclusion

There are many different approaches to calculating accessibilities; most focus on a particular component of accessibility, while other components influencing accessibility are represented only in a limited way. Accessibility computations used in transport planning, for instance, represent transport networks, and thus the transport component of accessibility very well, while they usually do not represent facility properties or temporal effects. As pointed out by Geurs and van Wee (2004), it would be optimal if an accessibility computation considered all accessibility components (i.e., transport, land-use, temporal, and the individual component) well. The accessibility extension of MATSim could be an approach to achieve this.

First, transport system dynamics are represented by the accessibility computation integration with the MATSim dynamic traffic simulation. Second, land use is represented in a very disaggregate way; single facilities' locations and attributes are taken into account. Third, the temporal dimension can be observed by representing facilities' opening times and time-dependent travel times on the network; these are given as a MATSim dynamic traffic simulation output. Finally, individual characteristics can be taken into account; in the MATSim simulation, each individual is represented by its own software object, i.e., an agent, whose properties could be considered in the accessibilities calculation.

Actual accessibility values calculated by the MATSim accessibility extension take the form of *potential accessibility measure*, as originally defined by Hansen (1959). The specific selection of the measure's mathematical form allows results to be interpreted as logsum values, making them suitable for utilization in economic evaluations like benefit-cost analyses. Because the MATSim accessibility extension can rely solely on publicly and freely available data, e.g., data from OSM, it is highly portable. By distinguishing activity facilities along various potential dimensions, many different analyses can be conducted. In the code example given (see `http://matsim.org/javadoc` → accessibility → `RunAccessibilityExample`), for instance, accessibilities for different land uses, i.e., different types of activity opportunities, are calculated. Being grid- instead of zone-based (which most other accessibility tools are), avoids certain problems associated with zones. At the same time, computations are still within reasonable ranges, partly due to a runtime optimization that reuses computational steps for locations sharing the nearest network node.

CHAPTER 36

Emission Modeling

Benjamin Kickhöfer

36.1 Basic Information

Entry point to documentation:

`http://matsim.org/extensions` → emissions

Invoking the module:

`http://matsim.org/javadoc` → emissions → `RunEmissionToolOnlineExample` class, `RunEmissionToolOfflineExample` class

Selected publications:

Hülsmann et al. (2011); Kickhöfer et al. (2013); Kickhöfer and Nagel (2011, 2013); Hülsmann et al. (2013); Kickhöfer (2014); Kickhöfer and Kern (2015)

36.2 Introduction

This chapter presents the emission modeling tool developed and tested by Hülsmann et al. (2011) and further improved by Kickhöfer et al. (2013). The text in this chapter is a slightly updated version of the emission modeling tool description in Kickhöfer (2014). The tool calculates warm and cold-start exhaust emissions for private cars and freight vehicles by linking MATSim simulation output to the detailed "HBEFA (Handbook on Emission Factors for Road Transport)" database, available for many European countries.

The chapter is structured as follows: Section 36.3 reviews literature for other attempts to model transport-related emissions. Section 36.4 presents an overview of the "EMT (Emission Modeling Tool, see Chapter 36)" and Section 36.5 shows how the tool is embedded in MATSim's software structure.

How to cite this book chapter:

Kickhöfer, B. 2016. Emission Modeling. In: Horni, A, Nagel, K and Axhausen, K W. (eds.) *The Multi-Agent Transport Simulation MATSim*, Pp. 247–252. London: Ubiquity Press. DOI: http://dx.doi.org/10.5334/baw.36.

36.3 Integrated Approaches for Modeling Transport and Emissions

Over the last two decades, the modeling of transport-related environmental externalities has received increasing attention in transportation science. The following paragraphs briefly present some recent work in the exhaust emission modeling area; additionally, they highlight differences to the EMT, which will then be described in subsequent sections.

Creutzig and He (2009) and Michiels et al. (2012) use very aggregated figures to estimate air pollution in Beijing and Belgium, respectively. Neither approach mentions any particular underlying transport model. It seems that transport related emissions are based on aggregated origin-destination matrices or aggregated demand functions. These two studies are on a very different level of aggregation than the EMT, and a comparison does not seem constructive.

Beckx et al. (2009) use a sophisticated activity-based model to simulate activity schedules for roughly 30% of all households in the Netherlands. Traffic assignment for passenger cars is performed by using an aggregated "all-or-nothing" assignment approach, resulting in hourly aggregated traffic flows on the network. Based on the average speed for a trip, the MIMOSA (Modélisation Isentrope du transport Méso-échelle de l'Ozone Stratosphérique par Advection) model then calculates emission and fuel consumption rates, possibly dependent on vehicle category. The idea of using an activity-based model to simulate time-dependent emissions is similar to the EMT. In contrast to the latter, the underlying transport in Beckx et al. (2009) does not account for congestion effects and different traffic states. Additionally, similar macroscopic emission models are typically unable to capture certain microscopic behavior accurately (see, e.g., Ahn and Rakha, 2008).

Hirschmann et al. (2010) link the microscopic traffic flow simulator VISSIM (Verkehr In Städten – SImulationsModell) with the instantaneous emission model PHEM (Passenger Car and Heavy-duty Emission Model).[1] At first glance, this approach seems very promising, as it also builds the basis for the HBEFA database. In contrast to the EMT, it is not suitable for large-scale scenarios due to the computational complexity of VISSIM (Verkehr In Städten – SImulationsModell). In Kraschl-Hirschmann et al. (2011), the same authors attempt to develop a parametrization of fuel consumption based on average speeds of vehicles. Such parametrization could be helpful—in the future—to replace time-consuming lookups in large databases (e.g., HBEFA). However, the model would need to allow for more input variables (e.g., vehicle category, traffic state, etc.) and provide more differentiated outputs, e.g., different emission types.

In a similar study, Song et al. (2012) couple VISSIM (Verkehr In Städten – SImulationsModell) with the emission modeling tool MOVES. They find that the VISSIM (Verkehr In Städten – SImulationsModell)-simulated, vehicle-specific power distribution for passenger cars deviates significantly from the observed distribution, meaning that the estimated emissions also contain significant errors. Here again, the proposed model cannot be used for large-scale scenarios. Additionally, it seems questionable whether such detailed modeling will prove to be superior to less detailed models as the EMT.

Wismans et al. (2013) compare passenger car emission estimates of static and dynamic traffic assignment models. They claim that little research has been done in connecting macroscopic or meso-scopic dynamic traffic assignment models with emission models. According to the authors, static assignment models predict congestion on the wrong locations and ignore spillback effects. They argue that emission hotspots are, in consequence, also predicted at the wrong locations and/or with the wrong amplitude. To counter these disadvantages, they couple a static and a dynamic traffic assignment model with the exhaust emission model ARTEMIS. Large differences in air pollutant emissions are found and hotspot locations differ.

---

1 The PHEM (Passenger Car and Heavy-duty Emission Model) model uses speed trajectories as input and was tested against the output of the EMT by Hülsmann et al. (2011).

Hatzopoulou and Miller (2010) develop a methodology for calculating exhaust emissions, using MATSim as transport model. The approach is therefore similar to the EMT. In contrast to that study, the EMT does not assume fixed exhaust emissions per time unit. It uses a more detailed calculation of emissions based on the two different traffic states: "free flow" and "stop&go". It is, thus, able to capture congestion effects that emerge, as well as the time spent in traffic jam. Furthermore, the EMT calculates exhaust emissions for passenger cars *and* for trucks. Finally, since the methodology is based on HBEFA, it can be transferred to any scenario in Europe.

36.4 Emission Calculation

Air pollution is caused by different contributions of road traffic: Warm emissions are emitted while driving and are independent of the engine's temperature. Cold-start emissions also occur during the warm-up phase and depend on the engine's temperature when the vehicle is started. Warm emissions differ with respect to: *driving speed*, acceleration/deceleration, *stop duration*, road gradient, and *vehicle characteristics* consisting of vehicle type, fuel type, cubic capacity, and European Emission Standard Class (André and Rapone, 2009). Cold emissions differ with respect to: *driving speed, distance traveled, parking time*, ambient temperature, and *vehicle characteristics* (Weilenmann et al., 2009).

Currently, the emissions contribution to MATSim considers all differentiations above marked in *italic*. Road gradient and ambient temperature are not considered; gradient is always assumed to be 0 %, and ambient temperatures are assumed to be HBEFA average. In addition to warm and cold-start emissions, evaporation and air conditioning emissions also result from road traffic. At the moment, these are not considered in the emission modeling tool, because they contribute little to the overall emission level.

The calculation of warm emissions is composed of two steps:

1. deriving *kinematic characteristics* from the simulation, and
2. combining this information with vehicle characteristics to extract emission factors from the HBEFA database.

In the first step, driving speed, as well as stop duration (and possibly an approximation of acceleration/deceleration patterns), is captured by a mapping of MATSim's dynamic traffic flows to HBEFA traffic states. These traffic states, namely "free flow", "heavy", "saturated", and "stop&go", have been derived from typical driving cycles, i.e., time-velocity profiles. A parametrization of these profiles led to the definition of these traffic states, which depend on speed limit, average speed, and road type. Thus, typical emission factors for a specific traffic state on a specific road segment can be looked up in the HBEFA database. In MATSim, neither the location on a road segment, nor the exact driving behavior of an agent is known (see Section 1.3). It is quite straightforward to extract agents' travel times on the road segment which, thanks to the queuing model, also includes interactions with other agents and spillback effects. The average speed of an agent on a certain road segment is thus used to identify corresponding HBEFA traffic states, and to assign emission factors to the vehicle. As of now, the emission modeling tool considers only two traffic states: free flow and stop&go.[2] Each road segment is divided into two parts representing these two traffic states. The distance l_s that a car is driving in stop&go traffic state is determined by the following equation:

$$l_s = \frac{l\, v_s(v_f - v)}{v(v_f - v_s)}, \tag{36.1}$$

[2] Simplified because the difference between traffic states—free flow, heavy, and saturated—emission factors are only marginal. In contrast, emission factors for stop&go are roughly twice as high.

where l is the link length in kilometers from the network, v_s is the stop&go speed in km/h for the HBEFA road type, v_f is the free flow speed in km/h from the network, and $v = \frac{l}{t}$ is the average speed on the link for the vehicle, t being the link travel time of the vehicle in the simulation. For the derivation of Equation (36.1), please refer to Kickhöfer (2014). The distance that the car is driving in free flow traffic state is then simply the remaining link length $l_f = l - l_s$. The interpretation of this approach: Cars drive in free flow until they have to wait in a queue. Stop&go traffic state applies only in the queue. According to the MATSim queue model presented in Section 1.3 , a queue emerges if demand exceeds capacity of a road segment, which can also result in spill-back effects on upstream road segments. The length of the queue is, thus, approximated by Equation 36.1, where the average speed v on a link is the only exogenous variable.

For the second step, agent-specific vehicle attributes are needed. They are usually obtained from survey data during the initial population synthesis. The vehicle attributes typically comprise: vehicle type, age, cubic capacity and fuel type. Because MATSim keeps socio-demographic information throughout the simulation process, it can be used at any time for reference in the detailed HBEFA database. Additionally, the emission modeling tool is designed in such way that fleet averages are used, whenever no detailed vehicle information is available.

The calculation of cold-start emissions is, again, composed of two steps:

1. deriving *parking duration* and *accumulated distance* from the simulation, and
2. combining this information with vehicle characteristics in order to extract emission factors from the HBEFA database.[3]

Parking duration refers to the time a vehicle is not moved *before* cold-start emissions are produced. It is calculated by subtracting an activity's start time from the same activity's end time and by checking if the trip to and from the activity is performed by car. Emission factors in HBEFA are differentiated by parking duration in one hour time steps from 1 hour to 12 hours. After 12 hours, the vehicle is assumed to have fully cooled down. The accumulated distance refers to the distance a vehicle travels *after* a cold start. According to HBEFA, there are different cold-start emissions for short trips less than 1 kilometer and and for longer trips equal to or greater than 1 kilometer. In reality, cold-start emissions are emitted along the route after a cold start; at this time, the emission modeling tool maps the short trip emissions to the road segment where the engine is started, and, if applicable, additional emissions to the road segment where the accumulated distance exceeds the first kilometer. Overall, cold-start emission factors increase with parking duration and accumulated distance; they also depend on vehicle attributes. The lookup for this information is identical to the one described for warm emissions.

In order to further process warm and cold-start emissions, so-called *emission events* are generated during the simulation in a separate events stream. The definition of emission events follows the MATSim framework that uses events for storing disaggregated information in XML format. The following section provides more information on the EMT's software structure.

36.5 Software Structure

The information in this section refers to code that can be found in the MATSim repository. In the following, the software structure of the EMT at revision 30 058 is described. For information on how to use the tool, please use the entry points listed at the beginning of this Chapter 36.

[3] Please note that HBEFA provides cold-start emission factors only for passenger cars. Freight traffic therefore only produces cold-start emissions of passenger cars.

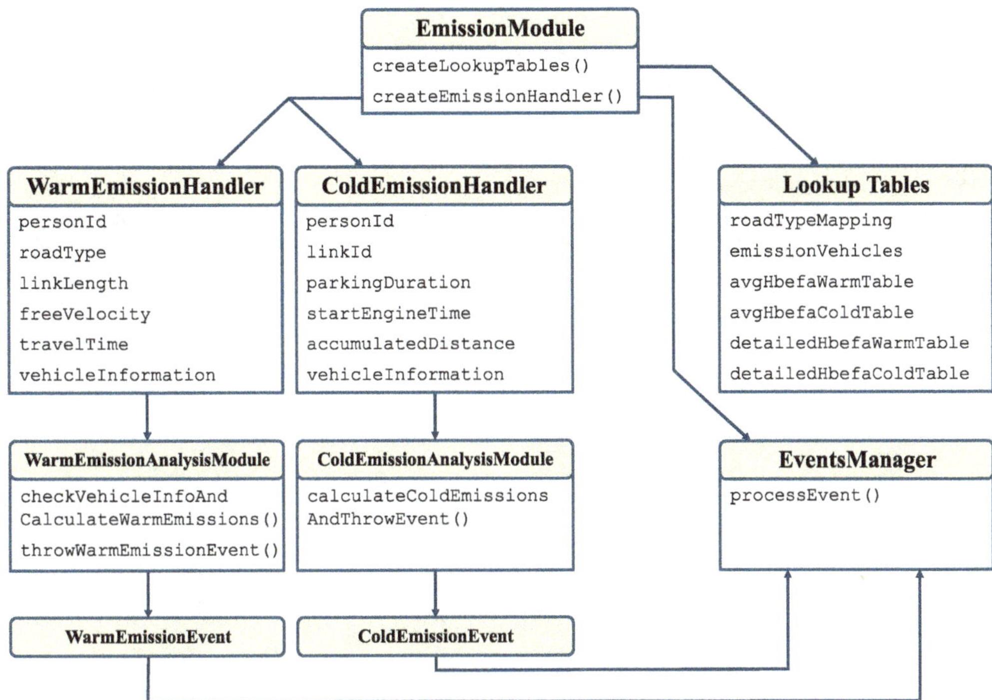

Figure 36.1: Software structure of the emission modeling tool.

Figure 36.1 shows the simplified software structure of the EMT. The core of the tool is the `EmissionModule` which needs to be created before the simulation starts. There are also two public methods that must be called: `createLookupTables()` and `createEmissionHandler()`.

The former creates lookup tables from input data that has to be exported from the HBEFA database. The path to these input files can be configured in the `EmissionsConfigGroup`. Mandatory input are files for the creation of `roadTypeMapping`, `emissionVehicles`, `avgHbefaWarmTable`, and `avgHbefaColdTable`. The first lookup table maps road types from the MATSim network to HBEFA road types. For this mapping, it is necessary to classify the network road types into HBEFA categories; this requires some transport engineering knowledge. The second lookup table defines the vehicle attributes of every owner in the population. It should therefore be generated during the population synthesis process. If no detailed information is available, the vehicle lookup table still needs to specify whether the vehicle is a car or a truck. The current implementation uses the MATSim vehicle interface `Vehicles` as container for storing the relevant data in `VehicleType`.[4] The last two mandatory lookup tables (`avgHbefaWarmTable` and `avgHbefaColdTable`) provide warm and cold emission factors in g/km, respectively. The data is stored using a unique key. For the construction of this key, information from `roadTypeMapping` and `emissionVehicles` is needed, as well as information derived from the simulation as described in Section 36.4. The latter information is depicted in Figure 36.1 as variables of the two classes `WarmEmissionHandler` and `ColdEmissionHandler`. These two handlers implement several MATSim `EventHandler` interfaces to extract necessary information from the simulation. After gathering this information, the `WarmEmissionHandler` asks its `WarmEmissionAnalysisModule` to reconstruct the key and look up the emission factors in the

[4] Please note that vehicle information provided to the `EmissionModule` is *only* used for storing data on individual vehicle characteristics and other information will be omitted by the simulation.

respective table. Similarly, the `ColdEmissionHandler` asks the `ColdEmissionAnalysisModule`. These analysis modules then create `Warm/ColdEmissionEvents`, which follow the MATSim `Event` interface definition. Finally, the resulting events stream is written in a joint emission events file by a separate `EventsManager`.

For the calculation of emissions dependent on agent-specific vehicle characteristics, `emissionVehicles` must contain that specific information, the corresponding flag in the `EmissionsConfigGroup` needs to be switched on, and detailed emission factor tables also need to be exported from HBEFA and provided to the `EmissionModule` with two additional input files: `detailedHbefaWarmTable` and `detailedHbefaColdTable`.

CHAPTER 37

Interactive Analysis and Decision Support with MATSim

Alexander Erath and Pieter Fourie

37.1 Basic Information

Entry point to documentation:

`http://matsim.org/extensions` → travelsummary

Invoking the module:

`http://matsim.org/javadoc` → travelsummary → `RunEventsToTravelDiaries`

Selected publications:

This chapter is largely based on work in Erath et al. (2013), where the interested reader will find references for further reading.

37.2 Introduction

Agent-Based Simulation Means Lots of Data Agent-based transport demand models require managing and integrating data sources several orders of magnitude larger than traditional aggregate models. In a truly disaggregate demand description, as seen in our MATSim implementation for Singapore, spatial data represents individual buildings and land parcels, not zones; travel demand takes the form of a full activity diary with connecting trips for every individual, based on their personal demographic attributes, instead of an aggregate number of trips from zone to zone for a specific time period. For this reason, input data for an aggregate four-step (or related) demand model can generally be edited on a laptop, using standard spreadsheet software, whereas agent-based modeling requires the manipulation and synthesis of large stores of structured, hierarchical data, frequently exceeding most personal computer capacity.

How to cite this book chapter:

Erath, A and Fourie, P. 2016. Interactive Analysis and Decision Support with MATSim. In: Horni, A, Nagel, K and Axhausen, K W. (eds.) *The Multi-Agent Transport Simulation MATSim*, Pp. 253–258. London: Ubiquity Press. DOI: http://dx.doi.org/10.5334/baw.37.

How MATSim Stores Data MATSim stores and retrieves data from XML, because XML reflects objects' hierarchical structure in the simulation and is readable. However, performing general exploratory analysis of large XML data stores is usually poorly supported by most data analysis software packages, especially GIS-based systems. To perform analyses, expert knowledge of XML querying technologies like XPath and XQuery is required (or Java, if one performs more specialized analysis on the objects themselves). In our experience, this specialized knowledge is lacking in transport and urban spatial planning practice. Therefore, in most MATSim applications so far, authorities, and other interested parties, must formulate their desired analysis in advance and have expert consultants perform the analysis. Any queries resulting from the analysis require another consultation cycle and the client's perceived value declines, due to both lack of interactivity and model ownership feeling. We believe this lack of a broadly supported exploratory data analysis interface, and the customer experience the interface can create, presents a considerable barrier to entry for many authorities and operators interested in using MATSim.

How Customers Interact With Data: Relational Databases, GUI-Driven Interaction Most transport and urban spatial planning customers rely on mature, GUI-driven software, such as ArchGIS (ESRI, 2011), EMME/3 (INRO, 2015), the PTV (PTV, 2009) transport planning suite, or even Microsoft Excel; all of these connect to relational databases and perform queries on large data sets. Many analysts can explicitly query databases using the SQL (Structured Query Language); the ODBC (Open Database Connectivity) standard allows software to connect to any relational database regardless of the actual technology driving it. Importantly, many interactive exploratory data analysis software suites, like Tableau, Tibco Spotfire, SAS and the open source R project, support relational databases and ODBC.

37.3 Requirements of a Decision Support Interface to MATSim

The event stream produced by the MATSim mobility simulation represents the transport simulation process at the atomic level. It could be fed into a relational database; an analyst fluent in procedural languages could process it in arbitrary ways. But we expect more general use case scenarios, where most analysts will perform general tasks that can be standardized. To this end, we set about compiling requirements specifications for potential audiences and their use case scenarios, to come up with a general interactive analysis framework and decision support to satisfy most requirements. We developed a set of Java classes to process MATSim input and output, producing tables in a relational database, and an entity relationship diagram that should be intuitive and useful to a large user audience.

37.3.1 Users

This chapter presents a decision support tool geared to decision makers and researchers in the fields of transport planning and operations, spatial planning and spatial economics and geography. Generally speaking, it should serve professionals interested in mobility and spatial analysis, who understand transport modeling principles, but do not have the expertise to operate an agent-based transport simulation directly. Currently, we envision the following stakeholders and some hypothetical questions for a decision-support system—a non-exhaustive list that, we expect, will grow with time:

Transport planners: How many trips occur where, when and what is the activity purpose? What are the socio-demographic characteristics of people performing these trips and activities?

Urban Planners: What are the temporal usage patterns of buildings and the surrounding neighborhood? What is the flow from public transport stops to surrounding buildings?

Policy-Makers: What are the costs and benefits of a new public transport service? Who are the winners and losers when constructing a new road?
Public Transport Operators: What is the breakdown of specific bus lines' ridership?
Service Industry: Which customers are in catchment areas, separated by mode?

37.3.2 Functional Requirements

The decision support framework should facilitate classic transport appraisal methods, such as cost/benefit analysis and evaluation of transport infrastructure spatial impact and policy measures. The framework should allow any sort of spatial analysis, on the finest granularity level provided by the transport model; usually, individual buildings or parcels, as well as public transport stops and selected links, like count stations or tolled road segments. However, these geographic features should be indexed against transport zones, or other geographic areas of interes,t to allow customized results aggregation. Furthermore, it should capture all temporal aspects of the simulation; full temporal dynamics are a crucial part of the agent-based approach.

37.4 General Framework for Decision Support

Figure 37.1 shows the general framework as we envision it: data from various sources feeds into a spatially-enabled database, with all geodata transformed to use the same spatial reference system (ideally, using the same projection used for MATSim coordinates, allowing for simple distance calculations). Simple Java programs using the MATSim API and JDBC (Java Database Connectivity) produce XML input data for MATSim scenarios; events from these scenarios are fed back into the database. Analysts query the database to produce "data cubes", which are aggregations and queries across many database tables. These are designed for specific purposes, such as calibration and validation, location analysis, winner/loser analysis or other application-specific purposes.

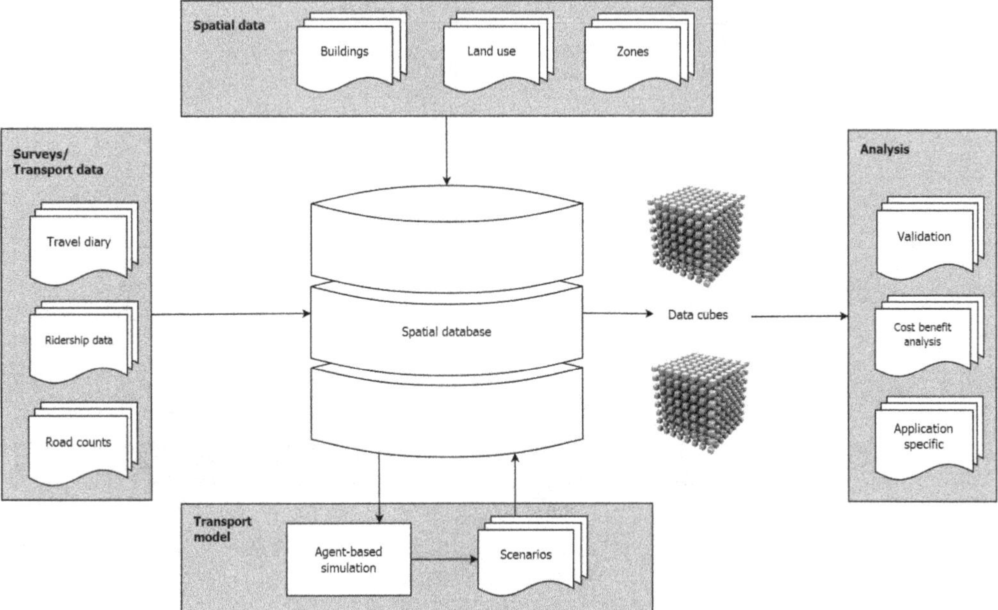

Figure 37.1: General framework of the decision support system.

37.4.1 Entity Relationship Diagram (ERD) for General Purpose Analysis

For entity relationships, we decided that a travel diary format is most suitable for the usual types of analysis, but works especially well for comparison with other data sources when validating simulation output. Most travel surveys take the form of a diary, recording travel time, purpose and mode, as well as aspects of the journey like number of stages, transfer walking and waiting time and in-vehicle time. Routines can be developed to transform survey data and public transport smart card records into the same format with consistent coding. Figure 37.2 shows the ERD (Entity Relationship Diagram) we propose, along with the primary/foreign key relationships between tables that facilitate aggregation and joining of e.g., personal/household attributes, such as income, with travel time experienced in the simulation.

37.4.2 Interactive Analysis Using Business Analytics Software

Modern business analytics software, like Tableau (Tableau Software, 2013), provide interactive data aggregation and visualization from relational databases. While basic analysis of individual tables

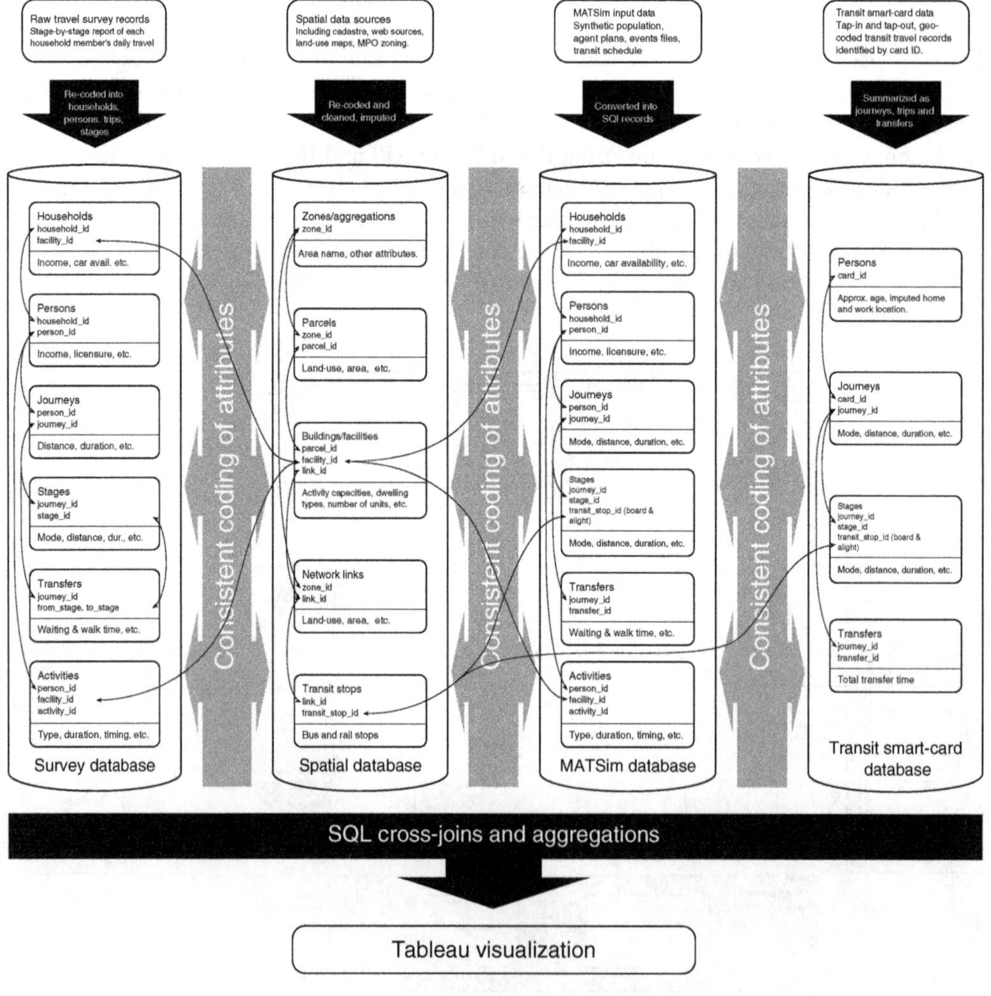

Figure 37.2: Simplified entity relationship diagram showing shared keys across tables.

in our proposed ERD could already provide valuable insight to MATSim simulations, much richer analysis is possible when tapping relationships between different tables in the database. With the help of graphical query building software, little or no knowledge is required to construct SQL scripts that create customized data cubes. These cubes are fed into the business analytics software, which is designed with a relatively programming-agnostic audience in mind. Relying on the familiar paradigm of drag-and-drop interaction in a simple, well-documented GUI, the user constructs "dashboards" summarizing information and allowing interactive aggregation, or drilling-down across multiple dimensions.

Figure 37.3 shows a Tableau visualization comparing public transport ridership from a MATSim simulation to actual smart card data records (transformed into the travel diary format specified in the ERD). Figure 37.4 shows the SQL query used to produce the data frame driving the Tableau analysis. The query exploits the primary/foreign key relationships in the database to perform rapid joins between the different tables.

37.5 Diaries from Events

In the package `contrib.analysis.travelsummary` (Chapter 38), the reader can find a set of classes that will transform their MATSim simulation results into a set of travel diary tables, like those discussed in the preceding section. The package contains a simple GUI class that can be run to specify input data XML files, the location to save output CSV (Comma-Separated Values) files and other information such as a subscript appended to the end of file names to identify different scenarios. These CSV files can be read into a relational database of choice, or directly queried in Tableau, or other interactive analysis software.

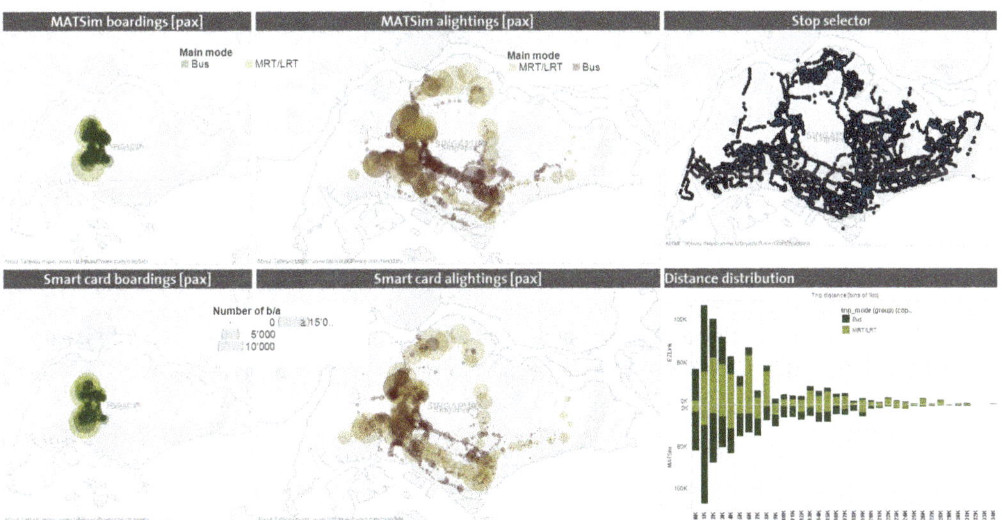

Figure 37.3: Tableau visualization of public transport ridership from a MATSim simulation compared against actual smart card data records in Singapore.

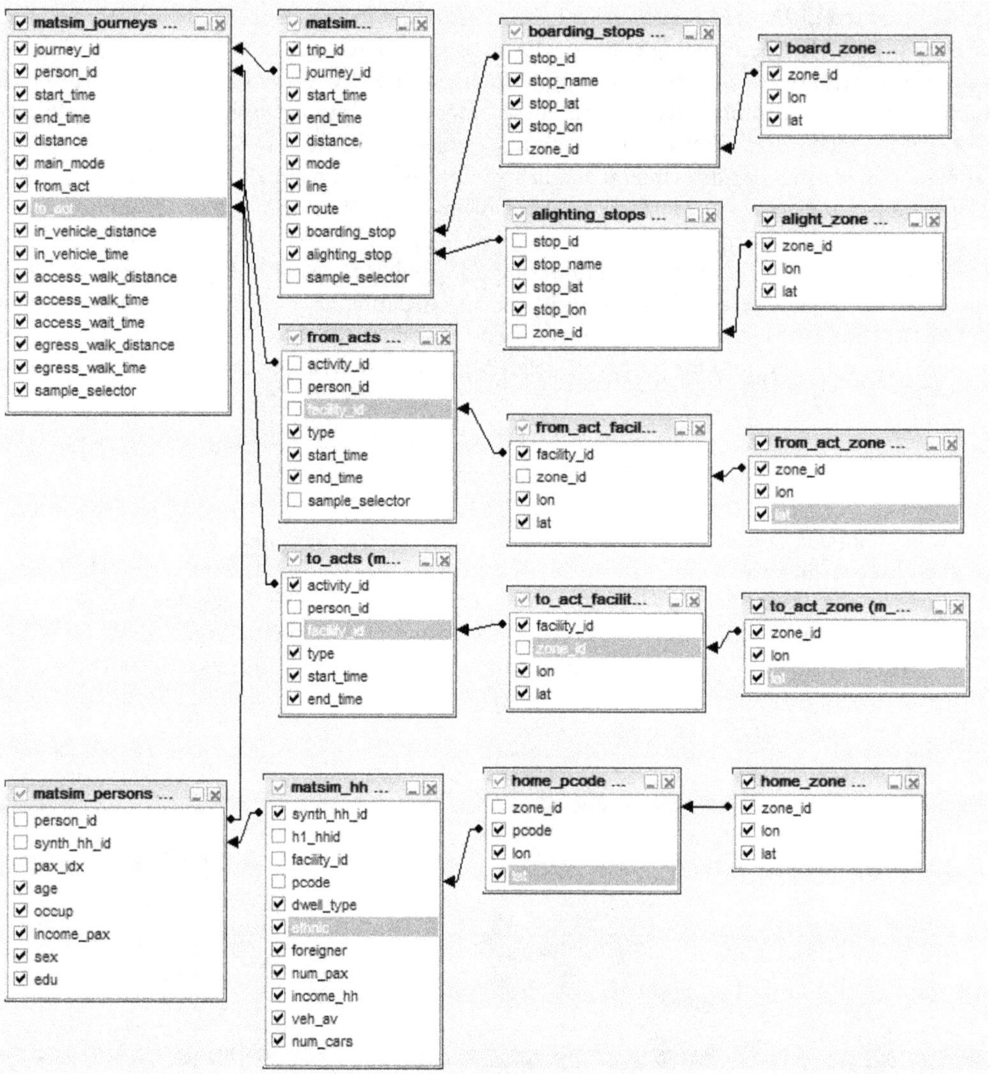

Figure 37.4: A diagram showing how the tables from Figure 37.2 are joined together for visualization in business analytics software, e.g., Tableau, as shown in Figure 37.3.
Source: (Erath et al., 2013)

CHAPTER 38

The "Analysis" Contribution

Kai Nagel

38.1 Basic Information

Entry point to documentation:

`http://matsim.org/extensions` → analysis

Invoking the module:

No standard invocation. See `http://matsim.org/javadoc` → analysis → `RunKNEventsAnalyzer` class for intuition.

Selected publications:

–

38.2 Summary

This contribution collects various analysis tools for MATSim output.

One important reason for having this in a contribution rather than in a playground is the Apache Maven layout of the repository: Contributions can use material from other contributions, but not from the playgrounds. In consequence, analysis tools that are needed in a contribution need to be in a contribution themselves. The analysis contribution is a possible place where to put them.

How to cite this book chapter:

Nagel, K. 2016. The "Analysis" Contribution. In: Horni, A, Nagel, K and Axhausen, K W. (eds.) *The Multi-Agent Transport Simulation MATSim*, Pp. 259–260. London: Ubiquity Press. DOI: http://dx.doi.org/10.5334/baw.38.

SUBPART ELEVEN

Computational Performance Improvements

CHAPTER 39

Multi-Modeling in MATSim: PSim

Pieter Fourie

39.1 Basic Information

Entry point to documentation:

`http://matsim.org/extensions` → pseudosimulation

Invoking the module:

`http://matsim.org/javadoc` → pseudosimulation → `RunPSim` class

Selected publications:

Fourie et al. (2013)

39.2 Introduction

MATSim's major current performance limitation is the network loading simulation, i.e., the mobsim, for example QSim or JDEQSim; this chapter focuses on QSim. As shown earlier, QSim is repeatedly executed in the MATSim loop for the entire agent population (Section 1.2).

With the multi-modeling approach (Fourie et al., 2013), shown in Figure 39.1, a MATSim run periodically replaces QSim for a number of iterations with a simplified meta-model or PSim (Pseudo-Simulation), running approximately one hundred times faster. In risk analysis, these models are called "surrogate models" (Sudret, 2012). PSim uses travel time information from the preceding QSim iteration to estimate how well an agent day plan might perform, allowing multiple iterations of mutation and evaluation between QSim iterations to more rapidly explore the agents' solution space, producing better performing plans in a shorter time.

How to cite this book chapter:

Fourie, P. 2016. Multi-Modeling in MATSim: PSim. In: Horni, A, Nagel, K and Axhausen, K W. (eds.) *The Multi-Agent Transport Simulation MATSim*, Pp. 263–266. London: Ubiquity Press. DOI: http://dx.doi.org/10.5334/baw.39.

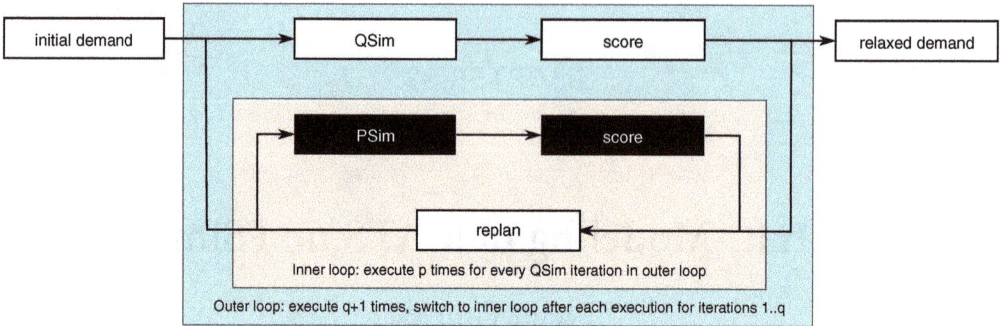

Figure 39.1: Operation of a MATSim run implementing pseudo-simulation.
Source: Fourie et al. (2013), Figure 1, p. 69

39.3 Basic Idea

PSim exploits classes that record various network performance aspects during queue simulations and uses them as an approximate meta-model of QSim. It fires the same sequence of events for car and public transport passengers that is produced during a QSim mobility simulation, except that event timings are approximate values expected at the time of day they occur.

For private vehicle traffic, it calls the

```
getLinkTravelTime(Link link, double time, Person person, Vehicle vehicle)
```

method of classes implementing the `TravelTime` interface to fire `LinkEnterEvents` and `LinkLeaveEvents` at appropriate times for all car route links. For public transportation, the events sequence generated for a passenger traveling on a particular service relies on a meta-model of stop-to-stop travel times (`interface StopStopTime`) and waiting times at stops (`interface WaitTime`); both concepts were developed by Sergio Ordòñez at the Future Cities Laboratory (package `playground.singapore.transitRouterEventsBased`).

PSim plans are scored using the same function as QSim iterations and are compatible with most replanning modules in MATSim. Following a series of PSim iterations, a plan is selected for each agent, in the usual fashion, and a QSim iteration is run to start a new cycle. The various classes used in PSim are updated with the latest network performance information and the process repeats.

39.4 Performance

Initial tests on the Zürich scenario (described in Chapter 56) have shown a dramatic decrease in computation times, compared to the default QSim-only approach; performance improves linearly with an increasing number of computational cores. Figure 39.2 compares the PSim-approach, in two configurations, against the existing approach, for a 10 % sample of private vehicle traffic in Zürich. All simulations were run until they reached a target score, i.e., the score reached after running the standard approach for 100 iterations. The first PSim-implementing configuration uses the same rate of plan mutation as the QSim-only approach, where 30 % of agents are selected for plan mutation (replanning) after each iteration, whether it is a QSim or PSim iteration. The new approach requires fewer QSim iterations to reach a target score, but requires more time for replanning. Replanning is fully multi-threaded, with no synchronization between cores required, so its performance increases linearly, with increasing number of cores; times improve more dramatically with the new approach than the standard approach. In the second configuration, the mutation rate

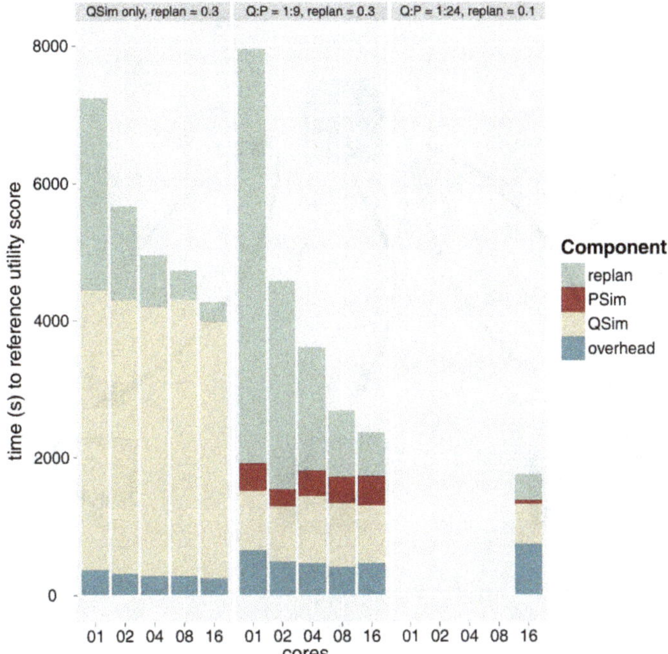

Figure 39.2: Computation time contributions vs. number of computational cores for QSim-only (0.3 replanning rate), 9 PSim iterations per QSim iteration at 0.3 replanning rate, and 24 PSim iterations per QSim iteration at 0.1 replanning rate.

Source: Fourie et al. (2013), Figure 4, p. 73

is reduced and the number of PSim iterations between QSim iterations increased to 24 for each QSim iteration. The system now tests many more combinations of different mutation operations (four in this case: activity timing, mode choice, secondary activity location choice, and re-routing), to reach the target state much faster, even though it produces a smaller expected number of mutated plans per agent between QSim iterations (three for configuration 1, 2.5 plans for configuration 2).

This last point raises an interesting issue; namely, that the distribution of mutation operation numbers can be dramatically spread out with the PSim approach, because increasing the number of iterations is relatively cheap. This should make the approach preferable, especially with random mutation-producing replanning strategies, where a large number of mutations are needed to produce a relaxed simulation state.

For a detailed discussion of the meta-modeling approach and the results of applying this method to the Zürich scenario, refer to Fourie et al. (2013).

39.4.1 Distributed Computing

Because PSim executes plans independently from each other, requiring no coordination of computational processes, it is possible to distribute it across multiple nodes, with no need of shared memory, as illustrated in Figure 39.3. To this end, we (Fourie and Ordòñez, FCL (Future Cities Laboratory)) are implementing a simple messaging protocol to transmit network performance objects to PSim slave nodes from a master node running QSim only. Slave nodes perform replanning operations and evaluate plans in a pre-determined number of PSim iterations per cycle. At the start of each QSim iteration, a single plan for each agent is transmitted back to the master from

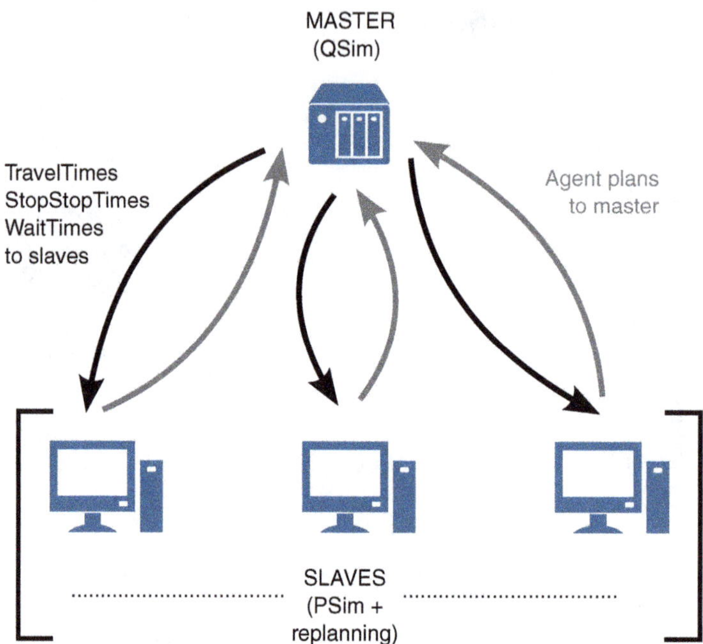

Figure 39.3: Master-slave configuration for running PSim in distributed mode, across many slave computer nodes in a local area network or in a cloud computational framework. The master runs selected plans in a full queue simulation and transmits updated travel time information to slave nodes after every iteration. In turn, slaves produce and evaluate new plans in repeated PSim/replanning cycles, sending the master a single plan for each agent at the start of a QSim iteration.

all the slaves, and updated `TravelTimes`, `StopStopTimes` and `WaitTimes` are rendered during the full mobility simulation, to be transmitted back to the slaves in the next cycle.

The approach yielded promising results, with a reduction in the number of QSim iterations, as in the previous work, as well as the potential for running large-scale simulations on much cheaper hardware than the current approach, that demands expensive shared memory servers. Most importantly, all replanning takes place in parallel with the QSim running on the master, so the time spent waiting for replanning operations can be reduced to nil. This performance increase is especially useful for large scenarios implementing public transportation, where the time spent replanning can be up to twice that of the queue simulation.

CHAPTER 40

Other Experiences with Computational Performance Improvements

Kai Nagel

MATSim has always had the simulation of large regions as its goal, and as such was always interested in high computational performance. The team had, when it started with the Java-based MATSim (cf. 46.2.1.4), considerable experience in parallel computing (Nagel and Schleicher, 1994; Rickert and Nagel, 2001; Nagel and Rickert, 2001; Cetin et al., 2003) as well as with more general message-based approaches (Gloor and Nagel, 2005) that resemble today's Protocol Buffers (Google Developers, 2015). However, the move to Java (see Section 46.2.1.4), a decision for faster conceptual progress and reduced maintenance effort, also had the consequence that the MPI (Message Passing Interface) approach to parallel computing could no longer be used and was thus given up. See Section 46.2.1.4 for details.

The behavioral modules of MATSim, such as route (Section 4.5.1.2) or destination (Chapter 27) innovation, are conceptually straightforward to parallelize by multi-threading, and that was implemented in MATSim from early on (Balmer et al., 2009b, see Section 4.2.3 how to use this). The remaining challenge then is to parallelize the mobsim, in which the parallel threads need to interact closely. For example, assume that we compute 24 hours of traffic in 120 seconds of computing time (cf. Table 40.1). With the 1 second time steps used in the QSim this means 720 update rounds per second, and thus 720 inter-thread interactions per second.

An attempt to use the CUDA (Compute Unified Device Architecture, a parallel computing platform and API by NVIDIA) for the C language (Strippgen and Nagel, 2009b,a; Strippgen, 2009) ran into the same problems as the earlier parallel DEQSim also written in C/C++ (Charypar et al., 2007a): The time necessary to transmit the necessary information back and forth between the Java-based MATSim and the C/C++-based external package used up all the performance gains. In consequence, the DEQSim was re-implemented as the so-called JDEQSim in Java (Waraich et al., 2015, also see Section 4.3.2). Before parallelizing the JDEQSim, however, it was decided to first accelerate the processing of the events since that was identified as the main bottleneck. Section 4.2.3 describes

How to cite this book chapter:

Nagel, K. 2016. Other Experiences with Computational Performance Improvements. In: Horni, A, Nagel, K and Axhausen, K W. (eds.) *The Multi-Agent Transport Simulation MATSim*, Pp. 267–268. London: Ubiquity Press. DOI: http://dx.doi.org/10.5334/baw.40.

how to use parallel events handling. The parallel version of the JDEQSim (Waraich et al., 2015) never made it into the MATSim main repository.

At the same time, the standard QSim was improved by other people, for example by keeping track of active links and not doing any computation on links without activity. Ch. Dobler made the QSim multi-threaded. He reported (Dobler, 2013, Chapter 5) close-to-linear speed-ups with large scenarios, but only small—if any—performance gains with small scenarios. That is, multi-threading helped greatly with overall computing times for large scenarios on large shared-memory computers, but little with with quick turn-around during experimentation. More recent hardware seems to have improved the situation also for small scenarios (Table 40.1) so that it was eventually decided to remove the single-threaded variant of the QSim and concentrate development on the multi-threaded variant only.

Lämmel et al. (2016) experiments with using Protocol Buffers (Protocol Buffers web page, accessed 2015) in order to couple two different mobsims.

The PSim (Chapter 39) addresses the problem from a different angle: Rather than accelerating the QSim itself, it attempts to make use of the fact that (1) adding or removing a small number of synthetic travelers does not change congestion patterns very much and thus alternative plans can be evaluated in parallel, and (2) the congestion patterns generated by the mobsim do not vary that much from one iteration to the next so that the mobsim does not have to be re-run every time after some synthetic travelers have moved to different alternatives.

Märki et al. (2014) and Dobler (2013) point out that the number of iterations to reach equilibrium can be reduced when the synthetic travelers perform within-day re-routing – this points into the same direction as Lu et al. (2015) who claim that equilibrium iterations will not be necessary at all with well-calibrated behavioral models and a realistic starting point.

MATSim needs, at least for large scenarios, a large amount of RAM. One could say that within the usual space-time tradeoff in computation,[1] in most situations MATSim rather consumes more memory in order to reduce the computation time. Memory-saving compressed routes are available as an option in the `<plans>` section of the config file. MATSim can be seen as an object-oriented database in RAM; attempts to provide a backing by a relational database were not successful when they were tried (Raney and Nagel, 2004, 2006, ; also see Section 46.2.1.3).

To summarize: (1) The behavioral parts of MATSim parallize easily; the main challenge is the mobsim. (2) The main challenge with parallelizing the mobsim is not so much the pure performance improvement, but to achieve this in a way that it remains integrated with the MATSim main development, and at little or no additional maintenance effort.

Computer	population size	1 thread	4 threads	6 threads	8 threads
laptop 2010	1% = 23 500	432 sec	(X)	(X)	(X)
laptop 2014	1% = 23 500	110 sec		57 sec	55 sec
laptop 2014	10% = 235 000			200 sec	

"(X)" means that the laptop was no longer useful for secondary tasks.

Table 40.1: Computing times of the mobsim for the Gauteng scenario (see Chapter 69) with 523 000 links for different computers, different population sizes, and different numbers of threads. "laptop 2010" refers to a high end Mac Pro laptop from 2010, "laptop 2014" refers to a high end Mac Pro laptop from 2014. We can see a speed increase close to a factor of four from 2010 to 2014, and then in 2014 an additional factor of two with multi-threading. These results were shown at several seminars, but never published elsewhere.

[1] See https://en.wikipedia.org/wiki/Space-time_tradeoff.

SUBPART TWELVE

Other Modules

CHAPTER 41

Evacuation Planning: An Integrated Approach

Gregor Lämmel, Christoph Dobler and Hubert Klüpfel

41.1 Basic Information

Entry point to documentation:

`http://matsim.org/extensions` → evacuation

Invoking the module:

`http://matsim.org/javadoc` → evacuation → `RunEvacuationExample` class

Selected publications:

Lämmel (2011); Lämmel et al. (2009)

This chapter presents an integrated approach for performing evacuation simulations with MATSim using the evacuation contribution. The approach comprises all workflow steps for performing an evacuation analysis: i.e., selecting the evacuation area and defining the population, specifying behavioral parameters (i.e., pre-movement time distribution and mode of evacuation—car or pedestrian) and analyzing the simulation output. These steps can all be performed within one graphical user interface. Additionally, two extensions of MATSim for simulating public transport and changing the network during simulation (i.e., network change events) are accessible from the GUI. In this chapter, the steps for performing such an integrated analysis are described and illustrated based on the Hamburg-Wilhelmsburg example. A detailed case study based on this scenario is given in Chapter 71, as well as in Durst et al. (2014); Hugenbusch (2012).

41.2 Related Work

Simulation of evacuation processes has attracted much attention in recent decades; reasons include increases in frequency and severity of natural hazards jeopardizing various populations

---

How to cite this book chapter:

Lämmel, G, Dobler, C and Klüpfel, H. 2016. Evacuation Planning: An Integrated Approach. In: Horni, A, Nagel, K and Axhausen, K W. (eds.) *The Multi-Agent Transport Simulation MATSim*, Pp. 271–282. London: Ubiquity Press. DOI: http://dx.doi.org/10.5334/baw.41. License: CC-BY 4.0

and regions, as well as (social) disasters (Rodríguez et al., 2006). Another factor is the availability of large-scale, fast simulation models and tools. Lämmel (2011) discusses such a model employed as a contribution to MATSim. Basically, this model implements the same iterative learning approach as that applied to "regular" transport scenarios. In the first instance, the cost function comprises only travel times, albeit a combination of travel time and travel distance; as a cost function has been investigated as well (Lämmel et al., 2009). Artificial agents represent evacuees trying to improve their evacuation plans from iteration to iteration, by creating new evacuation plans more responsive to the evolving situation. A typical simulation run comprises 500-1 000 iterations. The model is applied to a tsunami-related evacuation of the City of Padang in Indonesia (e.g., Taubenböck et al., 2013; Goseberg et al., 2013); some scenario details are discussed in Chapter 76.

Additional evacuation simulation related work in MATSim is presented by Dobler (2013). The main difference to this chapter's approach is that agents are allowed to adapt their plans spontaneously, using MATSim's within-day replanning framework (Dobler et al., 2012) (Chapter 30). Based on a behavioral model, agents coordinate their actions on a household level. If a household is, e.g., not complete when the evacuation starts, each member estimates time needed to return home, as well as the time required to leave the actual evacuation area. Then, the household decides whether meeting at home and leaving together is preferable to each member leaving on its own. Since the behavioral model is implemented on an agent, respectively household level, individual attributes such as children present in the household, or availability of a car, can be taken into account. In contrast to regular MATSim simulations, only a single iteration is performed. Since the agents can optimize their plans continuously using real time information, no further replanning is necessary. As a result, agents do not foresee future events like traffic jams caused by people leaving the threatened area.

An independent evacuation scenario, not using the evacuation, is presented in Chapter 60.

The remainder of this chapter is organized as follows: Section 41.3 gives a brief description on how to set up and run evacuation. A short start guide for evacuation is presented in Section 41.4. Obtaining the required input data is discussed in Section 41.5. Detailed instructions on how to use evacuation's `ScenarioManager`, running simulations and analysis is given in Section 41.6. This chapter concludes with a brief outlook in Section 41.7.

41.3 Download MATSim and Evacuation

Although the MATSim version `0.6.0-SNAPSHOT` is referred to here, the package should also work with later versions of MATSim.

1. Download the current nightly build of MATSim and evacuation from `http://matsim.org/files/builds/`.
2. Unzip the `Matsim_rxxxxx.zip`, `Matsim_libs.zip` and `evacuation-0.6.0-SNAPSHOT-rxxxxx.zip`.
3. Move the `evacuation-0.6.0-SNAPSHOT-rxxxxx.jar` and `libs` folder from the `evacuation-0.6.0-SNAPSHOT-rxxxxx` directory one level up, i.e., to the directory, where `MATSim_rxxxxx.jar` is located.

Test configuration by invoking

```
java -cp evacuation-0.6.0-SNAPSHOT.jar;MATSim_rxxxxx.jar
   org.matsim.contrib.evacuation.scenariomanager.ScenarioManager
```

(It is advisable to copy that command to a file `evacuation.bat`—or `evacuation.sh`, if using a Unix-like operation system. One can then run that file instead of typing the command.)

41.4 The Fifteen-Minute Tour

For just a quick impression, the following steps can be performed within a few minutes:

OSM Go to `http://www.openstreetmap.org`, search for the desired place and download a (small) OSM file. Please choose a small area, e.g., 500 meters by 500 meters; this is sufficient to begin and size of the exported area is limited. For larger areas, a direct download from sites like `http://www.geofabrik.de` is preferable (see next section).

Run the `ScenarioManager` as described in the previous section.

Create a scenario by clicking the leftmost button first and then `New`. Go to the directory where the designated project should be saved and name the project file (e.g., `london.xml` or `scenario.xml`).

Specify the path of the OSM file (by clicking `Set` next to network) and the output directory. Leave area and population file as it is, evacuation will handle this. **This step must be performed only once. After the scenario-file has been saved, one can open it in the `ScenarioManager`.**

Sample size Set the sample size to 0.1, using the mouse or the cursor buttons on your keyboard.

Departure Specify the departure time distribution. Plausible values are: normal distribution, μ and σ 600 seconds (10 minutes), earliest 300 seconds, latest 1200 seconds (20 minutes).

Save the scenario file.

Area Switch to the area tab. One can define the circular evacuation area by keeping the left mouse button pressed and defining the center and radius. Do not forget to save changes.

Population Switch to the population tab and define the population (handling is similar to area). Do not forget to save changes.

Convert Switch to the next tab and convert the scenario to MATSim input files by clicking the `run` button. The MATSim files will be stored in the output directory specified in the beginning.

Run the MATSim simulation by skipping the next two tabs/buttons (road closures and buses) and switching to the simulation tab (with the "M" for MATSim on the computer screen). Click `run`. This will take a while. **If an output directory (e.g., from a previous run) already exists, it will be renamed.**

Analyze your simulation results by switching to the final tab after the simulation is finished.

41.5 Input Data (any Place and any Size)

The only external input necessary for performing an evacuation analysis with `org.matsim.contrib.evacuation` is an OSM file. In this tutorial, we will use the file for Hamburg, Germany. Please go to `http://download.geofabrik.de/europe/germany/hamburg.html` and download the `hamburg-latest.osm.bz2` file. This is the only initial preparation needed. Everything else can be done with the `ScenarioManager` of the GUI.

41.6 Scenario Manager

The scenario setup, evacuation simulation, and analysis are handled by the `ScenarioManager` from the MATSim contribution package `org.matsim.contrib.evacuation`.

41.6.1 Scenario Configuration

At startup, the `ScenarioManager` offers the option to either: define a new scenario configuration or open an existing one from a XML file, which then can be modified. Figure 41.1 shows a screenshot of a scenario configuration in the `ScenarioManager` and the corresponding XML file, respectively.

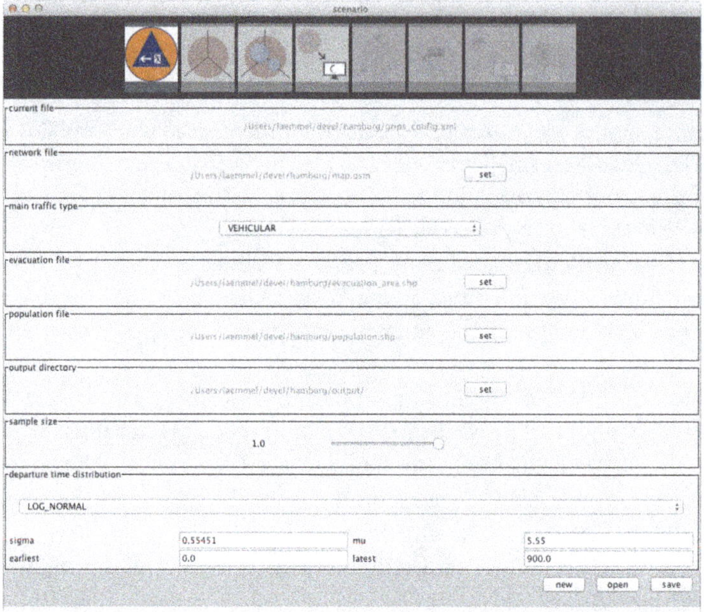

(a) ScenarioManager.

```
<?xml version="1.0" encoding="UTF-8" standalone="yes"?>
<grips_config xsi:schemaLocation="http://www.matsim.org/files/dtd http
://matsim.org/files/dtd/grips_config_v0.1.xsd" xmlns:xsi="http://www.w
3.org/2001/XMLSchema-instance">
    <networkFile>
        <inputFile>/Users/laemmel/devel/hamburg/map.osm</inputFile>
    </networkFile>
    <mainTrafficType>vehicular</mainTrafficType>
    <evacuationAreaFile>
        <inputFile>/Users/laemmel/devel/hamburg/evacuation_area.shp</i
nputFile>
    </evacuationAreaFile>
    <populationFile>
        <inputFile>/Users/laemmel/devel/hamburg/population.shp</inputF
ile>
    </populationFile>
    <outputDir>
        <inputFile>/Users/laemmel/devel/hamburg/output/</inputFile>
    </outputDir>
    <sampleSize>1.0</sampleSize>
    <departureTimeDistribution>
        <distribution>log-normal</distribution>
        <sigma>0.55451</sigma>
        <mu>5.55</mu>
        <earliest>0.0</earliest>
        <latest>900.0</latest>
    </departureTimeDistribution>
</grips_config>
```

(b) XML file.

Figure 41.1: Illustration of a configuration opened in the ScenarioManager and as XML file.

The evacuation scenario is specified by the following parameters:

- The path to the network file covering the evacuation area: Currently, OSM XML files are supported (*.osm).
- The main traffic type for the simulation: This can either be: VEHICULAR or PEDESTRIAN. Depending on the choice, a vehicular specific (the MATSim default) or a pedestrian-specific

(as discussed in Lämmel et al. (2009); Lämmel (2011)) simulation network will be generated by setting free speed, number of lanes and flow capacity for all links in the network.

- The path to a ESRI shape file describing the extent of the evacuation area, depicted by a simple polygon. This file does not have to be in place right from the beginning; it can be produced manually by the `ScenarioManager` itself, as discussed later.
- The path to an ESRI shape file detailing the size and distribution of the affected population. This file comprises a set of simple polygons; each polygon has an additional attribute for the number of persons residing at a location inside that polygon. The evacuation area file can be produced with help of the `ScenarioManager`.
- The path to the output directory where the simulation output and MATSim scenario files will be stored.
- The sample size for the MATSim simulation. A smaller sample size increases the simulation performance, while a larger size might increase accuracy of the results. Typical values are 1.0, 0.1, or 0.01, depending on the scenario and available computing resources.
- Departure time distribution defines the distribution from which departure times for the simulation will be drawn, based on the premise that, in real evacuation situations, all participants probably do not start evacuation simultaneously. People tend to perform pre-evacuation activities before they start, including: picking up relatives, packing food, clothes, valuable belongings, etc. Since number and duration of these activities differs by individual, population departure times are unknown quantities. The `ScenarioManager` supports three different distributions: (Dirac-delta, normal, and log-normal). If the user chooses the Dirac-delta distribution, then all evacuees will start simultaneously, which might be the worst case (Lämmel and Klüpfel, 2012). By choosing the normal distribution, departure times for individuals are drawn from a normal distribution with mean μ and standard deviation σ, where the parameters μ and σ are given in seconds. As an example, setting $\mu = 1800$ and $\sigma = 900$ will result in a departure time distribution where, on average, after 30 minutes 50 % of the population has departed and 68.3 % of the population departs in time intervals of 30 minutes $\pm$ 15 minutes. If the user chooses log-normal as the distribution, departure times are drawn from a log-normal distribution, where μ and σ are the parameters of the associated normal distribution (a discussion on this matter is given below). The parameters *earliest* and *latest* determine the earliest and latest possible departure time. The normal and log-normal departure time distribution are truncated accordingly.

The departure time distribution is perhaps the most tenuous parameter to set; the authors found no holistic research into this matter. In general, it seems reasonable to assume that many people start evacuating at the same time, or soon after the evacuation order has been issued and as time proceeds, fewer and fewer people are left to depart. This requires a departure time distribution that has a probability density function beginning with a steep positive gradient, leveling out slowly after a peak. The probability density function of a log-normal distribution produces this kind of curve; log-normal and normal distributions are closely related. If the random variable Y is normal distributed, then $X = \exp(Y)$ is log-normal distributed. The expected value $E[X]$ and the variance $Var[X]$ are

$$E[X] = \exp(\mu + \frac{\sigma^2}{2}) \tag{41.1}$$

and

$$Var[X] = \exp(2(\mu + \sigma^2)) - \exp(2\mu + \sigma^2). \tag{41.2}$$

Conversely, if the expected value and variance is given, μ and σ of the associated normal distribution can be obtained as follows:

$$\sigma = \sqrt{\log(1 + \frac{Var[X]}{(E[X])^2})} \tag{41.3}$$

and

$$\mu = \log(E[X] - \frac{1}{2}\sigma^2). \tag{41.4}$$

If users wish to generate a population with departure times following a log-normal distribution, it is hard to see how σ and μ will determine the outcome. It is much more convenient to consider expected value and variance. Given Equation (41.3) and Equation (41.4), a conversion from expected value and variance to σ and μ is straightforward.

41.6.2 Evacuation Area

The `ScenarioManager` integrates modules for the evacuation area definition and distribution of the affected population. The so-called evacuation area selection module allows the user to define the evacuation area by drawing either a simple polygon or circle on a map. The application can make use of either a WMS-provider or a tile map provider (e.g., OSM) as background map renderer. Zooming and panning is restricted to the bounding box of the OSM network file provided in the scenario configuration. An illustration of the evacuation area selector is given in Figure 41.2. In addition to defining a new evacuation area, a pre-existing one can also be loaded into the `ScenarioManager`. The requirements for a pre-existing evacuation area file are:

- It has to be provided as a ESRI shape file.
- The evacuation area must be defined as a simple polygon or a multi-polygon containing one, and only one, simple polygon.
- The coordinate reference system for polygon in the ESRI shape file must be set correctly.

Due to the high likelihood of error, this approach is recommended for experienced users only.

Later in the process, the `ScenarioManager` takes the evacuation area to cut out an evacuation network. However, after cutting out the evacuation net, there is no particular node as a target for the route calculation, as evacuees have more than one safe place as a destination. Instead, in the underlying domain, every node outside the evacuation area is a possible destination for an evacuee seeking an escape route. Thus, the evacuation problem is, in general, a multi-destination problem. To resolve this, the standard approach (e.g., Ford and Fulkerson (1962); Lu et al. (2005)) is to extend the network in the following way: All exit links (i.e., links that originate inside the evacuation area and terminate outside the evacuation area) are connected, using virtual links with very high (essentially infinite) flow capacity and equal length, to a super-node; all evacuation routes are routed to the super-node. This way, the problem is reduced to a multi-source single-destination problem. And thus, finding the shortest path from any node inside the evacuation area to this super-node and, in consequence, to safety, can efficiently be solved. For technical reasons, a super-link is added to the super-node and the evacuees are routed to that link (see the image at the beginning of this chapter).

41.6.3 Evacuation Demand

The process of defining the population distribution is similar to that of the evacuation area, differing because population is distributed over circles drawn on the map. The user can draw

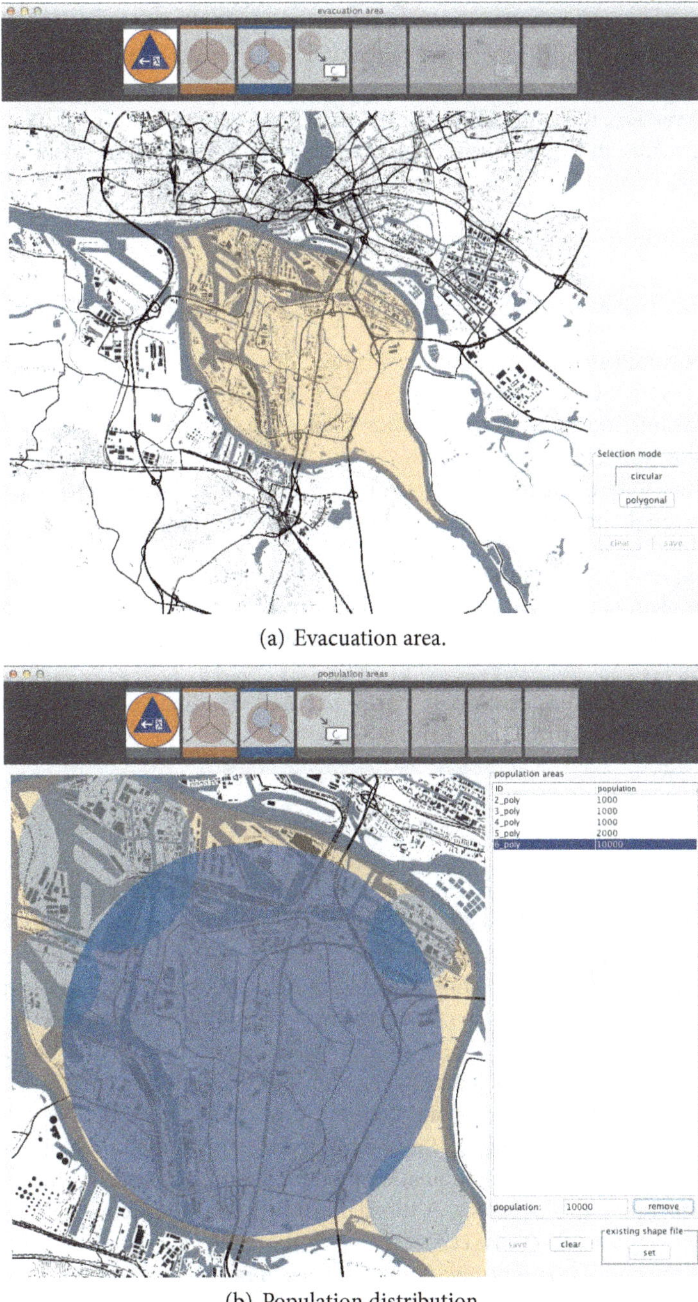

(a) Evacuation area.

(b) Population distribution.

Figure 41.2: The evacuation area and the population distribution can be defined with an integrated GIS application.

an arbitrary number of those circles and define population figures individually for each circle. Figure 41.2 illustrates the population editor. The population editor offers only basic functionality to define a population distribution. For every circular area, the `ScenarioManager` produces as many agents as required and assigns each agent a random coordinate inside the circular area. However, in MATSim agents depart on links, so the `ScenarioManager` calls the `getNearestLink()` method defined in `NetworkImpl`. Thus, agents will depart on links inside and possibly near the circular areas.

In the current version, it is impossible to use a predefined demand for the simulation. Extending the simulation package in this way would be straightforward, but is out of this work's scope.

41.6.4 Road Closures

In real situations, some evacuation roads might not be available for the evacuation, because:

- They might be impassable due to the event (often the case in flooding-related evacuations).
- The authorities might want to keep roads open only for action/help traffic.
- In some situations, like hurricane evacuations, lane direction on motorways might be reversed to increase flow capacity in one direction.
- The authorities have detailed evacuation plans in place, with pre-planned evacuation routes; road closures might be necessary to force evacuees onto certain routes.

The actual planning of road closures can be a complex undertaking; not all attributes can be integrated into a simple tool for rapid evacuation planning. Nevertheless, the `ScenarioManager` offers a tool to create time-dependent road closures. An illustration of the road closures editor is given in Figure 41.3(a).

Road closures are stored as `NetworkChangeEvents` and handled as time-dependent network attributes in MATSim (Lämmel et al., 2010).

41.6.5 Bus Stop Editor

Usually, not everyone has access to a private car. In the event of an evacuation, those people often rely on public transport. In regions prone to natural disasters, local authorities normally have detailed evacuation plans in place, probably including evacuation by public transport. Consequently, it is important to have a tool available to help integrate public transport into to the simulation scenario. The `ScenarioManger` offers this possibility by defining bus stops and bus schedules in the interactive GUI. Figure 41.3(b) gives an example of the bus stop editor. In addition to location, the user can define when the first bus will serve a bus stop, how many buses overall will serve this particular bus stop and these buses' capacity. The `ScenarioManager` transforms the inputs made into the GUI into a MATSim compatible transport schedule, enriching the scenario while using the same simulation model. Details about public transport simulations with MATSim are given in Chapter 16. A tutorial can be found on the MATSim webpage `http://matsim.org/docs/tutorials/transit`.

Limitations of the public transport evacuation approach in this project are:

- Each bus serves one and only one bus stop, perhaps a realistic assumption.
- Buses always take the shortest path from their designated bus stops to the safe area. As the shortest path is not necessarily the fastest, this approach might lead to avoidable delays. Some newer research investigates optimization of bus lines with respect to traffic demand and traffic conditions (Neumann, 2014). Implementing such optimization techniques in the evacuation context is a topic of future research.

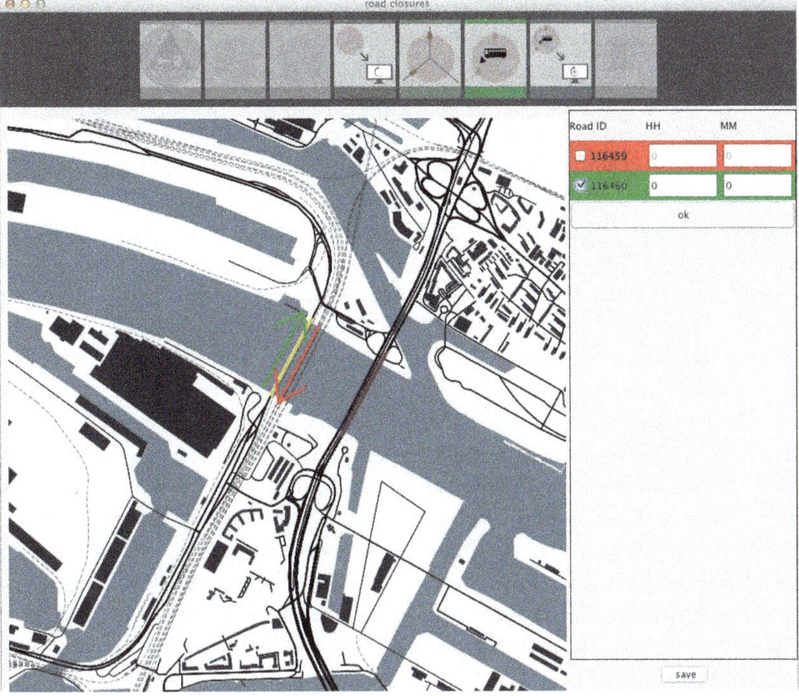

(a) Road closures.

(b) Bus stop locations and schedules.

Figure 41.3: Top: Road closures can be edited by an integrated GIS application. For every link the direction and the time of closure can be defined. Bottom: Tool that let the user define bus stop locations and schedules.

41.6.6 Running the Scenario

The `ScenarioManager` runs the evacuation simulation in a way similar to other transport simulation studies with MATSim. At the beginning, an evacuation plan is assigned to each evacuee. An evacuation plan describes how the evacuee intends to reach the safe area. If the evacuee leaves by car or on foot, the plan is essentially comprised of a route (typically the shortest) from home to the safe area. For evacuees who are depart by public transportation, the plan can be much more complex. All these evacuation plans will be executed in the mobility simulation; after this terminates, all plans are scored by travel time. The shorter a plan's travel time is, the higher is the score it receives. After this step, evacuees' plans are revised; some will receive new plans, while others continue with the current ones. This step is called re-planning. Mobility simulation, scoring, and re-planning are repeated in a loop for a predefined number of iterations; evacuees' individual performance improves over the iterations. In general transport studies, this approach emulates real-world travelers' behavior when they perform their daily commutes and try to find better travel alternatives. Evacuations, however, are singular events where such day-to-day re-planning would not occur. We argue here that the chosen iterative learning approach could be seen as the evacuees' anticipation of the conditions expected during an evacuation. People familiar with their surroundings would probably avoid roads that obviously constitute bottlenecks during an evacuation. Nevertheless, far more research is needed to definitively answer how people choose evacuation routes, or how many learning iterations are required to realistically reflect assumed anticipation skills adequately. As a rule of thumb, running 100 learning iterations are usually sufficient to achieve results constituting a lower evacuation times boundary.

41.6.7 Analysis

After the last iteration has finished, the `ScenrioManager` enables the analysis module. The analysis model evaluates the performed simulation run, using a number of different methods.

- The cumulative arrival curve tells the user the number of persons evacuated over time. From this curve, the user can, for example, learn at what time 50 % of the population has reached a safe destination.
- The GIS-based evacuation time analysis draws a grid over the evacuation area and computes, for every grid cell, average evacuation time. The evacuation times are indicated by different colors; the analysis modules run a quantiles-based clustering analysis for each cell. The size of cells can be varied by the user.
- The GIS-based clearance time analysis is performed in the same way as the evacuation time analysis. The clearance time of a cell is the time when the last evacuee leaves that cell. This evacuee is not necessarily the one who also started his/her evacuation inside the corresponding cell, but might also be one who crosses that cell somewhen during the evacuation.
- A similar quantiles-based clustering approach is used for the link utilization analysis. The link utilization analysis results help the user to identify the major evacuation routes.

The analyses can be run for every single iteration for which the MATSim `Controler` has dumped an events file (every 10th iteration by default). An overview of the analysis module is given in Figure 41.4

41.7 Conclusion

This chapter demonstrates how rapid evacuation planning can be performed with help of the evacuation contribution. The evacuation contribution provides an interactive GUI to perform this task. The only required external input is a network file extracted from OSM, thus a simple scenario

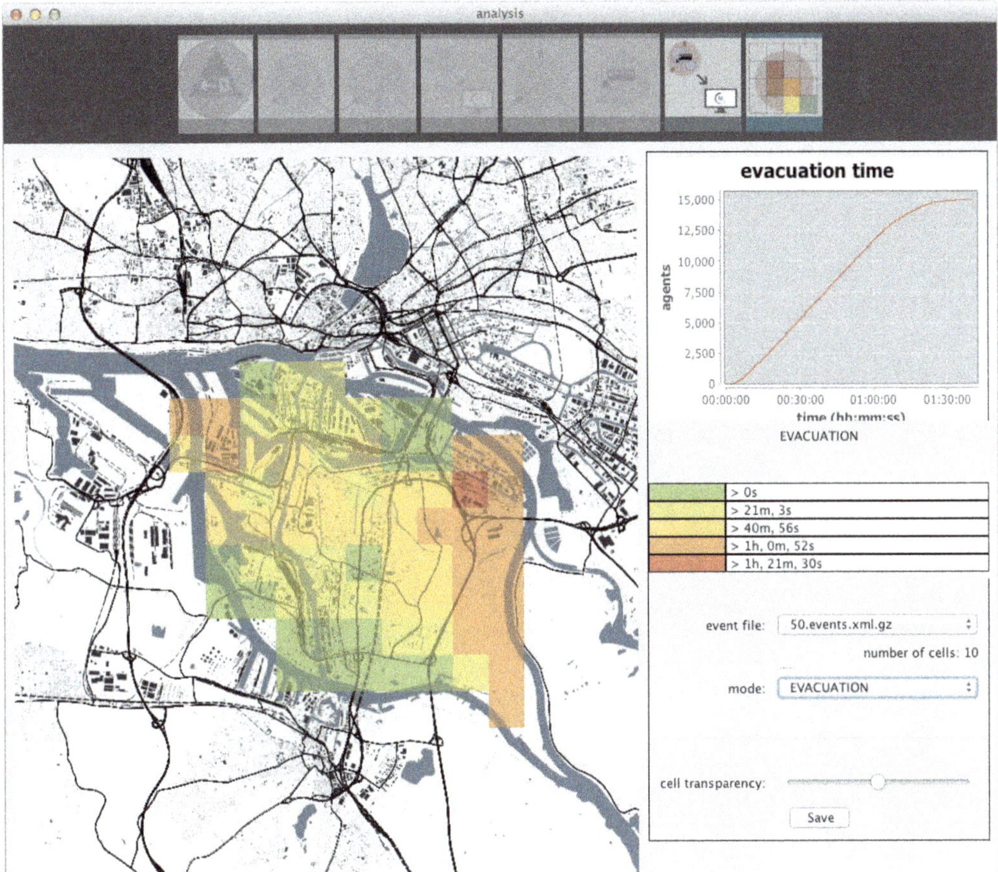

Figure 41.4: Screenshot of the analysis module showing GIS-based evacuation time analysis and the evacuation curve.

can be setup, simulated, and analyzed in less than an hour. Obviously, for an in-depth evacuation analysis of a certain area, a sort of expert knowledge is needed that a simple GUI can not supply. Still, for a rapid appraisal and for demonstration purposes, evacuation offers a powerful and easy-to-use tool. In the future, we plan to integrate a more advanced public transport planning tool based on Neumann (2014). Work is also ongoing to develop a more sophisticated pedestrian simulation model based on the theoretical framework given in Flötteröd and Lämmel (2015).

CHAPTER 42

MATSim4UrbanSim

Kai Nagel

42.1 Basic Information

Entry point to documentation:

`http://matsim.org/extensions` → matsim4urbansim

Invoking the module:

The module is invoked from a live UrbanSim implementation.

Selected publications:

Nicolai et al. (2011); Nicolai and Nagel (2014); Nicolai and Nagel (2015)

42.2 Summary

"MATSim4UrbanSim" is an adapter package for using MATSim as a travel model plug-in to UrbanSim, a well-known land use simulation (e.g., Waddell et al., 2003, see `http://www.urbansim.org`). UrbanSim has, for example, submodels for residential location choice, commercial location choice, or development and building construction, thus creating synthetic potential urban or regional development scenarios under various conditions and constraints. Traffic infrastructure plays a significant role in such developments; for example, very accessible areas are more attractive as residences and for commercial activities. Since accessibility is reduced by congestion, and congestion can only be realistically modeled through a sophisticated model of demand and supply interaction, UrbanSim does not have its own travel model, but delegates that task to external models, such as MATSim.

To use MATSim4UrbanSim, one first needs to have a running UrbanSim installation. From there, one can add MATSim to that installation; see the documentation mentioned above for more

How to cite this book chapter:

Nagel, K. 2016. MATSim4UrbanSim. In: Horni, A, Nagel, K and Axhausen, K W. (eds.) *The Multi-Agent Transport Simulation MATSim*, Pp. 283–284. London: Ubiquity Press. DOI: http://dx.doi.org/10.5334/baw.42.

information. Basic MATSim parameters are configured from the UrbanSim configuration file by adding an appropriate section; again, see the documentation mentioned above for more information. It is possible to add a standard MATSim config file allowing use of the extended MATSim features, including those added after the adapter package was designed.

The module was applied by Cabrita et al. (2015) and by Zöllig Renner (2014).

CHAPTER 43

Discontinued Modules

Kai Nagel and Andreas Horni

This chapter lists modules that were important for several projects in the past, but which are no longer being developed.

43.1 DEQSim

DEQSim was used for project *Westumfahrung* (Balmer et al., 2009a). It was a queue-based, event-based parallel simulation written in C++ (Charypar et al., 2007b; Charypar, 2008). This simulation included handling of reduced capacities due to traffic lights in an aggregate manner (Charypar, 2008, p.139 ff). It also supported modeling of gap back propagation at junctions (Charypar, 2008, p.98 ff).

Events were written do file by DEQSim and subsequently read by MATSim. This represented a major framework performance bottleneck. DEQSim was therefore replaced by a Java version, the JDEQSim (see Section 4.3.2).

43.2 Planomat

Chapter 45 explains how MATSim can be extended. One long-standing extension point is the `PlanStrategy` extension point (Section 45.2.9). It allows the addition of "innovative" strategy modules (see Section 4.5), above and beyond those available by default.

One such replanning model was Planomat (Meister et al., 2006; Meister, 2011). It replaced the randomizing modules for (departure) time innovation (Section 4.5.1.1) and for mode innovation (Section 4.5.1.3), with a module that computed a joint best reply for both choice dimensions internally, using a Genetic Algorithm. Thus, it evaluated not just one random alternative per iteration, as standard MATSim would do, but multiple alternatives within one single iteration, to obtain an (at least locally) optimal solution. Planomat was successfully applied in the project

How to cite this book chapter:
Nagel, K and Horni, A. 2016. Discontinued Modules. In: Horni, A, Nagel, K and Axhausen, K W. (eds.) *The Multi-Agent Transport Simulation MATSim*, Pp. 285–286. London: Ubiquity Press. DOI: http://dx.doi.org/10.5334/baw.43.

"KTI Frequencies" for time and mode innovation for sub-tours (Balmer et al., 2010, p.10). Unfortunately, there were three interacting problem complexes with Planomat:

- Any strategy module generating best reply plans must be able to compare plans and select a better one, at least along the considered choice dimensions. This is typically achieved by giving such a module an objective function which needs to be optimized. For example, a fastest path router minimizes the travel time; a generalized cost router minimizes the generalized travel cost.

  All best reply modules here face the challenge that they cannot run the full mobsim (= network loading = synthetic reality) every time they need such information. As a result, all best reply modules are forced to build some internal synthetic reality model.

  Planomat did this by running its plans through a simplified mobsim of its own. This mobsim was a re-implementation of the most important aspects for the core mobsim. Unfortunately, however, this meant that Planomat would not automatically pick up any change or addition to the core mobsim. Consequently, Planomat's idea of a good plan often diverged from MATSim's, especially when MATSim extensions were used. In other words, any addition to the MATSim system: e.g., tolls, or opening/closing time restrictions, or differentiating link travel times by turning directions, would have to be mirrored inside Planomat.
- Planomat always tended to return the same solution: understandable from a best-reply module, but it becomes a problem when what the module thinks is a best reply starts to differ from what the MATSim core thinks.

  While an innovative strategy that deliberately generates diversity can be useful even when not fully consistent with the MATSim core (Nagel et al., 2014), this cannot function with a non-diverse innovative strategy, since it then insists on returning only suboptimal plans.
- In addition, Planomat used the MATSim core router in a way that hindered further software development of the core router. Essentially, Planomat used MATSim classes and methods that were not designed for re-use, but just happened to be public.

  It was thus an obstacle for a major MATSim core router re-design, undertaken by T. Dubernet (see Section 45.2.7).

The combination of these three issues meant that Planomat was eventually discarded: Moving it to the new router infrastructure would have entailed a major piece of one-time work. After that, maintaining Planomat's best-reply capability would have been a permanent work-intensive obligation. It was thus decided instead to invest our scarce resources in the design of a better core, allowing extensions to survive without much manual intervention. Although this will always be work in progress, Chapter 45 explains our substantial progress toward pluggable extensibility.

However, it must be noted that the improved software architecture does not resolve the general conceptual problem; best reply modules somehow need to follow core system development. Chapter 39 discusses a newer alternative that re-uses mobsim output for plan evaluation without having to run the full mobsim every time. An alternative approach, based on plan diversity, is investigated by Nagel et al. (2014). Additionally, Chapter 49 discusses aspects of diversity in plan set generation.

43.3 PlanomatX

PlanomatX was based on Planomat. It extended it by performing activity choice and adopting a Tabu Search approach (Feil, 2010). To cope with the curse of dimensionality (due to the added choice dimension), PlanomatX introduced schedule recycling, basically a warmstart concept. Because of problems when using the standard MATSim logarithmic utility function for activity choice, PlanomatX also derived an alternative utility function from Joh (2004). Rough estimates for its parameters based on an MNL exist, but turned out to be problematic, as shown in Section 97.4.3.

PlanomatX, derived from Planomat, suffered from the same maintenance problems and was eventually abandoned for the same reasons.

SUBPART THIRTEEN

Development Process & Own Modules

CHAPTER 44

Organization: Development Process, Code Structure and Contributing to MATSim

Marcel Rieser, Andreas Horni and Kai Nagel

This chapter describes how new functionality enters MATSim. It describes the MATSim team and community, the different roles existing in the MATSim project, the development drivers and processes, and the tools used for integration. The goal is to provide an overview of the development process so that one quickly finds access to the MATSim community and is able to efficiently contribute to MATSim, based on one role or another.

44.1 MATSim's Team, Core Developers Group, and Community

The **MATSim team** currently consists of three research groups and a spin-off company:

- the VSP (VerkehrsSystemPlanung und Verkehrstelematik – The Transport Systems Planning and Transport Telematics group at TU Berlin) group at the ILS (Institut für Land- und Seeverkehr – Institute for Land and Sea Transport Systems), TU Berlin, led by Prof. Dr. Kai Nagel,
- the VPL (VerkehrsPLanung) group at the IVT, ETH Zürich, led by Prof. Dr. Kay W. Axhausen,
- the recently founded Mobility and Transportation Planning group at the FCL, based in Singapore and led by Prof. Dr. Kay W. Axhausen, and
- Senozon AG, based at Zürich with a subsidiary in Germany, founded by former PhD (Philosophiae Doctor – Doctor of Philosophy) and research students.

As is common in research, the university groups' composition changes frequently. Over the last decade more than 50 people, as listed earlier, contributed to MATSim.

How to cite this book chapter:

Rieser, M, Horni, A and Nagel, K. 2016. Organization: Development Process, Code Structure and Contributing to MATSim. In: Horni, A, Nagel, K and Axhausen, K W. (eds.) *The Multi-Agent Transport Simulation MATSim*, Pp. 289–296. London: Ubiquity Press. DOI: http://dx.doi.org/10.5334/baw.44.

A small group of the MATSim team defines the **MATSim core developers group**, maintaining MATSim's core as defined below in Section 44.3.2.

In addition, there is a **MATSim community** composed of closely connected research groups in other cities, e.g., Stockholm, Pretoria, Poznan, and Jülich, as well as more loosely connected external users coming together, e.g., at the annual MATSim User Meeting (see Figure 44.1).

MATSim is open-source software under the GPLv2 (GNU General Public License version 2.0). You are also very welcome to contribute to the code base as described in Section 44.6. New contributors are mentored in the beginning to become familiar with the project and the coding conventions.

44.2 Roles in the MATSim Community

The MATSim community includes the following roles:

- The **MATSim user** uses the official releases or nightly builds and runs the MATSim core with the config file (Section 5.1.1). He or she does not write computer code. Part I of the book is dedicated to the MATSim user. On the web page, he or she finds relevant information in the *user's guide* section and in the user's mailing list `users@matsim.org`.[1] There is also a list of questions and answers under `http://matsim.org/faq`.

  Users should also remember to consult the files `logfileWarningsErrors.log` and `output_config.xml.gz`, as also explained in Section 2.3. The former file is an extract from `logfile.log`, but only contains the warnings and errors. The latter is a complete dump of the currently available configuration options, including comments to most options.
- The **MATSim power user** is a MATSim user with knowledge on how to use the additional modules presented in the book's Part II. He or she does *not* program but knows how to use MATSim scripts-in-Java prepared by others or her/himself, as shown in Section 5.1.1. Parts I and II of the book are helpful to the MATSim power user. Information about extensions can be found under `http://matsim.org/extensions`. Most extensions come with an example script-in-Java. Again, questions and answers are under `http://matsim.org/faq`.
- The **MATSim developer** extends MATSim by programming against the MATSim API (Section 5.1.1). He or she also finds his or her information in Part II of the book, in particular, in Chapter 45, on the web page in the Developer's Guide, and in the mailing list `developers@matsim.org`.
- There are relatively few **MATSim core developers** in the MATSim team. These persons make necessary modifications of the core (as defined in Section 44.3.2), usually after having discussed them in the issue tracker (`http://matsim.org/issuetracker`), in the MATSim committee, or at a developer meeting (see below).

44.3 Code Base

The various pieces of MATSim are delineated by Apache Maven projects and sub-projects. The Apache Maven layout corresponds to the layout of the Git repository.[2] Note that the Java package structure does *not* directly correspond to the Apache Maven/Git layout.

---

[1] During the writing of this book, the information that had so far been contained in the User's Guide was moved to this book. Therefore, the User's Guide section on the web page is currently essentially empty, and may be removed.

[2] MATSim is currently at GitHub under `https://github.com/matsim-org/matsim`. The exact path name may change in the future, e.g., because of changes at GitHub.

Figure 44.1: MATSim events and community.

Source: ©Dr. Marcel Rieser, Senozon AG

44.3.1 Main Distribution

The "MATSim main distribution" corresponds to the "matsim" part of the Git repository. It is the part of the code that the MATSim team feels primarily responsible for. At the time of writing, the MATSim main distribution contains following packages:

- `org.matsim.analysis.*`, containing certain analysis packages that are added by default to every MATSim run.
- `org.matsim.api.*`, see Section 44.3.2.
- `org.matsim.core.*`, see Section 44.3.2.
- `org.matsim.counts.*`, see Section 6.3.
- `org.matsim.facilities.*`, see Section 6.4.
- `org.matsim.households.*`, see Section 6.5.
- `org.matsim.jaxb.*`, containing automatically or semi-automatically generated adapter classes to read XML files using JAXB (Java Architecture for XML Binding).
- `org.matsim.lanes.*`, see Chapter 12.
- `org.matsim.matrices.*`, containing (somewhat ancient) helper classes to deal with matrices, in particular, origin-destination-matrices.
- `org.matsim.population.*`, mostly containing a collection of algorithms that go through the population and modify persons or plans.
- `org.matsim.pt.*`, see Chapter 16.
- `org.matsim.run`, see Section 44.3.2.
- `org.matsim.utils.*` containing various utilities such as the much-used ObjectAttributes (see Section 45.2.2).
- `org.matsim.vehicles.*`, see Section 6.6.
- `org.matsim.vis.*`, containing helper classes to write MATSim information, in particular from the mobsim, to file. This has to a large extent been superseded by the Via visualization package (see Chapter 33).
- `org.matsim.visum.*`, containing code to input data from VISUM.
- `org.matsim.withinday.*`, see Chapter 30.
- `tutorial.*`, containing example code of how to use MATSim, referenced throughout this book.

44.3.2 Core

The core is part of the main distribution (see the previous Section 44.3.1) and contains material that is considered basic and indispensable, and resides in the packages

- `org.matsim.api.*`
- `org.matsim.core.*`
- `org.matsim.run.*`

The MATSim core is maintained by the MATSim Core Developers Group.

44.3.3 Contributions

The idea of the contributions part of the repository is to host community contributions. Historically, most contributors are from the MATSim team, but this is not a requirement.[3] The

[3] It is currently at GitHub under `https://github.com/matsim-org/matsim/tree/master/contribs`.

code is maintained by the corresponding contributor. Code in this section of the repository is considered more stable than code in playgrounds. The Java packages often have the root `org.matsim.contrib.*`, but this is not mandatory.

At the time of writing, there are the following contributions (= extensions which are in the "contrib" part of the repository), listed in alphabetical order:

- `accessibility`, presented in Chapter 35.
- `analysis`, presented in Chapter 38.
- `cadytsIntegration`, presented in Chapter 32.
- `common` is not a true contrib, i.e., it does not provide additional functionality by itself. Instead, it is a place where code used by several contribs, which has not yet made it into the main distribution is located. It also contains some long-running integration tests that are run at each build (i.e., more often than those contained in the `integration` contrib described below).
- `dvrp`, presented in Chapter 23.
- `emissions`, presented in Chapter 36.
- `freight`, presented in Chapter 24.
- `freightChainsFromTravelDiaries`, presented in Chapter 26.
- `grips`, presented in Chapter 41.
- `gtfs2matsimtransitschedule`, presented in Chapter 18.
- `integration` is not a true contrib, i.e., it does not provide additional functionality. Instead, it is a place where integration tests that should run daily or weekly (instead of as often as possible) can be committed.
- `locationchoice`, presented in Chapter 27.
- `matrixbasedptrouter`, presented in Chapter 20.
- `matsim4urbansim`, presented in Chapter 42.
- `minibus`, presented in Chapter 17.
- `multimodal`, presented in Chapter 21.
- `networkEditor`, presented in Chapter 10.
- `otfvis`, presented in Chapter 34.
- `parking`, presented in Chapter 13.
- `roadpricing`, presented in Chapter 15.
- `socnetgen`, presented in Chapter 29.
- `socnetsim`, presented in Chapter 28.
- `transEnergySim`, presented in Chapter 14.
- `wagonSim`, presented in Chapter 25.

44.3.4 Playgrounds

Another element of the MATSim repository is the "playgrounds". These are meant as a service to programmers. They have grown historically from the fact that MATSim's object classes and in consequence the interfaces between them have evolved and grown over time, and thus a stable API was not available. Regular code-wide refactorings, along the lines discussed, e.g., by Fowler (2004), were thus the norm for many years.

At this point, the extension points described in Chapter 45 should be somewhat stable and development against them should be possible without major changes from release to release. Anybody who needs tighter integration with the project should still apply for a playground.

44.3.5 *Contributions and Extensions*

Congruent with the structure of this book, the MATSim code structure contains a core which allows to run basic MATSim using the config file, a population and a network. Packages going beyond this basic functionality are extensions, where three different kind of extensions exist:

- extensions in the main distribution,[4]
- extensions contributed by the MATSim community known as contributions, and
- any code written anywhere published or unpublished extending the MATSim core.

Extensions are listed at `http://matsim.org/extensions`.

44.3.6 *Releases, Nightly Builds and Code HEAD*

Releases, nightly builds and the code head can be obtained from `http://matsim.org/downloads`.

MATSim releases are published approximately annually. Usually, MATSim users and MATSim power users as defined above in Section 44.2 work with releases.

MATSim uses continuous integration and, thus, nightly builds are available without stability guarantee under `http://matsim.org/downloads/nightly`. MATSim API developers that depend on a very recent feature might use Nightly builds.

Both Apache Maven releases and Apache Maven snapshots are available, see `http://matsim.org/downloads` for details.

MATSim API developers or core developers often work on the code's HEAD version that can be checked out from GitHub.

Nightly builds and maven snapshots are only generated when the code compiles and passes the regression tests. They are, in consequence, somewhat "safer" than the direct download from the HEAD.

44.4 Drivers, Organization and Tools of Development

Important drivers of the MATSim development are the projects and dissertations of the MATSim team. New features are developed as an answer to requirements of these dissertations and projects, where projects range from purely scientific ones—often sponsored by SNF (Schweizerischer Nationalfonds) or DFG (Deutsche Forschungsgemeinschaft)—via projects for governmental entities and projects where science and industry contribute equally—e.g., CTI (Commission for Technology and Innovation) projects—to purely commercial projects, which are managed by Senozon AG in the majority of cases. A significant number of innovations are also introduced by the collaboration with external researchers.

Systematic code integration is mainly performed by the Berlin group and by Senozon AG. This includes continuous code review and integration upon request of the community, but also comprehensive code refactorings to clean up code and to improve modularity. Refactorings are discussed and documented in the MATSim issue tracker (`http://matsim.org/issuetracker`).

The development process is supported by a MATSim standing committee discussing software and sometimes conceptual issues on a regular basis (`http://matsim.org/committee`). Another element that brings in innovation as well as organization are the annual meetings. Right from the beginning, there have been a MATSim developer meetings focused on coding issues. Later, a user meeting offering insights into current work by the community has been added, sometimes

[4] At the time of writing it is unclear if these extensions might one day become contributions, shrinking the MATSim main distribution to its core.

combined with a tutorial. Finally, a conceptual meeting is now held every year, concentrating on issues that go beyond pure software engineering. The developer meeting and the conceptual meeting together establish the road map that guides development for the remainder of the year.

MATSim development makes use of a large number of tools, hopefully leading to better software quality. Historically, many of those tools ran from automated scripts and were made available at `http://matsim.org/developer`. Nowadays, most of them are automatically available from the build server (see `http://matsim.org/buildserver`) and/or from the repository (`https://github.com/matsim-org/matsim`), so that many of them are scheduled for removal from `http://matsim.org/developer`. Some of these tools are: a change log; an issue tracker; the javadoc documentation; static code analyses performed by *FindBugs* and *PMD*; test code coverage analysis; copy paste analysis; code metrics; Apache Maven dependencies; and information about the nightly test results. These nightly test results are generated by the MATSim build server based on the Jenkins software.

Furthermore, there is a MATSim benchmark at `http://matsim.org/files/benchmark/benchmark.zip`. For results see `http://matsim.org/benchmark`.

Most MATSim developers use Eclipse as an IDE. The MATSim documentation is tailored to this IDE. Team development is currently based on Git as revision control system. External library dependencies are managed by Apache Maven.

44.5 Documentation, Dissemination and Support

The main documentation is now this book. Additional information, including tutorials, can be found under `http://matsim.org/docs`. Code documentation in form of javadoc can be found unter `http://matsim.org/javadoc`.

For fast application of MATSim, some small-scale example scenarios are provided in the code base (`folder: examples`), where recently an extended version of the well-known benchmark scenario for the City of Sioux Falls has been added (Chakirov and Fourie, 2014) (Chapter 59). Additional example datasets, including Berlin datasets, can be obtained via `http://matsim.org/datasets`.

Further information is disseminated at the afore-described annual user meetings and MATSim mailing lists, see `http://matsim.org/mailinglists`. Support is provided by the MATSim team via these mailing lists and via `http://matsim.org/faq`, both on a best effort basis. Many components of MATSim are documented by the numerous papers published in international journals and presented at worldwide conferences. Information about such publications can, e.g., be obtained from `http://matsim.org/publications` and from this book.

44.6 Your Contribution to MATSim

The technical details, i.e., the MATSim extension points, on where to hook with MATSim are detailed in Chapter 45. Here, the different ways of contributing to MATSim according to the roles presented in Section 44.2 are introduced.

As a MATSim user, power user, or API developer, you are warmly welcome to make an important impact by reporting your achievements, needs and problems with, or bugs of, the software via the users mailing list, the issue tracker, the FAQ, or at the annual MATSim user meeting.

If you would like to directly contribute to the code base of MATSim, you are welcome to become part of the contributions repository.

If you are the type of person that likes to change the core system, you can, although it is a long way, become a member of the MATSim core developers group. Core developers are usually picked from the MATSim team. Prerequisites are a strong computer scientist background, several years of experience with MATSim and a deep understanding of large software projects.

CHAPTER 45

How to Write Your Own Extensions and Possibly Contribute Them to MATSim

Michael Zilske

Notes

Documentation for the concepts described in this chapter can be found under `http://matsim.org/javadoc` → main distribution, by going to the corresponding class and interface documentation entries. These should also point to examples.

For programming against the MATSim API, we recommend `https://github.com/matsim-org/matsim-example-project` as a starting point; in particular, this should clarify how MATSim can be used as an Apache Maven plug-in.

45.1 Introduction

The three main elements of the MATSim cycle, execution, scoring, and replanning (Section 1.4), operate on what is essentially an in-memory, object-oriented data base of `Person` objects (Raney and Nagel, 2006). These three elements are the main elements to configure MATSim:

Execution The mobsim can be replaced, either by an internally available alternative, or by a fully external mobsim.

Scoring The scoring can be replaced, by possibly giving each individual agent a different recipe to compute its score.

Replanning Arbitrary implementations of type `PlanStrategy` can be added to the replanning; these either generate new plans from scratch or mutate existing ones, or they select between plans.

The simulation's behavior can be further configured by using `ControlerListeners`.

How to cite this book chapter:

Zilske, M. 2016. How to Write Your Own Extensions and Possibly Contribute Them to MATSim. In: Horni, A, Nagel, K and Axhausen, K W. (eds.) *The Multi-Agent Transport Simulation MATSim*, Pp. 297–304. London: Ubiquity Press. DOI: http://dx.doi.org/10.5334/baw.45.

The mobsim generates a stream of events. These are primarily used in two places:

- The scoring uses events to track each agent's success at executing its plan, and computes the scoring value based on this.
- `PlanStrategy` modules use events to build approximate models of the world in which they operate. For example, the router obtains time-dependent expected link travel times from a `TravelTime` object, which in turn listens to link enter and link exit events.

Additionally, one can write any sort of event handlers for analysis during the iterations or after a run by evaluating the events file.

Some modules are so large that fully replacing them in order to adapt the simulation system to one's needs is too much work. These are, in particular,

- the QSim, which is the default implementation of the `Mobsim` interface, and
- the router.

As a result, it is possible to add additional executable code into the execution flow of the QSim by `MobsimListeners` in a similar way as this is possible with the `ControlerListeners` mentioned above. The router, in contrast, is most importantly configured by replacing the definition of the generalized travel cost.

45.2 Extension Points

This section describes what could be called the SPI (Service Provider Interface) of MATSim. Historically, the main entry-point for writing a MATSim extension has been to literally extend (in the Java sense, i.e. to inherit from) the `Controler` class. Essentially, one would override the methods calling the mobsim, the scoring, and/or the replanning, as explained in Section 45.1. This is now discouraged. While this pattern worked when a each member of the team was working on extending the MATSim core by a different aspect, it fails when it comes to integrating those aspects to a single product: There is nothing one can do with a `PublicTransportControler`, an `EmissionsControler`, a `RoadPricingControler` and an `OTFVisControler`, if one wants to combine them to visualize the emissions of buses on toll roads. Also see Section 46.2.1.5.

45.2.1 Config Group

The configuration of a MATSim run is a grouped list of key-value pairs, stored in XML format in the config file (see Section 45.2.1).

At runtime, the entire configuration is stored in an instance of `Config`, from which instances of `ConfigGroup` can be accessed by their name. Config groups that are not in the main distribution need to be explicitly loaded; an approximate example is the following:

```
MyExternalConfigGroup myConfig
   = ConfigUtils.addOrGetModule(controler.getConfig(),
     MyExternalConfigGroup.GROUP_NAME,
     MyExternalConfigGroup.class);
```

The author of an extension can subclass the `ConfigGroup` class to provide named accessors for the parameters. A possibly better way is to subclass from `ReflectiveConfigGroup`, which you can use if you want to define the mapping of named parameters to accessors using Java annotations.

See `http://matsim.org/javadoc` → main distribution → `RunReflectiveConfigGroup` for an example.

45.2.2 ObjectAttributes and Customizable

MATSim operates on data types such as links, nodes, persons, or vehicles. Many of these data types have attributes, such as free speed (for links) or coordinates (for nodes). Rather often, one would like additional information for certain data types, such as "slope" for links, or "age" for persons. In order to not modify the data types every time this becomes necessary, but still allow experimentation, a helper container called `ObjectAttributes` is available. It essentially attaches arbitrary additional information to objects *that have an ID*, by a syntax of type

```
attribs.putAttribute( id, attribName, attribValue ) ;
```

where `id` is the object's ID, `attribName` is the name (type) of the attribute to be stored (e.g., "age"), and `attribValue` is the value of the attribute (e.g., "24").

Importantly, the package provides readers and writers for such attributes. It is thus possible for additional code to, say, generate additional attributes by preprocessing, write them to file, and read them back for every run. That approach is used, for example, by the Gauteng scenario (Chapter 69) to pre-allocate e-tag ownership to persons.

Note that there is currently no simple way to similarly attach information to data types that do not have an ID. This is, for example, the case with plans, activities, or legs, which are contained inside a data type with an ID (the person data type), but which do not possess an ID of their own and are therefore not addressable by `ObjectAttributes`. Some of these non-identifiable data types implement the Java interface `Customizable`, to which additional material can be attached by a syntax of type

```
plan.getCustomAttributes.put("myAttribName",myAttribValue) ;
```

For additional information, see the `Customizable` interface under `http://matsim.org/javadoc` → main distribution. Note that information contained in `Customizable` is not considered standard information by MATSim. It is not written to file when writing the corresponding container, it is in consequence not read from file, and it is undefined if it is copied when copying the data object (e.g., when cloning plans for the evolutionary algorithm). This is the status quo; the MATSim team is thinking about better solutions.

Please check the documentation of `ObjectAttributes` (see `http://matsim.org/javadoc` → main distribution) for more details and pointers to examples.

45.2.3 Scenario Element

The object-oriented, in-memory database which holds the `Person` objects with their plan memories is accessible via the `Population` interface. The `Network` interface gives access to the traffic network graph, consisting of links and nodes. There is a `TransitSchedule` interface which represents the public transit schedule. Your own modeling tasks may need an additional data container like these. We call them scenario elements. The freight carrier population of the freight extension described in Section 24.2 is a typical example.

`Scenario` is the interface which ties all scenario elements together. You can add your own named scenario element to the `Scenario` at startup, for example in a `StartupListener`. All standard scenario elements are populated from XML files at startup, but your own scenario elements could just as well be interfaces to an external relational database.

See `http://matsim.org/javadoc` → main distribution → `RunScenarioElementExample` for an example. Note, however, that in the meantime, the injection framework may have become a better alternative.

45.2.4 ControlerListener: Handling Controler Events

`Controler` remains the main user-facing class of MATSim, but please do not subclass it. Rather, use its setter methods to plug in your own code.[1]

`ControlerListeners` are called at the transitions of the MATSim loop (Figure 45.1), where so-called `ControlerEvents` are fed to the listeners.

The following `ControlerListeners` are currently available: `StartupListener`, `IterationStartsListener`, `BeforeMobsimListener`, `AfterMobsimListener`, `ScoringListener`, `IterationEndsListener`, `ReplanningListener`, and `ShutdownListener`. An up-to-date list can be obtained from `http://matsim.org/javadoc` → main distribution → `ControlerListener` interface.

A sample listener might look as follows.

```
public class MyControlerListener implements StartupListener {
    @Override
    public void notifyStartup(StartupEvent event) {
        ...
    }
}
```

`ControlerListeners` are called in undefined order, meaning that `AControlerListener` may only rely on the computation of `BControlerListener` if `BControlerListener` makes that computation in an earlier transition. For instance, if `BControlerListener` is a `StartupListener` and loads data into a `Map` on start-up, `AControlerListener` can be an `IterationStartsListener` and use that `Map`. But do not write two `IterationStartsListeners` where the first puts some data into a `Map` and the second expects to find it there, they may be called in any order.

Please check the documentation of `ControlerListener` (see `http://matsim.org/javadoc` → main distribution) for more details and pointers to examples.

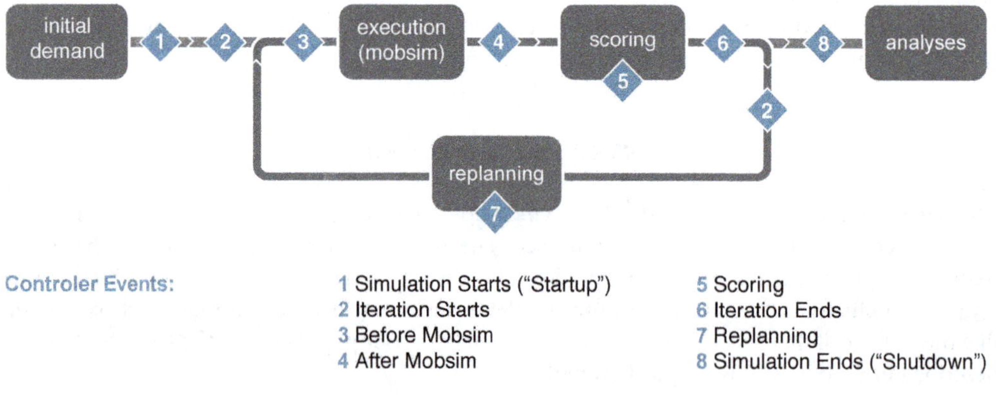

Figure 45.1: Controler events.

[1] Again, in the meantime, the injection framework may have become a better alternative altogether. The general structure, however, remains the same.

45.2.5 *Events*

The mobsim moves the agents around in the virtual world according to their plans and within the bounds of the simulated reality. It documents their moves by producing a stream of events. Events are small pieces of information describing the action of an object at a specific time. Examples of such events are (also see Figure 2.2):

- An agent finishes an activity.
- An agent starts a trip.
- A vehicle enters a road segment.
- A vehicle leaves a road segment.
- An agent boards a public transport vehicle.
- An agent arrives at a location.
- An agent starts an activity.

Each event has a timestamp, a type, and additional attributes required to describe the action like a vehicle id, a link id, an activity type or other data. In theory, it should be possible to replay the mobsim just by the information stored in the events. While a plan describes an agent's intention, the stream of events describes how the simulated day actually was.

As the events are so basic, the number of events generated by a mobsim can easily reach a million or more, with large simulations even generating more than a billion events. But as the events describe all the details from the execution of the plans, it is possible to extract essentially any kind of aggregated data one is interested in. Practically all analyses of MATSim simulations make use of events to calculate some data. Examples of such analyses are the average duration of an activity, average trip duration or distance, mode shares per time window, number of passengers in specific transit lines and many more.

The scoring of the executed plans makes use of events to find out how much time agents spend at activities or for traveling. Some replanning modules might make use of events as well: The router for example can use the information contained in events to figure out which links are jammed at certain times and route agents around that jam when creating new plans.

Handling Events MATSim extensions can watch the mobsim by interpreting the stream of events. This is done by implementing the `EventHandler` interface and registering the implementation with the framework. The lifecycle of an `EventHandler` can be chosen by the developer. Normally, an `EventHandler` lives as long as the simulation run. It is notified before the beginning of each new iteration so that its state can be reset to listen to a new iteration. This pattern can be used to collect information over all iterations. But if the purpose of an `EventHandler` is to make a calculation based on one single iteration, it may be more natural to create a new `EventHandler` instance for each iteration, query it for its result and discard it after the iteration finishes. This can be done in a `ControlerListener`.

See `http://matsim.org/javadoc` → main distribution → `EventHandler` for pointers to coding examples.

Producing Your Own Events One can extend the MATSim event model by extending the `Event` class to define own event types. Events can be produced from all places in the code which are executed during the running mobsim, and in particular from other `EventHandler` instances. Assume for example you want to analyze left-turns. A good starting point would be to specify a `LeftTurnEvent` class, and produce an instance of this class whenever a vehicle does a left-turn. You may do this from a class which is a `LinkLeaveEventHandler` as well as a `LinkEnterEventHandler`. A `LinkLeaveEvent` is produced every time a vehicle leaves a road segment, and a `LinkEnterEvent` is produced when it enters the next road segment. Pairing each `LinkLeaveEvent` with the next

`LinkEnterEvent` for the same vehicle gives a model for a vehicle crossing a node. At this point, your code would look at the road network model to determine if this was a left-turn, and if so, produce a `LeftTurnEvent`.

See `http://matsim.org/javadoc` → tutorial → `RunCustomScoringExample` for an example.

45.2.6 Mobsim Listener

A `MobsimListener` is called in each simulation timestep. This can, for example, be used to produce a custom event which is not triggered by another event. For example, if you wanted to include a model of weather conditions into the simulation, you could use this extension point to decide in every time step if it should start or stop raining on a certain road segment, and produce custom events for this. You would then calculate rain exposure per agent by adding an `EventHandler` which handles `LinkEnterEvent`, `LinkLeaveEvent` and your custom rain events.

Note that `EventHandler` and `MobsimListener` instances may be run in parallel by the framework. It is generally not safe to share state between them. The framework guarantees that the methods of an `EventHandler` instance are called sequentially, but two different instances may run on different threads of execution. Access to shared data must be synchronized externally. Whenever possible, different `EventHandler` instances should only communicate through events.

Example See `http://matsim.org/javadoc` → main distribution → `MobsimListener` for more details and pointers to examples.

45.2.7 TripRouter

A `TripRouter` is a service object providing methods to generate *trips* between locations, given a (main) mode, a departure time and a `Person`. A trip is a sequence of plan elements representing a movement. It typically consists of a single leg (`Leg` object), or of a sequence of legs with "*stage activities*" in between. For instance, public transport trips contain `pt interaction` activities, representing changes of vehicles in public transport trips.

Using the Router A `TripRouter` instance provides a few methods to work with trips, namely

- compute a route for a given mode and O-D pair, for a `Person` with a specific departure time,
- identify the *main mode* of a trip. For instance, a trip composed of several *walk* and *pt* legs should be identified as a public transport trip.
- Identify which activities are *stage activities*, and, by extension, identify the trips in the plan: A trip is the longest sequence of consecutive `Legs` and *stage activities*

Please check the documentation of the `TripRouter` class (see `http://matsim.org/javadoc` → main distribution) for more details and pointers to examples.

Configuring the Router The `TripRouter` computes routes by means of `RoutingModule` instances, one of which is associated with each mode. A `RoutingModule` defines the way a trip is computed, and is able to identify the *stage activities* it generates.

The association between modes and `RoutingModule` instances is configurable. You can even provide you own `RoutingModule` implementations. Do this if your use-case requires custom routing logic, for instance, if you want to implement your own complex travel mode.

Example Please check the documentation of `TripRouter` (see `http://matsim.org/javadoc` → main distribution) for more details and pointers to examples.

45.2.8 *Mobsim*

Alternative mobsim in Java Besides adding `MobsimListener` implementations to enrich the standard mobsim, it is also possible to replace the entire mobsim by a custom implementation. A mobsim is basically a `Runnable` which is supposed to take a scenario and produce a stream of events. This allows you to use the co-evolutionary framework of MATSim while replacing the traffic model itself.

Example for alternative mobsim in Java Please check the documentation of `Mobsim` (see `http://matsim.org/javadoc` → main distribution) for more details and pointers to examples.

Alternative mobsim in another programming language Your implementation need not be written in Java. The framework includes a helper class to call an arbitrary executable which is then expected to write its event stream into a file. This pattern has been used successfully many times, see, e.g., Section 43.1, or the CUDA implementation of the mobsim by Strippgen (2009). Note that we have found consistently that an external mobsim does *not* help with computing speed: Whatever is gained in the mobsim itself is more than lost again by the necessary data transfer between MATSim and the external mobsim. As of now, we cannot yet say if newer data exchange techniques, such as Google Protocol Buffers, may change the situation. Until then, the external interface should rather be seen as the option to inject a different, possibly more realistic, mobsim into MATSim.

45.2.9 *PlanStrategy*

Replanning in MATSim is specified by defining a set of weighted strategies. In each iteration, each agent makes a draw from this set and executes the selected strategy. The strategy specifies how the agent changes its behavior. Most generally, it is an operation on the plan memory of an agent: It adds and/or removes plans, and it marks one of these plans as selected.

Strategies are implementations of the `PlanStrategy` interface. The two most common cases are:

- Pick one plan from memory according to a specified choice algorithm.
- Pick one plan from memory at random, copy it, mutate it in some specific aspect, add the mutated plan to the plan memory, and mark this new plan as selected.

The framework provides a helper class which can be used to implement both of these strategy templates. The helper class delegates to an implementation of `PlanSelector`, which selects a plan from memory, and to zero, one or more implementations of `PlanStrategyModule`, which mutate a copy of the selected plan.

The maximum size of the plan memory per agent is a configurable parameter of MATSim. Independent of what the selected `PlanStrategy` does, the framework will remove plans in excess of the maximum from the plan memory. The algorithm by which this is done is another implementation of `PlanSelector` and can be configured.

The four most commonly used strategies shipped with MATSim are:

- Select from the existing plans at random, which are weighted by their current score.
- Mutate a random existing plan by re-routing all trips.
- Mutate a random existing plan by randomly shifting all activity end times backwards or forwards.
- Mutate a random existing plan by changing the mode of transport and re-routing one or more trips or tours.

Routes are computed based on the traffic conditions of the previous iteration, which are measured by means of an `EventHandler`. Using the same pattern, your own `PlanStrategy` can use any data which can be computed from the mobility simulation. The source code of the standard PlanStrategy implementations can be used as a starting point for implementing custom behavior.

Re-routing as a building block of many replanning strategies is a complex operation by itself. It can even be recursive: For example, finding a public transport route may consist of selecting access and egress stations as sub-destination, finding a scheduled connection between them, and finding pedestrian routes between the activity locations and the stations. With the TripRouter interface, the framework includes high-level support for assembling complex modes of transport from building blocks provided by other modules or the core.

Please check the documentation of PlanStrategy (see http://matsim.org/javadoc → main distribution) for more details and pointers to examples.

45.2.10 Scoring

The parameters of the default MATSim scoring function (Chapter 3) are configurable. The code, which maps a stream of mobsim events to a score for each agent is placed behind a factory interface and replaceable. However, replacing it means replacing the entire utility formulation. There is currently no mechanism for composing a utility formulation from contributions by different modules. For instance, a module which simulates weather conditions would probably calculate penalties for pedestrians walking in heavy rain, and the Cadyts (Chapter 32) calibration scheme already uses utility offsets in its formulation. A modeler who wishes to compose a scoring function from the Charypar-Nagel utility, the rain penalty and the calibration offset needs to do this manually, in code, accessing the code of all three modules contributing to the score and (for instance) summing up their contributions. As of the writing of this chapter, this makes scoring in a way the least modular part of MATSim: It has to be re-defined, in code, for every combination of modules which contribute to the utility.

Keep in mind that score and replanning are not inherently coupled or automatically consistent with each other. Consider a scoring function which penalizes left-turns. This is straight-forward to program: You would iterate over every route an agent has taken. Looking at the Network, you would calculate for each change of links if you consider it a left-turn, and if so, add a (negative) penalty to the score. However, this would not by itself lead to a solution where routes are distributed according to this scoring function. The reason is that the default replanning only proposes fastest routes, in other words, least-cost paths with respect to travel time. By default, the plan memory of an agent will only ever contain routes which have in one iteration been a fastest route. The behavior of the router is, in this case, inconsistent with the utility formulation.

Please check the documentation of ScoringFunction (see http://matsim.org/javadoc → main distribution) for more details and pointers to examples.

PART III

Understanding MATSim

CHAPTER 46

Some History of MATSim

Kai Nagel and Kay W. Axhausen

46.1 Scientific Sources of MATSim

As sketched earlier (Section 1.1), MATSim derives from the following research streams:

Microscopic Modeling of Traffic Microscopic modeling was a basis for traffic flow theory from the start (e.g., Herman et al., 1959; Seddon, 1972; Wiedemann, 1974), but the work was limited to individual links, or small sequences of links and could thus not address equilibrium, as aggregate assignment models could from the 1970's onward (see Sheffi, 1985; Ortúzar and Willumsen, 2011). The expansion to whole and large networks came with the increasingly powerful computers in the 1980's, as well as fast and sufficiently accurate flow models (e.g., Schwerdtfeger, 1984; Nagel and Schreckenberg, 1992; Daganzo, 1994; Gawron, 1998).

Computational Physics For MATSim, this development was aided by insights from computational physics, which often adopts simple and very fast models of physical processes and has performed simulations with 10^8, and more, particles since the 1980's (for a contemporary review see Beazley et al., 1995). It was thus clear from the beginning that urban or regional systems with 10^7 or 10^8 persons or vehicles could be simulated microscopically; the research then focused on where necessary compromises would have to be made.

Microscopic Behavioral Modeling of Demand/Agent-Based Modeling According to Russel and Norvig (2010, p. 53), an agent is "anything that can be viewed as perceiving its environment through sensors and acting upon that environment through actuators". In that sense, both the models of Seddon (1972) and of Wiedemann (1974) can be classified as agent-based; this holds even for the simple cellular automata models of Nagel and Schreckenberg (1992), since here driver-vehicle units perceive the distance to the vehicle ahead and act by adjusting their velocity.

Agent-based behavior can also be found at the demand modeling level, where aggregate models, such as the gravity model (Wilson, 1971), can be replaced by person-centric formulations.

How to cite this book chapter:
Nagel, K and Axhausen, K W. 2016. Some History of MATSim. In: Horni, A, Nagel, K and Axhausen, K W. (eds.) *The Multi-Agent Transport Simulation MATSim*, Pp. 307–314. London: Ubiquity Press. DOI: http://dx.doi.org/10.5334/baw.46.

In that sense, agent-based modeling of travel demand had been developed in Germany since the 1970's (see the references in Axhausen and Herz, 1989), as well as in English-speaking countries, as described in Jones et al. (1983)'s seminal book. While anglophone authors focused on sample enumeration methods to estimate total demand with their activity-based demand models (see Bradley and Bowman (2006) for North American, mostly discrete choice, model-based, developments and Arentze and Timmermans (2000) for an alternative Dutch approach), the simpler German approach was linked to an integral mesoscopic traffic flow simulation in Axhausen (1989), but not used for equilibrium search. It already had, however, a simple description of daily schedule total utility.

Complex Adaptive Systems/Co-Evolutionary Algorithms Nash-equilibrium-like approaches had been developed in transport assignment since the formative Wardrop (1952) paper. These aggregate, flow-based approaches were expanded to account for user perception errors and the social optimum (see Daganzo and Sheffi, 1977). In the late 1990's, transport science addressed the process of learning within the context and new possibilities of "intelligent transport systems", using various smoothing techniques to integrate data from iteration to iteration, reflecting the field tradition. Examples include Chang and Mahmassani (1989); Kaufman et al. (1991); Hatcher and Mahmassani (1992); Smith et al. (1995); Axhausen et al. (1995); Nagel (1995, 1996); Gawron (1998); Mahmassani and Liu (1999); Polak and Oladeine (2002); Arentze and Timmermans (2004). These approaches translated Nash equilibrium logic into co-evolutionary search schemes, which efficiently identified the optima of each agent's daily schedule.

46.2 Stages of Development

46.2.1 *Kai Nagel's Perspective*

46.2.1.1 *Fast Microscopic Modeling of Traffic Flow (University of Cologne/Los Alamos National Laboratory)*

Kai Nagel originally wanted to do his PhD (Philosophiae Doctor – Doctor of Philosophy) in meteorology. When funding did not come through, he began exploring alternatives and applied for a position in insurance modeling with Prof. A. Bachem at the University of Cologne. Instead, he was offered a position in operations research, solving problems like dynamic vehicle routing with time windows.

Having some background in computational statistical physics, he soon became skeptical whether it made sense optimizing up to the last second of a time window, while simultaneously facing a highly stochastic transport system. Using his training, he embarked on building a microscopic model of the transport system, in particular single-lane (Nagel and Schreckenberg, 1992; Nagel, 1999) road traffic on long links, as well as combining such links to large-scale network-based simulations, where each vehicle follows its own individual route (Nagel, 1996), including adaptive dynamics, being influenced most heavily by Arthur (1994). That paper already (Nagel, 1996) describes what is still the main MATSim architecture, where agents have many different plans, keep trying them out and eventually settle on the best option. In contrast to the current approach, in that paper, all plans were pre-computed; i.e., there was no innovation during iterations. This was possible because the network was much coarser than what we use today, making pre-computing route plans with enough diversity easy.

46.2.1.2 *TRANSIMS (Los Alamos National Laboratory/Santa Fe Institute)*

Some of the above PhD work was done during Kai's tenure as a Graduate Research Assistant at LANL (Los Alamos National Laboratory). After his PhD, he moved to LANL, where he worked with the TRANSIMS (see, e.g., Smith et al., 1995) team, under the leadership of Chris Barrett. The TRANSIMS project used some of the design described above, most notably the cellular automata approach to road traffic modeling, which was thus extended to multi-lane traffic (Nagel et al., 1998), to intersections (Nagel et al., 1997) and to massive parallel computing (Nagel and Rickert, 2001).

In terms of software design, TRANSIMS was a collection of stand-alone modules, coupled by a script. For example, the population synthesizer would generate a population file, the activity generator would take the population file as input and generate an activities file as output, etc. Iterations were done by running the traffic microsimulation (called mobsim in MATSim) based on plans and outputting average link travel times and then running the router based on link travel times and outputting plans.

46.2.1.3 MATSim in C++ (ETH Zürich Computer Science)

Kai Nagel moved to ETH Zürich Computer Science in 1999. It was difficult there to continue with TRANSIMS, partly because TRANSIMS was not under an open-source license at that time and also because TRANSIMS fell under U.S. technology export restrictions for some time. As a result, MATSim was started.

MATSim was different from TRANSIMS from the beginning in two important ways: (1) it tried to be more lightweight, i.e., running much faster, specifically by using the queue model (Gawron, 1998), rather than the cellular automata model for network loading and (2) other than TRANSIMS, agent properties such as demographic data, activity patterns or routes were no longer distributed across multiple files, but contained in one hierarchical XML file.

Another difference later appeared, which went back to the Nagel (1996) approach, but this time really followed Arthur (1994) by giving each individual agent its own memory (Raney and Nagel, 2006). After experimentation with relational databases such as MySQL (MySQL, accessed 2014) or Oracle (ORACLE www page, accessed 2005), it was eventually decided to implement MATSim as an object-oriented database in memory, i.e., by first reading in all XML files, modifying the data in computer memory RAM during a run lifetime and writing the data back from memory to XML files at the end of the run. The decision was based on the observation that the MATSim data model was described much better by XML files and that conversion to the relational format was impractical, prone to errors, and too slow if not kept in memory during iterations.

46.2.1.4 MATSim in Java (TU Berlin Transport Engineering)

Michael Balmer wrote his dissertation at ETH (see below) about demand modeling for MATSim, i.e., about the upstream process that leads to initial plans (Balmer, 2007). That project, different from the main MATSim code at that time, was written in Java. Along with the assessment that Java would be the better language than C++ to continue development, it was decided to use Michael Balmer's code as starting point for a Java version. Arguments for Java included:

- Java is more restrictive. For example, in Java, objects are always passed by reference, [1] while in C++, one has the choice between passing a pointer, a reference, or a deep copy of the object. Since standards are difficult to enforce in academic environments, a more restrictive language seemed (and still seems) the better choice.
- Java runs well on many platforms. This allowed (and still allows) us to let people work on their favorite platforms, be it Linux, Microsoft Windows, or Mac.
- There is good non-commercial support for Java; for example, the Eclipse IDE and numerous powerful libraries.

[1] We abstract from the notion that Java "passes object references by value".

- The Java compiler is easier to handle. For example, there is no extraction of header files and the Java compiler sorts out, by itself, the sequence in which modules need to be compiled.
- For our applications, Java was consistently *not* slower than C++. This assessment was based on several years of teaching a MATSim class at ETH Zürich, where computer science students implemented simple versions of MATSim in a programming language of their choice. Typically, while the fastest C++ code may have been 30 % faster than the fastest Java code, the slowest C++ code normally was a factor of 3 slower than the slowest Java code. In other words, while C++ gives more opportunities for optimization, it also gives more opportunities for very serious performance degradation. This assessment is corroborated somewhat in the literature (Prechelt, 1999), where, in one example, it is demonstrated that interpersonal differences within the same language are of the same magnitude as differences between languages.

  In addition, it seems that the gap between C++ and Java has narrowed further since then. Important differences remain in numerical applications, also partly because C++, other than the Java, allows operator overloading.[2] However, MATSim's agent-based approach means that complex objects are handled much more frequently than true numerical computations.
- One reason for using C++ was that it could be combined with MPI, which is a reliable message passing standard for parallel computing. Parallel computing was necessary both for performance reasons and to be able to run simulations that needed more than about 4 GB of memory—the maximum that could be addressed with the 32 bit architecture standard at that time. MPI is also available for Java, but it is much less well maintained.

  With the advent of the 64 bit architectures, the second reason for parallel computing became obsolete. In addition, with Kai Nagel now at a transport engineering department, it seemed that making conceptual progress was more important than keeping the parallel computing edge, especially since the maintenance of parallel code *permanently* consumes additional resources.

  With the decision to give up on parallel computing, it was no longer necessary to maintain compatibility with MPI; thus, the move to Java was facilitated.

In terms of language, C# might have been an alternative to Java. However, C# depends much more on the Microsoft Windows platform, and community support is not as good as it is for Java.

Clearly, the code by Michael Balmer already had all the necessary data classes, readers and writers. The code was used as a starting point to re-implement MATSim in Java. Nevertheless, many important elements like mobsim, events architecture, scoring, routing, and co-evolutionary architecture had to be re-implemented. It took about two years from making that decision to the first plausible run of MATSim in Java.

Important early steps with MATSim in Java were to add time choice (Balmer et al., 2005b) and mode choice (Rieser et al., 2009) as additional choice dimensions beyond route choice. A summary of the status around 2008 was written by Balmer et al. (2009b).

46.2.1.5 Code Reorganization

The C++ version of MATSim was, similar to the original TRANSIMS, a collection of stand-alone executables coupled by scripts. For example, the router would read plans and events and replace some of the plans by other plans with modified routes. The program flow was organized with shell scripts and makefiles. Later, it was possible to start all modules simultaneously where they used messages to interact (also see Gloor and Nagel, 2005), but the file-based and scripted interaction always remained available.

---

[2] See http://en.wikipedia.org/wiki/Operator_overloading.

That approach had, in consequence, very clearly defined interfaces, i.e., the files. Exchanging information not included in the files meant changing the readers and writers on *both* sides, which was, in consequence, rarely done; stand-alone modules instead tried to work with the information they had.

When MATSim was re-implemented in Java around 2006/07, it was re-implemented as one system. Now, everything could interact with everything. For example, a router could modify the network, compute routes on the modified network and then modify it back. Clearly, it could make an error in the process, thus erroneously modifying the network. In this way, any module could modify any data of MATSim, greatly increasing the scope for misunderstandings and errors.

What created even more problems, however, were extensions to the program flow. The program flow was, as it still is, organized by the `Controler` class.[3] Originally, everybody who wanted to change the program flow and insert his or her own research modules, would inherit from `Controler`, override some methods and insert his or her own instructions. This however, meant that it was impossible to *combine* the extensions without possibly massive manual interventions, illustrated as follows.

For example, assume the core program flow as

```
class Controler {
    void run() {
        ...
        aMethod() ;
        ...
    }
    void aMethod() {
        doA() ;
        doB() ;
    }
}
```

Also assume an extension called `MyControler` from one researcher and another extension called `YourControler` by another researcher:

```
class MyControler extends Controler {
    @Override
    aMethod() {
        doA() ;
        doMyStuff() ;
        doB() ;
    }
}
```

```
class YourControler extends Controler {
    @Override
    aMethod() {
        doA() ;
        doYourStuff() ;
        doB() ;
    }
}
```

If you wanted to combine both approaches, you could neither say `YourControler extends MyControler` nor `MyControler extends YourControler`, since either way one of the two extensions would get lost. In this simple case, one could possibly address the problem through manual

[3] Mis-spelled since its inception.

intervention, but in more complicated situations this would no longer possible without extensive additional testing.

Therefore, in 2008, a decision was made to make MATSim more modular. The first step in that direction was a decision to submit the whole MATSim repository to frequent refactorings, i.e., to *not* leave the code alone as much as possible, instead forcing the community to get used to frequent changes of code, while maintaining functionality. To facilitate that approach, coverage by automatic regression tests on the build server was hugely increased and all developers were encouraged to write automatic regression tests for their own code and projects.

The changes since then are too numerous to be listed here. They include, in particular, fairly restrictive data classes no longer extended or modified by every scientific project, and well-defined extension points in both the iterative loop and inside the mobsim. See Chapter 45 for currently existing extension points.

46.2.2 Kay W. Axhausen's Perspective

46.2.2.1 ORIENT/RV: Parking in Travel Demand Models (Karlsruhe University)

In 1984, Kay Axhausen returned to Karlsruhe University[4] after two years doing an MSc degree at the University of Wisconsin, to start his PhD (Philosophiae Doctor – Doctor of Philosophy) at the IfV (Institut für Verkehrswesen/Institute for Transport Studies). At that time, the IfV already had a long tradition of traffic flow analysis (Leutzbach, 1972) and agent-based traffic flow simulation, as pioneered by Wiedemann (1974) (see also Leutzbach and Wiedemann, 1986). In this environment, Sparmann and Leutzbach (1980) had implemented a sample enumeration-based simulation of traffic demand in the spirit of Poeck and Zumkeller (1978). This approach took the daily schedule of the traveler and simulated it activity-by-activity, including the necessary travel. Neither the traffic flow nor travel demand simulations aimed for equilibrium, but, in line with discussions at the time, both were more interested in the underlying behaviors (e.g., Jones et al., 1983).

Faced with a project to simulate parking as an extension of Sparmann's ORIENT approach, it became clear to Axhausen that sample enumeration approaches could not account for the temporal and spatial competition for parking spaces, but that the event-oriented approaches of the traffic flow model naturally could. Merging the two approaches was the natural solution and he then designed it for ORIENT/RV (Axhausen, 1989). Given the need to model the flow of traffic on the roads as part of the daily dynamics, the approach of Schwerdtfeger, an IfV colleague, was a natural and computationally-efficient choice. Schwerdtfeger (1984) had developed a mesoscopic simulation of traffic flow, which retained the agent-resolution, but employed macroscopic link-performance functions to calculate link speeds.

The work of Swiderski (1983), a second IfV colleague, started Axhausen thinking about the need to account for the constraints imposed by travelers' mental maps. As a full implementation of a mental map is impossible, even with today's computers, he chose to condition travelers' route choices on their travel time expectations, which were based on shortest-paths over an initially empty network. The agents reconsidered their routes at every junction if the experienced travel time deviated beyond an adaptive threshold from expected travel times. In this case, the route was recalculated with the current speeds. The framework was used to iterate (Axhausen, 1990) the expectations via shortest-paths based on stored mean travel times from the last iteration, but no formal tests of equilibrium were conducted, nor was the number of iterations extensive.

In the MATSim context, the competition for facilities was taken up by Horni et al. (2009). Reconsidering routing decisions while already being en-route was taken up by Dobler (2013),

4 Now: Karlsruhe Institute of Technology (KIT).

Number and type of activities
Sequence of activities

- Start and duration of activity
- Composition of the group undertaking the activity
- Expenditure division
- Location of the activity
    - Movement between the sequential locations
        - Location of access and egress from the mean of transport
        - Parking type
        - Vehicle/means of transport
        - Route/service
        - Group traveling
        - Expenditure division

Source: Axhausen (2014, 2006, 2009)

Figure 46.1: Behavioral dimensions to be included in a fuller scheduling model.

where he showed that such an approach can approximate the equilibrium in a small number of iterations.

46.2.2.2 From EUROTOPP to MATSim (Karlsruhe, Oxford, London, Innsbruck, Zürich)

The first framework program of the European Union offered a chance to continue with the work in a larger context; unfortunately, this extended version of ORIENT/RV never went beyond the design stage (Axhausen and Goodwin, 1991). The EUROTOPP approach was later implemented in a changed form at the IfV, again by Zumkeller, who also had been one of the partners of the first framework project (Schnittger and Zumkeller, 2004), and his students.

Moving to Oxford, London, Innsbruck and then Zürich in rapid succession kept Axhausen from initiating serious work on a large-scale simulation system. The focus switched to data collection and choice modeling and collaboration on travel demand simulation with Kai Nagel began when he also joined ETH in 1999. While this was initially low key, Michael Balmer and David Charypar's move to Kay Axhausen's group after Kai Nagel's departure to TU Berlin jump-started further work, which is now documented in this book.

46.2.2.3 "Best Response" and Further Choice Dimensions (ETH Zürich Transport Engineering)

Departure time, mode and route choice are the heart of the transport modeling enterprise and were addressed in MATSim almost from the start (Raney and Nagel, 2004; Balmer et al., 2005b; Rieser et al., 2009). Work in Zürich addressed further behavioral dimensions, as shown in Figure 46.1. earch or past studies, which did not produce stable enough code for general use. It is clear that there are more dimensions to consider. Those listed in the figure are only the more obvious examples: for example, rail travel service class or activity engagement intensity are not addressed .

Today, MATSim takes the activity chain and schedule, as given from the initial demand generation process, as input; modern "activity based-models" make it sensitive to accessibility, understood as the logsum term of the included destination and mode choice model (route choice is generally excluded in those models) (see Ben-Akiva et al., 1996, for an early example).Computational overhead costs of calculating non-chosen alternatives sets becomes prohibitive at the scale for which MATSim is designed, so alternative approaches were explored. Meister developed a **genetic algorithm on a household-basis to find optimal schedules for all members simultaneously** (reported in Meister et al., 2005), but only its time-of-day choice element was used in later scenarios (Meister et al., 2006). Feil set about finding a **best-response, but computationally fast approach to the optimization of the number and sequence of activities into a schedule** (Feil, 2010). While he made substantial progress using a tabu search and a cloning approach, it is still too slow as it currently stands. Fourie's PSim (see Chapter 39) might remove that constraint.

While Meister and of Feil's approaches, as well as the standard MATSim routing algorithm, attempt to directly provide best response solutions, the standard MATSim evolutionary algorithm also moves in the direction of good or best response (also see Section 97.3.1). With these approaches, it is impossible to directly model **destination choice**, since the best response destination would just be the closest possible destination (Horni et al., 2009). The problem: destinations similar from the analyst's point of view are quite different from each person's point of view: for example, allowing different types of leisure activity. As further explained in Chapter 27, the problem was addressed by attaching randomness directly to each person-alternative-pair (also see Horni et al., 2012b).

The need to address **parking** is obvious and even more so when considering electric vehicles and their current need to be recharged during the course of a day. Waraich addressed both aspects by integrating a local search into the overall MATSim iteration scheme to identify preferred parking spaces near the final destination (Chapter 13). Dobler's approach (Dobler, 2013) to evacuation is similar, but does not iterate, since that is not relevant for evacuation modeling. Waraich's local search can be extended with personalized walking time values.

The **group composition for joint travel and joint activities** is essential for making progress on a number of fronts, but especially to understand destination choice and activity generation. Gliebe and Koppelman (2005) or Zhang et al. (2005), for example, have proposed discrete choice models for household activity allocation. However, these approaches cannot be easily integrated into MATSim because of their computational costs. They are also too restrictive, with their exclusive focus on the household. Based on parallel empirical work on social networks (see Larsen et al., 2006; Kowald et al., 2013), Dubernet is currently exploring new game theoretic approaches to co-ordinate the timings and activities of households and wider social networks. These social networks are generated using the approach of Arentze et al. (2013), which was estimated against Swiss data for leisure social contact (Kowald and Axhausen, 2012) so as to reproduce measured characteristics of the real network, such as homophily, clustering and average number of leisure social contacts.

The **expenditure division** question is a promising research avenue (Section 97.6) not yet explored by transport planning and clearly interacting with joint activity participation and travel.

CHAPTER 47

Agent-Based Traffic Assignment

Kai Nagel and Gunnar Flötteröd

47.1 Introduction

This chapter presents MATSim from a DTA perspective. The following material is an abridged and edited version of Nagel and Flötteröd (2012).

The traffic assignment problem, whether macroscopic or microscopic, static or dynamic, trip-based or agent-based, is to identify a situation where travel demand and travel supply (network conditions) are consistent with each other. Travel demand results from a demand model that reacts to conditions in the network; these are the output of a supply model (network loading model) using travel demand as its input. A solution of the traffic assignment problem describes an equilibrium between travel demand and travel supply.

Possibly, the most intuitive mathematical formulation of this problem is defined by a fixed point: Find a demand pattern generating network conditions that, in turn, cause the same demand pattern to re-appear. This formulation is operationally important because it motivates a straightforward way of calculating an equilibrium by alternately evaluating the demand model and the supply model. If these iterations stabilize, a fixed point is attained that solves the traffic assignment problem.

The remainder of this chapter places MATSim into this DTA framework. Section 47.2 starts out from the static and macroscopic assignment of route flows and incrementally enriches this formulation into a dynamic and fully disaggregate agent-based assignment problem. Section 47.3 then turns to the problem of how to simulate (solve) this model system, with a particular focus on MATSim's coevolutionary approach. Section 47.4 concludes the presentation.

How to cite this book chapter:

Nagel, K and Flötteröd, G. 2016. Agent-Based Traffic Assignment. In: Horni, A, Nagel, K and Axhausen, K W. (eds.) *The Multi-Agent Transport Simulation MATSim*, Pp. 315–326. London: Ubiquity Press. DOI: http://dx.doi.org/10.5334/baw.47.

47.2 From Route Swapping to Agent Plan Choice

The following details an increasingly comprehensive specification of the traffic assignment problem, starting from the classical static user equilibrium model and ending with a fully dynamic model that captures arbitrary travel demand dimensions at the individual level. Computationally, the iterative fixed point solution procedure is carried throughout the entire development. Deliberately, this solution method also has a behavioral interpretation as a model of day-to-day replanning; see also Section 97.3.5.

We start by considering route assignment only. The generalization towards further choice dimensions will turn out to be rather straightforward.

47.2.1 *Static Traffic Assignment*

Consider a network of nodes and links, where some, or all, of the nodes are demand origins, denoted by o, and/or demand destinations, denoted by d. The constant demand q^{od} in an O-D relation od splits up among a set of routes K^{od}. Denote the flow on route $k \in K^{od}$ by r_k^{od}, where $\sum_{k \in K^{od}} r_k^{od} = q^{od}$.

Most route assignment models either specify a UE (User Equilibrium a.k.a. Wardrop's first principle) or an SUE (Stochastic User Equilibrium). A UE postulates that r_k^{od} is zero for every route k of non-minimal cost (Wardrop, 1952):

$$c(k) = \min_{s \in K^{od}} c(s) \quad \Rightarrow \quad r_k^{od} \geq 0 \tag{47.1}$$

$$c(k) > \min_{s \in K^{od}} c(s) \quad \Rightarrow \quad r_k^{od} = 0 \tag{47.2}$$

where $c(k)$ is the cost (typically delay) on route k.

An alternative, frequently-used approach is to distribute the demand onto the routes such that an SUE is achieved, where users have different perceptions of route cost and every user takes the route of *perceived* minimal cost (Daganzo and Sheffi, 1977). Mathematically, this means that the route flows fulfill some distribution

$$r_k^{od} = P_k^{od}(c(x(\{r_k^{od}\}))) \cdot q^{od} \tag{47.3}$$

where the route splits P_k^{od} are a function of the network costs $c(x)$, which depend on the network conditions x, which, in turn, depend on all route flows $\{r_k^{od}\}$.

In either case, the model needs to be solved iteratively, which typically involves the following steps (Sheffi, 1985):

Algorithm 47.1 Macroscopic and static route assignment

1. **Initial conditions:** Compute some initial routes (e.g., best path on empty network for every O-D pair).
2. **Iterations:** Repeat the following many times.
   (a) **Network loading:** Load the demand on the network along its routes and obtain network delays (congestion).
   (b) **Choice set generation:** Compute new routes based on the network delays.
   (c) **Choice:** Distribute the demand between the routes based on the network delays.

Defining the network loading as more on the "physical" side of the system, the behaviorally relevant steps are choice set generation and choice (Bowman and Ben-Akiva, 1998).

Choice set generation: Often, the new routes are best paths based on the last iteration ("best reply" or "best response" choice set generation). The routes are generated within the iterations because an a priori enumeration of all possible routes is computationally unfeasible.

Choice: Usually, demand is shifted among the routes to improve consistency with the route choice model, assuming—in the simplest case—constant network delays: In a UE, the flow on the current best routes is increased at the cost of the other route flows ("best reply" or "best response" choice), whereas for an SUE, flows are shifted towards the desired route choice distribution (often a version of multinomial logit, e.g., Dial, 1971; Cascetta et al., 1996; Ben-Akiva and Bierlaire, 1999). For stability reasons, this shift is typically realized in a gradual way that dampens the iteration dynamics. See below for more discussion on convergence issues.

The **iterations** are repeated until some stopping criterion is fulfilled, indicating that a fixed point is attained. In the best reply situation, the fixed point implies that no shift between routes takes place, i.e., what comes out as the best reply to the previous iteration is either the same, or at least of the same performance, as what was used in the previous iteration. Since, in this situation, no O-D pair can unilaterally improve by switching routes, the system is at a Nash equilibrium (e.g., Hofbauer and Sigmund, 1998). In the SUE situation, the fixed point means that a route flow pattern $\{r_k^{od}\}$ is found that leads to exactly those network conditions the travelers (the O-D flows) perceived when choosing their routes, giving no incentive to re-route.

Destination choice and elasticity in the demand are behavioral dimensions beyond route choice that can be captured by a static model. However, no technical generality is lost when discussing only route choice; both additional choice dimensions can be rephrased as generalized routing problems on an extended network ("supernetwork"; see, e.g., Sheffi, 1985; Nagurney and Dong, 2002).

47.2.2 Dynamic Traffic Assignment

The process above also works for *dynamic* traffic assignment (DTA; see Peeta and Ziliaskopoulos, 2001), where both demand and network conditions are time-dependent and the time-dependent travel times in the network define a physically meaningful progression of a demand unit through the network.

The algorithm structure does not change. The individual steps now look as follows:

Algorithm 47.2 Macroscopic and dynamic route assignment

1. **Initial conditions:** Compute some initial routing (e.g., best path on empty network for every O-D pair and departure time).
2. **Iterations:** Repeat the following many times.
   (a) **Network loading:** Load all demand items on the network according to their departure times, let them follow their routes and obtain network delays (congestion).
   (b) **Choice set generation:** Compute new routes based on the network delays.
   (c) **Choice:** Distribute the demand between the routes based on the network delays.

Once more, *if* the new routes are best replies (i.e., best paths based on the last iteration), *if* demand is shifted towards these new routes and *if* these iterations reach a fixed point, then this is then a dynamic UE since best reply dynamics mean that no traveler (no O-D flow) can unilaterally deviate to a better route. The SUE interpretation carries over in a similar way.

Destination choice and elasticity in demand apply naturally to the dynamic case as well. Beyond this, the dynamic setting also enables the modeling of departure time choice. Again, the sole consideration of route choice does, at least technically, not constitute a limitation because departure time choice can be translated into route choice in a time-expanded version of the original network (van der Zijpp and Lindveld, 2001).

47.2.3 Individual Travelers

In both the static and dynamic case, it is possible to re-interpret the algorithm in terms of individual travelers. In the static case, for every O-D pair, one needs to assume a steady (= constant) flow of travelers entering the network at the origin at a constant rate, corresponding to that O-D flow. A solution to the static assignment problem corresponds to the distribution of different travelers onto possibly different paths.

In the dynamic case, one needs to generate the appropriate number of travelers for every O-D pair and every time slot and distribute them across the time slot. From then on, the triple (origin, destination, departure time) is fixed for every simulated traveler; its goal is to find an appropriate path. Arguably, this re-interpretation is behaviorally more plausible in the dynamic case.

In a trip-based context, there are two major motivations to go from continuous flows to individual travelers:

- Traffic flow dynamics in complex network infrastructures are difficult to model as continuous flows (e.g., Flötteröd and Rohde, 2011), but are relatively straightforward to simulate at the individual vehicle level (TSS Transport Simulation Systems, accessed 2015; Quadstone Paramics Ltd., accessed 2015; Caliper, accessed 2015; PTV AG, accessed 2015; DynusT, accessed August 2014; Zhou and Taylor, 2014). Disaggregating an O-D matrix into individual trip-makers allows the assignment of one vehicle to every trip-maker in the microscopic traffic flow simulation.
- It is computationally inefficient to capture demand heterogeneity through a large number of commodity flows, but the sampling of trip-makers with different characteristics is fairly straightforward. For example, every vehicle can be given an individual route to its individual destination.

For a finite population of heterogeneous travelers, each traveler constitutes an integer commodity and the **choice** step thus must be changed from "gradually shift the route flows towards something consistent with the behavioral model" into "for a *fraction* of travelers, assign a *single* behaviorally plausible route to each of these travelers". The gradual shift that helps stabilize iterations in the continuous assignment carries over here to an equally stabilizing "inert shift"; not all travelers change their routes at once. This is a consistent reformulation; if one reduces the traveler size from one to $\varepsilon \to 0$ and increases the number of travelers by a factor of $1/\varepsilon$, a 10 % chance of changing routes in the disaggregate case carries over to shifting 10 % of all flows to new routes in the aggregate case ("**continuous limit**").

Apart from this, the iterations do not look much different from what was shown before:

Algorithm 47.3 Microscopic and dynamic route assignment

---

1. **Initial conditions:** Compute some initial routing (e.g., best path on empty network for every traveler).
2. **Iterations:** Repeat the following many times.
   (a) **Network loading:** Load all travelers on the network according to their departure times, let them follow their routes and obtain network delays (congestion).
   (b) **Choice set generation:** Compute new routes based on the network delays.
   (c) **Choice:** Assign every traveler to a route (which can be the previously chosen one) based on network delays.

UE and SUE notions carry over to the disaggregate case if the notion of an O-D pair (or a commodity) is replaced by an individual **particle** (= microscopic traveler).

A **particle UE** may be defined as a system state where no particle can unilaterally improve itself. This definition is consistent with definitions in game theory, which normally start from the discrete problem. It should be noted, however, that this makes the problem combinatorial, which means that even a problem that had a unique solution in its continuous version may have a large number of solutions in its discrete version. Further, the complexity of finding a solution increases similarly to the situation where linear programming jumps to being NP-hard when the variables are required to be integers.[1] The particle UE is hence deliberately not searching for an integer approximation of the continuous solution.

Situations may occur where mixed strategy equilibria exist; these are equilibria, where participants draw between different fixed strategies randomly. This implies that the opponents need to interpret the outcome of the game probabilistically: Even if they themselves play fixed strategies, they need to maximize some expectation value.

For a **particle SUE**, the continuous limit assumption of the macroscopic model is discarded in that the choice fractions $P_k^{od}(c(x(\{r_k^{od}\})))$ in (47.3) are now interpreted as individual-level choice probabilities $P_k^n(c(x(\{r_k^n\})))$ where r_k^n is a binary variable that indicates if traveler n takes route k or not. This implies that the individual-level route flows r_k^n are now random 0/1 variables; consequently, the cost structure—based on individual choices made—becomes probabilistic as well (Balijepalli et al., 2007; Cascetta and Cantarella, 1991; Cascetta, 1989).

A particle SUE is defined as a system state where travelers draw routes from a stationary choice distribution and where the resulting distribution of traffic conditions re-generates that choice distribution.

An operational **particle SUE** specification results if one assumes that travelers filter out the random fluctuations from what they observe and base their decisions only on average route costs:

$$P_n(k) = P_n\left(k \mid \mathrm{E}\{c(x(\{r_k^n\}))\}\right) \tag{47.4}$$

where $P_n(k)$ now is the probability that trip-maker n selects route k and $\mathrm{E}\{\cdot\}$ denotes the expectation. This approach incorporates some generality; it can be shown that choice distributions based on expected network conditions coincide, up to first order, with the stationary choice distributions based on fluctuating network conditions (Flötteröd et al., 2011).

The resulting route flows r_k^n represent not only mean network conditions but also their variability, due to the individual-level route sampling. Alternatively, one could use the particles merely as a discretization scheme of continuous O-D flows and distribute them as closely as possible to the macroscopic average flow rates (e.g., Zhang et al., 2008). The latter approach, however, does not lend itself to the subsequently developed behavioral model type.

No new behavioral dimensions are added when going from commodity flows to particles. However, the microscopic approach allows simulation of greater behavioral variability within the given choice dimensions because it circumvents the computational difficulties of tracking a large number of commodity flows. This will be discussed in more detail in Section 47.2.5.

47.2.4 Stochastic Network Loading

The network loading can be deterministic or stochastic. With deterministic network loading, given time-dependent route inflows, one obtains one corresponding vector of network costs. With stochastic network loading, given the same input, one obtains a *distribution* of vectors of network costs.

---

[1] See http://en.wikipedia.org/wiki/Linear_programming_relaxation.

The macroscopic SUE approach of Section 47.2.1 assumes a distribution of choices but converts choice probabilities into choice fractions before starting the network loading. That is, one effectively performs NetworkLoading(E{Choices}). It is, however, not clear that this is the same as E{NetworkLoading(Choices)}; in fact, with a non-linear network loading, even when it is deterministic, the two are different (Cascetta, 1989). Any Monte Carlo simulation of the particle SUE makes this problem explicit: If, at the choice level, one generates draws from the choice distribution, it makes sense to *first* perform the network loading and *then* do the averaging, rather than the other way around. This is especially true if day-to-day replanning is modeled, in which case draws from the choice distribution have a behavioral interpretation as the actual choices of the trip makers on a given day (but see also Section 97.3.5).

This, however, makes the output from the network loading effectively stochastic since the input to the network loading is stochastic. In consequence, any behavioral model that uses the traffic conditions as input needs to deal with the issue that these inputs are stochastic. *Thus, using a stochastic instead of a deterministic network loading makes little difference.* Making the network loading stochastic simplifies the implementation of certain network loading models. In particular, randomness is a method to resolve fractional behavior in a model with discrete particles.

With stochastic network loading, additional aspects of the iterative dynamics need to be defined. For example, a "best reply" could be against the last stochastic realization or against some average.

47.2.5 Extending the Route Assignment Loop to Other Choice Dimensions

Given the above behavioral interpretation, it is now straightforward to extend the assignment loop to other choice dimensions. For example, the "best reply" can include optimal departure time choice (e.g., de Palma and Marchal, 2002; Ettema et al., 2003) or optimal mode choice. This becomes easiest to interpret (and, in our view, most powerful in practice) if one moves from the concept of "trips" to daily plans. MATSim plans maintain the structure of DTA in terms of the triple (origin, departure time, destination); see Section 2.2.2.3 for an example. However, different from DTA, all activities are chained together.

This widens the behavioral modeling scope dramatically; all choice dimensions of an all-day travel plan can now be jointly equilibrated. This increases the degrees of freedom that need to be modeled but also carries a set of natural constraints along, which again reduce the solution space. Most notably, the destination of one trip must be the origin of the same agent's (synthetic traveler's) subsequent trip and an agent must arrive before it departs. Also, constraints such as Hägerstrand's space-time prisms (Hägerstrand, 1970) are automatically enforced when the agents need to return to their starting locations.

There is so significant conceptual difference between the network loading of a route-based and a plan-based model.

The notion of a particle (S)UE can now be naturally extended to agents that execute complete plans.

An **agent-based UE** implies individual travelers (Section 47.2.3), additional choice dimensions (Section 47.2.5) and possible stochastic network loading (Section 47.2.4). Corresponding to the particle UE, it is defined as a system state where no agent can unilaterally improve its plan.

An **agent-based SUE** implies individual travelers (Section 47.2.3), additional choice dimensions (Section 47.2.5) and, normally, stochastic network loading (Section 47.2.4). Corresponding to the particle SUE, it is defined as a system state where agents draw from a stationary choice distribution and where the resulting distribution of traffic conditions re-generates that choice distribution.

If the iterations aim at an agent-based UE, then choice set generation and choice should implement a "best reply" logic; some 'optimal' plans are calculated and assigned to the agents. This is anything but an easy task.

The disaggregate counterpiece of an SUE implies that every agent considers a whole choice set of (possibly suboptimal) plans and selects one of these plans probabilistically, which can lead to huge data structures.

Summarizing, we have now arrived at a fully disaggregate dynamic DTA specification that accounts for arbitrary behavioral dimensions.

47.3 Agent-Based Simulation

The conceptual validity of the agent-based traffic assignment model is fairly intuitive. However, since it comes with a substantial computational burden of solving the model, it presents entirely new challenges on the simulation side.

On the demand side, there is, in particular, the combinatorial number of choice alternatives that must be considered. For example, random utility models rely on an a-priori enumeration of a choice set that represents options each traveler considers when making a choice (Ben-Akiva and Lerman, 1985). This choice set is huge in an agent-based simulation (Bowman and Ben-Akiva, 1998). While there are sampling-based approaches to the modeling of large choice sets that aim at reducing this computational burden, they have not yet been carried over to the modeling of all-day-plan choices (Ben-Akiva and Lerman, 1985; Frejinger et al., 2009b; Flötteröd and Bierlaire, 2013). See also Chapter 49.

As long as household interactions are not included, the demand modeling problem can be decomposed by agent once the network conditions are given—a great computational advantage. The supply model, on the other hand, deals with congestion, which is, by definition, a result of all travelers' physical interactions. Modeling large urban areas requires dealing with millions of travelers, and an operational supply simulation must be able to load all of these travelers on the network in reasonable computation time .

The following sections describe solutions for these problems. Concrete examples of much of this material are implemented within MATSim.

47.3.1 Agent-Based UE; One Plan per Traveler

The simulation of an agent-based UE is possible through the following implementation of the behavioral elements.

Choice set generation: For every agent, generate what would have been best in the previous iteration. This does not concern just the route but all considered choice dimensions, e.g., departure times and/or mode choice.

Choice: Switch to the new plan with a certain probability.

The choice set generation implements a "best reply" dynamic. This now requires identification of an optimal all-day plan for given network conditions. While the calculation of time-dependent shortest paths for UE route assignment is computationally manageable, the identification of optimal plans is far more difficult (Recker, 2001). This is an important technical motivation to switch to an agent-based SUE, where optimality is not required (see below).

Even in the manageable cases of, e.g., shortest paths, any best reply computation is an approximation. Time-dependent routing algorithms require knowledge of every link's travel time as a function of the link entrance time. In computational practice, this information exists only in an average, interpolated way. Thus, such computations become more robust if plan performance is directly taken from the network loading instead of relying on the best reply computation prediction; an agent sticks with a new plan only if it *performs* better than the previous plan (Raney and Nagel, 2004). However, to keep run times manageable in computational practice, multiple agents must make such trial-and-error moves simultaneously. This is, therefore, not an exact best reply algorithm.

For the choice, a useful approach is to make the switching probability from current to best reply solution proportional to the expected improvement, i.e.,

$$P(\text{old} \rightarrow \text{best}) = \min[1, \mu \cdot (S_{\text{best}} - S_{\text{old}})]$$

where S_{best} and S_{old} are the (expected) scores of the best reply and the old plan, respectively, and the min takes care of the fact that a probability should not be larger than one. Truncation at zero is not necessary because the term $S_{\text{best}} - S_{\text{old}}$ cannot become negative. Chapter 3 gives an example of what a scoring function for all-day plans could look like. Note how the decreasing switching *fraction* of the continuous case is replaced by a decreasing switching *probability* (leading to a switching *rate*).

Clearly, any fixed point of such iterations is a UE since no switching takes place at the fixed point, meaning that the best reply plan has the same score as the already existing plan. Stability of the fixed point depends on the switching rate slope at the fixed point, in the above formulation on the μ: All else equal, making μ smaller makes the fixed point more stable but slows down convergence. These observations hold not only in transportation (e.g., Watling and Hazelton, 2003) but quite generally in the area of "evolutionary games and dynamical systems" (Hofbauer and Sigmund, 1998). In addition, in the context of traffic assignment, the existence of physical queues allowing for spillback across many links has apparently been shown to be an inevitable source of multiple Nash equilibria (Daganzo, 1998).

Alternatively, some MSA-like scheme may be used (Liu et al., 2007). One disadvantagewith MSA is that the switching rate does not depend on the magnitude of the expected improvement, which could mean slow(er) convergence. An advantage of MSA is that one does not need to find out a good value for the proportionality factor (μ in the above example).

Yet another approach would be to use a "gap" function measuring the distance of the current assignment from an equilibrium and to infer the switching rate from the requirement that this function must be minimized (Lu et al., 2009; Zhang et al., 2008). However, we are not aware of any operational gap function that applies to all-day plans.

The biggest criticism of agent-based UE is its lack of behavioral realism. In a UE, every agent is assumed to react with a best response according to a model of its objectives, which implies that real travelers are able to compute best responses despite their combinatorial nature and high dimension (Bowman and Ben-Akiva, 1998). Furthermore, as in a pure route assignment, it is reasonable to assume that (i) the behavioral objective is imperfectly modeled and that (ii) explorative travel behavior leads to—more or less—random variations in what real travelers do. While (ii) explicitly introduces stochasticity, (i) calls for it as a representation of imprecision in the behavioral model.

These considerations do not only lead naturally to the agent-based SUE; they also stimulate an additional behavioral component capturing real travelers' explorative learning. Similar to the symmetry between day-to-day replanning and the traffic assignment problem's iterative solution, an explorative learning algorithm can be interpreted either as a model of real learning or as a computational method to solve a stochastic assignment problem. The following section presents a possible implementation of such an algorithm.

47.3.2 Agent-Based SUE; Multiple Plans per Traveler

This section discusses MATSim's co-evolutionary algorithm for simulating plan choices. Chapters 49 and 51 provide an alternative perspective on MATSim's plan choice mechanisms in terms of mainstream discrete choice theory (Ben-Akiva and Lerman, 1985).

It is possible to approach every agent's daily planning problem as a population-based search algorithm. Such a search algorithm maintains a *collection* (= population) of possible solutions to a problem instance and obtains better solutions via that collection's evolution. This is a typical machine-learning (e.g., Russel and Norvig, 2010) approach; the best-known population-based search algorithms (also called **evolutionary algorithms**) are **genetic algorithms** (e.g., Goldberg, 1989).

It is important to note that "population" here refers to the collection of solutions for a single individual, as opposed to the population of travelers. Every individual uses a population-based algorithm to "co-evolve" in the population of all travelers (also see Balmer, 2007).

A **population-based search algorithm** typically works as follows:

Algorithm 47.4 Population-based search

1. **Initiation:** Generate a collection of candidate solutions for a problem instance.
2. **Iterations:** Repeat the following many times.
   (a) **Scoring:** Evaluate every candidate solution's "score" or "fitness".
   (b) **Selection:** Decrease the occurrence of "bad" solutions. There are many ways how this can be done.
   (c) **Construction of new solutions:** Construct new solutions and add them to the candidate solutions collection.

For the construction of new solutions, two operators are often used in genetic algorithms: **Mutation**—which takes a candidate solution and performs small modifications to it; and **crossover**—which takes *two* candidate solutions and constructs a new one from those. Since mutation takes one existing solution and crossover takes two, it makes sense to also move in the opposite direction and define an operator that takes zero solutions as input, i.e., generates **solutions from scratch**—a "best-reply to last iteration" would, for example, be such an operator.

Like what has been said before, we typically have a situation where multiple travelers evolve simultaneously: a *population of persons* where every person has a *population of plans*. The result is a co-evolutionary dynamic, where each person evolves according to a population-based **co-evolutionary algorithm**. The overall approach reads as follows (see, e.g., Hraber et al., 1994; Arthur, 1994, for similar approaches):

Algorithm 47.5 Co-evolutionary, population-based search

1. **Initiation:** Generate at least one plan for every agent.
2. **Iterations:** Repeat the following many times.
   (a) **Selection/Choice:** Select one of the plans for every agent.
   (b) **Scoring:** Obtain a score for every agent's selected plan by executing all selected plans simultaneously in a simulation and attaching some performance measure to each executed plan. Clearly, what was previously the network loading has now evolved into a full-fledged agent-based simulation of daily activities. See Section 47.3.2.4 for more detail on scoring.
   (c) **Generation of new plans (innovation)/Choice set generation:** For some of the agents, generate new plans; for example, as "best replies" or as mutations of existing plans (e.g., small departure time changes).

Note that this approach is really quite congruent with the SUE approach: Each person has a plan collection, which may be interpreted as the choice set. As in SUE, the choice set can be generated while the iterations run or before the iterations start. Each person selects between the plans, where one can attach to every plan a score-based probability to be selected, which becomes in the end similar to Equation (47.3). Clearly, a relevant related research topic is to specify an evolutionary dynamic that can be shown to converge to choice sets that are generated consistently with discrete choice theory requirements; see Chapter 49 and Section 97.3.

The following subsections give examples for the different elements of this approach.

47.3.2.1 Selection (Choice)

A possible choice algorithm is the following: For persons with unscored plans, select an unscored plan. For all other persons, select between existing plans with some SUE model, e.g., a logit model, i.e.,

$$P(i) = \frac{e^{\mu S_i}}{\sum_j e^{\mu S_j}} \tag{47.5}$$

where S_i is the score of plan i and μ models the travelers' ability to distinguish between plans with different scores. This is implemented in MATSim by `SelectExpBeta`.

In practice, we have found that it is much better to not use Equation (47.5) directly but instead use a switching process that *converges* towards Equation (47.5). This can, for example, be achieved by using a switching probability from i to j of the form

$$T(i \to j) = \gamma\ e^{\beta(S_j - S_i)/2} \tag{47.6}$$

where i is the previous plan, j is a randomly selected plan from the same person and γ is a proportionality constant that needs to be small enough so that the expression is never larger than one (since it denotes a probability). This works because the logit model (47.5) fulfills the detailed balance condition

$$P(i)\, T(i \to j) = P(j)\, T(j \to i) \tag{47.7}$$

for these $T(i \to j)$ (e.g., Ross, 2006).[2] This is implemented in MATSim by `ChangeExpBeta`.

The "switching approach" has additional advantages, including the following:

- Equation (47.6) can be behaviorally interpreted as the probability of switching from plan i to plan j. Plausibly, this probability increases with the magnitude of the improvement. For certain applications, one might want a more involved approach, e.g., an *expected* score of j, which then initiates the switch.
- One could replace Equation (47.6) by a threshold-based dynamics, i.e., a switch to a better solution will only take place if the improvement is above a certain threshold. One loses some of the mathematical interpretation, but it may be more consistent with discussion of project appraisal, where small improvements may not lead to a change in behavior.

Although not performed systematically in past work, it is possible to include formulations such as path-size logit (Ben-Akiva and Bierlaire, 1999) in the choice model.

---

2 Assume that, after a number of iterations, there is no more innovation, i.e., the choice set for every agent is fixed and that the scores are updated by MSA. On convergence of the iterations, all agents draw their plans from a fixed choice set based on constant score expectations, cf. (47.4). This means that all agents make their choices independently (and that all interactions are captured in the scores). The switching logic (47.6) then defines an ergodic Markovian process, which converges to the unique steady state probabilities (47.5).

47.3.2.2 *Score Convergence*

The assumption that the scores eventually converge to some constant value intuitively means that the scores cannot display spontaneous reactive behavior to a certain iteration. For example, a particular iteration might display a "network breakdown" (Rieser and Nagel, 2008). Converged scores would not trigger a next-day reaction to that breakdown. In practice, this can be achieved by averaging the scores over many iterations, which is somewhat similar to fictitious play (Monderer and Shapley, 1996; Garcia et al., 2000). Once more, MSA is an option (Section 3.3.4), with the same advantages and disadvantages discussed before (Section 47.3.1). An alternative is to use a small **learning rate** α (Section 3.3.3) in

$$S_i^{\text{new}} = (1 - \alpha)\, S_i^{\text{old}} + \alpha\, \tilde{S}_i \tag{47.8}$$

where S_i^{new} and S_i^{old} are the agent's memorized scores for option i, and $\tilde{S}_i$ is the most recent actual performance with that option; also see Chapter 49. The issue, in the end, is the same as the stable-vs-unstable fixed points (cf. Section 47.3.1): If the system is well-behaved (corresponding to a stable fixed point), it will converge benignly to constant scores and thus to the detailed balance solution. If the system is not well-behaved, one can still force it to such a solution with MSA, but the meaning is less clear (also see Sections 3.3.4 and 47.3.2.2).

As stated before, stochastic network loading makes no additional conceptual difference since there is already stochasticity caused by choice behavior.

47.3.2.3 *Innovation (Choice Set Generation)*

So far, this leaves open the question on choice set *generation*, i.e., the part that generates new plans or modifies existing ones.

One computationally simple technique not requiring a choice set enumeration is to simulate randomly disturbed link costs and run best response based on these costs. This, however, can yield unrealistic results if one does not get the correlation structure of the noise right.

An alternative is to calculate separate best responses after every network loading. Since the process is stochastic, this will generate different solutions from iteration to iteration. An advantage is that the correlations will be generated by the simulation—and are, presumably, realistic. Chapter 49 relates this to random utility modeling; see also Chapter 97.

Beyond that, there are many different algorithms that could be used here. Besides the previously-mentioned "mutation" (Balmer et al., 2005b) or "crossover" (Charypar and Nagel, 2005; Meister et al., 2006), there are also many possibilities for constructive algorithms, such as "agent-based" construction (Zhu et al., 2008). One attractive option, clearly, is to use a regular activity-based demand generation code (e.g., Bowman et al., 1998; Miller and Roorda, 2003), although we have found that this may not be as simple as it seems (Rieser et al., 2007b); in practice, activity-based models are often constructed with O-D matrices in mind. A successful integration is described by Ziemke et al. (2015).

47.3.2.4 *Adjusting the "Improvement Function" from Shortest Time to Generalized Utility Functions*

This chapter takes an inductive approach and argues that one can make the network assignment loop more general by including additional choice dimensions beyond routing. Clearly, for this to work, the scoring needs to take the effects of these additional choice dimensions into account (also see Balmer, 2007). Given evolutionary game theory, it is quite obvious how to do that: One has to extend the cost function used for routing to a general scoring function for complete daily plans.

That is, the performance of a daily plan needs to be scored. An established method to estimate scoring functions for different alternatives is random utility theory (e.g. Ben-Akiva and Lerman, 1985), which is why in the following "scoring" will be replaced by "utility". For a utility function for daily plans, the following arguments may serve as starting points:

- A heuristic approach, consistent with wide-spread assumptions about travel behavior, is to give positive rewards to performing an activity and negative rewards to traveling.
- For the activities, one should select functions where the marginal reward of doing an activity decreases over time.
- Without additional effects, such as opening times or time-varying congestion, the marginal utilities of all performed activities should be the same.

MATSim has, in the past years, gained some experience with the approach described in Chapter 3 and with more theory in Chapter 51; this then closes the loop.

47.4 Conclusion

Starting from regular route assignment, this chapter explains how one can extend the iterative solution procedure of static or dynamic traffic assignment to include additional behavioral dimensions such as time adaptation, mode choice or secondary activity location choice. This is somewhat similar to the so-called supernetworks approachbut argues from the viewpoint of the iterative solution procedure rather than the problem definition.

To address the combinatorial explosion of commodities caused by the expansion of the choice dimensions, a move to individual particles is suggested. This allows an interpretation of the solution procedure as behavioral day-to-day learning but maintains a connection to the SUE definition by interpreting synthetic travelers' behavior as random draws from individual choice sets.

Most of this chapter discusses simulation/computer implementation issues. From the definition given above, progress can be made by using methods from machine learning and co-evolutionary search algorithms. The SUE problem of random selection between different alternatives can be cast as a so-called population-based optimization algorithm, where each synthetic traveler randomly selects between the different members of the population of possible solutions. At the same time, the *population of the travelers* co-evolves towards a stationary distribution of choices.

Overall, this chapter has worked out the structural similarity between the "classical" DTA problem and the more recent agent-based assignment problem.[3] The presentation has focused on the algorithmic issue of how to find solutions to these problems. This is complemented by the subsequent Chapters 48 to 50, which mostly discuss modeling (descriptive) aspects of MATSim.

---

[3] It is, in fact, possible to run MATSim in DTA mode, by converting each trip into a dummy person, with dummy activities at the beginning and end of the trip. The class RunExample5Trips (see http://matsim.org/javadoc → main distribution) runs an example; the class itself points to a configuration file, which in turn points to examples /equil/plans100trips.xml. A dummy person that denotes a trip from link 1 to link 20, departing at 6 am, is coded as

```
<person id="1">
   <plan>
      <act type="dummy" link="1" end_time="06:00" />
      <leg mode="car" />
      <act type="dummy" link="20" dur="00:10" />
   </plan>
</person>
```

CHAPTER 48

MATSim as a Monte-Carlo Engine

Gunnar Flötteröd

48.1 Introduction

"Agents" that "learn" in a "synthetic reality" is a common term in Artificial Intelligence (Russel and Norvig, 2010) and/or Multi-Agent simulation (Ferber, 1999), but it does not belong to the standard terminology of transport modeling. This chapter explains the functioning of MATSim in terms of modeling and simulation concepts that are more established in the transportation field.

It is important to distinguish between a model and a simulation. A model describes certain aspects of a system; a simulation evaluates a model. For instance, a simple route choice model may state that route A is selected with 25 % probability and route B with 75 % probability. A simulation of this model then draws one or more realizations (route choices) from this distribution. One always needs a model before one can simulate. Possible feedback from simulation to modeling comprises (i) new insights into emergent model properties and (ii) computational constraints that prohibit overly complex model specifications. In MATSim, both kinds of feedback are strong drivers of the modeling.

Consider Figure 48.1, displaying MATSim as a model system comprising a (travel) demand model and a (network) supply model. The travel demand model predicts travelers' behavior, given their information about the network conditions. The network supply model predicts these network conditions using a certain travel behavior chosen by all travelers in the system. This is complemented by the modeling assumption that demand and supply are mutually consistent in the sense that the network conditions resulting from a certain travel behavior are statistically equal to the network conditions that caused this behavior.

Simulation addresses the question of how to identify this state of mutual demand/supply consistency, i.e., it solves the model. The model system shown in Figure 48.1 is complicated—it is nonlinear, stochastic and extremely high-dimensional. The only known operational technique to solve it exploits an additional modeling assumption that justifies the real occurrence of

How to cite this book chapter:
Flötteröd, G. 2016. MATSim as a Monte-Carlo Engine. In: Horni, A, Nagel, K and Axhausen, K W. (eds.) *The Multi-Agent Transport Simulation MATSim*, Pp. 327–336. London: Ubiquity Press. DOI: http://dx.doi.org/10.5334/baw.48.

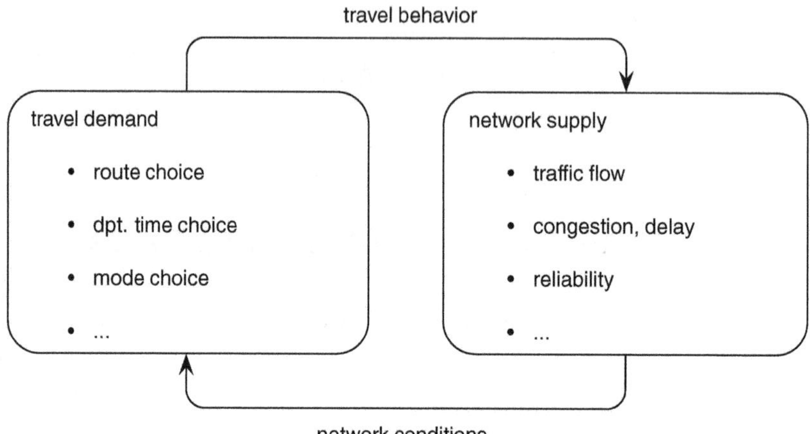

Figure 48.1: Demand/supply perspective on MATSim.

Algorithm 48.1 Iterative scheme to reach demand/supply consistency

1. Create a synthetic agent population.
2. Create a synthetic environment.
3. Iterate:
   (a) All agents choose some planned travel behavior.
   (b) All agents execute their travel plans.
   (c) All agents see the resulting network conditions.

demand/supply consistency: travelers adjust their behavior for their own benefit and only stop doing that when further improvement is insubstantial. Demand/supply consistency characterizes the outcome of this process jointly for all travelers.

Now, consider Algorithm 48.1, which displays the high-level simulation logic of MATSim. This is indeed a logic that iteratively adjusts travel demand. If this logic adjusts the simulated behavior of the simulated travelers until further simulated improvements are insubstantial, then this logic should approach a state of demand/supply consistency. That is, Algorithm 48.1 may be a valid solution method for the model system shown in Figure 48.1. However, that model system does not specify how demand and supply become consistent; it merely specifies that this eventually happens. The only modeling assumption made is that some process of this type exists. The purpose of Algorithm 48.1 is not to mimic this (unspecified) process; it only identifies the final outcome of that process.

The fact that Algorithm 48.1 mimics real, urban, day-to-day dynamics invites misleading interpretations of the underlying model system. In particular, it is a misconception that there is more than a superficial resemblance between the "learning agents" in MATSim and the (hardly understood) learning processes of real humans. If the notion of "learning" has to be used at all when interpreting Algorithm 48.1, it should be understood as "moving a MATSim model closer to its solution point". Also see Section 97.3.5.

The remainder of this chapter phrases these statements more technically and explains their implications for the interpretation of MATSim outputs. This presentation is in parts a more technical reformulation of Chapter 47.

48.2 Relaxation as a Stochastic Process

48.2.1 *Probabilistic Model Components*

Algorithm 48.1 can be written more formally. Denoting the iteration index by k, the following happens in every iteration:

1. All agents choose some planned travel behavior, resulting in the travel demand D^k of the entire agent population.
2. All agents execute their travel plans, resulting in the (time-of-day dependent) network conditions C^k.
3. All agents see the resulting network conditions C^k. As a result, the information Z^k is now available to all agents.

The variables D and Z apply to the population as a whole, comprising all agents. Similarly, the variable C represents network conditions for an entire day and for the entire physical system. Given MATSim's high level of detail, one can think of D, C and Z as placeholders for arbitrarily large and complex data structures. Under MATSim's standard conditions, D corresponds to the set of selected plans, C to the collection of all events and Z to the full plans file including the scores.

Step 1 evaluates the (stochastic) travel behavior model of each agent. Technically, this comprises (i) an optional update of the plan choice set and (ii) the choice of one plan to be executed. Symbolically, this is written as

$$D^k \sim P(D \mid Z^{k-1}), \tag{48.1}$$

meaning that the travel demand of iteration k follows a probability distribution that is conditional on the information Z^{k-1} available to the agents at the end of iteration $k-1$.

Step 2 runs the (stochastic) mobility simulation that moves all agents jointly through the network. In symbols, this becomes

$$C^k \sim P(C \mid D^k), \tag{48.2}$$

meaning that the network conditions of iteration k follow a probability distribution that is conditional on the demand D^k.

Step 3 updates the (possibly stochastic) information available to all agents using the new network conditions C^k. This is written as

$$Z^k \sim P(Z \mid C^k, Z^{k-1}). \tag{48.3}$$

That is, the new information Z^k is not only a transformation of the current network conditions C^k but may also be based on the previously available information Z^{k-1}.

The conditional distributions Equation (48.1)–(48.3) are detailed elsewhere in this book: Chapter 49 describes the plan selection mechanisms leading to $P(D \mid C^{k-1})$, Chapter 50 explains the physical processes underlying $P(C \mid D^k)$, and Chapter 3 specifies at least some of the information update logic behind $P(Z \mid C^k, Z^{k-1})$. A greater level of detail is, however, not necessary in this chapter.

48.2.2 *Markov Chain Perspective*

Algorithm 48.1 constitutes a discrete time stochastic process. "Discrete-time" because it evolves in stages (from iteration to iteration), stochastic because it evaluates stochastic models. Further, one iteration of this process requires only information about the previous iteration's outcome. This allows the expression of Algorithm 48.1 in terms of a "Markov chain" (Ross, 2006).

In symbols, let X^k be the Markov chain's stochastic state during stage k, and let $P(X^k = x)$ be the probability that the chain is in the concrete state x. Further, let T_y^x be the probability that the chain enters state x in its next stage given that it is currently in state y. The transition from one stage to the next can then be expressed as follows:

$$P(X^{k+1} = x) = \sum_{y} P(X^k = y) \cdot T_y^x. \tag{48.4}$$

Each argument of the sum expresses the probability of the chain being in one particular state y and then entering x. The overall probability of arriving in x results from summing up these probabilities.

Markov chains tend, under certain assumptions sketched in the next section, to stabilize after sufficient iterations, in the sense that a long-term probability $\Pi(x)$ of encountering the process in state x exists. This stationary distribution satisfies

$$\Pi(x) = \sum_{y} \Pi(y) \cdot T_y^x, \tag{48.5}$$

which essentially results from removing the k-indices from Equation (48.4). Intuitively, removing the stage-indices k means that Equation (48.5) now applies, in the long term, for any stage k.

Given that the long-term behavior of Algorithm 48.1 shapes the predictions made with MATSim, and updated information its characterization in terms of the stationary distribution of a corresponding Markov chain is of interest. To obtain a Markov chain representation of Algorithm 48.1, one needs to specify (i) what variables in MATSim represent the states of that chain and (ii) what transition distribution underlies the MATSim simulation logic.

A state variable must provide sufficient information to simulate a process further into the future. Candidates for MATSim's state space arethe demand D, the network condition C and the information Z. Of these, only the information Z qualifies as a state variable: If one knows Z^k, it is possible to draw the next day's travel demand D^{k+1} based on Equation (48.1), to insert this demand into Equation (48.2) and obtain the network conditions C^{k+1} and to finally use both C^{k+1} and Z^k to obtain an updated Z^{k+1} through Equation (48.3). This last step is what disqualifies D and C as state variables because an evaluation of Equation (48.3) is impossible without having Z in the state space.

Letting $X^k = Z^k$, the transition distribution hence needs to express how the information Z^k available to the population in iteration k carries over to the information Z^{k+1} available in iteration $k+1$. This relationship is given by

$$T_y^x = \sum_{c} \sum_{d} P(Z^{k+1} = x \mid C^k = c, Z^k = y) P(C^k = c \mid D^k = d) P(D^k = d \mid Z^k = y). \tag{48.6}$$

Each argument of the double sum represents the probability of one particular sequence of given information y, resulting travel demand d, resulting network conditions c and updated information x. The double sum over all possible travel demand realizations d and network conditions c then accounts for the fact that there are many different such sequences through which one can start out at y and end up at x.

This completes the representation of MATSim in terms of a Markov chain. The next section illustrates practical uses of this representation.

48.3 Existence and Uniqueness of MATSim Solutions

The long-term (stationary) behavior of a Markov chain can be derived from its transition function. This also leads to useful insights for MATSim, despite of the complexity of its transition function Equation (48.6).

Two key properties are aperiodicity and irreducibility. Informally, a Markov chain is aperiodic if all of its states can be visited at irregular times; Figure (48.2) provides an example. It is irreducible if it can reach any other state from any given initial state with one or more transitions; see Figure (48.3) for an example. Aperiodicity and irreducibility are essential when it comes to long-term predictions, where (i) aperiodicity guarantees that the concrete iteration in which one evaluates the simulation does not play a role and (ii) irreducibility ensures that every possible future system state can be reached (predicted) by the simulation. If both properties are given, the Markov chain has the following properties (Ross, 2006):

1. A unique stationary distribution exists. The simulation process attains this distribution after many iterations, independently of its initial state.
2. It is feasible to compute statistics of the stationary distribution from a single simulation run, meaning that it is not necessary to run replications.

With respect to MATSim, the following holds:

- Periodicity is already broken if a nonzero probability of staying in the same state exists. This is likely to be the case in MATSim, for instance because the following sequence of events may occur by chance: (i) No agent uses plan innovation, (ii) all agents select the same plan as in the previous iteration, (iii) the mobility simulation creates identical congestion and travel time patterns as before, meaning that Z^k from Equation (48.3) remains the same as Z^{k-1}. Practically, this means that all plan scores stay unchanged.
  More intuitively: Even if the system returns multiple times exactly to a state where it has been before, it unlikely that it does so in the same number of steps.
- With plan innovation (see Sections 4.5, 4.5.3 and 47.3.2.3) switched on, irreducibility cannot be postulated:
  Every time a new plan is added somewhere, the previous state space subspace where the plan was not available *cannot* be reached any more until that plan is removed; similarly, every time a plan is removed, the previous state space subspace where the plan was available cannot be

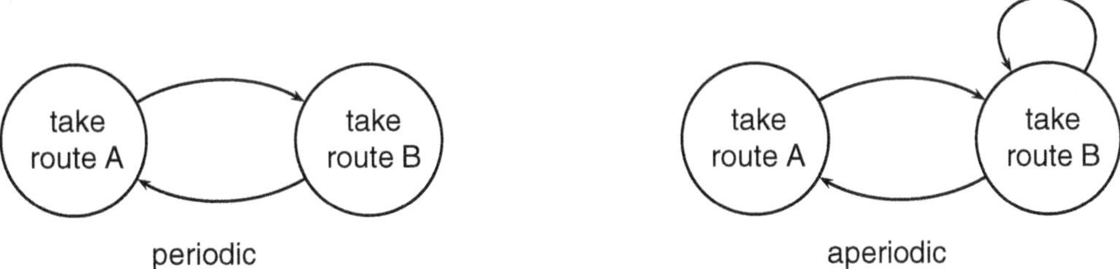

Figure 48.2: Example of (a)periodicity

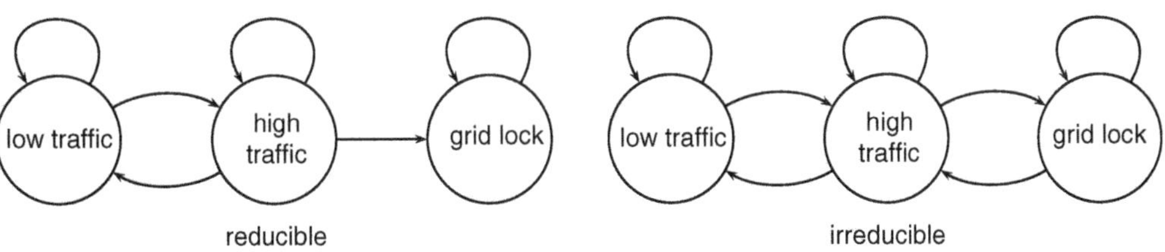

Figure 48.3: Example of (ir)reducibility

reached any more until the plan is re-created again. (Chapter 49 discusses this in greater detail and also suggests a solution for this problem.)
Even if plans creation and removal could be modeled such that irreducibility was guaranteed, the resulting process dynamics would be slow due to the state space size.

- With plan innovation switched off, MATSim in its standard configuration is likely to be irreducible. This is only "likely", because the notion of a "standard configuration" itself is not rigorously specified here. Arguments behind this follow Cascetta (1989), who presents a related result for a much simpler, trip-based traffic simulation that only allows for route choice. Observing that travel plans are, technically, paths in a rather complicated decision network, one can then carry this result over to MATSim. See also Nagel et al. (2000) and Flötteröd et al. (2011).
- When scores are additionally forced to their expected values (Section 3.3.4), the system eventually draws agent behavior from fixed choice distributions, thus varying independently from one iteration to the next.
If config option `ChangeExpBeta` is used, some correlation is maintained between choices in subsequent iterations, even though the long-term choice distributions remain unchanged.

In summary, a mathematical framework exists allowing a rather rigorous characterization of the outcome of MATSim's relaxation process. It turns out that MATSim, in its current form, is not necessarily a "well-behaved" stochastic process; however, casting it into this framework enables a structured approach to developing the simulation logic further. An example of how to go about this is given in Chapter 49.

48.4 Analyzing Simulation Outputs

Many of the models used in MATSim are stochastic. Examples are the discrete choice models used for plan selection or the randomized selection of the next vehicle to enter a congested downstream link in the mobility simulation. The reason for this randomness is that real mobility and transportation processes are not completely understood. The insertion of randomness represents the uncertainty remaining in the modeling.

This uncertainty may apply to both (i) model inputs, meaning that random variables are computed once before a simulation run and then kept fixed (for instance, the random generation of a synthetic population) and to (ii) processes, meaning that random variables are computed throughout the simulation (for instance, the repeated evaluation of discrete choice models). Technically, if a MATSim scenario is simulated R times with different random seeds one obtains $r = 1 \dots R$ independent simulation outputs y_r. Note that, while the raw outputs are plans and event files, the actual quantities for which y_r stands here are numerical in the majority of applications.

Given that one has used different random seeds, $y_1, \dots, y_R$ constitute independent draws from a distribution $\Pi(Y)$. This means that if one performed a huge number of simulation runs and plotted a (possibly multidimensional) histogram of the y values, then this histogram would eventually attain the shape of $\Pi(Y)$. It is important to acknowledge that stochastic simulation outputs are a desirable consequence of stochasticity inserted elsewhere in the simulation; just as a deterministic model output is a truthful representation of its input consequences, a stochastic model output contains a truthful representation of the prediction uncertainty resulting from uncertainties in its input and its process specification.

To help intuition, one may think in the following of y_r as a large vector containing travel times on all links in all one-hour time bins as observed during the last iteration of the rth simulation run. Questions like these may then be asked:

- What travel times can one expect on average?
- What is the travel time variability?
- How probable are travel times beyond some threshold θ?
- ...

This list can be arbitrarily continued. It turns out that most (if not all) of these questions can also be expressed symbolically. For instance:

- What travel times can one expect on average?

$$E\{Y\} = \sum_y y \cdot \Pi(Y=y) \tag{48.7}$$

  This asks for the expected value of the simulation output distribution.
- What is the travel time variability?

$$VAR\{Y\} = \sum_y (y - E\{Y\})^2 \cdot \Pi(Y=y) \tag{48.8}$$

  This asks for the variance (or, for multidimensional outputs, the variance-covariance matrix).
- How probable are travel times beyond some threshold θ?

$$Pr(Y \geq \theta) = \sum_y \mathbf{1}(y \geq \theta) \cdot \Pi(Y=y) \tag{48.9}$$

  This expression merely sums up the probabilities of all simulation outputs that exceed the threshold.
- ...

This enumeration of symbols reveals a common structure. The mathematical formulation of each question can be written in the form

$$\sum_y m(y) \cdot \Pi(Y=y) \tag{48.10}$$

with different specifications of $m(y)$ (see Table 48.1).

By definition, Equation (48.10) is the expectation $E\{m(Y)\}$ given that Y is distributed according to its stationary distribution $\Pi(Y)$. Combining this with the observation that the mean over a

quantity of interest	corresponding $m(y)$
$E\{Y\}$	y
$VAR\{Y\}$	$(y - E\{Y\})^2$
$Pr(Y \geq \theta)$	$\mathbf{1}(y \geq \theta)$
...	...

Table 48.1: Examples of m functions.

sample converges to its expectation as the number of samples grows (the Law of Large Numbers), one obtains

$$\mathrm{E}\{m(Y)\} = \sum_y m(y) \cdot \Pi(Y = y) \tag{48.11}$$

$$= \lim_{R\to\infty} \frac{1}{R} \sum_{r=1}^{R} m(y_r) \tag{48.12}$$

$$\approx \frac{1}{R} \sum_{r=1}^{R} m(y_r) \quad \text{for a finite } R, \tag{48.13}$$

where the simulation outputs y_r, $r = 1 \ldots R$, are independent draws from $\Pi(Y)$.

Now recall that initially certain questions about simulation outputs were asked. The Equation (48.11)(first row) represents exactly these questions in a formal way–and Equation (48.13) (last row) provides a simple method for computing answers to these questions. It reads as follows:

1. Define the function $m(y)$ that represents the question of interest.
2. Perform R independent simulation runs and obtain the outputs $y_1, \ldots, y_R$.
3. Compute $m(y_r)$ for all $r = 1 \ldots R$ and average these numbers.

Returning to the example questions, one thus obtains the following:

- What travel times can one expect on average?

$$\mathrm{E}\{Y\} \approx \frac{1}{R} \sum_{r=1}^{R} y_r \tag{48.14}$$

  Not surprisingly, this turns out to be the mean value over all simulated travel times.
- What is the travel time variability?

$$\mathrm{VAR}\{Y\} \approx \frac{1}{R} \sum_{r=1}^{R} (y_r - \mathrm{E}\{Y\})^2 \tag{48.15}$$

  This is the empirical variance of the simulated travel times. (Note that in practice $\mathrm{E}\{Y\}$ needs to be replaced by its estimator.)
- How probable are travel times beyond some threshold θ?

$$\Pr(Y \geq \theta) \approx \frac{1}{R} \sum_{r=1}^{R} \mathbf{1}(y_r \geq \theta) \tag{48.16}$$

  This divides the number of times the threshold was exceeded by the total number of experiments, i.e. it yields the frequency of the event of interest.
- ...

Revisiting Section 48.3, it may be possible to make these computations more efficient. If (i) there is no uncertainty in the model inputs and (ii) the simulation uses fixed choice sets, then it could be feasible to compute the above statistics by averaging over many stationary iterations of a single simulation run instead of having to run a large number of replications to convergence.

Practically, all of this is just a starting point. Important questions, such as how precise these estimates are, how many runs one needs to obtain a certain level of precision, etc. are not answered here; Ross (2006) is a good starting point for further reading.

48.5 Summary

This chapter attempted to clarify certain mechanisms underlying MATSim's iterative solution scheme. The specification of MATSim's model (components) was distinguished from MATSim's iterative solution algorithm. It was stressed that the behavioral day-to-day interpretation of MATSim is not to be taken literally; realism can only be expected from the long-term process behavior.

This long-term behavior was then related to the properties of the iteration logic using the the Markov chain formalism. MATSim was phrased as such a chain, with its state space comprised of the information available for replanning. This representation was exploited to observe that the long-term distribution of MATSim is likely to exist and be unique if the plan choice sets are a priori fixed.

It further was explained that (i) there are good reasons for the stochasticity both in MATSim's inputs and outputs and that (ii) instead of avoiding stochasticity where it constitutes a truthful representation of uncertainty, one should access adequate statistical techniques to make sense of it.

CHAPTER 49

Choice Models in MATSim

Gunnar Flötteröd and Benjamin Kickhöfer

This chapter attempts to reconcile MATSim's mechanisms of plan "mutation", "selection" and "execution", borrowed from evolutionary computation, with a discrete choice modeling perspective.

Discrete choice theory originates in work by Luce and Suppes (1965) and McFadden (1975); Ben-Akiva and Lerman (1985) and Train (2003) are the two standard textbooks in this area. The theory is mainly used to describe individual choices among mutually exclusive alternatives. Discrete choice models typically do not predict individual choices with complete accuracy. Luce and Suppes (1965) distinguishes between two possible interpretations of this phenomenon: (1) People choose randomly among their alternatives, rendering their behavior inherently unpredictable. (2) The choice only *appears* to be random since the model does not perfectly capture the decision process and its relevant decision variables. Both perspectives lead to the same result, the introduction of probabilistic choice models.

Let $\mathcal{U}_n$ be the universal set of all plans that may ever be considered by agent n and let C_n denote that agent's concrete plan choice set. The choice set independent probability that agent n selects plan i for execution can then be written as

$$P_n(i \mid \mathcal{U}_n) = \sum_{C_n \subset \mathcal{U}_n} P_n(i \mid C_n) \cdot P_n(C_n \mid \mathcal{U}_n), \tag{49.1}$$

explained as follows. Selecting a plan requires a plan choice set. The term $P_n(C_n \mid \mathcal{U}_n)$ represents the probability that this concrete choice set is C_n, which must be a subset of $\mathcal{U}_n$. Technically, the MATSim plan *innovation* modules draw from this distribution. The term $P_n(i \mid C_n)$ represents the probability that agent n selects plan i given that its concrete choice set is C_n. Technically, the MATSim plan *selection* modules draw from this distribution. The product of these terms thus

How to cite this book chapter:

Flötteröd, G and Kickhöfer, G. 2016. Choice Models in MATSim. In: Horni, A, Nagel, K and Axhausen, K W. (eds.) *The Multi-Agent Transport Simulation MATSim*, Pp. 337–346. London: Ubiquity Press. DOI: http://dx.doi.org/10.5334/baw.49.

represents the joint probability that choice set C_n is available and that plan i is chosen from that set. The probability of selecting plan i independently of the concrete choice set then results from summing up the probabilities of selecting it in the presence of all possible choice sets $C_n \subset \mathcal{U}_n$.

It is evident in Equation (49.1) that an agent's behavior depends on both the choice model $P_n(i \mid C_n)$ and the way the choice set is generated through $P_n(C_n \mid \mathcal{U}_n)$. The following two sections will look at each step in more detail.

49.1 Evaluating Choice Models in a Simulated Environment

This section's discussion focuses on the choice distribution $P_n(i \mid C_n)$ for given choice sets. In MATSim, a plan is evaluated and selected based on the score as the sole property of the plan. This is only a technical specification; the scoring and selection protocols are responsible for representing adequate perceptional and behavioral mechanisms. The notions of "choice" and "selection" are subsequently used interchangeably (cf. Section 4.5.2).

The usual selection protocol of MATSim resembles a MNL choice model. Letting S_{ni} be the score of plan i of agent n, one has

$$P_n(i \mid C_n) = \frac{e^{\mu S_{ni}}}{\sum_{j \in C_n} e^{\mu S_{nj}}} \tag{49.2}$$

with μ controlling the preference for higher scores. It is set to one in the remainder of this section.

49.1.1 Case 1: Score is or Converges Towards a Deterministic Value

If the score of a plan was a deterministic number representing an expected value, then Equation (49.2) would constitute a plain MNL choice model with μ taking the role of a scale parameter (see, e.g., Train, 2003, p.45). Such behavior can be approximated in MATSim by the following configuration settings:

- A fixed choice set C_n is eventually obtained by setting the configuration option `fractionOfIterationsToDisableInnovation` below one, meaning that innovation (see Section 49.2) will be switched off for the remaining fraction of iterations beyond the configured value.
- Score convergence to its expectation value can be achieved by setting the configuration option `fractionOfIterationsToStartScoreMSA` below one, meaning that scores will be averaged according to MSA (Method of Successive Averages) for the remaining fraction of iterations.

49.1.2 Case 2: More General

Without the particular configuration mentioned in the previous section, things are somewhat more complicated. Assume that the attribute vector $\mathbf{x}_{ni}$ of the alternatives in Equation (49.2) is defined through (a transformation of) the network conditions observed during the last iteration(s). Assume further that the score is a linear function of these attributes:

$$\begin{aligned} S_{ni} &= \boldsymbol{\beta}^T \mathbf{x}_{ni} && (49.3) \\ &= \boldsymbol{\beta}^T (\mathrm{E}\{\mathbf{x}_{ni}\} + \boldsymbol{\eta}_{ni}) && (49.4) \end{aligned}$$

where $\boldsymbol{\beta}$ is a coefficient vector, superscript T denotes the transpose and η_{ni} is a zero mean random vector. In the general case of S_{ni} being a random variable and not just an expected value, one obtains a mixture-of-logit model with the choice distribution

$$P_n(i \mid C_n) \quad = \quad \int \frac{\exp\left(\boldsymbol{\beta}^T \mathrm{E}\{\mathbf{x}_{ni}\} + \boldsymbol{\beta}^T \eta_{ni}\right)}{\sum_{j \in C_n} \exp\left(\boldsymbol{\beta}^T \mathrm{E}\{\mathbf{x}_{nj}\} + \boldsymbol{\beta}^T \eta_{nj}\right)} p(\eta_n) d\eta_n \tag{49.5}$$

where $p(\eta_n)$ is the probability density function of $\eta_n = (\eta_{ni})_i$, i.e., the joint probability density function of the random disturbances of all alternatives of individual n (Train, 2003, Section 6). This formulation comprises most, if not all, MATSim configurations currently used. It represents the `ExpBetaPlanSelector` and the equivalent `ExpBetaPlanChanger`. It also comprises the `BestPlanSelector`, because that is equivalent to the `ExpBetaPlanSelector` with a very large (infinite) μ. Arbitrary score averaging schemes are also included; this only leads to different instances of $p(\eta_n)$.

Mainstream applications of mixture-of-logit models attempt to combine the tractability of closed-form logit models with the flexibility of simulating arbitrary $p(\eta_n)$ distributions. The distribution of η_n is often as simple as a multivariate normal because this already allows for the introduction of rich correlation structures into the underlying random utilities. In MATSim, however, the simulated error term η_n is extremely complicated. Revisiting Equation (49.4), it defines the variability of the scores resulting from the fact that the simulated network conditions are stochastic. The distribution from which these network conditions are drawn is defined implicitly through the mobility simulation. It is not available in closed form; one can only draw from it.

Additional complexity results from the simulated network conditions being, in turn, the consequence of simulated travel behavior that is again defined through Equation (49.5). Just as a representation of the mutual demand/supply dependency is essential in transport planning, the circular definition of the η_n terms adds realism to MATSim:

1. Assume one could somehow make the simulated network conditions more realistic. The result would be a more realistic distribution $p(\eta_n)$ of the simulated error terms.
2. All else equal, increasing the realism of $p(\eta_n)$ in Equation (49.5) would also increase the realism of the resulting choice distribution.
3. This, in turn, would lead to the selection of more realistic travel plans, meaning that their execution would result in even more realistic network conditions.

However, this positive feedback only applies to the extent to which the error terms in the behavioral model are indeed mobility simulation outputs. Simulated travel time (variability) is such a case. Unobserved preferences of the decision maker, however, are not an output of the mobility simulation and hence need to be differently captured.

It is by no means obvious how the randomness of the simulated network conditions should enter η_n. The notion of "learning" again enters the picture, cf. Chapter 48. However, if the simulation iterations really represented simulated days then a real human learning model would be needed to combine a sequence of past network conditions into an instantaneous η_n realization. Without a sound instance of such a learning model, a learning-based interpretation of Equation (49.5) cannot be given.

Another perspective on this problem is possible, continuing the arguments of Chapter 48. It is stated there that the purpose of MATSim's iterative mechanism is merely to attain a realistic stationary distribution. If so, then the sole purpose of the simulated η_ns is to yield a realistic stationary choice distribution. To illustrate this perspective, consider the following moving-average score updating rule:

$$\bar{S}_{ni}^{k+1} = \begin{cases} \alpha S_{ni}^{k} + (1-\alpha)\bar{S}_{ni}^{k} & \text{if } n \text{ chose plan } i \\ \bar{S}_{ni}^{k} & \text{otherwise} \end{cases} \tag{49.6}$$

where $\bar{S}^k$ is the filtered score of iteration k and S^k is the concrete score observed in that iteration. The learning rate α controls how strongly the filtered score is smoothed out, thus controlling the variability of η_n. MATSim enables this mechanism through the `learningRate` parameter.

Assuming – for simplicity – that the unfiltered stationary score S^{∞} fluctuates in stationary conditions independently from iteration to iteration around its expected value, one can derive the following (as demonstrated in this chapter's appendix):

$$\mathrm{E}\{\bar{S}_{ni}^{\infty}\} = \mathrm{E}\{S_{ni}^{\infty}\} \tag{49.7}$$

$$\mathrm{VAR}\{\bar{S}_{ni}^{\infty}\} = \frac{\alpha}{2-\alpha}\mathrm{VAR}\{S_{ni}^{\infty}\}. \tag{49.8}$$

This means that the filtered score is unbiased with respect to the underlying score process and that its variance is in the interval from zero to the variance of the unfiltered score, depending on the chosen α. There is no need to justify this through a learning process. One has merely constructed a parametrization of the distribution $p(\eta_n)$. In the resulting mixture model Equation (49.5), α should be estimated from real data, just like any other model parameter. Even though this apparently has not yet been attempted, techniques necessary for such an endeavor are, in principle, available (Gourieroux et al., 1993).

49.1.3 *Expected Maximum Utility*

The expected maximum utility of Equation (49.5) is relevant to the microeconomic interpretation of MATSim outputs. A recipe for its computation is described next. Let

$$U_i = V_i + \eta_i + \varepsilon_i \tag{49.9}$$

using the shortcuts $V_i = \boldsymbol{\beta}^T \mathrm{E}\{\mathbf{x}_{ni}\}$, $\eta_i = \boldsymbol{\beta}^T \boldsymbol{\eta}_{ni}$, letting ε_i be the Gumbel error assumed by the multinomial logit model and dropping the n index for brevity. Following this notation, Equation (49.4) is rewritten as

$$S_i = V_i + \eta_i. \tag{49.10}$$

One needs to distinguish between the score of a plan when it is selected and its updated score after it has been executed. To start, it is assumed that the agent receives an expected maximum utility depending on the scores at the time of plan selection, not after plan execution. The expected maximum utility of Equation (49.5) could then be expressed as follows:

$$\mathrm{E}\left\{\max_{i\in C_n} U_i\right\} = \mathrm{E}\left\{\max_{i\in C_n} V_i + \varepsilon_i + \eta_i\right\} \tag{49.11}$$

$$= \mathrm{E}_\eta\left\{\mathrm{E}_\varepsilon\left\{\max_{i\in C_n} V_i + \varepsilon_i + \eta_i \mid \eta\right\}\right\} \tag{49.12}$$

$$= \mathrm{E}_\eta\left\{\ln \sum_{i\in C_n} e^{V_i + \eta_i}\right\}. \tag{49.13}$$

where the law of total expectation is used and E_ε and E_η represent expectations with respect to ε and η, respectively. The remaining argument of the expectation is the expected maximum utility of a multinomial logit choice model given the systematic utilities $V_i + \eta_i$. This expression can be numerically approximated by averaging over many realizations of η_i (i.e. over simulation iterations):

$$E_\eta \left\{ \ln \sum_{i \in C_n} e^{V_i + \eta_i} \right\} \approx \frac{1}{R} \sum_{r=1}^{R} \ln \sum_{i \in C_n} e^{V_i + \eta_i^r} \tag{49.14}$$

where η_i^r is the realization of η_i in iteration r. This expression holds regardless of the functional form of the mobsim-generated mixture distribution.

Now, one needs to account for the fact that agents can only evaluate *past* information when making a choice leads to a *future* score payoff. Recalling that score variability is represented by the η_i variables in Equation (49.5),

$$\eta_i = \hat{\eta}_i + \gamma_i \tag{49.15}$$

is written with η_i contributing to the score actually received Equation (49.10), $\hat{\eta}_i$ being the agent's prediction of that and γ_i being a random variable capturing the difference between the two.

To express the expected maximum *experienced* utility, one hence needs to add (an estimator of) the expectation of γ_i to Equation (49.14). Using Equation (49.15) and Equation (49.10), one obtains

$$\begin{aligned} \gamma &= \eta_i - \hat{\eta}_i && (49.16) \\ &= (\eta_i + V_i) - (\hat{\eta}_i + V_i) && (49.17) \\ &= S_i - \hat{S}_i && (49.18) \end{aligned}$$

where $\hat{S}_i$ can be interpreted as the agent's prediction of the selected alternative i's score. The expectation of this quantity can again be approximated by averaging, resulting in the following estimator of the expected maximum experienced utility, with $i(r)$ indicating the alternative that was selected in iteration r:

$$E\left\{ \max_{i \in C_n} U_i^{\text{experienced}} \right\} \approx \frac{1}{R} \sum_{r=1}^{R} \ln \sum_{i \in C_n} e^{\hat{S}_i^r} + \frac{1}{R} \sum_{r=1}^{R} \left(S_{i(r)}^r - \hat{S}_{i(r)}^r \right). \tag{49.19}$$

The second sum of this expression estimates a "cost of uncertainty"; the less predictable the network conditions (and thus the selected plan's future score), the worse off an agent is on average. The usefulness of this expression depends on the simulation's ability to create realistic network condition variability, for instance along the lines of the last paragraphs of Section 49.1.2. Section 51.2.5.5 discusses this a bit further.

49.2 Evolution of Choice Sets in a Simulated Environment

49.2.1 Overview

The choice set of agents can in principle be computed a priori and then held fixed during a MATSim simulation run. However, the pre-computation would have to be done for every relevant system state (e.g., before and after a policy change). Alternatively, MATSim can be used to generate agents' choice sets within the iterative loop (Section 1.2).

As Equation (49.1) shows, the generation of the choice sets affects the simulated choices. The simplest illustration of this mechanism isthat alternatives that never appear in the choice set cannot be chosen. Similarly, including certain alternatives with a low (high) probability in the choice set decreases (increases) their probability of being chosen, given that the choice model is not changed. When a policy study's synthetic choice sets are very different from the alternatives considered in the real world, it is unlikely that the simulation will display correct aggregated quantities or useful sensitivities for policy measures.

These types of biases are well-known in the discrete choice community, even though the focus is there, arguably, more on estimation than simulation. The problem is particularly acute in route choice modeling because the combinatorial size of the universal route choice set prohibits its enumeration. Drawing further from the discrete choice literature (specifically Frejinger and Bierlaire, 2010), different interpretations can be given to "plan mutation" and "plan innovation" in MATSim.

An interpretation of mutation and innovation as *perceptional models* of travel plan choice set formation is hindered by the need to validate them against real and unobservable choice sets. Alternatively, one may assume that travelers consider the universal choice set and that the choice of unfeasible alternatives is impeded by correspondingly low utility values. In this setting, mutation and innovation constitute *sampling techniques* serving the *computational* purpose of reducing the universal choice set to a small, representative subset. However, one still faces the problem from above that the concrete sampling protocol has a concrete effect on the simulated behavior. The cure when *estimating* choice models is to correct for the sampling based on known sampling probabilities (e.g. Ben-Akiva and Lerman, 1985, Chapter 9), even though these probabilities can be rather difficult to obtain (Flötteröd and Bierlaire, 2013; Frejinger et al., 2009a). The problem appears to be less explored when it comes to *simulation*.

MATSim's currently implemented mutation and innovation procedures constitute concrete, yet heuristic, approaches to the choice set generation problem, aiming at valid predictions at the system level. Possible biases induced by these procedures can, however, be difficult to quantify. For example, the current MATSim implementation might, under certain conditions, yield incomplete choice sets and correlated alternatives (also see Chapter 51). To mitigate the effect of strong correlations between alternatives within the choice set, so-called diversity increasing re-planning modules have been tested (see, e.g., Nagel et al., 2014). In the same context, Grether (2014, Chapter 6) and Neumann et al. (2013) have tested path size logit approaches (see, e.g., Daganzo and Sheffi, 1977; Frejinger and Bierlaire, 2007) to maintain diversity in the choice set by penalizing similar alternatives. Still, these approaches are–as of now–ad-hoc solutions, with little theoretical foundation in the simulation context.

It thus seems worthwhile to revisit the plan choice set generation problem from a statistical perspective. The goal of the following presentation is more to establish a corresponding mindset than deliver a complete solution.

49.2.2 *Towards Unbiased Choice Set Generation*

To make the simulated long-term (stationary) plan choice independent of the plan choice set generation, one may require the following stationary choice distribution:

$$P_n(i \mid \mathcal{U}_n) = \frac{e^{\mu S_{ni}}}{\sum_{j \in \mathcal{U}_n} e^{\mu S_{nj}}}, \tag{49.20}$$

meaning that plans are selected from the universal choice set $\mathcal{U}_n$.

Denoting by $P(C_n \rightarrow C'_n)$ the probability that plan mutation/innovation turns the choice set C_n into C'_n, it is possible to enforce the long-term choice distribution Equation (49.20) through an application of the MH (Metropolis-Hastings) algorithm (Hastings, 1970, see also Flötteröd and Bierlaire (2013) for a related approach to a similar problem).

The MH algorithm specifies the transition distribution of a Markov chain so that a desired stationary distribution of that chain is reached. Given that Chapter 48 has established a formulation of MATSim as such a chain, the MH machinery can hence be inserted into the MATSim iterations. A simplification made in the following is that the choice distribution of agent n is considered independent of all other agents.

To make this concrete, let the state space of the algorithm be the tuple $(C_n, i \in C_n)$ consisting of choice set and resulting choice. During each (MATSim) iteration, one first draws a new choice set C'_n, then draws a new choice $i' \in C'_n$ according to the usual model (49.2) and finally accepts the new state (C'_n, i') with probability

$$\phi[(C_n, i), (C'_n, i')] = \min\left\{\frac{P_n(i' \mid \mathcal{U}_n)}{P_n(i \mid \mathcal{U}_n)} \cdot \frac{P(C'_n \to C_n)P_n(i \mid C_n)}{P(C_n \to C'_n)P_n(i' \mid C'_n)}, 1\right\} \tag{49.21}$$

and rejects it otherwise (meaning that the original choice set C_n and choice $i \in C_n$ are maintained).[1] Intuitively, the first fraction introduces a preference for states comprising a more probable choice and the second fraction corrects for the way transitions between states are proposed.

Assume that the plan innovation yields exactly one new plan i_{in} through a against the last iteration. Let the corresponding plan innovation distribution be approximated by $e^{\mu_{\text{inno}} S_{ni}} / \sum_{j \in \mathcal{U}_n} e^{\mu_{\text{inno}} S_{nj}}$ with a very large μ_{inno}. Assume further that i_{in} replaces exactly one uniformly selected plan i_{out}, which implies that the choice set size J is constant and exclude for simplicity the case that the best response innovation reconstructs the removed plan exactly. This leads to

$$P(C_n \to C'_n) = \frac{1}{J} \cdot \frac{e^{\mu_{\text{inno}} S_{ni_{\text{in}}}}}{\sum_{j \in \mathcal{U}_n} e^{\mu_{\text{inno}} S_{nj}}} \tag{49.22}$$

$$P(C'_n \to C_n) = \frac{1}{J} \cdot \frac{e^{\mu_{\text{inno}} S_{ni_{\text{out}}}}}{\sum_{j \in \mathcal{U}_n} e^{\mu_{\text{inno}} S_{nj}}}. \tag{49.23}$$

Inserting this as well as Equation (49.2) and Equation (49.20) into Equation (49.21), one obtains

$$\phi[(C_n, i), (C'_n, i')] = \min\left\{\frac{e^{\mu S_{ni'}}}{e^{\mu S_{ni}}} \cdot \frac{e^{\mu_{\text{inno}} S_{ni_{\text{out}}}} \dfrac{e^{\mu S_{ni}}}{\sum_{j \in C_n} e^{\mu S_{nj}}}}{e^{\mu_{\text{inno}} S_{ni_{\text{in}}}} \dfrac{e^{\mu S_{ni'}}}{\sum_{j \in C'_n} e^{\mu S_{nj}}}}, 1\right\} \tag{49.24}$$

$$= \min\left\{e^{\mu_{\text{inno}}(S_{ni_{\text{out}}} - S_{ni_{\text{in}}})} \cdot \frac{\sum_{j \in C'_n} e^{\mu S_{nj}}}{\sum_{j \in C_n} e^{\mu S_{nj}}}, 1\right\} \tag{49.25}$$

$$\overset{\mu_{\text{inno}} \to \infty}{=} \begin{cases} 1 & \text{if } S_{ni_{\text{out}}} \geq S_{ni_{\text{in}}} \\ 0 & \text{otherwise.} \end{cases} \tag{49.26}$$

---

[1] The acceptance probability $\phi(X \to X')$ in MH sampling is calculated as

$$\min\left(\frac{w(X') \cdot p_{\text{propose}}(X' \to X)}{w(X) \cdot p_{\text{propose}}(X \to X')}, 1\right),$$

where $p_{\text{propose}}(. \to .)$ is the probability that a certain transition is proposed, and $w(X)$, $w(X')$ are the relative weights of the respective states. It is important to note that w does not have to be normalized; it is sufficient if $w(X)/w(X') = p(X)/p(X')$. $P(C \to C')P(i'|C')$ is the probability that the choice set transitions from C to C' and that i' is selected from the resulting choice set.

Some care is needed when evaluating this expression because it assumes $S_{ni_{\text{in}}}$ and $S_{ni_{\text{out}}}$ to be independent random variables, whereas $S_{ni_{\text{in}}}$ is (due to the best response) always maximal among all alternatives given the most recent iteration. One should thus evaluate this expression by computing either score from the network conditions of a different, randomly selected stationary iteration.

This would allow the selection of plans according to (49.20) from an unconstrained choice set, even though one enumerates only a small subset of the full choice set, which is updated through a computationally efficient best-response mechanism.

In summary, one does the following for each agent in each iteration:

1. Randomly select a given plan for removal and compute a new best-response plan against the last iteration.
2. Is the new plan better than the one selected for removal, based on network conditions from two randomly selected stationary iterations?
   - Yes: Keep the previously selected plan and the previous choice set.
   - No: Remove the randomly selected plan from the choice set, add the newly generated plan and select a new plan from the new choice set.

This (at first glance perhaps counter-intuitive) logic can be explained as follows: Best-response creates new plans that are by chance better than any other plan in a given iteration. Best-response is thus corrected for by accepting the new plan only if it is by chance worse than a randomly selected alternative plan, with both plans being evaluated in randomly selected stationary iterations.

Note that the accuracy of this approach depends on the ability of the best-response plan innovation to create sufficiently variable plans, in the sense that the plan choice set innovation process is irreducible (Ross, 2006, see also Section 48.3 for an intuitive definition of irreducibility).

49.3 Summary

This chapter attempted to phrase MATSim's mechanisms of plan scoring, innovation, mutation and selection in the more mainstream terminology of discrete choice modeling. The implications of evaluating stochastic scores when selecting a plan were explained. The chapter also addressed how simulated choices depend on the way the underlying plan choice sets are generated, and different ways to address this problem were described.

The chapter clearly brought up more issues than it resolved. The take-away message, if any, is probably that even though MATSim agent behavior is roughly based on discrete choice modeling, one needs to be careful when assuming full consistency with a particular discrete choice model.

Appendix: Derivation of Filtered Score Statistics

Writing out the expection:

$$\mathrm{E}\{\bar{S}_{ni}^{k+1}\} = P_n(i)\mathrm{E}\{\alpha S_{ni}^k + (1-\alpha)\bar{S}_{ni}^k\} + (1-P_n(i))\mathrm{E}\{\bar{S}_{ni}^k\} \tag{49.27}$$

$$\Leftrightarrow \mathrm{E}\{\bar{S}_{ni}^{k+1}\} - \mathrm{E}\{\bar{S}_{ni}^k\} = \alpha P_n(i)(\mathrm{E}\{S_{ni}^k\} - \mathrm{E}\{\bar{S}_{ni}^k\}). \tag{49.28}$$

From $\lim_{k\to\infty} \mathrm{E}\{\bar{S}_{ni}^{k+1}\} - \mathrm{E}\{\bar{S}_{ni}^k\} = 0$ then follows

$$\lim_{k\to\infty}\left(\mathrm{E}\{S_{ni}^k\} - \mathrm{E}\{\bar{S}_{ni}^k\}\right) = 0. \tag{49.29}$$

Proceeding in a similar way for the second moment:

$$\begin{aligned} \mathrm{E}\{(\bar{S}_{ni}^{k+1})^2\} &= P_n(i)\mathrm{E}\{(\alpha S_{ni}^k + (1-\alpha)\bar{S}_{ni}^k)^2\} \\ &\quad +(1-P_n(i))\mathrm{E}\{(\bar{S}_{ni}^k)^2\} \qquad (49.30) \\ \Leftrightarrow \lim_{k\to\infty} \mathrm{E}\{(\bar{S}_{ni}^{k+1})^2\} - \mathrm{E}\{(\bar{S}_{ni}^k)^2\} &= P_n(i)\alpha\Big[\alpha\mathrm{E}\{(S_{ni}^k)^2\} \\ &\quad +2(1-\alpha)\mathrm{E}\{S_{ni}^k\}^2 - (2-\alpha)\mathrm{E}\{(\bar{S}_{ni}^k)^2\}\Big] \qquad (49.31) \end{aligned}$$

From $\lim_{k\to\infty} \mathrm{E}\{(\bar{S}_{ni}^{k+1})^2\} - \mathrm{E}\{(\bar{S}_{ni}^k)^2\} = 0$ then follows

$$\lim_{k\to\infty}\mathrm{E}\{(\bar{S}_{ni}^k)^2\} = \frac{\alpha}{2-\alpha}\mathrm{E}\{(S_{ni}^k)^2\} - \frac{2-2\alpha}{2-\alpha}\mathrm{E}\{S_{ni}^k\}^2. \qquad (49.32)$$

The limiting variance then results from inserting of Equation (49.29) and Equation (49.32) into

$$\begin{aligned} \lim_{k\to\infty}\mathrm{VAR}\{\bar{S}_{ni}^k\} &= \lim_{k\to\infty}\Big[\mathrm{E}\{(\bar{S}_{ni}^k)^2\} - \mathrm{E}\{\bar{S}_{ni}^k\}^2\Big] \qquad (49.33) \\ &= \frac{\alpha}{2-\alpha}\mathrm{VAR}\{(S_{ni}^k)^2\}. \qquad (49.34) \end{aligned}$$

CHAPTER 50

Queueing Representation of Kinematic Waves

Gunnar Flötteröd

50.1 Introduction

MATSim comes with a number of mobsims (cf. Sections 4.3, 43.1); the most important are the so-called QSim and JDEQSim. These differ from the implementation perspective (time-stepping vs. event-based, degree of parallelism), but all are (at least approximate) solvers of the same underlying traffic flow model. The purpose of this chapter is to relate MATSim's mobsims to the existing traffic flow theory. There are other simulation packages rooted in the same underlying modeling concepts (Tian et al., 2007; Zhou and Taylor, 2014).

The flow-density relationship (also called FD (Fundamental Diagram)) shown in Figure 50.1 is at the heart of MATSim's traffic flow model. Given a long, homogeneous road, it predicts average flow q (in vehicles per time unit) through any cross-section of that road, given an average vehicle density ϱ (in vehicles per length unit) on that road.

The FD is defined as the minimum of a sending function $S(\varrho)$ (solid) and a receiving function $R(\varrho)$ (dashed), resulting overall in a triangular curve parametrized by free flow speed v, maximum density $\hat{\varrho}$ and backward wave speed w. The maximum velocity is an observable parameter that can be set in the network file (`freespeed` attribute of the `link` element). The maximum density equals one over the length of a vehicle (`effectivecellsize` attribute of the `links` element) for a single-lane link and needs to be multiplied with the number of lanes (`permlanes` attribute of the `link` element), otherwise. The backward wave speed turns out to be the (negative of the) ratio of vehicle length to the safety time gap adopted by drivers in congested conditions. This parameter is fairly constant; a vehicle length of 7.5 meters and a time gap of 2 seconds leads to a value of (minus) 13.5 kilometers per hour. The backward wave speed can be set in the JDEQSim through the `gapTravelSpeed` parameter; it cannot currently be set in the QSim.

The considered FD alone applies only in stationary conditions, where it predicts that (i) flow increases linearly with density at low densities (i.e., in uncongested conditions); (ii) flow decreases

---

How to cite this book chapter:
Flötteröd, G. 2016. Queueing Representation of Kinematic Waves. In: Horni, A, Nagel, K and Axhausen, K W. (eds.) *The Multi-Agent Transport Simulation MATSim*, Pp. 347–352. London: Ubiquity Press. DOI: http://dx.doi.org/10.5334/baw.50.

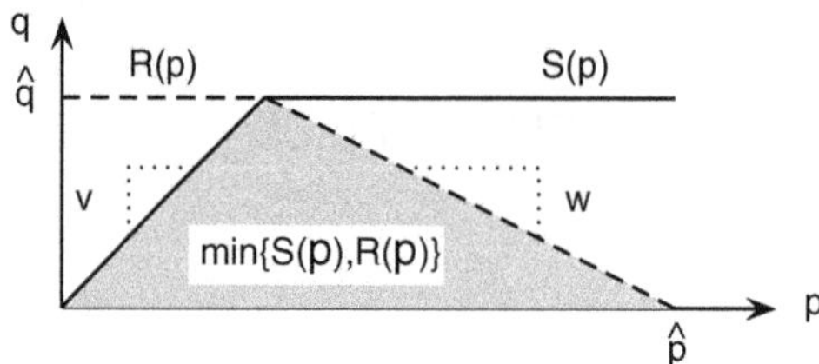

Figure 50.1: Fundamental diagram.

linearly with density at high densities (i.e., in congested conditions); and (iii) in between, it attains a maximal value constituting the flow capacity

$$\hat{q} = \frac{vw\hat{Q}}{v+w} \tag{50.1}$$

of the link. This parameter represents the maximum throughput of the link in the absence of any other flow constraint (such as downstream traffic lights or other bottlenecks, which are discussed further below).

A realistic representation of non-stationary traffic flow (where density and flow change over space and time) is possible by inserting the FD into a continuity equation (which intuitively models vehicle conservation, in the sense that vehicles cannot vanish or spontaneously appear on a road segment without on- and off-ramps). This leads to the KWM (Kinematic Wave Model) of traffic flow (Lighthill and Witham, 1955; Richards, 1956), where the sending and receiving function receive an intuitive interpretation: The instantaneous flow across any interface, possibly with different densities prevailing and FDs applying up- and downstream of that interface, is defined by (i) inserting the density upstream of the interface into the upstream sending function, (ii) inserting the density downstream of that interface into the downstream receiving function and (iii) taking the minimum of these two quantities (Daganzo, 1994; Lebacque, 1996). Intuitively: The flow is limited by what can be sent from upstream and what can be received downstream, but otherwise it is maximized.

The remainder of this chapter expresses MATSim's link model (Section 50.2) and its node model (Section 50.3) in terms of the sending and receiving function framework of the KWM. Some technical detail is omitted from the presentation for the sake of readability; pointers to the literature are provided.

50.2 Link Model

To compute flows entering and leaving a link, one needs to know how much flow can maximally enter the link and how much flow can maximally leave the link. Both constraints depend on the internal (congestion) state of the link. In symbols, one is interested in the instantaneous receiving flow rate R of the link's upstream end and the instantaneous sending flow rate S of the link's downstream end. Multiplying these rates by the duration δ of a simulation time step then yields the maximum number of vehicles that can enter or leave the link during a time step.

MATSim also needs to compute these quantities; how it does so is rooted in Newell's "simplified theory of kinematic waves" (Newell, 1993), which provides a tracktable recipe for computing flow and density anywhere in a link, given that one keeps track only of the flows at the link's up- and downstream interface. In the continuum model (i.e., one that allows for real-valued flows and densities at real-valued locations and times) specified by Newell (1993), the cumulative in- and outflow of a link are defined as

$$N^{\text{in}}(t) = \int_0^t q^{\text{in}}(z)dz \tag{50.2}$$

$$N^{\text{out}}(t) = \int_0^t q^{\text{out}}(z)dz \tag{50.3}$$

where t denotes time, q^{in} and $q^{\text{out}}$are the instantaneous in- and outflow rates (in vehicles per time unit) of the link and an initially (at $t = 0$) empty link is assumed. From MATSim's vehicle-discrete perspective, cumulative inflow (outflow) at a given point in time hence represents the total number of vehicles having entered (left) the link up to that point in time.

Yperman et al. (2006); Yperman (2007) observe that if Newell's theory allows computation of instantaneous densities anywhere in a link, then it also allows computation of densities at the up- and downstream ends of that link. Inserting these densities in the link's sending and receiving function then allows expressing the sending and receiving flows as functions of time-shifted cumulative in- and outflows only, with the time-shifts specified according to the original Newell (1993) formula:

$$R(t) = \min\left\{\hat{\varrho}L - \left[N^{\text{in}}(t) - N^{\text{out}}(t+\delta - L/|w|)\right], \hat{q}\delta\right\} \tag{50.4}$$

$$S(t) = \min\left\{\left[N^{\text{in}}(t+\delta - L/v) - N^{\text{out}}(t)\right], \hat{q}\delta\right\} \tag{50.5}$$

where L is the link length and δ is the (small) discrete time step length. Yperman (2007) provides some intuition for this rather formal specification.

The connection to MATSim can now be made explicit by labeling the two bracketed terms in Equation (50.4) and Equation (50.5) as "upstream queue" (UQ) and "downstream queue" (DQ) (Osorio et al., 2011; Osorio and Flötteröd, published online in *Articles in Advance*):

$$\text{UQ}(t) = N^{\text{in}}(t) - N^{\text{out}}(t+\delta - L/|w|) \tag{50.6}$$

$$\text{DQ}(t) = N^{\text{in}}(0, t+\delta - L/v) - N^{\text{out}}(t). \tag{50.7}$$

These expressions can be given a recursive meaning. Evaluating $\text{UQ}(t) - \text{UQ}(t-\delta)$ yields $[N^{\text{in}}(t) - N^{\text{in}}(t-\delta)] - [N^{\text{out}}(t+\delta - L/|w|) - N^{\text{out}}(t - L/|w|)]$, which under the assumption that flow rates are held constant throughout a simulation time step simplifies into $\delta[q^{\text{in}}(t-\delta) - q^{\text{out}}(t - L/|w|)]$. From this (and symmetric operations for DQ), one obtains

$$\text{UQ}(t) = \text{UQ}(t-\delta) + \delta\left[q^{\text{in}}(t-\delta) - q^{\text{out}}(t - L/|w|)\right] \tag{50.8}$$

$$\text{DQ}(t) = \text{DQ}(t-\delta) + \delta\left[q^{\text{in}}(t - L/v) - q^{\text{out}}(t-\delta)\right]. \tag{50.9}$$

These recursive definitions turn out to be the continuum version of how the JDEQSim updates its link model: In every time step, all vehicles that have just left the link are taken out of the DQ and all vehicles that have entered the link L/v time units ago (corresponding to free-flow travel time) are inserted into the DQ. Similarly, all vehicles that have just entered the link are put into the UQ and all vehicles that have left the link $L/|w|$ time units ago are only now taken out of the UQ. Further, inserting (50.6) and (50.7) into (50.4) and (50.5) yields

$$R(t) = \min\left\{\hat{\varrho}L - \text{UQ}(t), \hat{q}\delta\right\} \tag{50.10}$$

$$S(t) = \min\left\{\text{DQ}(t), \hat{q}\delta\right\}, \tag{50.11}$$

which again corresponds to how JDEQSim evaluates the boundary conditions of a link: The amount of flow allowed to enter the link is limited by the space in its UQ and the amount of flow allowed to leave the link is limited by the number of vehicles in its DQ.

A mobsim that implements the rules Equation (50.8), Equation (50.9), Equation (50.10) and Equation (50.11) implements a KWM-consistent link model. This is almost the case for the JD-EQSim, which, in its implementation as of December 2014, exhibits the sole inconsistency of not limiting the link's inflow to its flow capacity. The QSim turns out to be a particular instance of the same model where backward wave speed is set to $|w| = L/\delta$. Inserting this into Equation (50.8) leads to

$$\mathrm{UQ}(t) \quad = \quad \mathrm{UQ}(t-\delta) + \delta\left[q^{\mathrm{in}}(t-\delta) - q^{\mathrm{out}}(t-\delta)\right], \tag{50.12}$$

which represents the total number of vehicles in the entire link. This corresponds to QSim behavior, where inflow to a link is limited only by the available space in the link as a whole. Letting $|w| = L/\delta$ means that the QSim behaves like a KWM with an extremely high backward wave speed, which physically means that a queue on the link does not dissolve from its downstream end but moves "en block" over the link.

50.3 Node Model

All mobsims in MATSim implement the same node model. Surprisingly, this node model can be traced back at least to (Cetin et al., 2003, under the name of "fair intersections"), while the literature establishing its consistency with the KWM is only a few years old (Tampere et al., 2011; Flötteröd and Rohde, 2011; Corthout et al., 2012).

Nodes in MATSim have no spatial dimension; they merely connect up- and downstream links. Tampere et al. (2011) specify a set of requirements for a (continuum) node model to be consistent with the KWM. They require that the flow through the node shall be maximized subject to the following constraints:

1. Flows are non-negative and conserved within the node. This means that vehicles cannot drive backwards and they must neither disappear nor appear within the node.
2. Flow ratios comply with exogenously specified turning fractions. For instance, if it is specified that 20 % of the outflow of link i shall turn into link j, then the amount of flow that actually advances from link i into link j shall indeed be 20 % of the flow that actually leaves link i.
3. Sending flows of upstream links and receiving flows of downstream links are not exceeded. This is explained in Section 50.2.
4. The invariance principle of Lebacque and Khoshyaran (2005) is satisfied. The most important intuitive implication of this principle is that the advancement of a queuing vehicle is not affected by the vehicles behind it.
5. A "supply constraint interaction rule" is satisfied. It defines how the limited receiving flow of a downstream link is shared by competing upstream links: in practical terms, a right-of-way specification.

Flötteröd and Rohde (2011) specify an "incremental node model" that satisfies these requirements and also provide an intuitive, computationally efficient solution algorithm. In each simulation time step, this algorithm incrementally (hence the name) moves flow from upstream links into downstream links. It does so such that all the previously enumerated constraints are satisfied anytime during the transfer, terminating only once no more flow can be moved. Thus, the ultimately moved flows also comply with all constraints and are maximal.

Now consider the code documentation of MATSim's `queuesim.QueueNode.moveNode` (as of December 2014):

```
Moves vehicles from the inlinks' buffer to the outlinks where
possible. The inLinks are randomly chosen, and for each link all
vehicles in the buffer are moved to their desired outLink as long as
there is space. If the front vehicle in a buffer cannot move across
the node because there is no free space on its destination link,
the work on this inLink is finished and the next inLink's buffer is
handled.
```

This is an informal description of how the incremental node model of Flötteröd and Rohde (2011) works, given that one adopts the conventions that the sending flow of a link is stored in its "buffer" and that the receiving flow of a link is labeled here as free space in (the upstream queue of) that link. A more detailed inspection of the underlying implementation reveals no inconsistencies with incremental node model specification.

There are two aspects of the MATSim node model that are not reflected by the above code comment.

- The sending flow of an upstream link may be limited by an outflow capacity below the flow capacity Equation (50.1) of that link; for instance, to approximate a capacity reduction resulting from a downstream traffic light. This is still consistent with the framework described above.
- The selection probability of "inLinks" is proportional to their flow capacity, meaning that links with higher capacity send, on average, more flow. This is again consistent with Flötteröd and Rohde (2011) and constitutes a concrete "supply constraint interaction rule", as required by Tampere et al. (2011).

The relative simplicity of MATSim's intersection logic may be refined in many ways. For instance, turning pockets may be added and conflicts within intersections may be modeled (cf. Chapter 12). However, some caution is needed when implementing such extensions. The present node model is, due to its simplicity, guaranteed to yield unique node flows. This property needs to be revisited when implementing more complicated specifications (Corthout et al., 2012).

50.4 Summary

This chapter demonstrated that MATSim's mobility simulation is already very close to implementing a particle-discretized instance of the KWM. For full consistency, one needs to (i) use the JDEQSim (or to implement a realistic backward wave speed in the QSim) and to (ii) limit the inflow of a link by its flow capacity (which corresponds to the maximum of its triangular FD).

CHAPTER 51

Microeconomic Interpretation of MATSim for Benefit-Cost Analysis

Benjamin Kickhöfer and Kai Nagel

This chapter explains how MATSim's agent-based framework can be interpreted from a microeconomic perspective and how it can be used for the economic evaluation of transport policies, e.g., for BCA (Benfit-Cost Analysis). The text of this chapter is partly taken from Kickhöfer (2014, Section 2.3).

Typically, the process of economic policy evaluation consists of three steps: First, forecasting changes in the system by modeling users' reactions to a policy (Section 51.1). Second, assigning some (potentially monetary) valuation to these changes (Section 51.2). And third, applying an appropriate aggregation rule (Section 51.3). As will be shown in the next sections, these steps are neither completely independent nor completely dependent on each other.

51.1 Revisiting MATSim's Behavioral Simulation

Estimating policy intervention benefits relies on a sound descriptive model able to predict individuals' related behavioral changes. As explained in Section 1.2, agents in MATSim optimize their mobility behavior over several iterations by reacting to the behavior of other agents. Even if one assumes homogeneous individual preferences in the behavioral parameters of their utility functions (see Section 3.4), activity locations and activity patterns of agents typically differ, meaning that the simulation deals with heterogeneous decision makers. It thus seems reasonable to interpret the simulation from a discrete choice modeling perspective (see Chapter 49). Another attractive reason to use this interpretation lies in the well-established approaches to estimate user benefits and system welfare changes in discrete choice models.

How to cite this book chapter:
Kickhöfer, B and Nagel, K. 2016. Microeconomic Interpretation of MATSim for Benefit-Cost Analysis. In: Horni, A, Nagel, K and Axhausen, K W. (eds.) *The Multi-Agent Transport Simulation MATSim*, Pp. 353–364. London: Ubiquity Press. DOI: http://dx.doi.org/10.5334/baw.51.

As shown by Nagel and Flötteröd (2012, also see Chapter 47 and Section 49.1.1), the MATSim choice model is equivalent to a standard MNL model under the following two conditions: first, valid choice sets have been found for all individuals; second, the score of each plan has converged to its expectation value (self-consistent state). An approximation of this can be reached by switching innovation off (Section 4.5.3) and forcing scores to convergence (Section 3.3.4, also see Section 49.1.1). Still, the following methodological issues remain:

1. **Choice set incomplete**: The maximum number of plans J in each agent's choice set is limited by memory constraints; the choice set for decision making is, hence, unlikely to be complete.
2. **Plans correlation from innovation**: Plans might be correlated. This is very likely if they are modified or replaced by best-response re-planning modules (e.g., the route choice module), since they always have a tendency to generate the same answer. However, random mutations, in general, also tend to result in correlated plans, since the concept of a mutation implies only a small move away from the parent. This violates the required IIA (Independence from Irrelevant Alternatives) property of the choice set necessary for a MNL model.
3. **Plans correlation from plans removal**: The current MATSim implementation has a tendency to retain similar, i.e., correlated, plans when the number of plans has grown beyond J, because the current default plans remover deletes the plan with the lowest score, which is also typically most different from other plans. As a result, normally only very similar plans—with very similar scores—remain in the choice set.

These three issues can lead to biased behavior, which would have consequences for economic evaluation. Possible solutions for these shortcomings are discussed in Section 49.2, and again, from a different angle, in Section 97.3. For the rest of this chapter, it will be assumed that the above issues are solved, and that a consistent solution has been found for the system states before and after the policy change. However, the following text briefly discusses possible impacts of the above issues on policy appraisal results, to facilitate better understanding.

51.2 Valuing Human Behavior at the Individual Level

Following de Jong et al. (2007), a major advantage of the agent-based approach is a seamless integration of (i) forecasting behavioral changes as a reaction to changes in the system, and (ii) the subsequent economic evaluation. In this section, it is shown how estimated agent-specific preferences, which determine behavior, can directly be used for deriving individual VTTSs and how they need to be modified for running a MATSim simulation to obtain individual utility differences resulting from a policy change. The next Section 51.3 will then focus on how these individual utility changes can be used to derive an indicator of overall welfare change for the considered population.

51.2.1 The Utility of Time

The MATSim scoring function of plan (= alternative) i consisting of $q = 0..N-1$ activities and trips has been introduced in Chapter 3 in the following form:

$$U = \sum_q U_{act,q}(t_{dur,q}, \ldots) + \sum_q U_{trav,q}(t_{trav,q}, \ldots) \, , \qquad (51.1)$$

where monetary payments (e.g., tolls) are included in $U_{trav,q}$ and the index i was dropped for notational convenience.[1]

An approximate argument about optimal time allocation can be made as follows: Assume the constraint $T - \sum_q t_{dur,q} - \sum_q t_{trav,q} = 0$, i.e., that the time per day is limited by $T = 24h$, during which all trips and activities need to be completed. Let us now also assume that all travel times are fixed; i.e., we ignore the possible optimization from departure time or mode switches and concentrate on the activity time allocation problem. Optimizing under this constraint leads to the Lagrangian

$$L = \sum_q U_{act,q}(t_{dur,q}, ...) + \sum_q U_{trav,q}(t_{trav,q}, ...) + \mu \cdot (T - \sum_q t_{dur,q} - \sum_q t_{trav,q}) , \qquad (51.2)$$

where μ is the Lagrangian multiplier corresponding to the time constraint.[2]

Solving the optimization problem leads to

$$0 \stackrel{!}{=} \frac{\partial L}{\partial t_{dur,q}} = U'_{act,q}(t_{dur,q}, ...) - \mu \qquad (\forall q) \qquad (51.3)$$

and the time constraint equation from above, where $U'_{act,q} := \partial U_{act,q} / \partial t_{dur,q}$. Equation (51.3) states that, at the optimum and without further constraints, the $t_{dur,q}$ need be selected for all activities q such that all $U'_{dur,q}(t_{dur,q}, ...)$ are the same and equal to μ.

Equation (51.2) can also be seen as a linearized version of the indirect utility function; for example, reducing travel duration by Δt_q affects not only $U_{trav,q}$, but will also lead to a utility change of $\mu \cdot \Delta t_q$ from the constraint, which can be interpreted as the linearized utility effect of spending that time otherwise.[3] In consequence, the **marginal utility of time spent traveling** reads

$$\frac{\partial L}{\partial t_{trav,q}} = U'_{trav,q}(t_{trav,q}, ...) - \mu \, . \qquad (51.4)$$

μ is the **marginal utility of time as a resource**—the marginal utility generated by increasing T, i.e., by making the day longer than 24 hours. The marginal utility of time spent traveling is thus determined by μ, modified by "any enjoyment or dislike of the travel itself" (Small, 2012).

To get a handle on the MATSim utility function in Equation (51.1), μ and $U'_{trav,q}$ need to be obtained separately: μ in order to calibrate $U'_{act,q}$ as in Equation (51.3) and $U'_{trav,q}$ to calibrate the direct utility of time spent traveling, the offset to the marginal utility of time as a resource. This will be further discussed in Section 51.2.4.

51.2.2 The Utility of Money

Time allocation theory (DeSerpa, 1971; Jara-Díaz and Guevara, 2003) makes a similar argument for money, with a budget constraint similar to the time constraint. Just as the time constraint leads to a marginal utility of time as a resource, the budget constraint leads to a marginal utility of money as a resource.

---

[1] Strictly speaking, at this point, it would make more sense to stay with the scores S that MATSim generates. Section 51.2.5 discusses the relation between MATSim scores S, systematic utility V and total utility U in more detail. However, since the following text uses terms like "marginal utility of time" or "marginal utility of money", equations are also noted using U instead of S.

[2] This should not be confused with the scale parameter from discrete choice theory; here, to be consistent with time allocation theory, μ represents the marginal utility of time as a resource and corresponds to β_{dur} in Chapter 3.

[3] A reminder: the indirect utility function describes utility as a function of the value of the constraint that emerges when, for each value of the constraint, utility is maximized.

However, MATSim does currently not include such a monetary budget constraint. It is also questionable whether it should be introduced: the typical theoretical argument assumes the possibility of increasing one's income by working more hours. It is questionable if this functions in European countries, where work contracts typically include a fixed number of working hours, which cannot easily be changed. Hence, an alternative derivation of the marginal utility of money is necessary.

Assume that $U_{trav,q}$ includes a change in the monetary budget, $\lambda \cdot \Delta m$, e.g., invoked by fares or tolls. Then

$$\frac{\partial U}{\partial m} = \frac{\partial U_{trav,q}}{\partial m} = \lambda, \tag{51.5}$$

that is, reducing the monetary budget by Δm reduces the utility by $\lambda \cdot \Delta m$. We will therefore interpret λ as the **marginal utility of money**.[4] Taking the first derivative of L with respect to m would lead to the same result.

In contrast to the marginal utility of time above, we do *not* break down the marginal utility of spending money for travel into a marginal utility of money as a resource, and an offset for spending money on a particular purpose (for an example of this decomposition, see, e.g. Munizaga et al., 2008). Because there is no monetary budget constraint, there is also no neutral Lagrange multiplier that would give the marginal utility of money as a resource.

This, however, leads to the problem that if there are multiple monetary channels, they may have different marginal utilities of money. For example, the marginal utility of toll payments is larger than the marginal utility of payments for fuel—i.e., people find it less irritating to pay for fuel than to pay tolls (see, e.g., Vrtic et al., 2008). That is, each monetary channel, such as fuel cost, toll, public transport fare, or a toll refund, may lead to different preference estimates.

To our knowledge, there is no best solution to this problem in the literature. For the time being, we work with forcing all alternatives' cost-related parameters to a uniform value in preference estimation. However, choice modelers typically avoid limiting the model's degrees of freedom in this way, since it suppresses some information contained in the data.[5] It is therefore often impossible to obtain necessary parameter estimates from the literature. Where raw data is available, the same model can be re-estimated with a uniform marginal utility of money across alternatives (see, e.g., Kickhöfer et al., 2011; Tirachini et al., 2014).

Also, Small (2012) points out that the "neutral" marginal utility of money as a resource is difficult to estimate; for example, it is *not* the marginal utility of income. As an alternative research avenue, we could hypothesize that a measure's monetary channels are included in the choice experiment. For example, a travel time improvement in a value-of-time study could come together with a hypothetical income tax increase, *or* with a hypothetical toll. A rudimentary version of this actually takes place in Switzerland, where large infrastructure investments are bundled with tax increases that pay for them before they are put to public vote (see, e.g., BAV, 2013).

51.2.3 Value of Time

The **VTTS of trip** q is now defined as the marginal utility of time spent traveling (Equation (51.4)), divided by the marginal utility of money (Equation (51.5)), i.e.,

$$VTTS_q = \frac{\partial L / \partial t_{trav,q}}{\partial L / \partial m}, \tag{51.6}$$

[4] This constant, potentially person-specific, implies that income effects (Herriges and Kling, 1999; Daly et al., 2008; Dagsvik and Karlström, 2005; Jara-Díaz and Videla, 1989) do not play a role, i.e., that changes in expenses resulting from transport policies are not strong enough to change λ. In microeconomic theory, λ is the usual variable for the marginal utility of money and corresponds to β_m in Chapter 3.

[5] J. de Dios Ortúzar, personal communication.

where we are using the indirect utility function since we assume that the traveler compares optimal allocations before and after the change.

With $\partial L/\partial t_{trav,q} = U'_{trav,q} - \mu$ from Equation (51.2) one obtains

$$VTTS_q = -\frac{U'_{trav,q}}{\lambda} + \frac{\mu}{\lambda}\,, \tag{51.7}$$

μ/λ is sometimes called the value of time as a resource.

51.2.4 From Estimated to MATSim Parameters

As stated above, most value of time studies do not separately estimate μ, λ, and $U'_{trav,q}$ $(\forall q)$. Assume that an MNL estimation of behavioral parameters from a mode choice survey between car and PT uses the following utility functions:

$$\begin{aligned} U_{car,q} &= && \hat{\beta}_{trav,car} \cdot t_{car,q} + \hat{\beta}_m \cdot \Delta m_{car,q} \\ U_{pt,q} &= \hat{\beta}_0 + && \hat{\beta}_{trav,pt} \cdot t_{pt,q} + \hat{\beta}_m \cdot \Delta m_{pt,q}\,, \end{aligned} \tag{51.8}$$

where $t_{car,q}$, $t_{pt,q}$, $\Delta m_{car,q}$ and $\Delta m_{pt,q}$ are, respectively, travel times and monetary costs in the different modes, and $\hat{\beta}_x$ are the corresponding parameter estimates As explained in Section 51.2.2, $\hat{\beta}_m$ (the same as λ above) is assumed to be the same for all modes, or more precisely, for all types of expenditure.

According to Equation (51.4), the marginal utility of time spent traveling needs to be split into two components:

1. The marginal utility of time as resource, which needs to be used for $U'_{act}(t_{dur,q},\ldots)$ $(\forall q)$ in Equation (51.3).
2. The direct marginal utility of time spent traveling, which needs to be used for $U'_{trav,q}(t_{trav,q},\ldots)$.

We do not know of any good way to perform this split; Kickhöfer et al. (2011) and Kickhöfer (2014) use the least negative $\hat{\beta}_{trav,mode}$ for μ (i.e., β_{dur}) and then re-calculate all other direct marginal utilities of travel time relative to that. As indicated in Section 3.4 of this book, this is currently the preferred procedure.

51.2.5 From Simulation Output to Evaluation

At the end of the simulation run, each agent n has a number of plans $i = 1..J$, each of them associated with a score $S_{n,i}$, computed according to Equation (51.1). For economic evaluation, the question arises how to aggregate these $S_{n,i}$ into an agent-value S_n, which can then be interpreted as a utility U_n. Possibilities include using:

- the logsum of the agent's plans scores, i.e., $\ln \sum_i e^{S_i}$
- the score of the agent's last executed plan,
- the average of the agent's plans scores, or
- the highest score of the agent's plans.

51.2.5.1 Using the Logsum of the Agent's Plans Scores

In literature, the logsum term

$$logsum_n = \ln \sum_i e^{V_i}$$

has been proposed for applied welfare analysis with Discrete Choice Models (Small and Rosen, 1981; de Jong et al., 2006; Kohli and Daly, 2006; de Jong et al., 2007). Under the assumption of a correctly specified model and choice set, the logsum term represents the EMU (Expected Maximum Utility) for a user with several options $i = 1..J$ in her choice set and the systematic utility of each option i is V_i. It is the expectation value, given that a random (Gumbel-distributed) ε_i is added to each V_i, and that the individual chooses the alternative with the highest $U_i = V_i + \varepsilon_i$.[6] In this interpretation, the (expected/average) MATSim score S_i is equated with the systematic part of the utility V_i.

However, as described in the previous Section 51.1, the use of MATSim as choice set generator yields issues with incompleteness of the choice set and with similarity of daily plans. In the current MATSim implementation, the maximum error occurs when all plans are copies of the best plan, rather than a diversity of plans. An upper bound of this error can be approximated as follows. Without loss of generality, assume that $i = 1$ is the plan with the largest systematic utility. Then

$$logsum_n = EMU_n = \ln \sum_{i=1}^{J} e^{V_i} \le \ln \sum_{i=1}^{J} e^{V_1} = \ln(J \cdot e^{V_1}) = \ln J + \ln e^{V_1} = V_1 + \ln J \,.$$

At the same time, obviously

$$logsum_n = EMU_n = \ln \sum_{i=1}^{J} e^{V_i} \ge \ln e^{V_1} = V_1 \,.$$

Overall,

$$V_1 \le logsum_n \le V_1 + \ln J \,.$$

That is, for a choice set with J alternatives, the true logsum value lies between the systematic utility of the best option, V_1, and $V_1 + \ln J$.

51.2.5.2 *Using the Score of the Agent's Last Executed Plan*

Using, for each agent, the logsum over the scores of all plans implies that all these plans are valid behavioral choices. An alternative would be to simply use the score of the last executed plan. The behavioral interpretation consistent with this procedure is that there is no additional relevant randomness beyond what MATSim generates intrinsically. There has been no systematic work in this direction in the MATSim context, but such an approach might be justified in conjunction with the idea of explicitly generating the missing $\varepsilon_{n,i}$ for each person-alternative-pair n, i, then always selecting the best plan, as described in Section 97.4.6.

51.2.5.3 *Using the Average of the Agent's Plans Scores*

In principle, it is also possible to use

$$S_n = \frac{1}{J} \sum_{i=1}^{J} S_{n,i} \cdot P_{n,i} \,, \tag{51.9}$$

where $P_{n,i}$ is the probability of plan i for agent n. This can, however, only be justified when the choice probabilities, $P_{n,i}$, are interpreted like mixed strategies from game theory, i.e., that sampling from these probabilities is the true agent behavior. In principle, we cannot see why such an

[6] At this point, we assume that V_i is absorbing the scale parameter.

interpretation should be plausible—except that it is statistically the same as Section 51.2.5.2 with the advantage of having less variance. Note, however, that the approach is intertwined with the choice model. If, e.g., $P_{n,i}$ is one for the plan with the highest score and zero for all other plans, then Section 51.2.5.2 and Equation (51.9) are identical.

51.2.5.4 *Using the Highest Score of the Agent's Plans*

Alternatively, one could simply use the highest score that the agent has in its plan. This would only make sense if true behavior is assumed to always select the plan with the highest score. Again, this should then also be expressed by the choice model, i.e., using the highest score only makes sense when the agent always selects the plan with the highest score, in which case the result becomes the same as Section 51.2.5.2 and 51.2.5.3.

51.2.5.5 *More Complicated Variants*

Section 49.1.2 discusses the idea that MATSim's typical choice model might be described by a mixture-of-logit model. In that model, ϵ_{ni} remains fixed per agent n and alternative i, but other attributes such as the network conditions vary from one iteration to the next. In Equation (49.5), η_{ni} denotes these random, but simulation-generated, deviations from the average conditions; let us add an index k for the iteration number, i.e., write η^k_{ni}. That is, it is postulated that a real person would know both ϵ_{ni} and η^k_{ni}, but the simulation only knows the latter (through the MATSim score). Equation (49.5) then just describes the resulting choice distribution from what MATSim often does, i.e., apply a logit model to scores that are not averaged.

At least for η^k_{ni} that are uncorrelated from one iteration to the next it is, however, clear that this will not result in optimal *average* agent behavior – the agent may be pushed towards some choice by a random fluctuation of the η^k_{ni}, but obtaining a much lower score from that choice in the average. Overall, the agent would be better off by first averaging the score of each alternative over many iterations, and then basing her choice on those scores. This goes back to the converged scores of Section 49.1.1.

Calculating benefits from a mixture-of-logit interpretation becomes thus rather involved: we postulate that the agent sees the full MATSim score, plus some private ϵs; that she optimizes based on the sum of these two; that the MATSim simulation, however, does not know the ϵs and thus has to sample from the logit model; but that the economic utility has to include the effect of the ϵs although we do not know them, as in Section 51.2.5.1. Overall, thus, assigning utility values to such behavior as described by Equation (49.5) requires a better understanding of underlying behavioral rationality. Section 97.4.6 discusses this further.

51.2.5.6 *Summary*

Overall, there seem to be two consistent strategies to aggregate various plan scores of an agent n into one value:

- If the choice model is a logit model, then using the logsum term over all plan scores as the agent's utility U_i is consistent with the choice model.
- If the choice model is such that the plan with the highest score is selected, then using that score as the agent's utility U_i is consistent with the choice model.

In both cases, the choice model needs to be consistent with the behavioral assumption about the agent, i.e., in the first case it needs to be assumed that the model does not know the true agent choice beyond the choice probabilities and the model system thus has to repeatedly sample from these probabilities. In the second case it needs to be assumed that the randomness has already been "frozen" into the score computation (see Section 97.4.6) and the agent thus selects the plan with the highest score.

In both cases, the calculated individual score differences that result from a policy measure can be directly used in order to identify winners and losers.

Some economists claim that the modeler's task of providing information for decision support ends at this point (Ahlheim and Rose, 1989). However, in practice, some (monetary) valuation of the resulting behavioral changes is often required. The next section reviews different possibilities to monetize and aggregate individual utility differences in the MATSim context.

51.3 Aggregating Individual Values

After having obtained the individual changes in terms of utility, it is often necessary to convert these utility changes into monetary terms for economic evaluation, e.g., in BCA. Unfortunately, no "correct" monetization or aggregation approach exists for individual utility differences. This is reflected by the ongoing discussion[7] between transport policy appraisal experts:

1. The first stream argues in favor of a consistency in values used in demand modeling and appraisal (Grant-Muller et al. (2001, p.255), Bickel et al. (2006, p.S4 and p.S8), and Proost[8]). Values from literature should only be used if behavioral model values are not available. These researchers are, however, aware that this procedure potentially limits the comparability of projects in different regions of the same state, or in different member states of the EU (European Union). In consequence, additional indicators such as absolute time savings per income group should also be reported to address equity issues.
2. In contrast to the above, Mackie and Worlsley (2013, p.12) state, that in the United Kingdom, "standard [VTTS] values per minute would be used across incomes, modes and regions. Therefore, their practice is to use behavioral information for modeling but standard values for appraisal." Also Daly (2013) distinguishes between "valuation", i.e., people's willingness-to-pay (or accept) for marginal changes, and "appraisal", i.e., what these changes are worth from a societal point of view.
3. Fowkes (2010), OECD (2006), and Gühnemann[9] argue slightly differently, but in the same direction: modeling and evaluation should be based on the best heterogeneous preferences available; in the evaluation, additional weights should be introduced, e.g., to counter the effect of decreasing marginal utilities of money, or increasing VTTS with income, respectively. These weights would, thus, define the underlying equity concept of the appraisal method.
4. However, as Ahlheim and Rose (1989) point out, no approach to empirically determine these weights is available without assuming some arbitrary a-priori specification. In consequence, every interpersonal comparison of utility changes requires some normative decision and the weights need therefore to be determined on a political level.

One goal of this section 51.3 is to show the impact of a possible integration between behavioral modeling and economic evaluation in the same agent-based framework. First, a conversion into *income equivalents*, and second, a conversion into *time equivalents* (possibly followed by some conversion into money terms).[10]The choice of the procedure depends on a (normative) decision whether one *EUR* or one *h* should be valued equally across individuals. It is, therefore, important

[7] A similar overview on this discussion is given by Börjesson and Eliasson (2014).

[8] S. Proost, personal communication.

[9] A. Gühnemann, personal communication.

[10] Kickhöfer (2014) shows that the choice of the monetization and aggregation procedure can have major impact on the results when heterogeneity is assumed in user preferences.

that decision makers and modelers who deal with economic evaluation understand the possible effects of that choice; simply going with the most common approach may not be advisable.

51.3.1 *Income Equivalents*

Basic Approach The most common approach used in welfare economics to convert utility changes into money terms is to calculate the monetary amount ΔY_n that one would need to give or take from individual n to offset the impact of the policy on the utility level ΔU_n. According to Equation (51.1), it is calculated as

$$\Delta Y_n = -\frac{\Delta U_n}{\lambda_n} . \tag{51.10}$$

Note that the marginal utility of money, λ_n, might be person-specific, e.g., dependent on the person's income.

The monetary amount $-\Delta Y_n$ from above represents individual Consumer Surplus. Its absolute value is, in the absence of income effects (see Footnote 4 in Section 51.2.1), equal to the Compensating Variation and the Equivalent Variation (Daly et al., 2008). The overall welfare change ΔW for the population with individuals $n = 1..N$ is then calculated by

$$\Delta W = -\sum_{n=1}^{N} \Delta Y_n . \tag{51.11}$$

Equity The above approach is often criticized for equity reasons: if the marginal utility of money is—in the behavioral model—assumed to decrease with income, and these values are directly (without additional weights) used in economic evaluation, rich people will have a stronger impact in the evaluation process than poor people. In turn, this might lead, e.g., to investments in expensive high-speed trains on major corridors rather than affordable train services for everyone. In terms of equity and public acceptance, such specification in the appraisal method might not be desirable. To counter this effect in economic evaluation, the use of standard or equity values is proposed in the literature. In this context, Jara-Díaz (2007, p.106ff) introduces the *social utility of money* and the *social price of time*. For a more general overview of possible solutions how to address equity issues, see Rizzi and Steinmetz (2015).

A rather ad-hoc but simple possibility is to replace the person-specific marginal utility of money, λ_n, with a population average,

$$\bar{\lambda} := \frac{1}{N} \sum_{n} \lambda_n , \tag{51.12}$$

and then

$$\Delta Y_n = -\frac{\Delta U_n}{\bar{\lambda}} . \tag{51.13}$$

Following the argument by Fowkes (2010), OECD (2006) and Gühnemann et al. (2011) mentioned above (Item 3), this would be one particular way to introduce the necessary weights. Alternatively, one could think of fixing the social weight of every person to 1.0, and derive the social price of all attributes included in the generalized costs from there (Jara-Díaz, 2007, p.108f).

51.3.2 *Time Equivalents*

Another option to derive a monetary measure of welfare changes is composed of two steps: First, a conversion of individual utility changes into *equivalent hours of time as a resource* (Jara-Díaz et al., 2008; Mackie et al., 2001). This would be the number of hours ΔT_n that one would need to give or take from individual n to offset the policy impact on the utility level ΔU_n. Second, a monetization of the resulting numbers through an arbitrary conversion factor, i.e., the monetary value of one hour for the individual or for society.

In the MATSim sense, one could first calculate the corresponding time equivalent by

$$\Delta T_n = -\frac{\Delta U_n}{\mu_n} \,. \tag{51.14}$$

Similar to the marginal utility of money, also the marginal utility of time as a resource, μ_n, might be person-specific.

One option would be simply provide time equivalents, i.e., the BCA would return time equivalents per invested monetary unit. In many situations, however, it is desirable to convert all impacts of a policy into monetary terms, i.e., to compute,

$$\Delta Y_n = \alpha_n \cdot \Delta T_n \,, \tag{51.15}$$

and to compare ΔY_n with investment or changes in external costs. The following options are then possibilities for α_n:

- The obtained time equivalents ΔT_n could be converted in monetary terms using the person-specific resource values of time, i.e.,

  $$\alpha_n = \frac{\mu_n}{\lambda_n} \,. \tag{51.16}$$

  This would obviously result in the same monetary amount as the income equivalent approach from Equation (51.10).
- Following Mackie and Worlsley (2013), one could argue that the resource value of time should be the same for every individual, and, thus, use some average value for monetization, e.g.,

  $$\alpha_n \equiv \overline{\alpha} = \frac{1}{N} \sum_{n=1}^{N} \frac{\mu_n}{\lambda_n} \,.$$

- As another alternative, one could average over the marginal utility of money only, i.e.,

  $$\overline{\lambda} = \frac{1}{N} \sum_n \lambda_n$$

  and then

  $$\alpha_n = \frac{\mu_n}{\overline{\lambda}} \,. \tag{51.17}$$

  This would highlight that some persons are more pressed for time than others, while, at the same time, using an equal value for the marginal utility of money. Clearly, this gives the same result as Equations (51.12) and (51.13). It does, however, lend itself to a clearer interpretation: first, all utility differences are converted to a comparable scale, i.e., time as a resource (Equation 51.14). Then, these times are converted to a monetary scale, using a conversion factor which includes the pressure for time (i.e., the person-specific μ_n) but assumes an average marginal utility of money.

In all cases, the overall welfare change ΔW for the population with individuals $n = 1..N$ is then calculated identically to Equation (51.11), i.e., by $\Delta W = -\sum_{n=1}^{N} \Delta Y_n$.

51.3.3 *Income vs Time Equivalents: Discussion*

The sections above show how to monetize and aggregate individual utility differences though income equivalents or time equivalents. To summarize:

- Income equivalents put emphasis on the individual willingness-to-pay, whereas time equivalents focus on time pressure.
- The aggregation of income equivalents yields the overall equivalent monetary cash flow that would be generated by the project for the population considered. That is, one *EUR* is valued equally across individuals.
- The aggregation of time equivalents yields the overall equivalent lifetime hours that would be generated by the project for the population considered. That is, one hour of lifetime is valued equally across individuals.
- A monetization of time equivalents using person- and activity-specific resource values of time leads to the same total benefit as directly aggregating income equivalents.
- A monetization of time equivalents using some average value of time as a resource, therefore generally leads to a different total benefit than directly aggregating income equivalents. Such an approach maintains the equal value for one h of lifetime.

51.3.4 *Conclusion and Recommendations*

Scoring Function A correct scoring function is central to correct MATSim functioning. The mathematics and understanding of that scoring function need to be derived from time allocation theory in economics. In particular, any marginal utility of travel time needs to be split into the marginal utility of time as a resource (μ in the text above, and β_{dur} in Section 3.4) and an additional direct marginal utility of time spent traveling ($U'_{trav,q}$ in the text above, and $\beta_{trav,mode,q}$ in Section 3.4).

Since most discrete choice models estimate the sum of these two, definition is required about how to split up this sum. A somewhat ad-hoc way to achieve this is to find the mode with the largest (= least negative) marginal utility of time and use that value for the marginal utility of time as a resource. That reference mode's direct marginal utility of time spent traveling is then zero; all other modes' direct marginal utilities of time spent traveling are relative to that of the reference mode.

If one is interested in monetization, i.e., converting utility values into monetary terms, then additionally the marginal utility of money as a resource (λ in the text above, and β_m in Section 3.4) needs to be known. Our current approach to obtain an approximation to λ is to force all monetary preferences in the estimation of a choice model to a unique value. If this is not possible, then one has to make a normative decision which monetary channel is considered most "neutral", i.e., most similar to an "unearned income" channel.

Choice Model and Score Aggregation MATSim agents normally have more than one plan; each plan has a score. There are two consistent approaches to come up with a utility value from those scores:

- Using a MNL choice model that makes probabilistic draws from those plans using their scores: The correct aggregation is then the logsum of all scores.
- Using a choice model that selects the plan with the highest score: The correct aggregation is then to use the score of that plan.

In both cases, the result is the total utility U of the choice set. In the first case, the logsum term includes an expectation value of the randomness, typically denoted by ε. In the second case, all randomness, if any, needs to be "frozen" into the alternatives, and included into the computation of the score.

Monetization Individual utility differences resulting from a change in the transport system can be converted into monetary terms by dividing them by λ_n. The result is the change in individual user benefit. Aggregating these individual benefits provides an indicator for the overall welfare change. Since λ_n may vary among agents, e.g., according to their incomes, such approach will put a higher weight on people with small λ_n, typically those with large incomes. An alternative is to use an average $\overline{\lambda}$ for this conversion, even when the behavioral model (= the scoring function) uses person-specific λ_n.

Acknowledgements

The authors are grateful to G. Liedtke (DLR Berlin) who provided very helpful and detailed feedback after reading two rather different versions of this chapter. The authors would also like to thank C. Winkler (DLR Berlin) for his useful comments, in particular on the use of the indirect utility functions for economic evaluation. Finally, the authors are very thankful for the discussions with F. B. Birke (DIW Berlin) who formalized the possible decomposition of the marginal utility of money. The responsibility of any remaining errors stays with the authors.

PART IV

Scenarios

CHAPTER 52

Scenarios Overview

Marcel Rieser, Andreas Horni and Kai Nagel

This last book part summarizes MATSim scenarios, as located on the map in Figure 52.1 and listed at `http://matsim.org/scenarios`.

Figure 52.1: Locations with known MATSim scenarios. Most of them are described in this book.

How to cite this book chapter:
Rieser, M, Horni, A and Nagel, K. 2016. Scenarios Overview. In: Horni, A, Nagel, K and Axhausen, K W. (eds.) *The Multi-Agent Transport Simulation MATSim*, Pp. 367–368. London: Ubiquity Press. DOI: http://dx.doi.org/10.5334/baw.52.

Although there are real-world scenarios based on free and public data such as the Santiago or Cottbus scenarios (Chapters 84 or 66), many scenarios are not public, due to data privacy issues. However, knowing about general methods and approaches adapted for scenario creation and understanding problems faced during these processes might significantly support and encourage the building of new scenarios. Each of the following chapters provides information on study area, population and demand generation, activity locations, network, simulated modes, calibration and validation, achieved results, and associated projects. Further topics involve where to find more information and where/when emphasis is put on certain scenario specialties—be it parsimonious data usage procedures, special modules used, or special modes simulated (such as the parataxis in the Gauteng scenario). Some scenarios have been used for years, with ongoing further development. We target the latest version when reporting.

Different levels of MATSim involvement are possible. For some regions and projects, MATSim is, for example, used only for traffic assignment, where for others, the complete demand is endogenously handled. Couplings with other forecasting models for transport demand generation have been successfully applied, like the coupling with TASHA (Travel Activity Scheduler for Household Agents) for Toronto, or the combination of MATSim with the Tel Aviv activity-based transport model.

CHAPTER 53

Berlin I: BVG Scenario

Andreas Neumann

The BVG is Berlin's main public transport company, running virtually all services, with the exception of the S-Bahn urban rail system. This includes bus services, the subway network, the largest tram network in Germany and ferry services. The bus network consists of 149 different lines, 6468 directed stops and a vehicle fleet of 1316 buses (BVG, 2012). In total, about 937 million trips were served by BVG in 2012, 41% of them by bus.

With the opening of the new Berlin and Brandenburg BER international airport, Berlin expects major travel demand changes; importantly, the existing airport Tegel, now exclusively served by BVG-operated buses, will close. BVG thus had substantial interest in a new Berlin area transport model. To deal with these changes, the model not only had to provide a base for future regional transport system planning, but also had to supply detailed information about different user groups' passenger flows. Such user group-specific analyses were very important for BVG in providing a platform for their future business strategies; thus, an agent-based model was specifically requested. Two scenarios were required, one for the year 2008 (actual state), and one for the year 2015 (prediction). To meet the above needs, PTV (2013), Senozon (2013) and VSP (2012) at TU Berlin offered a combined model consisting of both a static macroscopic model built with VISUM, as well as an integrated, activity-based demand and dynamic traffic assignment model, built with MATSim. During the project, efforts were made to base both models on the same data sources and to ensure that both modeling processes interacted with each other to allow data exchange.

The model contains about 115 000 links, about 15 000 directed stops, about 6 million agents, and 539 public transport lines operated by BVG and other Berlin and Brandenburg state companies. Besides motorized individual traffic and public transport, the model also considers biking and walking. For a more in-depth description of the model, its generation and its calibration, the reader is referred to the work of Neumann et al. (2014). The model has extensively been used by Neumann (2014, Ch. 7/8) for the development of the minibus module presented in Chapter 17.

How to cite this book chapter:

Neumann, A. 2016. Berlin I: BVG Scenario. In: Horni, A, Nagel, K and Axhausen, K W. (eds.) *The Multi-Agent Transport Simulation MATSim*, Pp. 369–370. London: Ubiquity Press. DOI: http://dx.doi.org/10.5334/baw.53.

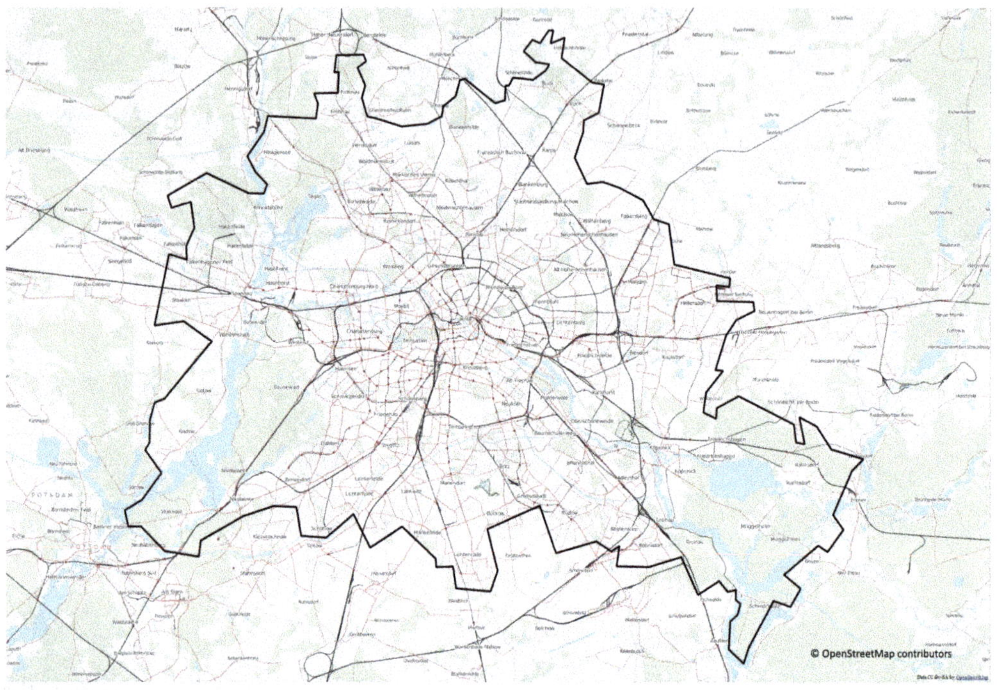

Figure 53.1: The city of Berlin and its transit network.

CHAPTER 54

Berlin II: CEMDAP-MATSim-Cadyts Scenario

Dominik Ziemke

To correctly model initial demand properties not included in MATSim iterations in specific studies (i.e., activity choice), suitable data are needed. Travel diaries containing departure times, mode choice decisions and activity locations are widely used. However, much of this data source content, particularly location information, is considered sensitive in terms of data privacy legislation and thus increasingly difficult to obtain and process in many areas (e.g., in Germany and the United States; Ziemke et al., 2015).

The *Berlin II scenario* (also referred to as the *CEMDAP-MATSim-Cadyts scenario* according to the applied models in its setup), is the outcome of an alternative approach relying exclusively on freely available or easy-to-obtain input data. All of these data do not rely on individual trajectories, but instead on "anonymous" data that is aggregated so much that the data providers are no longer concerned about privacy issues.

The starting point for this scenario is a publicly available commuting matrix containing homes and workplaces of workers with social security on the municipality level. Based on this information, it is possible to model morning and evening commuting peaks.

To obtain a full-population demand representation, two further major modeling steps are required. First, in cases like the Berlin case, see below, where the commuter matrix spatial resolution is quite coarse, higher resolution O-D information is necessary. Second, a procedure is needed to model secondary activities, i.e., all other activities beyond home and work.

The importance of the first step becomes obvious when looking at the German case; here, the whole city of Berlin, with 3.4 million inhabitants, is represented by exactly one zone (Bundesagentur für Arbeit, 2010). In the United States, commuting matrices are typically available only on a county-to-county level. Since such location-aggregation-based matrices may become the rule, rather than the exception, in privacy-sensitive societies, a (generalizable) method to attain O-D information at a higher resolution is needed (Ziemke et al., 2015). The standard solution would be to estimate an activity location choice model. This, however, is difficult if no trip data to estimate

How to cite this book chapter:
Ziemke, D. 2016. Berlin II: CEMDAP-MATSim-Cadyts Scenario. In: Horni, A, Nagel, K and Axhausen, K W. (eds.) *The Multi-Agent Transport Simulation MATSim*, Pp. 371–372. London: Ubiquity Press. DOI: http://dx.doi.org/10.5334/baw.54. License: CC-BY 4.0

the model is available. O-D matrix estimation studies (van Zuylen and Willumsen, 1980) suggest that traffic counts may be used to make an initially rough O-D matrix more appropriate for a region. As MATSim is not based on O-D flows, but on full daily plans, the issue comes down to whether a procedure exists to update these initial full daily plans using traffic counts. In the approach used to create the Berlin II scenario, a procedure proposed by Flötteröd et al. (2011) and implemented in the software Cadyts—explained in Chapter 32—is applied for this task. Specifically, random draws of possible home and work locations within the home or work municipality given by the commuter matrix are made. Various MATSim plans, each containing one pair of home and work locations, are created for each agent. Then, the Cadyts calibration procedure is applied within the iterative MATSim simulation to select plans and locations more likely to occur with given traffic counts.

As stated above, however, full daily plans (as opposed to mere home-work-home commuting patterns) are needed. Therefore, the second modeling step, the modeling of secondary activities for each individual in the region, needs to be addressed. For the Berlin II scenario, CEMDAP (Comprehensive Econometric Microsimulator for Daily Activity-Travel Patterns (Bhat et al., 2008)) is used to generate initial complete daily plans for each individual. On one hand, however, no CEMDAP parameter set is available for Berlin. On the other hand, and more importantly, one major goal of the study creating the Berlin II scenario was to show its generalizability (Ziemke et al., 2015). So, the model parameters of CEMDAP estimated for the Los Angeles region (the estimation context) are retained and then used to generate initial plans for individuals in Berlin (the application context in the current paper), based on Berlin demographic data.

To sum up, home and work municipalities are taken from the commuter matrix. Within these municipalities, a set of (more precisely spatially defined) potential home and work locations are randomly chosen for each agent. Full daily plans incorporating the various potential locations of each agent are generated with CEMDAP, based on a parameter set from another region.

Then, the Cadyts calibration procedure is used to select those initial full daily plans most consistent with Berlin traffic count data. In other studies, Cadyts has already been applied to update route choice predictions, both for car (Flötteröd et al., 2011a) and for public transit (Moyo Oliveros and Nagel, in press). However, it has not been used to update full daily activity-travel plans, as it was in the procedure that created the Berlin II scenario.

The Berlin II scenario is thus an activity-plan-based MATSim transport model for Berlin based exclusively on freely, or readily, available data. If a commuter matrix, some basic population demographics, and traffic counts (or, theoretically, another suitable data source on which to run the calibration procedure) are available for a particular regional context, the approach used to create the Berlin II scenario can be transferred to that other context. In fact, the Berlin II scenario itself should be seen as a *transferred model*, because initial plans generated by CEMDAP are based on parameter estimates from another geographic region (the Los Angeles area).

Through a validation based on the Berlin 2008 SrV (System repräsentativer Verkehrsbefragungen (Ahrens et al., 2009)), an extensive, regularly-conducted travel survey, the created transport demand representation quality has been successfully tested. So far, the Berlin II scenario exists for a 1% and a 10% population sample of all persons, i.e., including workers without social security, as well as non-working people, aged 18 and above, for the study region. Currently, only motorized traffic is considered. Stability tests, showing that plausible agents' daily plans continue to be chosen when Cadyts calibration functionality is switched off, have been successfully carried out. This is a clear indication that the scenario is applicable and meaningful for policy studies.

Further improvements, like the addition of public transport and a more realistic representation of the population, are planned. Moreover, similar approaches to integrating activity-travel pattern generators (e.g., the FEATHERS model) with MATSim in transport simulation are planned.

CHAPTER 55

Switzerland

Andreas Horni and Michael Balmer

The Switzerland scenario was initially created for the project Westumfahrung (Balmer et al., 2009a) and serves as the base for the very frequently used Zürich scenario (Chapter 56).

Two main branches can be distinguished. The first, older one is based on a one-to-one translation of the Swiss population census (Swiss Federal Statistical Office (BFS), 2000); the second applies approaches from the IPF (Iterative Proportional Fitting) family, reported by Müller and Axhausen (2013, 2012); Müller (2011b,a, 2012) to generate the synthetic population.

The scenario's study area covered all of Switzerland. Due to administrative borders, no demand and supply data were available for adjoining countries, which leads to boundary effects; studies focusing on Swiss border areas are difficult.

The population was derived from the Swiss Census of Population 2000 (Swiss Federal Statistical Office (BFS), 2000). The complete Swiss population was modeled, resulting in around 7.5 million agents.

This population's home locations were given at hectare level and work locations were specified at municipality level from commuter matrices, a component of the Swiss Census of Population 2000 (Balmer et al., 2009a, p.35). A very good overview, in German, of the population generation, its initial individual demand and activity locations can be found in Meister et al. (2009). Further information is given by Ciari et al. (2008); Meister et al. (2010); Balmer et al. (2009a, 2010, 2009b).

Travel demand was basically taken from the 2000 and 2005 National Travel Surveys (Swiss Federal Statistical Office (BFS), 2006) (Swiss microcensus), although this sample substantially underestimated freight traffic and ignored cross-border traffic of non-Swiss residents. Freight traffic for Switzerland was missing at that time (except Zürich, see next chapter). Cross-border traffic was derived from mode-specific, hourly origin-destination matrices given by Vrtic et al. (2007). These were disaggregated to around 600 000 individual MATSim plans for the whole country, which contain the cross-border traffic originating *outside* Switzerland. Non-Swiss, cross-border traffic starting in Switzerland was supposed to be negligible.

How to cite this book chapter:

Horni, A and Balmer, M. 2016. Switzerland. In: Horni, A, Nagel, K and Axhausen, K W. (eds.) *The Multi-Agent Transport Simulation MATSim*, Pp. 373–374. London: Ubiquity Press. DOI: http://dx.doi.org/10.5334/baw.55.

The activity location data set, comprising home, work, education, shopping and leisure locations, was also derived from the 2000 Swiss Census of Population and the 2001 Federal Enterprise Census (Swiss Federal Statistical Office (BFS), 2001), providing hectare level information. Facility generation was described by Balmer et al. (2009a, p.33).

For car traffic, navigation networks from Teleatlas (Tele Atlas MultiNet, 2010) and NAVTEQ (NAVTEQ, 2011) were available. The most-used network was the planning network derived from from the Swiss National Transport Model (Vrtic et al., 2003).

The public transport simulation network was derived from the National Transport Model of the UVEK (Eidgenössisches Departement für Umwelt, Verkehr, Energie und Kommunikation), described by Vrtic and Fröhlich (2010).

The scenario simulated car and public transport; schedules for public transport were given at the municipality level. Fine-granular schedules were not available then, but were in preparation. The modes walk and bike were usually "teleported".

Calibration was mainly performed for modal split and distance distributions; utility function values were set accordingly.

For validation, count data on city level, cantonal level and national level (ASTRA, 2006) were available from various sources, resulting in 600 links measured for Switzerland. An average working day (Monday to Thursday, excluding public holidays) was used for comparisons in current projects.

CHAPTER 56

Zürich

Nadine Rieser-Schüssler, Patrick M. Bösch, Andreas Horni and Michael Balmer

The MATSim team frequently uses the Zürich scenario, based on the Switzerland scenario described above. The Zürich scenario, however, is more detailed; it was enhanced by data available only for the smaller region; e.g., traffic light data or freight demand data was only included for Zürich city and the canton. It is under continuous development, calibration and validation and has been applied in numerous projects, serving as a real-world research example.

Horni et al. (2011b) provide a technical overview of the first scenario branch; Balmer et al. (2009a) describe its generation for the "Westumfahrung" project.

The study area was delineated by a circle, with a 30 kilometer radius around Bellevue, a central and prominent Zürich location. This delineation led to two versions, the *Zürich diluted scenario* and the *Zürich cut scenario*. For the first, all agents crossing the study area during the simulated day were considered (Figure 56.1), resulting in almost two million agents. For the second, only agents remaining in this area the whole day were modeled. The *Zürich cut scenario* was employed as an experiment in Hackney (2009), but using the *Zürich diluted scenario* for production runs is preferable.

Demand was taken directly from the Swiss model; freight traffic was added to the Zürich scenario, as follows. Canton Zürich raw freight traffic data was taken from the KVMZH (Kantonales Verkehrsmodell Zürich), provided by Amt für Verkehr, Volkswirtschaftsdirektion Kanton Zürich (2011) and documented by Gottardi and Bürgler (1999). Zonal level matrices were disaggregated to single MATSim plans (Shah, 2010). Matrices for small delivery and heavy trucks were combined into one activity called *freight*. An additional 180 000 agents were generated for the Zürich region.

For the diluted Zürich scenario, all Swiss facilities, as described above, were used as activity locations and the networks were not thinned out. For public transport simulation, network and transport schedules were derived from the KVMZH. Walk and bike modes were "teleported".

Calibration was mainly done for modal split and distance distributions and utility function values set accordingly.

How to cite this book chapter:

Rieser-Schüssler, N, Bösch, P M, Horni, A and Balmer, M. 2016. Zürich. In: Horni, A, Nagel, K and Axhausen, K W. (eds.) *The Multi-Agent Transport Simulation MATSim*, Pp. 375–378. London: Ubiquity Press. DOI: http://dx.doi.org/10.5334/baw.56.

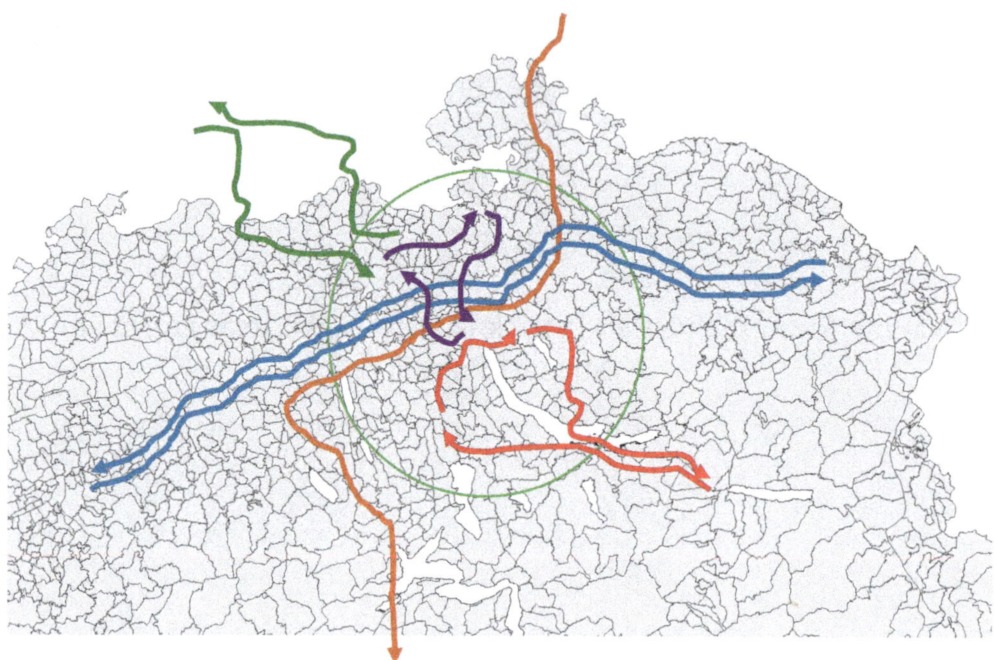

Figure 56.1: The diluted Zürich scenario

For validation, count data on city level, cantonal level and national level (ASTRA, 2006) were available from various sources, resulting in 123 links measured for the Zürich inner city, delineated by a 12 kilometer radius around Bellevue. The reduced count analysis radius was applied to reduce boundary effects resulting from demand reduction outside the 30 kilometer radius study area. An average working day (Monday to Thursday, excluding public holidays) was used for comparison in current scenarios.

Some traffic signal data was available for Zürich city (Stadt Zürich, Dienstabteilung Verkehr, 2008); this was integrated for the Westumfahrung project.

56.1 Studies Based on the Zürich Scenario

Besides its widespread use for the development of new MATSim functionality—e.g., the contributions for destination innovation (Chapter 27), joint decisions (Chapter 28), parking (Chapter 13), or electric vehicles (Chapter 14)—the Zurich scenario has also been used in policy studies. The most prominent one was the study Westumfahrung (Balmer et al., 2009a), where MATSim was used to estimate the effects of opening a new motorway section and different accompanying measures. In addition to classic evaluations such as link volumes and spider analyzes, the project focused on estimating who the winners and losers of the Westumfahrung were and where they lived. Other policy studies looked at the potential for Park & Ride, organized as well as informal ride sharing, the effects of a substantially improved public transport offer, and the influence of road capacity changes on transport behavior.

A more recent example for a study based on the Zürich scenario is described by Heyndrickx et al. (2016); Boesch et al. (2014); Heyndrickx et al. (2014); Pilli-Sihvola et al. (forthcoming); Boesch and Ciari (2014); Boesch (2014). It was conducted as a part of the EU project ToPDAd (Tool supported Policy Development for regional Adaptation). ToPDAd tried to find the best strategies for decision makers to adapt to the expected short and long term effects of climate change. The international project focused on the three potentially climate sensitive and important economic sectors Energy, Transport and Tourism.

For each sector different case studies were investigated to develop the tools required to find suitable adaptation strategies. In the transport sector the IVT together with the TML (Transport & Mobility Leuven), Belgium, conducted a study on the potential influence of extreme weather events, which are predicted to increase in frequency and intensity for Western Europe due to climate change, on the transport system.

The Zürich scenario was used to identify the transport system reactions on different, weather-induced disturbances. The number of trips, activities, and their durations were compared for different scenarios. The applied scenarios represented variations both on the supply side and on the demand side. On the supply side, next to the baseline scenario eight different scenarios were simulated. A medium and a high disturbance scenario, where the capacity and the free-flow speed on the entire network were reduced due to unfavorable weather conditions and a medium and high disruption scenario where certain, exposed street and public transport links were (temporary) blocked. These disturbances and disruptions occurred only in the peak hour or for the full day, resulting in the eight scenarios on the supply side. On the demand side the agents were allowed five different degrees of flexibility to react to this situation: 1. Worst case (no reaction allowed); 2. Rerouting; 3. Rerouting and modal change; 4. Rerouting, modal change and rescheduling; and finally 5. Rerouting, modal change, rescheduling and relocation.

It was found that rerouting and mode choice together have the highest impact in terms of reaction to the disturbances. If the public transport system is disrupted, the expected shift to car and slow modes is observed. The opposite, expected shift to increased pt-usage is also correctly observed if the transport system is disturbed by unfavorable weather conditions (e.g., rain or snow).

The results of these scenarios were used by TML to calculate the direct and indirect economic costs of extreme weather events through an impaired transport system. Extreme events with a return value of five to ten years are estimated to cause costs of up to 19 million EUR per event for the region of Zürich, while the more extreme events with a return value of only 50 to 100 years would cause costs of up to 100 million EUR per event. Compared to estimations for historic events these are relatively low values (costs of billions per event). One of the reasons for this difference is assumed to be in the inability of MATSim agents to drop activities. So, while in reality people would for example likely drop work activities in the case of severe floods and thus cause additional economic costs, MATSim agents will always try to find a way to get to their work location and to work { no matter how bad the circumstances. Current efforts at IVT try to overcome this limitation while still producing realistic simulation outcomes.

CHAPTER 57

Singapore

Alexander Erath and Artem Chakirov

The MATSim Singapore scenario (Erath et al., 2012) was implemented and is maintained at the FCL (Future Cities Laboratory), a research program of the SEC (Singapore-ETH Center for Global Environmental Sustainability) and part of Singapore's National Research Foundation CREATE (Campus for Excellence and Technological Enterprise). The scenario covered the whole Singapore area, with a population of approximately five million and included traffic to and from neighboring Malaysia. Singapore provides an excellent study case for an agent- and activity-based modeling approach: a fairly densely populated city, with an extensive public transport infrastructure and advanced transportation and pricing policies.

57.1 Demand

In the absence of a full-population census for Singapore, a synthetic population was generated based on data from the HITS (Household Interview Travel Survey) 2008 (Choi and Toh, 2010) and population breakdowns of Singapore's population census 2010. The synthetic population was derived using the fitting and sampling method (Müller and Axhausen, 2011), where a reference sample of household and person records was weighted, using an IPF technique, until the weighted sample matched marginal census control totals. In our case, the reference sample was from travel survey records; fitting technique was the entropy optimization method proposed by Bar-Gera et al. (2009) and implemented by Kirill Müller, IVT, ETH Zürich. Then, the reference sample records were replicated through weighted sampling until the population total was met.

Car ownership was modeled on a household level and driving licenses were assigned to individuals, using discrete choice methods. Given the high car tax in Singapore, the model reflected lower car ownership level than in other developed nations. The model presented by van Eggermond et al. (2012) included not only socio-economic, but also spatial variables and proved to be essential to the MATSim Singapore model, leading to accurate mode choice and mode share predictions.

How to cite this book chapter:
Erath, A and Chakirov, A. 2016. Singapore. In: Horni, A, Nagel, K and Axhausen, K W. (eds.) *The Multi-Agent Transport Simulation MATSim*, Pp. 379–382. London: Ubiquity Press. DOI: http://dx.doi.org/10.5334/baw.57.

Activity locations were defined on an individual building level, with information on building and facility types compiled from various sources: i.e., the land-use master plan (URA, 2008), government websites and online directories, as well as points of interest information provided by NAVTEQ. In the absence of a business census, an innovative approach for location identification and corresponding number of work places was developed, drawn from the full smart card data record of public transport journeys and enriched with information on land-use and estimates of building floor space. In a first step, a probabilistic model was applied to a daily public transport journey record to identify types of activities performed between two subsequent public transport trips. Estimated and calibrated using HITS 2008 records, the model combined variables such as time of day, activity duration and land-use around each stop or station to ensure an accurate differentiation between home, work, or other activities. After accounting for mode shares in 53 different zones, an optimization technique employing accessibility computation was applied to distribute work activities to individual buildings. More details on the newly developed methodology and its practical application were reported by Chakirov and Erath (2012) and Ordóñez Medina and Erath (2013a).

Assignment of households to buildings was performed using detailed information on residential developments; for public housing, which represented about 80 % of Singapore's residential building stock, information on distribution of different dwelling types was employed, while for privately owned condominiums, only information on number of apartments per building was available. Work locations were assigned using a zone-based gravity model using prior estimated number of work activities in each building as additional information for distribution of workplaces within each zone. Activity chains were assigned based on their observed frequency in HITS, taking into account key socio-demographic parameters like sex, age, occupation and income. Activity chains of type home – work – home were by far the most frequent, accounting for approximately 50 % of the trips. Freight and cross border traffic, as well as tourist travel demand, were derived based on a set of origin destination matrices provided by the LTA (Singapore Land Transport Authority). These matrices were converted into special daily plans. Information on the temporal distribution of freight trips was derived from loop detector data for freight and temporal attraction profiles of major tourist sites.

57.2 Supply

Using a semi-automatic map-matching algorithm (see Chapter 9), a high-resolution navigation network provided by NAVTEQ was map-matched to, and enhanced with, LTA's planning network lane and capacity information. Without access to traffic signal cycle time data, traffic lights were not specifically modeled. Extensive attention was paid to public transport modeling; interaction between private and public transport with Singapore's high density and limited space was very important. Simulating dynamic effects, such as bus bunching, was crucial for obtaining realistic travel times and mode shares. Public transport network and schedule data provided by LTA included bus and train routes, as well as stop and station location. This information was matched to the road network, using yet another map-matching algorithm presented by Ordóñez Medina and Erath (2011); Ordóñez Medina (2011b). Recently, the scenario was updated using public transport schedule data derived from public transport smart card data records (Fourie, 2014). Such schedule information provided actual vehicle dispatch frequencies and headways, which are continuously adjusted and, in some cases, can substantially deviate from published schedules. Additional features of public transport simulation in Singapore's model included advanced bus dwell time model (Sun et al., 2014b), as well as an approximation of the distance-based public transport fare scheme.

Other modes, specifically walking and cycling, were "teleported" with constant travel speeds without any interaction with other users.

57.3 Behavioral Parameters

Behavioral parameters specific to Singapore's context were borrowed from Land Transport Authority (2009) and used with the widely applied Charypar-Nagel function for activity scoring (Charypar and Nagel, 2005). Thus, the same parameters were used for all agents, ignoring user preferences, heterogeneity, and time values in the initial scenario implementation. Furthermore, no additional crowding penalties (impacting travelers' discomfort) were considered at this stage; public transport overcrowding effects were taken into account only with physical vehicle capacity limitations, as well as their implications for dwell time and the bus bunching phenomenon.

57.4 Policy

The MATSim model for Singapore also included ERP (Electronic Road Pricing) scheme, featuring time and vehicle-dependent road pricing. Based on two data sets, with location and time-dependent price levels, prevailing tolls were specified for 73 network links where toll gantries had been installed, as of February, 2012. To account for the numerous dedicated bus lanes, additional links attributed to exclusive bus use were added to the network. The existing links' capacity was reduced accordingly, even if, in some cases, dedicated exclusive bus lanes by buses existed only during peak hours. Such a simplified setup, insensitive to the time-dynamic operation of dedicated lanes, led to actual road capacity underestimation during periods when bus lanes were also open to other motorized traffic. However, as most links featuring bus lanes consisted of three or more lanes, the effect on modeled traffic conditions during off-peak hours appeared to be low.

57.5 Calibration and Validation

Road usage data is available for around 200 count stations at hourly intervals. Public transport smart card data availability provides an additional validation dimension. For the future, the opening of new MRT lines—since setting up the model in 2012—presents a unique opportunity for comparing observed ridership with predicted ridership in the model. However, systematic calibration and detailed validation have not yet been conducted.

CHAPTER 58

Munich

Benjamin Kickhöfer

The MATSim scenario for the Munich metropolitan area was set up during 2010.[1] The main goal was, and is, simulation of local air pollutant and global greenhouse gas emissions and how their levels change with different policy measures—on aggregated and spatially disaggregated levels. Thus the scenario was used for development and testing of the EMT (Emission Modeling Tool, see Chapter 36). For an example illustrating where overall NO_2 private car and freight vehicle emissions are produced over one day, see Figure 58.1.

Network information from VISUM was converted into MATSim format, resulting in a network of 17 888 nodes and 41 942 links. This transport supply was then linked to travel demand from different sources; an inner-urban traffic activity-based demand from survey data was created, based on MiD (Mobilität in Deutschland (MiD 2002, Follmer et al., 2004)). This synthetic population segment consisted of roughly 1.4 million individuals, with detailed vehicle information for every household. Commuters and reverse commuters were modeled with data provided by the German Federal Employment Office (Böhme and Eigenhüller, 2006). This part of the population consisted of approximately 0.5 million individuals, with 0.3 million commuting to Munich for work. The rest lived in Munich and commuted to their workplace outside the city. Freight traffic was also introduced into the model using data from the German Ministry for Transport (BVU Beratergruppe Verkehr + Umwelt GmbH und Intraplan Consult GmbH, 2007). This consisted of roughly 0.15 million freight vehicles, performing one commercial trip per day.

The scenario was used for several case studies: Hülsmann et al. (2011) used a single street corridor to validate simulated travel times and emission levels against actual data obtained from a test vehicle. Kickhöfer et al. (2013) investigated the relationship between the price elasticities of car travel demand and air pollutant emissions. Hülsmann et al. (2013) identified city areas with

[1] Detailed descriptions of the scenario can be found in Kickhöfer et al. (2013) and Kickhöfer (2014).

How to cite this book chapter:
Kickhöfer, B. 2016. Munich. In: Horni, A, Nagel, K and Axhausen, K W. (eds.) *The Multi-Agent Transport Simulation MATSim*, Pp. 383–384. London: Ubiquity Press. DOI: http://dx.doi.org/10.5334/baw.58.

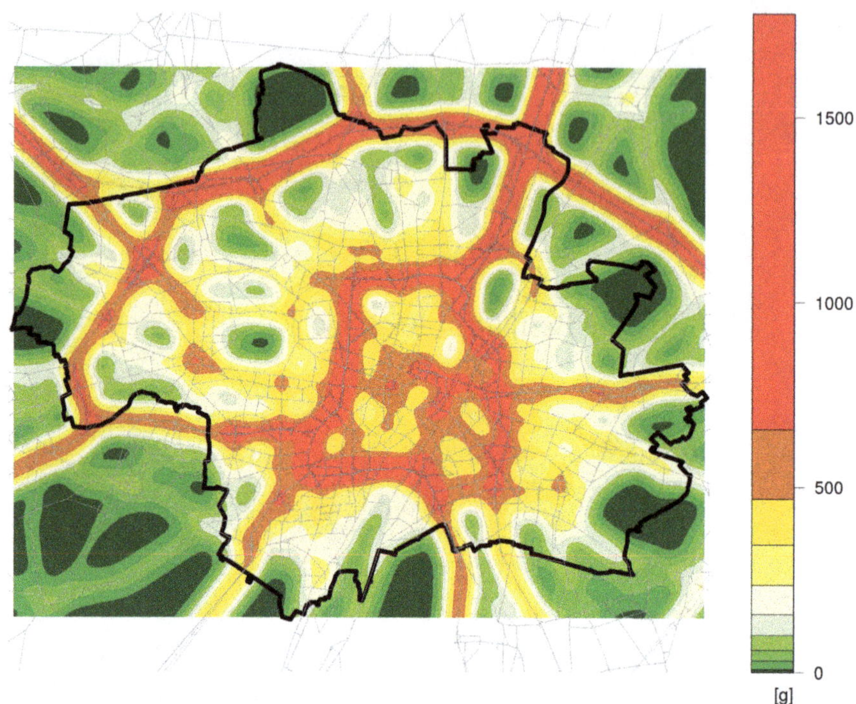

Figure 58.1: NO_2 emissions in Munich

high air pollution concentration. They defined these areas as "hotspots", exceeding the EU limits for NO_2 (Nitrogen Dioxide). The authors raised toll levels incrementally for vehicles passing these hotspots, until high pollution concentrations disappeared, to estimate true threshold value EU avoidance costs. Kickhöfer and Nagel (2013) derived time-dependent, vehicle-specific, first-best air pollution tolls to create a benchmark for real-world policy evaluation. Kickhöfer and Kern (2015) went one step further and calculated time-dependent, vehicle-specific air pollution *exposure* tolls.

CHAPTER 59

Sioux Falls

Artem Chakirov

The Sioux Falls scenario provided a convenient test-case, combining fully dynamic demand fitted with realistic socio-economic and demographic attributes with a small-scale road network including an integrated public transportation system. Based on the Sioux Falls road network commonly used for tests and demonstration purposes in transportation literature (Bar-Gera, 2013), it allowed quick and convenient experiments on new policy or software implementations with MATSim on a heterogeneous agent population, with a high degree of spatial resolution, but without significant computational requirements. However, it is important to stress that, despite the use of real world data for the generation of the enriched Sioux Falls scenario, it did not aim to replicate the real City of Sioux Falls in South Dakota, US and remains a fictitious test case. Detailed report on scenario generation and its characteristics is provided by Chakirov and Fourie (2014) and can also be found at `http://www.matsim.org/scenario/sioux-falls`.

59.1 Demand

A realistic, socio-economically and demographically diverse demand population—with heterogeneous use preferences—was crucial for unlocking the full potential of an agent-based simulation like MATSim. However, generation of a disaggregated demand description on individual and household levels close to reality was challenging; not only for trip origins and destinations, but also with respect to travel pattern relation and socio-demographic travelers' characteristics.

To address this challenge for the Sioux Falls scenario, and represent the household structure, demographic profile and income distribution as realistically as possible, a synthetic household population, using the Bar-Gera et al. (2009) entropy optimization technique, was generated. It matched the aggregate distribution of demographic attributes (age, sex and household income) recorded during the 2010 US Census. It contained census tracts inside, and adjoining, the city center of

How to cite this book chapter:

Chakirov, A. 2016. Sioux Falls. In: Horni, A, Nagel, K and Axhausen, K W. (eds.) *The Multi-Agent Transport Simulation MATSim*, Pp. 385–388. London: Ubiquity Press. DOI: http://dx.doi.org/10.5334/baw.59.

Sioux Falls and was composed of household and person records taken from the (anonymous) 5-year American Community Survey sample (2007-2011), covering 5 % of all households.

To keep the scenario accessible, as well as facilitating interpretation and understanding of possible effects on policies studied, only two simple activity chains were initially included: "home – work – home" and "home – other – home". Activity locations were identified using building stock data set provided by the City of Sioux Falls GIS division. Each household's home location was assigned randomly to a residential unit within the household's tract. Because no information on the real number and distribution of work places within the relevant area was easily accessible, the static O-D matrix from LeBlanc et al. (1975) was taken as a workplace attraction indicator for each zone. Then, assignment of work places to individual workers, as well as locations of secondary (other) activities, was performed using a parameter-free radiation model presented by Simini et al. (2012).

To exploit the full potential of disaggregated demand and add another degree of realism to the scenario, car ownership on the household level was modeled using an ordered probit model, presented by Giuliano and Dargay (2006) and based on the NPTS (US Nationwide Personal Transportation Survey) 1995. In addition to socio-demographic household characteristics (number of adults, children, pensioners, household income), the model used residential location attributes (population density, public transport access and dwelling type), which better described specific Sioux Falls scenario characteristics, as well as its area-wide bus network.

59.2 Supply

A realistic transportation test network should ensure sufficient complexity of travelers' choice dimensions while limiting computational effort. To this end, the Sioux Falls test network was introduced by Morlok et al. (1973) and later adapted as a benchmark and test scenario in many publications (see Chakirov and Fourie (2014) for overview). The network structure captured the major arterial roads of Sioux Falls, South Dakota, but was never intended to replicate the real city, or all characteristics of its transportation system, such as travel times or modal split. The original network was comprised of 76 arcs, 24 nodes and 552 O-D pairs. For this scenario, road capacities were adjusted according to values provided by the Highway Capacity Manual Transportation Research Board (2010) and other related research publications (e.g., Ng and Small, 2012). The public transportation network added to the scenario included five bus lines, as initially proposed by Abdulaal and LeBlanc (1979), with bus stops placed at regular intervals of 600 meters.

Due to the design of MATSim queue simulation, agents were handled only at the beginning and end of each network link and could not enter or leave a link along its length. Therefore, origins and destinations located along very long links led to spatial detail loss, as all origins and destinations along the length of the link were effectively assigned the same coordinates. Consequently, to improve spatial detail level, all links of the Sioux Falls network were evenly split into smaller links, with maximum length of 500 meters each. Following this operation, number of nodes was increased to 282 and number of links to 334, without changing effective network topology.

In addition to car and bus modes, walking as "teleported" mode, with constant travel speeds, and with no interaction with other users, is used as the non-motorized transportation mode.

59.3 Behavioral Parameters

Behavioral parameters used in utility functions were based on estimated demand model for Sydney by Tirachini et al. (2014). Before applying parameters in an activity-based context, time-related parameters had to be adjusted to account for utility gained from activity performance. Thus, to provide a value for marginal utility of performing an activity, the travel mode with smallest the disutility was set as a baseline, under the assumption that traveling with this mode was equivalent to idling/doing nothing. Corresponding parameters were split into opportunity costs of time and a mode-specific disutility of traveling, as has been done in previous MATSim-related publications (e.g., Kickhöfer et al., 2011).

59.4 Results, Drawbacks and Outlook

Sioux Falls scenario stability and performance was tested using two sets of activity timing constraints, as well as five different random seeds, which all delivered stable and realistic results. Chakirov and Fourie (2014) also investigated MFD (Macroscopic Fundamental Diagram) existence and hysteresis characteristics, as discussed in Geroliminis and Daganzo (2007, 2008); Geroliminis and Sun (2011).

However, recent experience has shown certain coarse network drawbacks; it represented only major arterial roads and neglected minor neighborhood and collector road links. With an elaborate synthetic population and high rush hour demand peaks, the network seemed to be sensitive to network breakdowns under high loading conditions.

Along with the coarse road network, the coarse public transport network level and the resulting low level of accessibility (for parts of the population) represented another drawback, particularly relevant to simulation and evaluation of policies sensitive to, or requiring, a certain share of public transport users.

Replacing the original Sioux Falls network with a finer network obtained from the crowd-sourced OSM and adding additional public transport lines would address the above-mentioned scenario weakness. However, this introduces a different set of drawbacks and would require further attention. First, the significantly larger number of network links and nodes increases time and resources for routing and dynamic queue simulation and could erase the advantages of a small-scale network. Extended simulation times can be tackled with the new pseudo-simulation methodology, currently developed by Fourie et al. (2013). Second, total network capacity increase leads to reduction or even disappearance of congestion during peak hours, although including freight and through traffic in the scenario can make it more realistic and address congested conditions during peak-hours.

CHAPTER 60

Aliaga

Pelin Onelcin, Mehmet Metin Mutlu and Yalcin Alver

Aliaga, in Turkey, is situated about 50 kilometers north of Izmir; it is one of the 30 Izmir province districts in the Aegean region of Turkey and is crucial to the national economy.

Aliaga is home to Petkim, one of the largest petrochemical enterprises of Turkey. In 2011, Petkim was ranked as the 12th largest company in Turkey's 500 top industrial list (Istanbul Chamber of Industry, 2012, accessed 03.07.12); the enterprise includes 14 plants and seven auxiliary units.

According to the Turkish Statistical Institute, the 2011 population of Aliaga was 68 432; 56 440 lived in central neighborhoods and 11 992 in surrounding villages (Turkish Statistical Institute, 2011).

Many chemical factories are located near residential areas. The evacuation zone was determined using a scenario developed for a chemical accident in one of the Petkim factories. Chemical substance elements and NFPA (National Fire Protection Association) (704) ratings, ranging from 1 to 4 for flammability, health and reactivity, were compared. The most dangerous substance was acrylonitrile (ACN), rated 3, 4 and 2 for flammability, health and reactivity, respectively.

Risk zone radii were found using Aloha software developed by the Office of Emergency Management and Emergency Response Division. The software divided the risk area into three zones, based on the chemical substance type, wind speed and wind direction. The wind data, obtained from Aliaga wind measurement station, showed that maximum wind speed was 17 meters per second (WolframAlpha, 2012, accessed 02.08.12) and the prevailing wind direction was WNW. Wind blowing from the west would be the most dangerous for Aliaga, carrying the smoke over residential areas and increasing the number of persons to be evacuated.

The evacuation zone was divided into 19 TAZs (Traffic Analysis Zones). Trips generated from these zones were directed to six destinations TAZs, three of which are health care centers and three gathering places. The Petkim area iwa divided into six zones; the first in the impact area. The evacuation planning zone is shown in Figure 60.1.

How to cite this book chapter:

Onelcin, P, Mutlu, M M and Alver, Y. 2016. Aliaga. In: Horni, A, Nagel, K and Axhausen, K W. (eds.) *The Multi-Agent Transport Simulation MATSim*, Pp. 389–392. London: Ubiquity Press. DOI: http://dx.doi.org/10.5334/baw.60.

Figure 60.1: Evacuation planning zone.

Number of evacuees was calculated considering both permanent residents and employees, who were classified as transients. The scenario was prepared assuming the following conditions:

- The explosion occured in the evening when there were no students in schools and people were awake.
- All employees in the first risk zone, and some in the second, were taken to the Aliaga state hospital in zone 30, as well as to other health care centers in zones 26 and 27. The first risk zone was the most vulnerable; thus, persons needing medical intervention in this area would be taken to hospital. The typical behavior pattern in Turkey is to flock to hospitals in emergency situations. When generating scenarios, this behavior was considered; in the first and second risk zones, health care centers were designated as destination zones.
- People in residential zones would self-evacuate. Since Aliaga is a small town, public transportation service is weak and in the evening, public transportation frequency is low. Therefore, public transportation was not considered in this study.
- Employees in Petkim and in Tupras worked in three shifts; factories were active 24 hours a day and employees were always present.

There were 3 883 employees in the area studied; number of employees to be evacuated from factories was computed using the following assumptions:

- The total employee figure was divided into three, as they worked in three shifts.
- The explosion did not occur during shift change.

Evacuations from residential buildings were calculated using these steps:

- Number of persons living in an evacuation zone neighborhood was divided into the number of neighborhood buildings, giving the mean number of persons living in one building.

- Number of buildings that remained in the evacuation zone was multiplied by the mean number of persons in one building.

To estimate the number of evacuation vehicles needed, car occupancy ratio rate was used. This rate was 1.57 in normal situations—as given in the Urban Transportation Plan of the Istanbul Metropolitan Area by the Istanbul Metropolitan Municipality Directorate of Transportation Planning—however, in emergency situations, it was expected to be higher. In this study, car occupancy ratio rate was taken as two, number of evacuees was computed as 14 472 and number of vehicles 7 236.

The Aliaga network was taken from OSM and converted to a shape file and MATSim network file with the tutorial's `PNetworkGenerator` class. Zones used in generating synthetic population for MATSim were created in QGIS.

Three different scenarios were identified for the evacuation simulation. In Figure 60.2, O-D matrices for each scenario can be seen. These three scenarios were selected based on destination zone location and traffic demand criteria; free spaces close to the risk zone were designated as gathering areas. The time required for evacuees to reach health care centers was very important in emergency situations like this. The traffic demand generated for health care centers was distributed between zones 26, 27 and 30 in the scenarios, though the first risk zone was always directed to Aliaga State Hospital, which had the most capacity of all health care centers; severely injured persons would be transferred to this state hospital. Evacuating vehicles departing from the second risk zone were directed to health care centers in zones 26 and 27. Changing the number of evacuating vehicles in any given zones resulted in different evacuation times; thus, these different scenarios enabled observation of traffic demand effect on traffic and whether this led to evacuation time reduction.

Initial demand refered to synthetic population derived from numbers and locations of evacuees to be transferred to health care centers or gathering-areas, sorted by distance. A starting place for

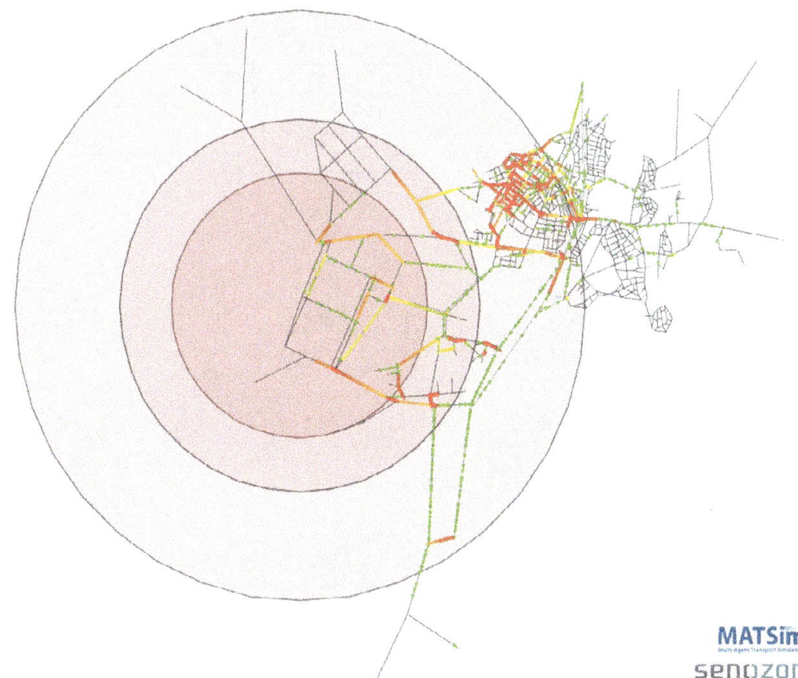

Figure 60.2: MATSim simulation snapshot.

initial demand generation was found in the tutorial's `DemandGenerator` class. Zones were modified according to both actual population density in given zones and road links where evacuees could start their trips at the time of a possible chemical accident to generate a relatively realistic scenario. Population zones were set for a group of origin zones, which were assigned for a predefined destination. For each agent, random activity coordinates were generated: home, work and leisure, or in this case, evacuation zones, hospitals and gathering areas. Agents' departures and arrivals took place on the nodes closest to activity coordinates and, in the first iteration, shortest path was calculated for route choice.

MATSim assigned trip start to the node closest to agent activity coordinates (i.e., home or work) for each agent. MATSim simulation results were analyzed by Senozon AG Via. Figure 60.3 shows a simulation snapshot.

Clearance time for three risk zones and total arrival time for three different scenarios were listed in Table 60.1. For the first scenario, evacuation times were 45 minutes for the first risk zone, 83 minutes for the second, and 86 minutes for the third. For the second scenario, evacuation times were 44 minutes for the first risk zone, 82 minutes for the second, and 91 minutes for the third. Finally, for the third scenario, evacuation times were 47 minutes for the first risk zone, 86 minutes for the second, and 88 minutes for the third. The third scenario yielded the best results, with minimum clearance time for the entire risk area. Scenario results confirmed that clearance times were insufficient for people to evacuate safely, especially from the first risk zone.

		Scenario 1						Scenario 2						Scenario 3					
		26	27	30	32	33	34	26	27	30	32	33	34	26	27	30	32	33	34
RZ 1	1			33						33						33			
	3			17						17						17			
	4			33						33						33			
	6			83						83						83			
	7			50						50						50			
RZ 1–2	10			221						221						221			
	13	16						16						16					
	14	90						50		40				50		40			
RZ 2–3	11				837						837						837		
	15		320					120	200						120			200	
	17				53						53						53		
RZ 3	16						660						660						660
	18					151						151						151	
	19					158						158						158	
	20					216						216						216	
	21					137						137						137	
	22					175	280						280						280
	23											175						175	
	24						88						88						88

Figure 60.3: OD matrices for evacuation scenarios.

	Risk zone 1	Risk zone 2	Risk zone 3
Scenario 1	45	83	86
Scenario 2	44	82	91
Scenario 3	47	86	88

Table 60.1: Risk zones evacuation times in minutes.

CHAPTER 61

Baoding: A Case Study for Testing a New Household Utility Function in MATSim

Chengxiang Zhuge and Chunfu Shao

61.1 Introduction

Baoding is a medium-sized city in Hebei Province, China. The Baoding case study—testing a new household utility function—proposed two scenarios to compare the performance of two utility functions: the household and individual utility functions. In Scenario 1, it was assumed that each household sought to maximize their overall household utilities when they scheduled; thus, family members' communication and coordination was communal in each household. In Scenario 2, the individual utility function—the default utility function in MATSim—was utilized to score plans; here, each agent tried only to maximize his own utilities without communicating with other family members.

Overall, Scenario 1 differed from Scenario 2 only in the utility function; other input data and parameters in these two scenarios were kept the same. The scenarios simulated only urban residents' travel behavior. In 2007, the study area population was 1 060 783, in 299 850 households, encompassing 355 465 privately owned cars.

61.2 Population and Demand Generation

Population The scenarios' agent population was created using a new population synthesis, which starts with initial household weights obtained from the 2007 Baoding Household Travel Survey. The final household weights, used for creating the population, were calculated by iteratively adjusting initial household weights in a directed way. Gender and household car ownership were also used as person- and household-level control variables, respectively. In the scenarios, only 20 % of Baoding's total population, approximately 212 000, was synthesized and used, to speed up the simulation.

How to cite this book chapter:
Zhuge, C and Shao, C. 2016. Baoding: A Case Study for Testing a New Household Utility Function in MATSim. In: Horni, A, Nagel, K and Axhausen, K W. (eds.) *The Multi-Agent Transport Simulation MATSim*, Pp. 393–396. London: Ubiquity Press. DOI: http://dx.doi.org/10.5334/baw.61.

Travel Demand Generation For initial demand generation, a GA (Genetic Algorithm), adopting utility maximization theory, was implemented. For Scenario 1, this GA used the new proposed household utility function as the fitness function; this was employed to generate initial individual daily plans for each household in the synthetic population. Specifically, in the GA, each chromosome represented a household's set of daily plans and each gene represented a family member's daily plan. During evolution (including mutation, crossover and selection), each chromosome was scored; only those with higher household utilities remained. Then, a set of daily plans with the highest household utility function were selected and allocated to the household. Similarly, other daily household plans in the synthetic population were generated, one by one. It should also be noted that the travel time in the initial daily plans was estimated. Therefore, elements like travel time and activity duration in the initial daily plans would be adapted (optimized) when executed in MATSim.

In Scenario 2, the GA incorporated the individual utility function to search for each agent's (family member's) plans.

61.3 Activity Locations, Network and Transport Modes

Activity Locations Five typical activity types, including work, home, leisure, education and shopping, were taken into account in the scenarios. The activity facilities numbers for these five types were: 1 647, 462, 246, 372 and 445, respectively.

Transport Network The scenarios contained two network types, including road and public transit networks. Figure 61.1 demonstrated Baoding's 2007 road and transit network. The road network was composed of 1 650 nodes and 539 links; the transit network contained transit routes and transit schedules, with 49 transit lines and (98 transit routes).

Transport Modes The simulated transport modes included car, public transport, bike and walk. Car drivers and public transport passengers used the road network and transit network. Because agents who traveled by bike or on foot had no access to the transport network, they were teleported from origin to destination and assigned no exact routes, but their travel time was calculated.

61.4 Historical Validation

Historic validation. composed of the following two steps, was carried out to assess MATSim's performance and applied to both scenarios.

Step 1: Comparison of both real and simulated car flows and comparison of real and simulated transit passenger flows were carried out in each scenario, to assess MATSim's performance for car and transit simulation. The MRE (Mean Relative Error), calculated by the equation (61.1), was employed to assess performance.

$$MRE = \frac{\|F_{simulated} - F_{real}\|}{F_{real}} \times 100\,\% \tag{61.1}$$

where, $F_{simulated}$ and F_{real} denotes the simulated and the real flow (car flow or passenger flow), respectively.

Step 2: Comparison of both scenarios' performance for car and transit simulation, based on results from step 1.

61.4.1 Comparison of Two Scenarios: Car Traffic

Car flow data on six road links (equal to 12 links in MATSim scenario) from 7 am to 9 am, was used for comparison of car simulation and was manually counted in 2007.

Figure 61.1: Road and transit network of Baoding in 2007.

Figure 61.2(a) demonstrated car simulation performance for both scenarios. Four dots were approximately located in the $y = x$ line and the other two dots, below the line, also were very close to it. Mean relative error margins of Scenario 1 and Scenario 2 were 44.8 % and 47.5 %, respectively. It can thus be concluded that the performance of Scenario 1 (using household utility function) was slightly better than Scenario 2 (using individual utility function).

61.4.2 Comparison of Two Scenarios: Transit Traffic

Data (passenger flow for nine transit lines from 7 am to 9 am) used for transit simulation comparison was also manually counted in 2007. Figure 61.2(b) illustrated both transit simulation scenarios' performance. Clearly, most dots did locate close to the $y = x$ line, however, two dots below the line were significantly distant from it. Also, mean relative errors of Scenario 1 and Scenario 2 were 38.7 % and 47.9 %, suggesting that Scenario 1 better represented transit passenger flows than Scenario 2.

61.5 Achieved Results

A proposed MATSim household utility function was tested comparing two scenarios using household and individual utility function. Historical validation confirmed that MATSim improved its own car and transit simulation performance by using the new utility function. However, more case studies are needed to further confirm this new proposed utility function's advantages.

More information on the Baoding scenario can be found in Zhuge (2014) (in Chinese).

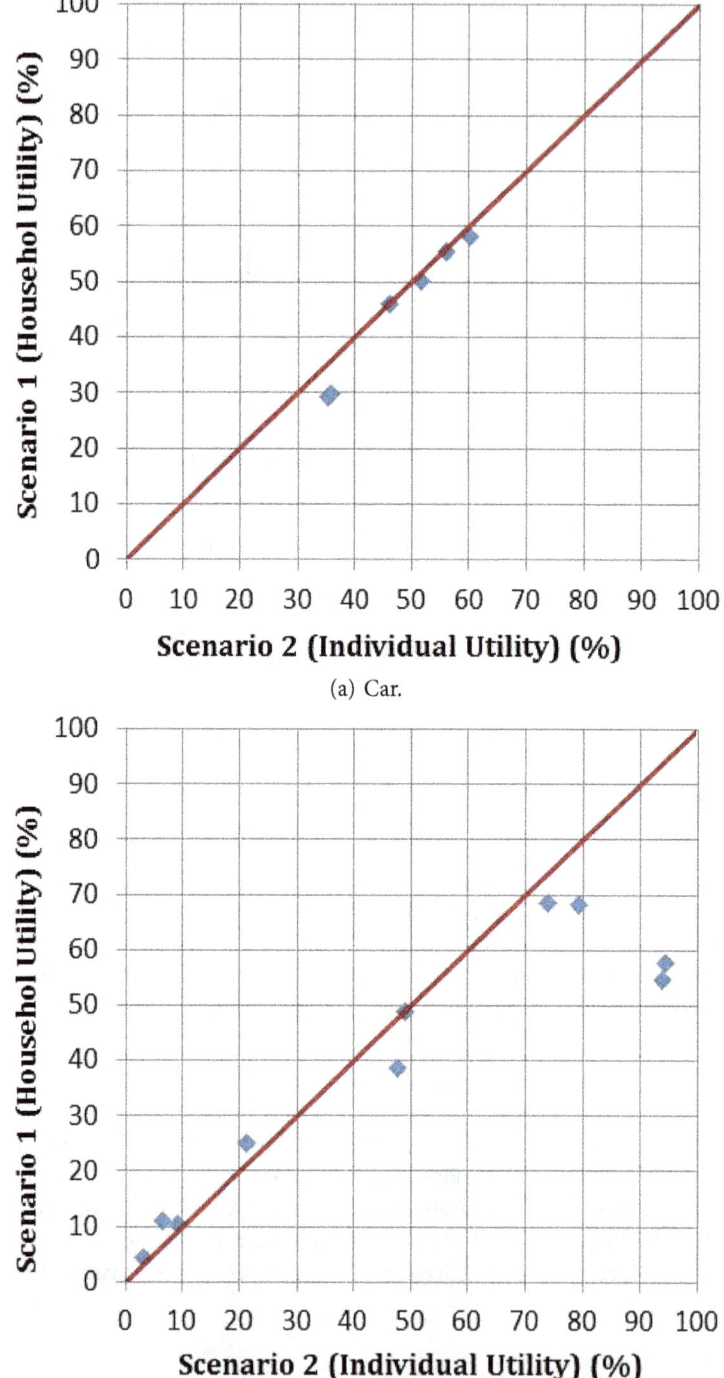

(a) Car.

(b) Public transit.

Figure 61.2: Performance comparison of Scenario 1 and 2.

CHAPTER 62

Barcelona

Miguel Picornell and Maxime Lenormand

The Barcelona scenario is one of the three case studies (together with London and Zürich) carried out under the framework of the EUNOIA (Evolutive User-centric Networks fOr Intraurban Accessibility) project. The main goal of the Barcelona case study was to evaluate the impact of different public bike-sharing schemes in the city. The study area covers the metropolitan Barcelona area, with special focus on the city center, where public bike-sharing stations are located. For this study a novel bike-sharing module was developed by ETH Zürich.

62.1 Transport Supply: Network and Public Transport

The road network was extracted from the TransCAD (Transportation Computer Assisted Design) model used by the city of Barcelona. Public transport supply was also considered, comprising: bus, underground, tram, train and bike-sharing. Information about stops and schedules was obtained from the public information available at the Barcelona Open Data platform, as well as from the Barcelona transport authority website.

62.2 Transport Demand: Population

Agent plans were defined using anonymised mobile phone registers CDRs (Call Detail Records). From mobile phone data, it is possible to identify places where agents perform activities and corresponding trips. Activities have been classified as "home", "work" and "other" (including as "other", "leisure", "shopping", etc.). A sample of around 15 % of the population was used in the simulation. Modes used in the simulation model include: walking, cycling, public transport and car.

How to cite this book chapter:

Picornell, M and Lenormand, M. 2016. Barcelona. In: Horni, A, Nagel, K and Axhausen, K W. (eds.) *The Multi-Agent Transport Simulation MATSim*, Pp. 397–398. London: Ubiquity Press. DOI: http://dx.doi.org/10.5334/baw.62.

62.3 Calibration and Validation

Different data sources were used to calibrate and validate the model. First, to validate agent plans obtained from mobile phone data, results were compared to EMEF (Enquesta de Mobilitat en dia Feiner), indicating that mobile phone data provides information similar to traditional surveys. Additionally, agents' utility function was calibrated using the modal split from EMEF and road counts.

62.4 Results and More Information

At the time this summary was written, the calibration process was still ongoing. More detailed information about the scenario and main results can be found at: `http://www.eunoia-project.eu/publications/` (project deliverables/Report on Case Study 3: Barcelona).

CHAPTER 63

Belgium: The Use of MATSim within an Estimation Framework for Assessing Economic Impacts of River Floods

Ismaï Saadi, Jacques Teller and Mario Cools

63.1 Problem Statement

With the history of river floods in Belgium and the significant probability that such events will again take place in the near future, assessment of both direct and indirect economic impact was deemed essential to allow formulation of an adequate policy program and efficient flood risk management. One proposal would assess flood risk at the micro-scale level: i.e., individual buildings for exposure analysis and direct economic damage estimation, individual companies for indirect economic damage estimation, 10 meter grid spacing for land-use modeling and individuals/vehicles for transportation models. To enable this assessment, an integrated modeling framework combining different simulation theories from a multidisciplinary perspective is being developed. Figure 63.1 describes the procedure to measure the annual flood risk. A more detailed description of the whole modeling chain is available in Dewals et al. (2015).

A basic modeling framework premise is that different spatial pattern 'families' might influence the damage intensity caused by river floods (e.g., land use change, transportation systems). In this chapter, we focus on how MATSim is being integrated into this overall framework, thus focusing on the TSA (Transport System Analysis) within the overall estimation procedure. For TSA, two configurations (freight and passenger model) are distinguished. For the passenger model, a MATSim scenario is developed on a national scale to simulate travel demand at base year 2010 and its evolution during the following years. The main objective is to study the effects of river floods on the transportation network and, consequently, on travel demand from an economic point of view. In addition, a freight travel demand model has been developed, to enable interactions between

How to cite this book chapter:

Saadi, I, Teller, J and Cools, M. 2016. Belgium: The Use of MATSim within an Estimation Framework for Assessing Economic Impacts of River Floods. In: Horni, A, Nagel, K and Axhausen, K W. (eds.) *The Multi-Agent Transport Simulation MATSim*, Pp. 399–404. London: Ubiquity Press. DOI: http://dx.doi.org/10.5334/baw.63.

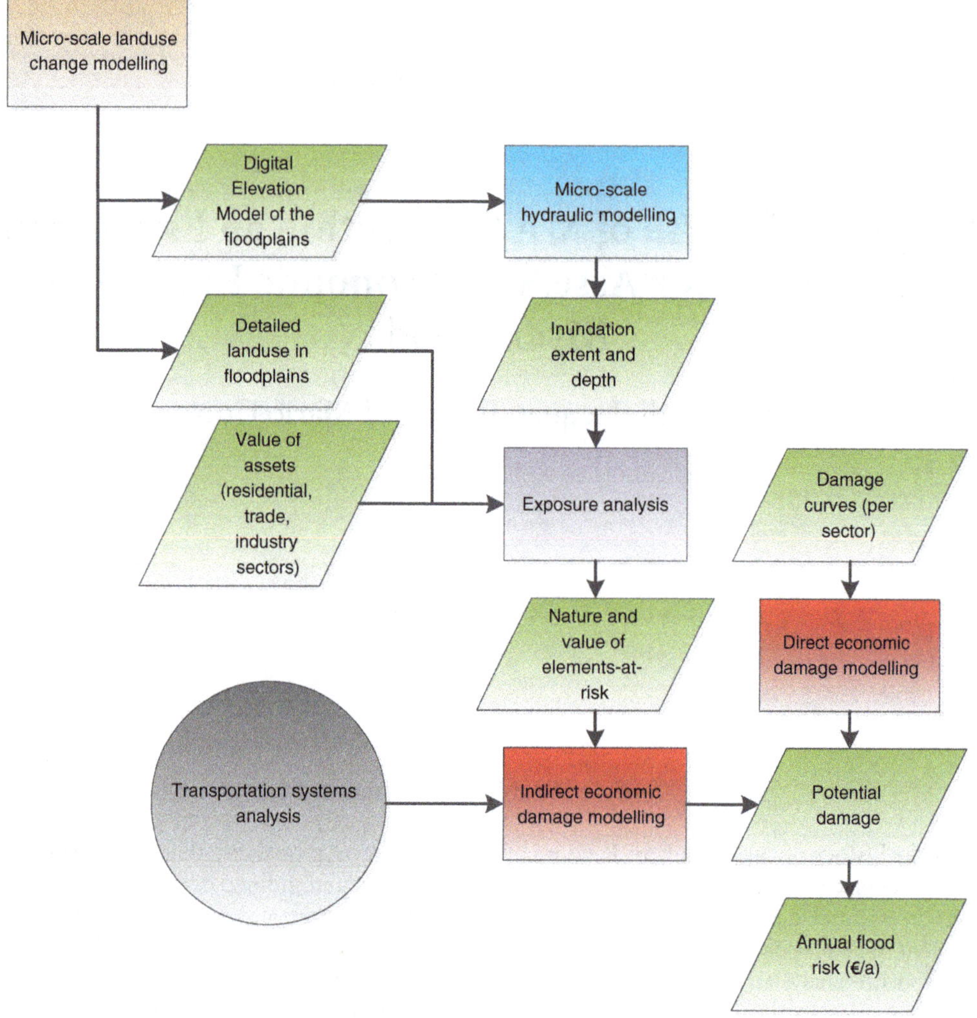

Figure 63.1: Economic impact estimation procedure.

passenger and goods flows. Note: at this time, this is still an aggregate four-step model, but development is ongoing to develop an agent-based model for the freight side.

63.2 Data Collection

As inputs, MATSim requires a synthetic population (or travel demand) file, as well as the related transportation network. Unfortunately, no recent census is available for the first input; the latest dates from 2001. To compensate, a synthetic population was derived from more recent travel surveys (e.g., Cornélis et al., 2012) by employing a Gibbs sampler (Farooq et al., 2013). The Belgian National Household Travel Survey (e.g., Cornélis et al., 2012) contains socio-demographics and activity travel diaries with a detailed description of activity start, end times and durations. Activity locations are also available, but at the municipality code level. They are generally accessed by using the new municipalities referencing system: LAU (Local Administrative Unit) level 2. For the transportation network, OSM network data has been used.

63.3 Input Preparation

63.3.1 Network

The network data of Belgium, downloaded in 2015, is available online from the OSM server. It consists of 100 467 nodes and 232 715 links. Network quality is generally acceptable, according to many MATSim users, even if manual adjustment is necessary for specific links.

63.3.2 Synthetic Population

Preparation of a synthetic population presents a significant challenge for this case study; only micro-data are available to enable population synthesis. From these partial views of the actual population, use of a Gibbs sampler enables the joint distribution (re-)construction. The outputs seem to be encouraging when comparing computed predictions to the reference dataset. Here, we propose testing the methodology by synthesizing some relevant variables for both transportation and urban systems simulations at the household level (see Figure 63.2).

63.3.3 Activity-Based Pattern Generation

After the synthetic population has been generated, *activity types*, *activity times* and *activity locations* are generated and associated to the agents, using an activity-based pattern generator. Using a combined set of machine learning techniques, daily activity planners are generated for each agent. As shown in Figure 63.3, the model suggests some promising first results. The activity-pattern generator is calibrated by using micro-data, such as activity travel diaries extracted from travel surveys.

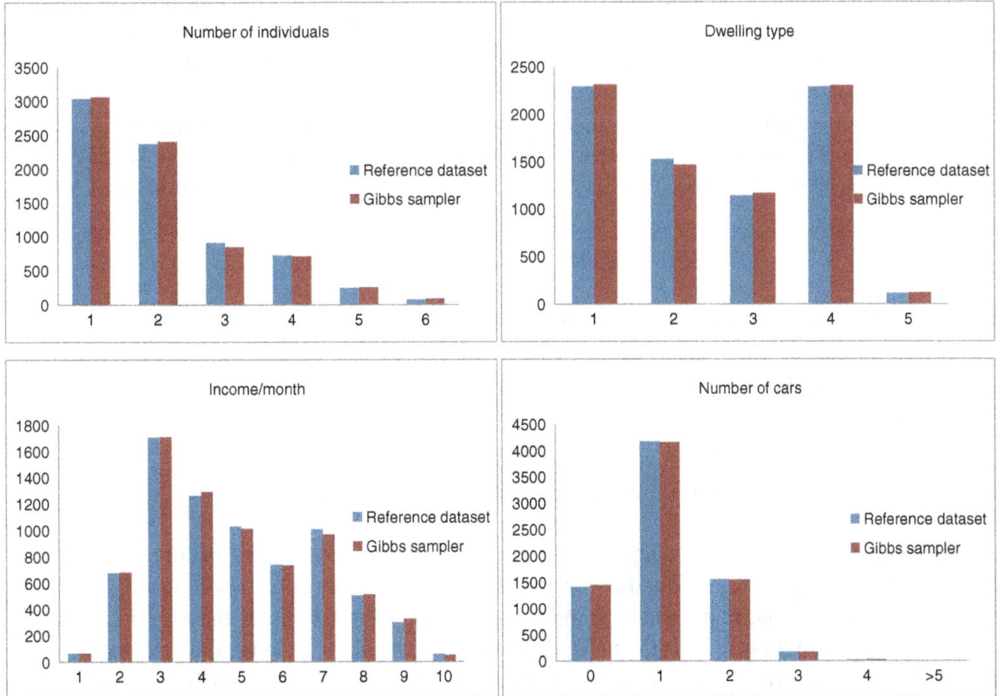

Figure 63.2: Examples of households synthesized attributes.

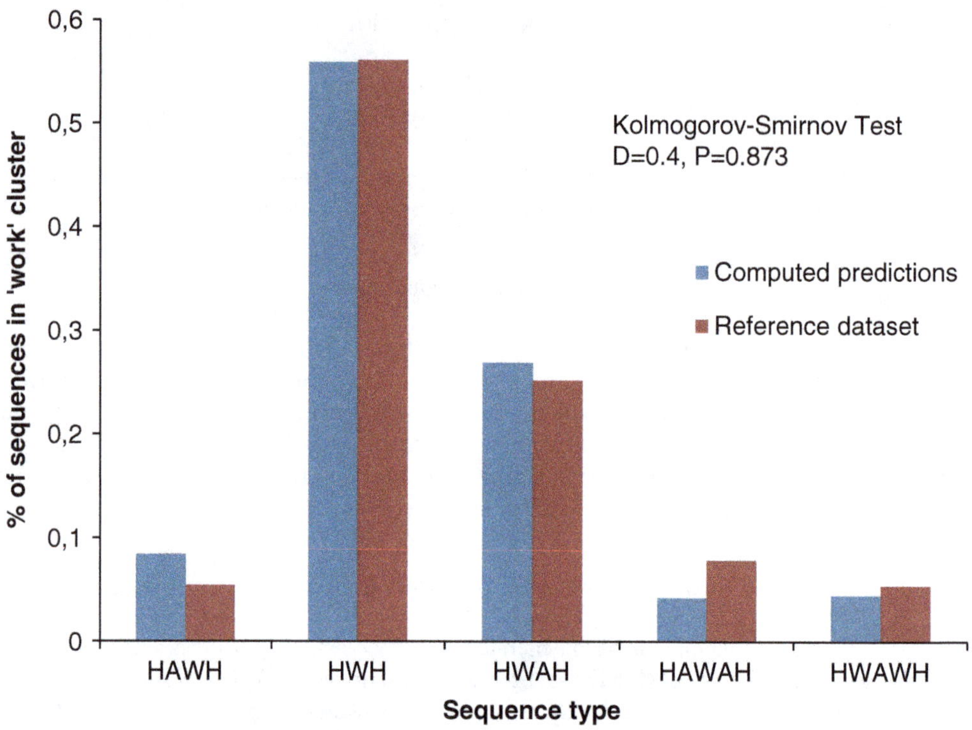

Figure 63.3: Activity chains generation.

Calibration quality will be measured after analyzing MATSim scenario outputs when traffic counts are compared. If the comparison between observed and simulated traffic counts suggests a significant deviation, a direct approach based on traffic counts (Cools et al., 2010) could work to adjust activity-based pattern generator parameters.

As outlined by Cools et al. (2011), uncertainties introduced by statistical distributions of random components in most activity-based models might be significant. Thus, some key indicators (e.g., sequences type proportions) will be investigated to measure micro-simulation error impact.

63.4 General Modeling Framework

In Saadi et al. (2014), the overall modeling framework is presented, as well as the integration of scheme components. This paper covers all concepts expected to be used in building the future MATSim scenario. Figure 63.4 is a partial view of the overall modeling framework being researched at the moment.

63.5 Modeling Network Disruption

As mentioned, this study also suggests modeling network inaccessibility occurring after river floods. This approach assumes that link capacities subjected to river floods are reduced, depending on flood intensity. Given that damage is mainly a function of water depth, the idea is to intersect a steady-state inundation map with the transportation network or, at least, the area impacted by floods (Saadi et al., 2014). Then, an analysis extension will be achieved by including a time series of river floods for a better understanding of dynamic effects: e.g., response to river floods propagation, return way and time to the new equilibrium point between transport supply and demand.

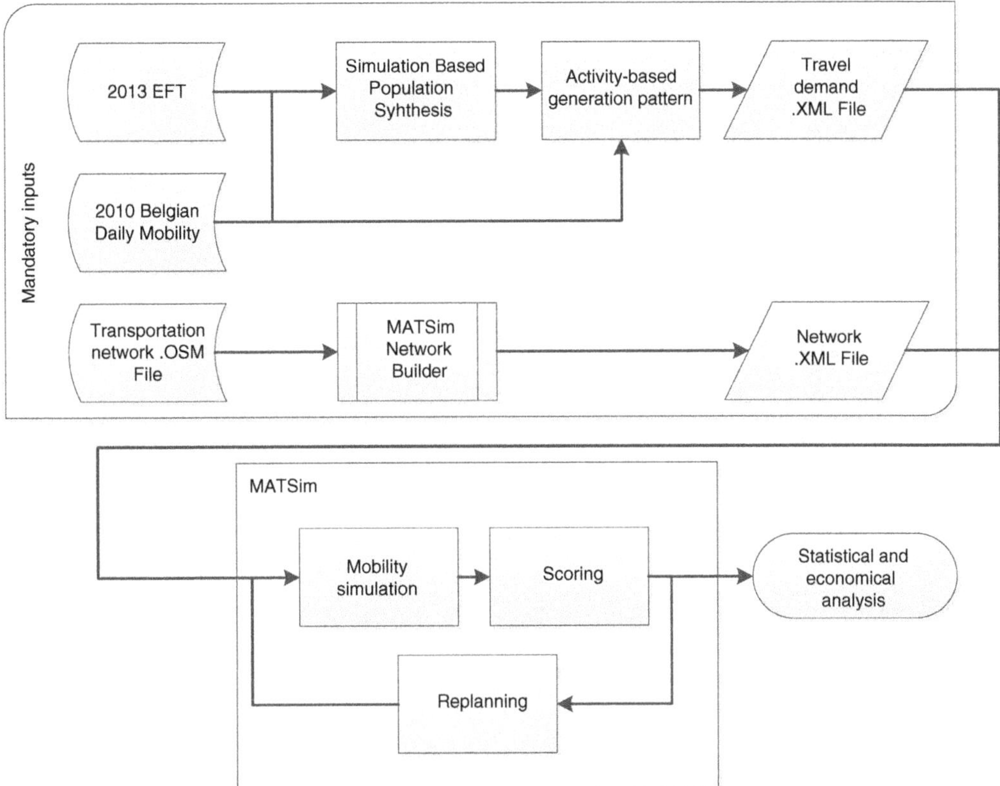

Figure 63.4: Partial modeling framework.

A similar problem was studied in a tsunami evacuation scenario simulation in the city of Padang (Lämmel et al., 2010) (Chapter 76) and was particularly interesting in terms of network dynamic evolution during the scenario simulation.

63.6 Next Development Steps

When the complete integrated agent-based transportation model is ready, combination with the land-use change CA (Cellular Automaton) based model proposed by Mustafa et al. (2014) will to allow more interactions between those two patterns. This connection will be the basis for an innovative micro-scale LUTI (Land-Use and Transport Interaction) model, allowing more accurate predictions about future river floods influenced by different micro-scale patterns.

CHAPTER 64

Brussels

Daniel Röder

The MATSim scenario for Brussels was developed as part of the SustainCity project. This project's goal was to couple an urban land use model, in this case UrbanSim, with the MATSim mobility simulation, to evaluate transport policy impact on urban land use and vice versa. A detailed description of this coupling is given by Nicolai (2013) and others. A detailed description of the scenario development is found in Röder et al. (2013).

The scenario covered the greater Brussels area in Belgium; input data was derived from two main sources. The population was directly generated from the UrbanSim model, covering a total of 860 214 persons. At home- and at work-locations (per person) were given and converted into a daily home-work-home plan. For computational reasons, a randomly-drawn population sample of one percent was used. OSM was sourced for the street network generation, which consisted of 10 861 nodes and 19 830 links, i.e., using mainly the trunk road network.

For the modeling of public transport, two different approaches were tested: first, the MATSim default approach for scenarios where no detailed transit schedule is available, based on either: beeline distance and average speed, or network-based freespeed travel times and a designated factor. The second approach was not part of the MATSim core during the project, but was available as a contribution (`matrixBasedPtRouter`, see Chapter 20). It was based on O-D travel time matrices between transit stops, i.e., travel times for all relations were computed in a pre-process. The travel times can based on a real-world-schedule or certain assumptions which can take spatial coverage into account. Advantages of this model are obvious; on one hand, it may depict spatial coverage with public transport supply—here, distance to the next transit stop influences travel time. On the other hand, it may depict the real network, i.e., routes and lines and possible waiting times for switching. Both approaches were compared against travel times and mode share measures from a SATURN (Simulation and Assignment of Traffic to Urban Road Networks) model. Since the matrix-based approach came closer to this model, further investigations were based on that.

How to cite this book chapter:

Röder, D. 2016. Brussels. In: Horni, A, Nagel, K and Axhausen, K W. (eds.) *The Multi-Agent Transport Simulation MATSim*, Pp. 405–406. London: Ubiquity Press. DOI: http://dx.doi.org/10.5334/baw.64.

To evaluate the model's sensitivity to certain policies, a cordon toll scenario was set up, where a toll is charged between 6 and 10 am every time a car passed a cordon border, i.e., every time a car entered a link crossing a cordon border defined by the Brussels freeway ring. Accessibility was calculated and compared for both scenarios. Röder et al. (2013) provides a detailed analysis.

CHAPTER 65

Caracas

Walter J. Hernández B. and Héctor E. Navarro U.

Capital of the country, Caracas is the largest city in Venezuela, with serious vehicle traffic issues. Its daily estimated circulation of 1.5 million units represents three times the load originally estimated for the city's growth. Despite the lack of official statistics, it is possible to estimate the amount of Caracas' traffic using other national figures, such as the *Time Travel Index* employed by the Federal Highway Administration (2013). This index estimates approximately 50 % longer than *free-flow travel* to traverse inner city circles and 75 % around metropolitan areas. This is in stark contrast to an average city in the US, which normally does not go beyond 35 %, even in the worst case.

Apart from obvious budget-related deficits these delays cause in work force productivity for companies and organizations, the country itself loses an estimated $2.1 billion per year. This includes the precious subsidies that have helped to maintain the country's world lowest prices of gas for decades (Wilson, 2008); $1 billion could be saved by reducing the average circulation time by just 30 minutes. Equally important, the accompanying significant reduction in CO_2 (Carbon Dioxide) emissions would help meet greenhouse targets for the country.

In recent years, several measures have been initiated to cope with increased traffic in Caracas:

- HOV (Highly Occupancy Vehicle) lanes (Turnbull, 1990), implemented in a *contraflow* fashion to increase traffic flow on central roads and highways,
- bus lanes for rapid bus trips and bicycle lanes, to stimulate use of alternative means of transport, and
- shifting job starting hours to non-peak times and increasing the number of at-home working hours for certain types of jobs, to cut back vehicle use and general costs to public transport.

In addition to the these measures, other mechanisms could be implemented, such as weekday circulation restrictions (e.g., based on license plate numbers) and smart traffic lights. Especially in the case of smart traffic devices and planning of special lanes, careful study and simulation of traffic

How to cite this book chapter:
Hernández B., W J and Navarro U., H E. 2016. Caracas. In: Horni, A, Nagel, K and Axhausen, K W. (eds.) *The Multi-Agent Transport Simulation MATSim*, Pp. 407–410. London: Ubiquity Press. DOI: http://dx.doi.org/10.5334/baw.65.

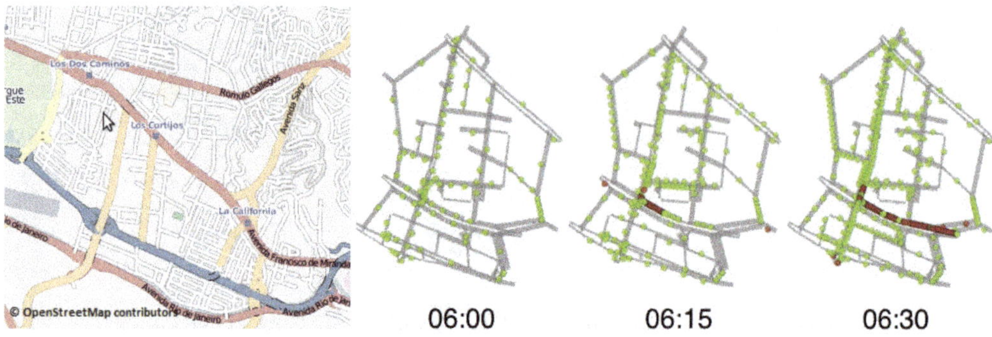

(a) Snapshot obtained from OSM. (b) Three snapshots of a simulation ran in MATSim between 6:00 and 6:30 am, depicting the increasing traffic in the zone.

Figure 65.1: An area of "Los Cortijos" in Caracas, Venezuela.

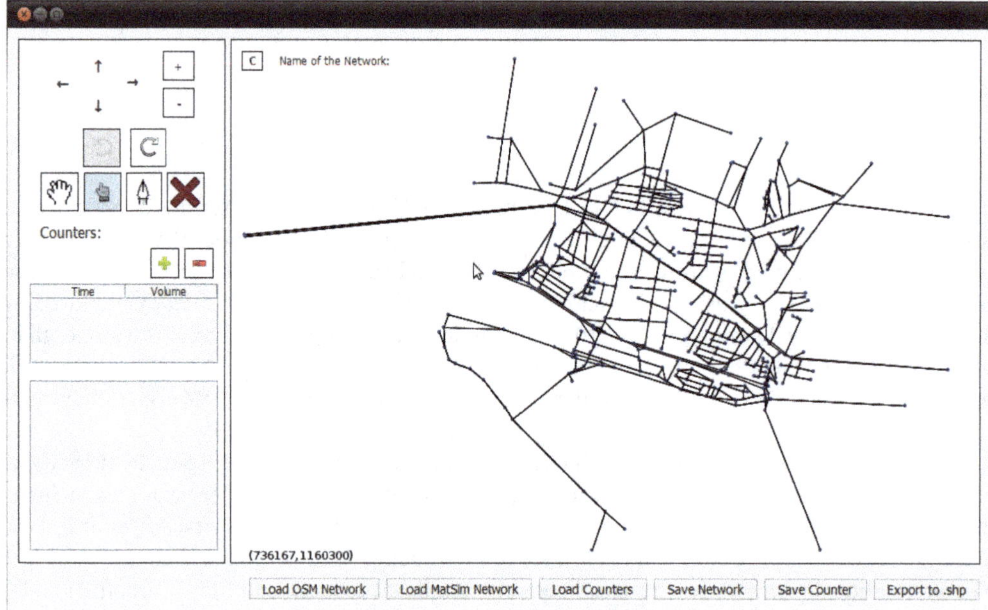

Figure 65.2: Interface of the software tool developed in Java showing the area studied. Blue dots over the roads in the map represent the counters positioned in the area to capture vehicle flow used as input for the simulation.

patterns must be undertaken. To help achieve this, a software tool was envisioned with the following objectives:

- to envision creation and editing of traffic networks on MATSim format and assign validation points to the network,
- to study and analyze simulation results, especially the traffic volumes assigned to roads,
- to translate data obtained in O-D format for input to MATSim, and
- to run the simulations and validate outputs in order to calibrate the parameters involved.

The tool was tested with real data traffic in a Caracas area "Los Cortijos" (Figure 65.1(a)), one of the most heavily traveled zones on the city's east side. The simulation model belongs to the *microscopic* category, made by Gartner et al. (2001), since only individual elements are taken into account (i.e., vehicles).

The network was created by using data from OSM, then manually modifying it (i.e., setting correct speed, capacity attributes) based on information delivered by a company conducting a study in the same area. Demand was given in a O-D matrix by the same company, but only for the morning period. As the area researched is mainly a consuming zone in the morning and a producing zone in the afternoon, values from the O-D matrix were used to create day-plans for the agents. An initial departure time around 7:30 am was assigned to the plans.

Several scenarios with different re-planning rates were run to test how much agents have to change their departure time in the morning to allow the network to accommodate all travel demand. Figure 65.1(b) shows how traffic jams builds up in the scenario where simulated demand best matches real-world traffic count.

Figure 65.2 shows the interface of the tool providing options to: load a map from OSM, load a network from MATSim, load counters (blue dots in the map image), save the map and export to a shape file (an open file format for GIS systems).

Figure 65.3(a) shows an image output of the same area after running a simulation generated with MATSim and including a vehicle-density color map. Figure 65.3(b) shows a sample score

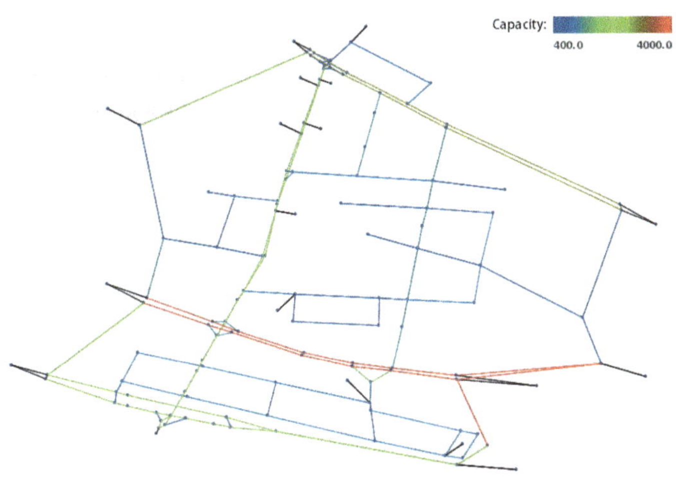

(a) A snapshot of the area at 7:00 am with a color map for vehicle flow.

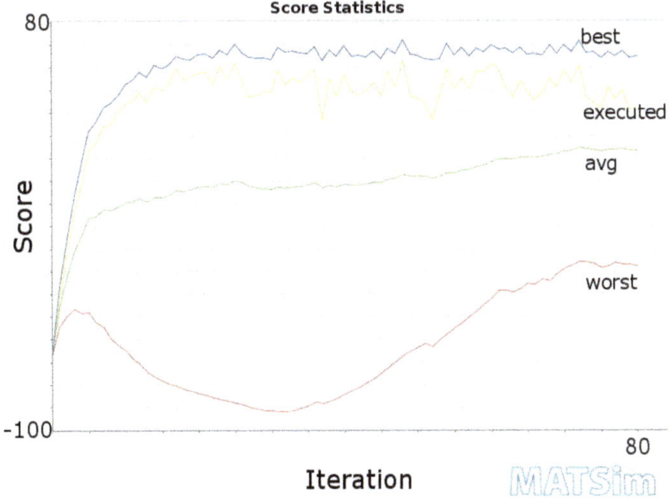

(b) MATSim sample score statistic for one of thescenarios defined.

Figure 65.3: Simulation results.

graph from one of the many tests run: *avg. trip time* of 02h:06m:38s and *avg. distance per agent* of 1727.83 meters; *total run time* of the simulation was 41 minutes 56 seconds.

In spite of differing approaches between the company's study and our own results, through simulations, the numbers were quite similar with a range of difference not exceeding 3 % (-0.42 % to 2.52 %). Examining these promising results, but also the limitations encountered, the following future lines of work were defined:

- Run a larger number of simulations and compare with real data, to fine-tune accuracy of results.
- Improve capacity to incorporate simulation plans from censuses and polls, among other alternative data sources different from O-D and develop a methodology allowing disaggregated collection of data.
- Include more options for network creation, such as generating links based on characteristics like zebra crossings, speed humps, curb extensions and/or a number of traffic signs.
- Create options to manage a simulation project incorporating the internal organization made by the tool, where all iterations of simulations are separated in folders with all outputs produced. This also implies the creation of a more refined reporting tool that could be used to support the decision making process of smart traffic devices, contraflow lanes, etc.

Acknowledgements The authors wish to acknowledge Daniel Ampuero Anca and Jesús Francisco Gómez Ortíz, for their Bachelor's degree final work at *Universidad Central de Venezuela*. Also to óscar Anzola, founder of *URVISA S.A.*, who provided the logistics for the capture of real traffic data used on the simulations.

CHAPTER 66

Cottbus: Traffic Signal Simulation

Joschka Bischoff and Dominik Grether

The Cottbus (Germany) scenario is used for traffic light simulation (see Chapter 12). It is explained by Grether (2014, pp. 87); this chapter briefly reviews the main points. The scenario data is generally available to the public, and can be found from `http://matsim.org/datasets`.

The network was derived from OSM data in summer 2010 (Bischoff, 2010), and covers all streets within the city boundaries, as well as main roads in the surrounding Spree-Neiße administrative district. It is designed as a 100 % sample. The population is based on the German federal employment agency commuter statistics for both Cottbus and Spree-Neiße (Wiethölter et al., 2010). As such, the population has only home-work-home plans spread over the usual commuting times, resulting in two peaks, including 33 479 agents traveling exclusively by car. The scenario is generally not very busy; the area does not usually have major congestion issues.

Figure 66.1(a) shows the network over the "Corine Land Cover" landuse (European Environment Agency, 2011), provided by European Environmental Agency. Woods and agricultural areas are depicted; most of the region is agricultural use area. Virtual persons in MATSim need a geographic coordinate for their activities. If this coordinate is drawn randomly (solely based on municipality borders), home and work activity locations are uniformly distributed over the area, i.e., most of them in woods and fields. Thus, activity locations are drawn randomly in combination with land use data. The coordinate must be in the municipality area and for home activity, it must be located in urban fabric areas; for work locations, industrial or commercial areas are also allowed. The resulting home activity locations are shown in Figure 66.1(b).

The scenario contains data for 22 traffic signals within the city center, based on the city's 2009 signal plans; junction layout is also modeled in detail. Fixed-time control data is taken from Köhler and Strehler (2010). Due to higher transport network resolution, several originally recorded fixed-time control schedules are invalid and were removed; data for 22 junctions is available. Figure 66.2 shows their transport network location.

Public transit, not part of the original scenario, is available based on 2011 schedules, although it is not currently used.

How to cite this book chapter:

Bischoff, J and Grether, D. 2016. Cottbus: Traffic Signal Simulation. In: Horni, A, Nagel, K and Axhausen, K W. (eds.) *The Multi-Agent Transport Simulation MATSim*, Pp. 411–412. London: Ubiquity Press. DOI: http://dx.doi.org/10.5334/baw.66.

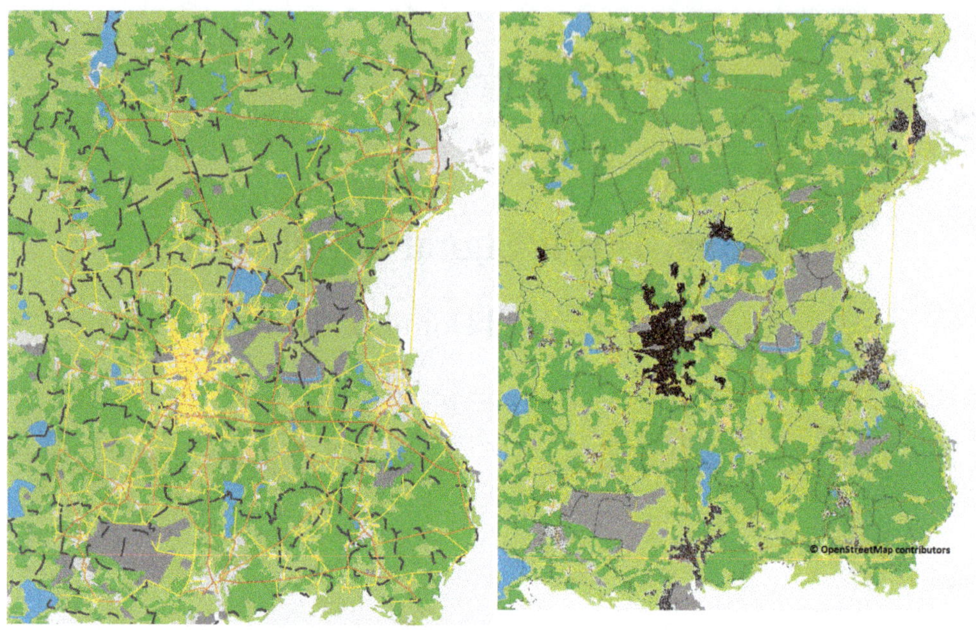

(a) Cottbus network and municipality borders.

(b) Synthetic population for the Cottbus scenario, geospatial location of home activities.

Figure 66.1: Cottbus scenario: Network and population.
Source: Grether (2014)

(a) Location within city of Cottbus.

(b) Signalized area in detail.

Figure 66.2: Cottbus scenario: Network, area with traffic signals within the city of Cottbus.
Source: Grether (2014)

CHAPTER 67

Dublin

Gavin McArdle, Eoghan Furey, Aonghus Lawlor and Alexei Pozdnoukhov

67.1 Introduction

To demonstrate a new spatial choice model, a microsimulation of urban traffic flows for the greater Dublin region was implemented using MATSim. The scenario simulated leisure activities and commuting trips completed by individuals using private cars over a twenty-four hour period. For commuting trips, detailed information from the Irish Census was used; a new spatial choice model, inspired by the radiation model, was developed for leisure trips. The effectiveness of the approach was validated using hourly data from count stations on the main motorways around Dublin City. The results show that the microsimulation accurately reproduced traffic volumes.

67.2 Study Area

County Dublin, in Ireland, covers an area of approximately 115 square kilometers and encompasses several administrative areas. Dublin is a coastal county with the Irish Sea lying to the east. To capture both intra-city and inter-city flows, the scenario considered individuals who live or work in Dublin, capturing those who commute to or out of Dublin, as well as those who live and work there.

67.3 Network

To capture the desired study area for the scenario, the network consisted of all roads in the greater Dublin region and major roads for the remainder of the country. The road network was a mix of motorway, national routes and local roads and was extracted from OSM, along with other information such as speed limits and number of traffic lanes. This OSM network was prepared for use in

How to cite this book chapter:
McArdle, G, Furey, E, Lawlor, A and Pozdnoukhov, A. 2016. Dublin. In: Horni, A, Nagel, K and Axhausen, K W. (eds.) *The Multi-Agent Transport Simulation MATSim*, Pp. 413–418. London: Ubiquity Press. DOI: http://dx.doi.org/10.5334/baw.67. License: CC-BY 4.0

MATSim. This study focused on private vehicles; the public transport network was not considered, but can be incorporated into the microsimulation in future studies.

67.4 Population Generation

The population for this scenario consisted of all car drivers who live or work in the greater Dublin region and was prepared from a variety of data sets. To obtain home and work locations, the 2011 Irish Census was used, particularly a census subset called POWSCAR (Place of Work, School or College Census of Anonymised Records). This provided home and work locations, mode of commuting transport used, time of departure for work or school and a variety of socio-economic data at an individual level. The individuals relevant to this scenario (drivers who live or work in Dublin) were extracted from the data set. In POWSCAR, home locations are anonymized by aggregating them into a statistical unit called the small area, consisting of 80 to 100 households. In the greater Dublin region, this represents a street or an apartment complex. We translated this to an individual address point by selecting a random address point within the small area. For this process, we used a commercial database of addresses and their coordinates in Ireland called Geodirectory. To account for non-workers, we used census statistics to generate the spatial distribution of the number of sick, unemployed and retired persons along with car ownership details to produce the non-working population for the greater Dublin region. These were also assigned to individual address points, providing us with a population of 600 000 agents for the scenario (see Figure 67.1).

67.5 Demand Generation

Individuals from the population were assigned work and school locations according to POWSCAR (Figure 67.1). In POWSCAR, work and school locations are given at a 250 meters grid level and we translated them into individual address points using Geodirectory. For school and collage locations, the address point was checked using NACE (from the French title 'Nomenclature générale des Activités économiques dans les Communautés Européennes') codes, to confirm its status as an educational institute. Departure times for work and school were assigned using a Gaussian curve centered at the declared 30 minute departure time from POWSCAR. INTS (Irish National Travel Survey) was used to create non-commuter demand for the road network. Through a survey, the INTS collected a 24 hour travel diary for an Irish population sample recording journey origin, destination, departure time and mode. We extracted the private car mode and combined the data with the commuter data to create a 24 hour activity chain for each individual in the population.

67.6 Activity Locations

A set of activity locations were obtained from an in-car navigation system's POIs (Points of Interest) database and augmented with additional POIs from OSM. While work locations were assigned from demand generation, locations for secondary activities, such as shopping and leisure, were not specified in the INTS and so had to be modeled to create spatial and temporal activity chains for the population. We developed a radiation model variant that applied emission-absorption ideas to compute interaction probabilities for a set of origins and destinations. The radiation model was parameter-free and distance decay was replaced by a ranked-based decay (Simini et al., 2012). While generally used for modeling movement between regions or cities, we used this approach to produce probabilities of selecting different locations capable of fulfilling a given activity. Where the radiation model uses known populations of locations to produce region ranking, we used attractiveness scores for areas and facilities that could fulfill an activity. A facility, venue or area's

Figure 67.1: The distribution of work (upper image) and home (lower image) locations for part of the Dublin scenario.

attractiveness was derived from venue size, which was calculated using domain knowledge and the model was calibrated with trip distribution patterns from social media check-in statistics. This radiation model variant was used to assign locations to secondary activities in the agents' day chains for the Dublin scenario demand.

67.7 Validation and Results

Network, population and demand data were prepared for use with MATSim. For efficiency reasons, a 25 % sample of the population was used for the simulation. The location choice model described above was used to generate the initial demand. On each interaction of the simulation, agents could be rerouted or rescheduled according to the MATSim default settings, but the locations defined in activity chains remained constant. The simulation reached a stable state after 350 iterations. The road volume data output was scaled according to the sample used, aggregated to an hourly count and compared to the observed count data from 6 count stations on motorways around Dublin. In order to compare the effect of the new location choice model, the simulation was re-run using the MATSim nearest neighbor algorithm for selecting secondary activities' locations.

67.8 Achieved Results

Aggregated hourly counts were compared with those observed at the 6 count stations which determine the number of vehicles traveling in two directions. A typical hourly distribution was obtained by averaging mid-week traffic volumes for a 3 month period. The results produced by the radiation model showed a stronger correlation between simulated and observed counts than those from the nearest neighbor approach. Figure 67.2 shows hourly observed and simulated count data for two count stations; the inset shows the relative percentage error for the two approaches being tested. The results indicate that both techniques are effective for estimating commuter traffic during morning and evening peaks. This was to be expected as the location of school and work activities were provided from real world data, but it did confirm the MATSim routing algorithm effectiveness. For daytime traffic, which consisted mostly of secondary activities, our variant of the radiation model outperformed the nearest neighbor approach; it included individuals who were willing to travel further for better opportunities, producing more accurate results.

67.9 Associated Projects and Where to Find More

The Dublin scenario validation results demonstrated the effectiveness of MATSim as a traffic simulation tool and also showed the power of our spatial choice model which adapted the radiation model to predict individual movement at a small spatial scale. In the future, the research will be expanded by considering a multimodal transport network and scaling the scenario from an urban simulation to a national one. Full details of the Dublin scenario can be found in McArdle et al. (2012) and McArdle et al. (2014).

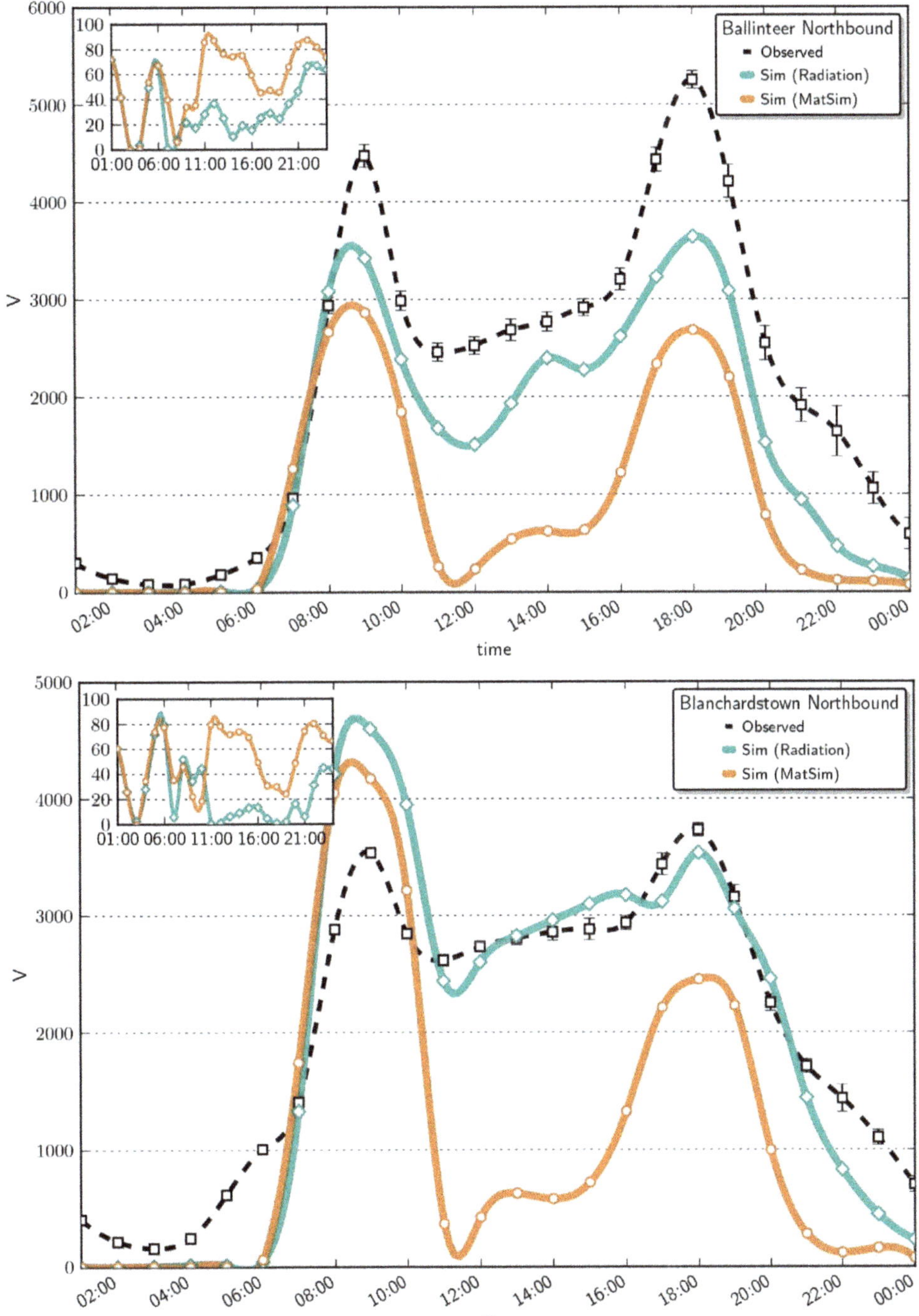

Figure 67.2: Hourly observed traffic volumes (dashed line) compared to the estimated traffic volumes produced by MATSim using the radiation model (green line) and nearest neighbor model (orange line).

CHAPTER 68

European Air- and Rail-Transport

Dominik Grether

This chapter discusses simulation of air- and rail-transport technology and passengers using MATSim. There is no great difference in overall travel times between middle-range rail and air transportation. Airports and railway stations are affected by capacity and opening time constraints. For passengers and goods, geospatial location is an important property. Both modes, but especially air transport, are faced with difficult capacity restrictions at certain departure times.

This chapter discusses how MATSim can be applied to capture these constraints and how interaction between passenger demand and constraints on technology supply can be modeled. The public transit model of MATSim (Chapter 16) is applied. Airports and aircraft are microscopically modeled the same way as bus stops and buses. Passengers are represented microscopically as multi-agent demand for air transportation. Their choices of transport mode, routes, and departure time are restricted by the air transport technology simulation model's capacity. The modeling of rail transport is based on teleportation. With appropriate data, the modeling approach for air transport could also be applied to rail transport (Quick, 2012).

The modeling of technology and demand is sketched in Section 68.1. On the basis of simulation results for a pure air transport model, rail transport is added and effects of mode choice are presented (Section 68.2). Section 68.3 then interprets simulation results and highlights some modeling aspects requiring further study. The choice set generation and plans removal algorithm of MATSim is discussed in detail; that is also the subject of Section 97.4. Modeling, results, and studies of this chapter present the highlights of Grether (2014, Chapter 6, pp. 119), in more detail.

How to cite this book chapter:
Grether, D. 2016. European Air- and Rail-Transport. In: Horni, A, Nagel, K and Axhausen, K W. (eds.) *The Multi-Agent Transport Simulation MATSim*, Pp. 419–428. London: Ubiquity Press. DOI: http://dx.doi.org/10.5334/baw.68.

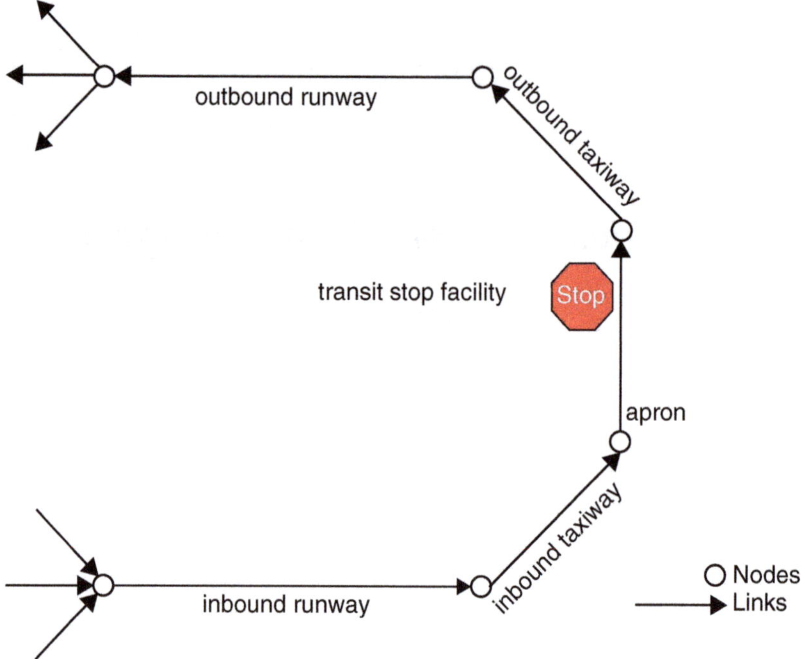

Figure 68.1: Layout of airports in the air transport network: In- and outbound runways are modeled by separate links connected by taxiways and a link representing the apron. There the transit stop facility is attached.
Source: Grether et al. (2013)

68.1 Air Transport Scenario

68.1.1 Modeling & Simulation of Air Transport Technology

The air traffic technology model uses data provided by OAG Aviation.[1] Relevant data for schedule and network generation is taken from the September 2009 OAG data, using all flights departing on a Tuesday, taking each specific flight number into account only once. This may not always result in complete flight cycles, e.g., when the outbound and inbound flight operate on different days of the week. Compared to using all flights of an entire week, the network may be incomplete, as certain destinations are only served on specific days.

The air network modeling aims at a simulation with MATSim. The network consists of airports, each showing an identical layout and point-to-point connections in between. Every runway is solely used either for inbound or outbound flights, with taxiways connecting the runways to the apron. The latter accommodates a transit stop, i.e., the terminal, where flight movements originate and terminate (Figure 68.1). Each airport pair is directly connected by airway links, one for each flight and direction of travel (Figure 68.2). Maximum speed on any of these links is calculated based on distance and flight duration provided by OAG. Times for taxi, take-off, and landing are also taken into account, i.e., flight duration is reduced by the time needed from push-back to airborne before the maximum speed for an airway link is calculated. Each flight has an individual link that could be interpreted as route, each possessing individual characteristics. Figure 68.3 shows parts of the network for European air traffic.

[1] `http://www.oagaviation.com`, last access 08.08.2012

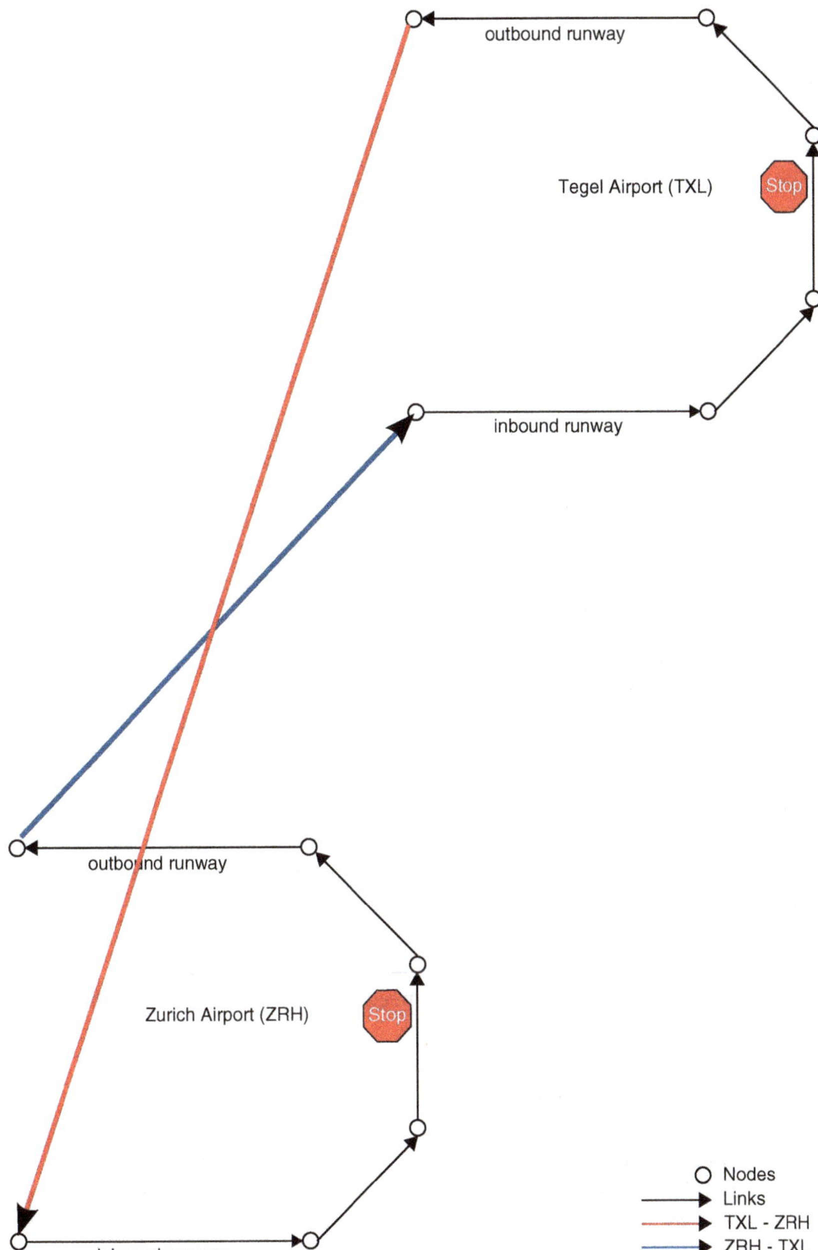

Figure 68.2: Layout of airways in the air transport network: each airport pair is directly connected by two airway links, one for each flight and direction.
Source: Grether et al. (2013)

Flight schedules are taken from the OAG data and translated to a MATSim transit schedule containing information about each line, route, and departure. For each airline offering a connection between two airports, a transit line is generated. A transit route, which represents the route on the air traffic network, is created for each flight offered by this airline. Mutual interferences of aircrafts en-route are not included in the studies presented in this chapter.

Figure 68.3: European air network with country borders in the background (country borders © http://www.openstreetmap.org).
Source: Grether et al. (2013)

To represent individual aircraft in the simulation, transit vehicles are created on the basis of OAG data. IATA aircraft codes, operating airlines, and seating capacities are reflected in the respective aircraft representation for every flight. Information about boarding times, i.e., passenger flow per door over time, is not available, but could be set for each aircraft type. One aircraft per flight is generated, thus delays resulting from a delayed incoming aircraft are not modeled. Accordingly, no aircraft rotations and vehicle trip chains are implemented at this time. The maximum velocity of each aircraft is set to twice sonic speed, since speed limitations are set for each network airway link.

68.1.2 Passenger Demand

As soon as the technology side of air transport is modeled, passenger demand simulation can begin. The passenger demand for trips in Germany created and used for the results of this section is based on O-D data of DESTATIS.[2] For each O-D pair and trip a virtual person is created. Each virtual person performs two activities, one at the origin and the other at the destination airport. Both activities are of same type, thus time spent performing both activities is accumulated before it is evaluated by the utility function according to Section 3.2. A typical duration, $t_{typ,q}$, of 21 hours is set for this activity type. The time virtual persons arrive at the origin airport and start waiting for a connection is drawn randomly from a uniform distribution in 4 am to 6 pm, UTC. This reflects estimated typical opening hours of European airports. No other time constraints are set, thus the only incentive for virtual persons is to reduce overall travel time and maximize time spent at the activity. A flight leg is scheduled between the two activities, connecting origin and destination. As

[2] Deutsches Statistisches Bundesamt, http://www.destatis.de, Fachserie 8 Reihe 6, last access 10.09.2012

usual, the demand does not specify if a direct flight from O to D is chosen or the virtual person is on a route containing one or more transfers. The synthetic population contains 51 832 virtual persons, 1 550 trips from the original data are neglected as origin and destination are equal.

68.2 Simulation Results

68.2.1 Air Transport

As a scenario for air transport technology, a coverage model from Europe to world wide destinations is used; with the synthetic population, it serves as input for the simulation. The assignment of flights to the desired O-D connection, i.e., the passenger routing, is calculated by MATSim's default public transit routing module.

Each simulation is run for 600 iterations. In each iteration, 10 % of the virtual passengers may shift their departure time randomly within a 2 hour interval. Another 10 % may seek a new route, i.e., a connection between origin and destination. Each passenger chooses from a set of 5 plans using an MNL. The outcome is stable after 500 iterations, then departure time choice and routing are switched off. For another 100 iterations only the MNL is used by passengers to select a plan.

Results are then taken from the output of the 600th iteration. Filtered by flights in Germany, Figure 68.4 depicts passengers in aircraft (red) and seats (black) over time of day and reveals passengers' tendency to depart early.

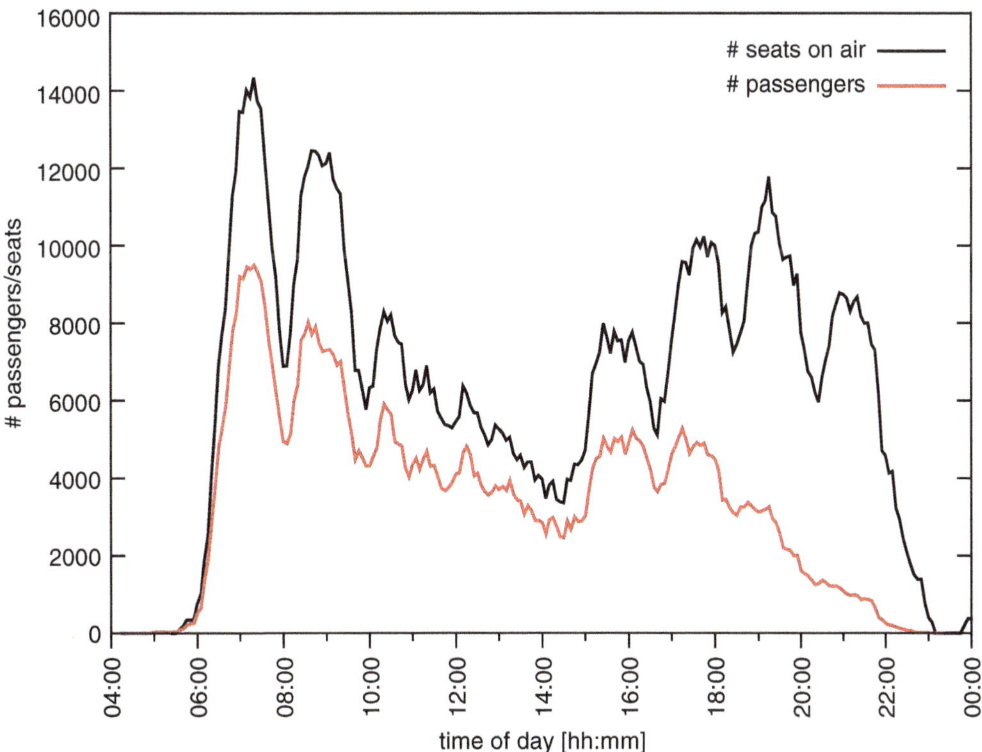

Figure 68.4: Passengers in aircraft and available seats over time in Germany: At any time, there are more seats than passengers. Air transport-only scenario based on O-D data for Germany, iteration 600.

Some passengers fail to reach their destination and are "stranded". This is unrealistic; only trips within Germany are modeled. These are usually completed within a few hours, with no requirement for an overnight airport stay. 320 passengers are stranded at the end of the day. Getting stranded is not a result of insufficient seating; at any time of day, there are more seats than demand. There are many reasons why passengers could be stranded in such a situation. Further analysis of the $c_{lineswitch} = 0$ scenario simulation results indicates:

- 92 passengers are stranded because there is no seat and no other flight on the same airline later that day, to which they could be shifted.
- 228 passengers are stuck at an airport because there is no connection after their departure time between that airport and their destination airport.

Behavioral aspects: neither departing early, nor getting stranded, are explicitly modeled.

68.2.2 Adding an Alternative Mode

To gain further insights, in the following a slightly different simulation setup is applied. A second option for mode choice is added. Each virtual passenger can now choose between the microsimulated air transport options and an alternative mode. The alternative mode has no capacity restrictions. Passengers traveling with the alternative mode can start directly at their randomly selected departure time. The travel time, tt, is computed by the microsimulation, with an estimation of the beeline distance between the O-D pair d and a velocity v, i.e., $tt = d/v$. This velocity is varied in several simulation runs, i.e., $v \in \{100, 150, 200, 250, 300\}[km/h]$. If the alternative mode is chosen, the (dis-)utilities for traveling are calculated accordingly in the scoring.

With this population, the simulation is again run for 600 iterations. As in the previous simulations 10 % of the virtual passengers may shift their departure times, while another 10 % seek a different route between origin and destination in the air transport network. Additionally, further 10 % of virtual persons may change mode, i.e., they can switch between the air traffic mode and the alternative mode. After 500 iterations all choice modules are switched off; thus, for the last 100 iterations, passengers use the logit model to select a plan.

Simulation results for the 600th iteration show that the increasing speed of the alternative mode affects the modal split. While for a $v = 100\,km/h$ the alternative mode is chosen by 1.2 % of the passengers, a mode alternative with a speed of 300 kilometers per hour attracts 15.69 % of travelers. The number of stranded passengers for the alternative mode with $v = 100$ kilometers per hour is substantially reduced, from approximately 320 to 67. Higher speeds of the alternative mode further reduce the number of stranded passengers. Slow speeds of the alternative mode imply dominance of the air transport mode. If there is a seat on a flight, travelers receive a higher score than when they use the alternative mode. However, travelers risk getting stranded, which can be hard to analyze and interpret. The implemented algorithm is also an open issue; if the number of plans per traveler exceeds a threshold of 5, the plan with the lowest score is removed from the plan database.

Instead of this deterministic plan removal, a probabilistic algorithm can be implemented: e.g., plans for removal can be selected based on a path size logit model. With this modification, simulation runs are repeated. Figure 68.5 shows the resulting travel patterns over time for alternative modes at speed 100 kilometers per hour and 300 kilometers per hour. Traveler distribution on the alternative mode over time of day is quite homogeneous. The alternative mode speed increase attracts more passengers, as reflected by the modal splits in Table 68.1. At most, one passenger is stranded at the end of day.

Simulation results are compared in more detail with DESTATIS data serving as a base for the virtual population. Synthetic population is generated based on O-D pairs that may contain transfers ($od_{transfers}$), while other DESTATIS data counts the number of passengers on actual direct flights

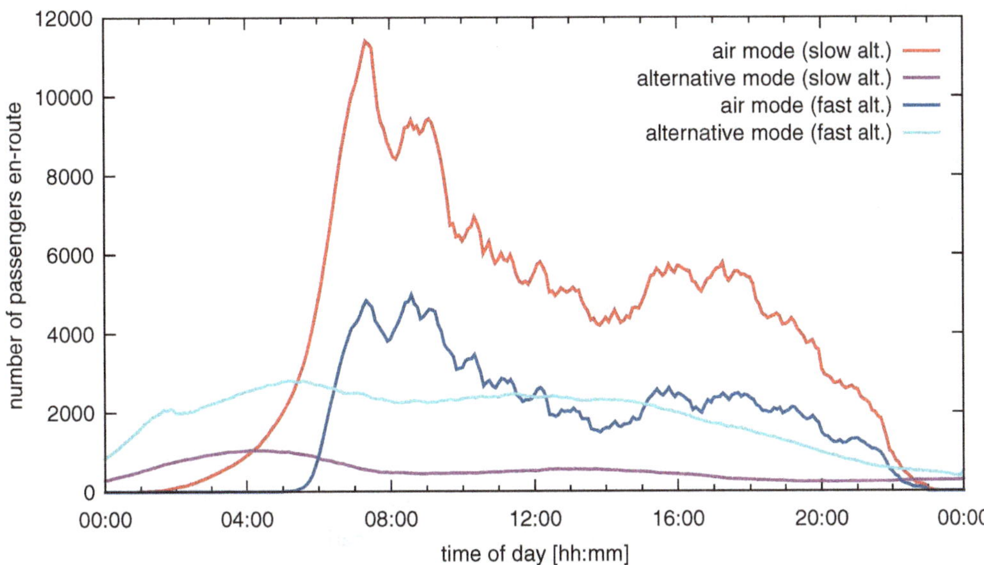

Figure 68.5: Passengers waiting for a flight, traveling by plane, or by alternative mode over time of day. Air transport and alternative mode scenario for Germany, iteration 600. Results with random selector for plan removal.
Source: Grether (2014)

v[km/h]	# air mode	# alt. mode	# stuck	air mode[%]	alt. mode[%]	stuck[%]
100	49280	2551	1	95.08	04.92	00.00
150	44835	6996	1	86.50	13.50	00.00
200	39929	11902	1	77.04	22.96	00.00
250	34332	17499	1	66.24	33.76	00.00
300	27270	24562	0	52.61	47.39	00.00

Table 68.1: Modal split for different speeds *v* of the alternative mode. Air transport and alternative mode scenario for Germany, iteration 600. Results with random selector for plan removal.

(od_{direct}). The latter is used to evaluate model accuracy. For comparison, number of passengers on direct flights is calculated for each O-D pair (sim_{direct}) from the simulation results. Based on these data sets, the mean square error and the mean relative error are calculated.[3]

Table 68.2 shows the outcome of these calculations. The first line is the comparison of two input data sets from DESTATIS.[4] This serves as reference, as it would assume that all demand is served by direct flights. All simulation runs explain the data better than that reference. Mean square error and variance increase with the speed *v* of the alternative mode; logical, as the demand covers only air transport trips.

[3] The mean square error σ^2 is computed as $\sigma^2 = \frac{\sum_{i \in OD}(sim_{direct}(i) - od_{direct}(i))^2}{|OD|}$, whereby $|OD|$ denotes the number of O-D pairs, $sim_{direct}(i)$ the simulated passengers on a direct flight between the O-D pair *i*, and $od_{direct}(i)$ the same, but retrieved from data. With the same values, the (unsigned) mean relative error for each O-D relation is calculated as mean rel error $= \frac{\sum_{i \in OD}|(sim_{direct}(i) - od_{direct}(i))|/od_{direct}(i)}{|OD|}$.

[4] In the calculation, sim_{direct} is replaced by $od_{transfers}$.

$v[km/h]$	σ^2	σ	mean rel error	stuck
$od_{transfer} - od_{direct}$	12640	112	1.75	-
100	10367	102	0.35	1
150	13820	118	0.43	1
200	18651	137	0.56	1
250	25291	159	0.68	1
300	36059	190	0.76	0

Table 68.2: Error calculations for different speeds v of the alternative mode. Air transport and alternative mode scenario for Germany, iteration 600. Results with random selector for plan removal.

68.3 Interpretation & Discussion

The alternative mode can be defined as a combination of train, bus, or car connection availability. Clearly, the results hinge on the assumption that the alternative mode is always available and not capacity-restricted. All passengers on the alternative mode travel at the same speed, but this assumption is too coarse for the scenario presented. For example, average speed and temporal availability of train connections depends on the O-D pair. In principle, the alternative mode could be refined by including O-D pairs' dependent average speed data. Alternatively, train, bus, and car can be simulated explicitly, featuring capacity restrictions and mutual interactions. Even considering these factors, a homogeneous velocity for the alternative mode seems to be more appropriate for the overall modeling approach illustration. Effects triggered by the alternative mode availability are illustrative. Data for the demand provides O-D pairs for air transport, but not for car, train or bus trips. For more plausible interpretations, further demand data for other modes is required.

All the presented modeling approaches explain passenger routing in more detail than technically possible from the input data. Most passengers use a direct connection, which is very plausible, considering the geospatial demand extent. Flying within Germany is often not worthwhile if the connection includes a transfer; empirically it is faster to travel by train, car, or bus. For further insights, the geospatial extent of the modeled demand could be increased; but this depends on data availability, not on the overall simulation approach.

Passengers are modeled without specific desired departure or arrival times. This study's input data does not contain any information about time distribution. The simulation approach can capture such individual time constraints and the information can be added, without too much effort, with some more data, thus resolving several departure time choice problems.

Stranded passengers are an unwanted product of the simulation. Without the alternative mode, the only available transport mode is a capacity-restricted flight connection provided in discrete, irregular time intervals. The number of stranded passengers is higher than for the simulation runs with the alternative mode. Passengers are more likely to get stuck in O-D pairs, where demand is higher than seat capacity, for extrinsic and intrinsic model reasons.

The quality of the simulation model's outcome hinges on the data available. For older studies of air transport passenger demand, DESTATIS data for 09-2011 was used, but the air transport technology model was created on an 09-2009 flight schedule. The number of flight starts within Germany increased slightly between 2009 and 2011 (DLR, 2012, p. 23). Assuming that the number of available seats increased accordingly, the simulation model provided too little capacity, at least on certain O-D pairs. As result, the number of passengers not reaching their destination (stranded) was much higher. With the availability of 09-2009 DESTATIS data, the overall quality of results improved. Replacement of the 2011 data with 2009 data reduced the number of stranded passengers significantly, from around 1 500 to 350 travelers.

Data is provided on a monthly basis, while the simulation model time horizon is one day. Number of trips per day is retrieved using the assumption that trips are uniformly distributed over all days of a month. The remaining 350 stranded passengers might be resolved by a more accurate distribution. Otherwise, a longer time horizon could be simulated.[5] This would also include flights not departing on a Tuesday. With these alternations,the issue of stranded travelers might be solved.

The problem of stranded passenger can be model-intrinsic. The algorithm removing plans is apparently critical to avoid stranded passengers. Replacing the deterministic formulation with the probabilistic resolves most of the stranded passenger problem. The applied path size logit modeling approach seems to be feasible, but requires further studies for parametrization and interpretation. In general, this modeling approach allows the generation of more heterogeneous choice sets, see also Section 97.3. With the deterministic plan, removal plans with a high score (but similar structure) dominate all other generated plans. In combination with capacity restrictions, lack of alternatives results in stranded passengers. All other approaches to simulate more heterogeneity—discussed on the following—should consider these effects.

In further studies, departure time choice and cost structures can be refined. If there is only one early connection to a hub per day, some passengers' departure times might be too late to make connections. The random departure time mutation may not be able to find a connection for all passengers. This has been ruled out for the current setup, but should be considered in further studies.

Alternatively, passengers could have a connection that works in theory, but are "crowded out" by other passengers arriving earlier at the gate; these passengers would reach their destination if they would take a different route. The current approach would not find such a solution, since passengers do not consider costs they impose on others; see Lämmel and Flötteröd (2009) for an approach taking that into account. The real-world solution, presumably, would be to raise prices on seats during congestion periods until a passenger re-routes. Currently, all passengers have homogeneous time values. For a more meaningful price modeling, additional heterogeneous passenger attributes can be included. As the present model is based on only O-D data, it does not include such a process. In principle, other data, e.g., Lorenz curves and median incomes, can be merged with the O-D data (Kickhöfer et al., 2011).

An alternative approach to improve heterogeneity is a router generating a greater route diversity for the same departure time. Such a router would be able to direct a passenger to a route where seats are available, without actually knowing about seat availability. That approach would, however, not address the issue that some passengers might need to switch their path to allow others to obtain a feasible path. In Graf (2013), a first prototype of such a router is tested in a different context, with first tests for the flight model revealing only slight improvement. As more diverse routes are dominated by the direct connection, they are removed by the algorithm similar to routes on slow alternative modes. After this general problem is solved, a more diverse routing should be reconsidered.

68.4 Conclusion

Overall, the results show that a microscopic, agent-based simulation of passenger demand for air transport is feasible. Most passengers are able to learn the constraints of air transport technology and arrive at their desired destination.

The technology modeling is similar to the Clarke et al. (2007) approach, although the level of detail is coarser. In the same way as Clarke et al. (2007), further models for, e.g., gates, taxiing, weather

[5] Note, that this requires some changes in the source code that may not be resolved by sole customizations of MATSim. Please ask the developers before running MATSim for a longer time horizon.

or airline operations can be added to the approach. As the open source code of MATSim comes with options for extension, more detailed models of the technology side hinge on the availability of data. In contrast, and going beyond Clarke et al. (2007), passengers are captured at all stages of their trip and passengers traveling on alternative transport modes can be simulated. The chapter discusses certain open general issues not specific to air transport systems. Interested users should support the MATSim team in solving these more general questions first, which will aid the model in achieving a more detailed picture of mid-distance travel patterns.

Clearly, potential applications of the proposed model depend on type and detail of information included. In general, application for policy planning allows a more detailed evaluation of mid-distance travel policy effects, including mode alternative consideration. The approach could also be useful for private companies' planning of flight-schedules and capacities to their connections. The impacts of these changes on customers can be assessed in close detail.

CHAPTER 69

Gauteng

Johan W. Joubert

Gauteng is a landlocked province in South Africa, with three main metropolitan areas: the city of Johannesburg, city of Tshwane (formerly Pretoria) and Eurhuleni. Although the province covers less than 3 % of the country's surface, it is the country's economic hub and contributes a third of the country's GDP (Gross Domestic Product). The 2011 census reported a population of 12.2 million inhabitants, a quarter of the South African population.

The first Gauteng scenario was developed in 2008/9 and appeared in Fourie (2009) and Fourie and Joubert (2009). The population was synthesized from 2001 census data and travel demand was inferred from the 2003 NHTS (National Household Travel Survey). Initially, the network was created from a proprietary source made available for research purposes this has been replaced with a much richer OSM network.

Early comparisons already showed that the Gauteng MATSim scenario provided far more detailed results than the four-step models available at the time (Fourie, 2010). The scenario was also extended to include freight vehicles (Joubert et al., 2010).

With the introduction of an open-road tolling scheme referred to as the GFIP (Gauteng Freeway Improvement Project), the scenario was used to study the diversion patterns of different road user groups. The population was extended to included background traffic, in the form of public transport (buses and minibus taxis) and external through-traffic. This data was taken from Saturn O-D-matrices made available by the sponsor, the SANRAL (South African National Roads Agency Limited). The impact of the tolling scheme, using vehicle-specific values of time, and a complex toll pricing regime was reported in Nagel et al. (2014).

The most recent update to the synthetic population generation for the Gauteng scenario is documented on MATSim's `https://matsim.atlassian.net/wiki/display/MATPUB/South+Africa` Confluence site.

How to cite this book chapter:

Joubert, J W. 2016. Gauteng. In: Horni, A, Nagel, K and Axhausen, K W. (eds.) *The Multi-Agent Transport Simulation MATSim*, Pp. 429–430. London: Ubiquity Press. DOI: http://dx.doi.org/10.5334/baw.69.

CHAPTER 70

Germany

Johannes Illenberger

The Germany scenario was developed by DB ML AG (DB Mobility Logistics AG), a subgroup of DB AG (Deutsche Bahn AG), the German state-owned railway company, in 2014. To help evaluate MATSim's applicability in the strategic planning process, as well as defining its compatibility in the traditional zone-based four-step process, this scenario has been constructed to establish a Germany-wide O-D matrix for private car travel. A solid understanding of the transport market for private car travel (rail's major competitor) is required for the strategic planning process at the DB ML AG. This scenario intentionally focuses just on road transport since, on one hand, there are already well-established models for rail transport at the DB ML AG and, on the other hand, this scenario is meant to be the first step towards MATSim's application.

Considering this scenario's objective, MATSim is used here as a tool to build a microscopic representation of the current transport market. Unlike the majority of MATSim studies, the focus is not to build and calibrate a behavioral model with forecasting power to answer the "what if" question. Hence, this study emphasizes reproducing empirical measurements, rather than on modeling plausible causalities and behavioral processes.

The final outcome of this exercise is a O-D matrix with average daily trip volumes. Although a higher temporal resolution is possible and also available from travel data, this dimension will not be considered during calibration and validation of the scenario. The matrix is based on a zonal structure with approx. 10 K TAZs, with a granularity comparable to municipalities (LAU 2) and a higher detail in large cities.

How to cite this book chapter:

Illenberger, J. 2016. Germany. In: Horni, A, Nagel, K and Axhausen, K W. (eds.) *The Multi-Agent Transport Simulation MATSim*, Pp. 431–436. London: Ubiquity Press. DOI: http://dx.doi.org/10.5334/baw.70.

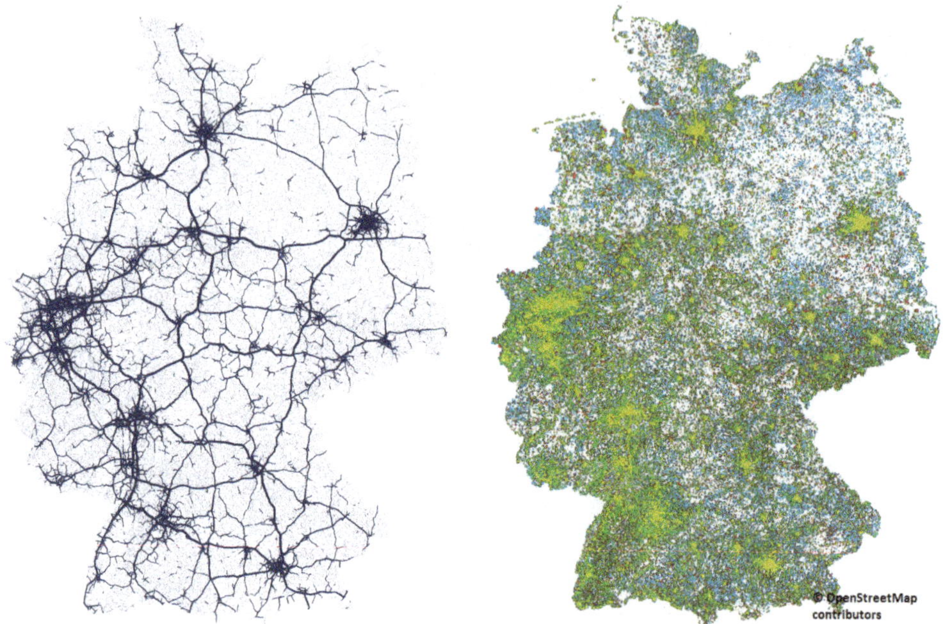

Figure 70.1: Left: Simulation road network including roads down to the level of major arterials. Right: Spatial distribution of activity facilities colored according to activity type.

70.1 Demand and Supply Data

The German national travel survey MiD from 2008 (Follmer et al., 2010) builds the foundation for the synthetic population and its travel plans. The survey features approx. 60 K person records with one-day travel episodes. A travel episode specifies trip sequences with mode, purpose, travel distance and day of reporting. Home locations are known at the level of states and municipality type (urban/rural). From each record, a synthetic person with one travel plan is generated. Initially, activity locations are set to random facilities and each person is cloned multiple times (according to the person's weight), so that in total a population of 4 M persons results.

The road network is extracted from OSM. The geographical resolution of the O-D matrix allows omission of minor roads (everything below "tertiary" in OSM terminology) and the network is further simplified so that connected nodes with a distance less than 50 meter are merged to one node. The resulting network consists of 126 K nodes and 360 K links (Figure 70.1).

Activity facilities are taken from OSM as well. A facility can be synthesized from a OSM node representing a point-of-interest (shop, restaurant, bar, etc.), a polygon representing a building and a polygon representing a region with specified land use. In the latter case, multiple facilities are generated proportional to the polygon's area. Activity options are inferred from meta information associated with the node or polygon. Given the huge amount of "home" facilities, only a 20 % subsample is used. This still yields a total of 5.6 M activity facilities.

70.2 Imputation and Calibration

The location of activities—origin and destination of trips—are required for building the O-D matrix. The travel survey does not provide any information on activity locations. Thus, the study's main task is to impute plausible activity locations, given the sequence of trip distances and underlying geographical distribution of activity facilities. The intermediate solution resulting from this imputation is calibrated against count stations and selected O-D flows from car navigation devices.

The imputation process implementation can be considered as a Monte Carlo Markov Chain simulation converging into a distribution where the activity locations configuration best fits the constraints imposed by trip distances, count stations and selected O-D flows. Solving this task in one simulation process would be congruent with theory. However, considering the scenario size and computational limitations, the process is split into three steps: (i) assigning "home" locations, (ii) generating an initial state with assigned "non-home" activity locations, and (iii) varying a subset of "non-home" activity locations to meet car volumes at count stations and selected O-D pairs' flows. Steps (i) and (ii) can be considered as imputation processes and are realized outside of the MATSim iteration framework. Step (iii) can be considered the calibration step and is realized with a MATSim-Cadyts setup configured as a Monte Carlo engine (see Chapter 48).

70.2.1 Imputation of Home Locations

Home locations are known at state and municipality type levels. The municipality type is divided into six categories by number of inhabitants. Initially, each person is placed on a random home facility, while inhabitants' geographical distribution is given at the TAZs level. A Monte Carlo simulation relocates persons to best meet their specified state and municipality. More formally:

1. Generate an initial configuration $\mathcal{P}_k$:
   (a) Randomly assign each person n a home facility.
   (b) Evaluate the configuration: $H(\mathcal{P}_k) = \sum_n \theta_1 \delta_n + \theta_2 \left| m_n - m_n^* \right|$, where δ_n is 0 if the person is located in the correct state and 1 otherwise, m_n denotes the category index of the current person's municipality and m_n^* its target category. Parameters θ_1 and θ_2 control how close the simulation converges to the target values.
2. Generate a new configuration $\mathcal{P}_{k+1}$ by switching the home facilities of two random persons.
3. Accept the new configuration $\mathcal{P}_{k+1}$ with probability $\pi_{k+1} = 1/\left(1 + \exp\left(H\left(\mathcal{P}_{k+1}\right) - H\left(\mathcal{P}_k\right)\right)\right)$, otherwise return to $\mathcal{P}_k$.
4. Repeat step 2 and 3 until the system reaches a steady state distribution.

Switching home facilities, instead of assigning a random facility, ensures that persons' spatial distribution remains constant. Running the simulation for 10^9 iterations results in a configuration where more than 90 % of persons are located in their correct state and, on average, three of four persons are located in their correct municipality; the forth persons is just one category index distant from its target index.

70.2.2 Imputation of Non-Home Locations

Activity locations (non-home) are assigned with an analogous process, like home locations. The simulation relocates activities so that resulting trip distances best meet their empirical target distances. In this imputation step, distance always refers to the beeline distance and thus avoids expensive route search.

About one forth of all trip chains are composed of more than two trips; that is, trips are not symmetrical. Accordingly, drawing a random activity location on a annulus centered at the origin activity location, with radius according to the target distance, does not necessarily fulfill the target distances of succeeding trips. The simulation process is specified as follows:

1. Generate an initial state $\mathcal{P}_k$:
   (a) Assign a random facility to each non-home activity by considering the activity type and the facility's activity options.
   (b) Evaluate the configuration: $H(\mathcal{P}_k) = \sum_n \sum_q \theta_3 \left| \frac{d_{nq} - d^*_{nq}}{d^*_{nq}} \right|$, where d_{nq} denotes the realized distance of n's trip to activity q and d^*_{nq} the target distance.
2. Generate a new configuration $\mathcal{P}_{k+1}$ by assigning a random facility to a random non-home activity (by considering activity type and activity options).
3. Accept the new configuration $\mathcal{P}_{k+1}$ with probability $\pi_{k+1} = 1/\left(1 + \exp\left(H\left(\mathcal{P}_{k+1}\right) - H\left(\mathcal{P}_k\right)\right)\right)$, otherwise return to $\mathcal{P}_k$.
4. Repeat step 2 and 3 until the system reaches a steady state distribution.

A configuration is evaluated based on the relative error or realized distances to target distances, so that short and long trips are treated equally. Parameter θ_3 controls the randomness. That is, if $\theta_3 = \infty$ each trip would exactly (if possible) meet its target distance, which, however, is not the desired solution. Rather, θ_3 is adjusted so that there is some randomness in realized trips distances, but without distorting global target distance distribution.

70.2.3 Calibration

The outcome of step 2 (Section 70.2.1) and 3 (Section 70.2.2) is a valid population with imputed home and activity locations. In the third step, the population is calibrated against measurements of count stations and flows of selected O-D pairs. This step is implemented in a "standard" MATSim-Cadyts combination. The scoring function accounts only for the "linear plan effect", agents are allowed to have only one plan and `SelectExpBetaForRemoval` is used for the `planSelectorForRemoval` parameter. All Cadyts parameters are left to their defaults.

During replanning, non-home activities that are not part of a complex trip chain are relocated. More specifically, an activity is valid for relocation if:

- The facility of the previous and succeeding activity is equal (round trip), or
- the activity is the origin of the first trip, or
- the activity is the destination of the last trip.

The above conditions ensure that complex trip chains that have been adjusted in step 2 (Section 70.2.2) are not distorted. New activity locations are randomly chosen within a distance of $\pm 10\,\%$ of the trip's target distance, so that global distance distribution is conserved.

70.2.3.1 Counts Calibration

The German Federal Highway Research Institute, BASt (Bundesanstalt für Straßenwesen), provides average daily vehicle volumes (distinguished in car and trucks) yearly, measured at about 1500 count stations on motor- and highways. After separating each station's data into both directions of traffic and validating measured vehicle volumes, about 2 500 link volumes are available for calibration. Empirical car occupancy rates (depending on trip purpose) are taken from the MiD to convert person volumes to car volumes.

70.2.3.2 O-D Calibration

O-D calibration is based on an O-D matrix representing car navigation device flows. Since occupancy rate of devices in cars is unknown, only the distribution of flows is used for calibration.

Further, comparison with other data sources shows that the O-D matrix is only valid for O-D pairs above a distance of 100 kilometers. This appears reasonable, considering that navigation devices are probably not switched on for short (likely commuter) trips. Accordingly, only O-D pairs above 100 kilometers distance and, from those, only the 6 000 pairs with the highest volumes are extracted into a reduced *calibration matrix*. The latter conditions ensures that only pairs with a sufficient sample size are used.

The reduced calibration matrix is normalized to the sum of all trips in a *reference matrix*. The reference matrix contains all trips from the initial population that correspond to all non-null O-D pairs (matrix entries) in the reduced calibration matrix. This yields a calibration matrix with valid absolute trip volumes.

For each O-D pair, a virtual link is inserted into the road network at runtime. A virtual count station with the corresponding count value from the reduced calibration matrix is attached to each virtual link. In the mobility simulation, after a person arrives at its destination, it then travels the virtual link corresponding to the traveled O-D pair. This travel, however, is only communicated towards the Cadyts calibrator by injecting additional `PersonDeparture`, `LinkLeave`, and `PersonArrival` events.

70.3 Simulation Results and Travel Statistics

The synthetic population of 4 M persons corresponds to approx. 8.5 % of the German population that conducts at least one car trip per day. This yields a scaling factor of 11.8 by which all simulation statistics need to be multiplied. The simulation produces an overall trip volume of 57.5 M trips, quite close to 57.2 M trips in the official statistics (DIW, 2014). Passenger mileage of 947 G person-kilometers is slightly overestimated compared to 917 G person-kilometers in the official statistics, yet still reasonable.

Figure 70.2 visualizes the calibration results in a scatter plot. Each dot represents a count station (left) or a O-D pair (right), respectively. On average, absolute value of the relative error yields 0.18, considering vehicle volumes at count stations and 0.16 considering O-D flows. A median relative

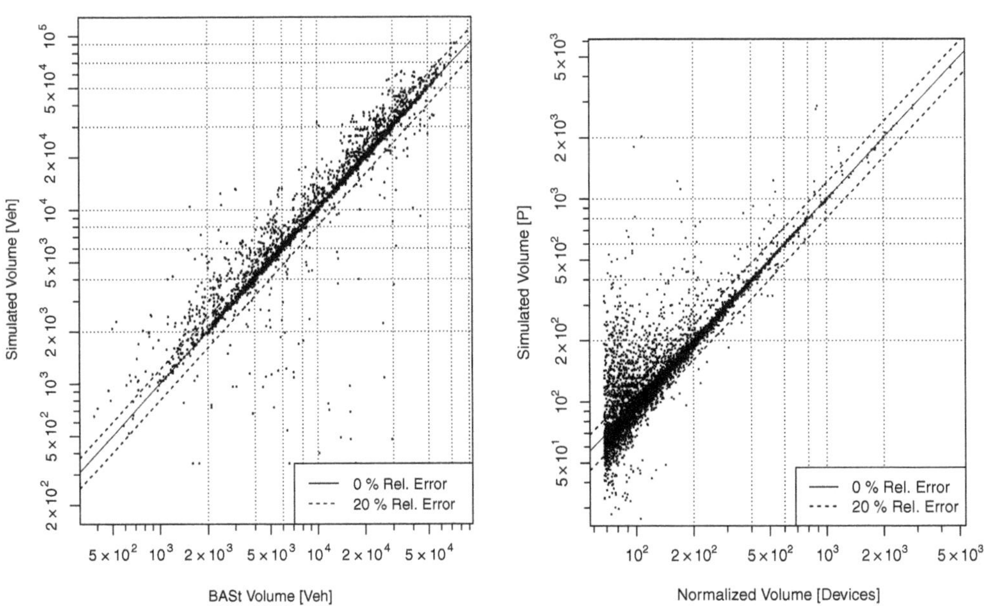

Figure 70.2: Comparison of simulation and empirical measurements. Left: count stations. Right: O-D pairs.

count station error of 0.12 indicates that there are a few count station that are significantly off. These two issues should be noted; first, there is only a description of the count station's location, which is not always unique and may be misinterpreted by the matching algorithm. This can yield a false assignment of count stations to network links. Second, there is no information on the count stations' measuring error, or on temporarily capacity reductions (for instance, caused by construction sites) modeled in the simulation network.

O-D flows error correlates with the O-D pairs' volumes; thus, pairs with low volume show, on average, a higher error. This is related to this scenario and MATSim's characteristics: a population sample size (8.5 %) and the discrete nature of MATSim. For instance, a real-world O-D flow with 118 individuals is represented by ten agents in the simulation. A variation during re-planning to this O-D pairs by, say, one agent already has a significant impact on the scaled real-world value. Averaging over multiple iterations reduced the variance but does not entirely remove the effect.

CHAPTER 71

Hamburg Wilhelmsburg

Hubert Klüpfel and Gregor Lämmel

The following describes the evacuation of Hamburg-Wilhelmsburg as a case study. The scenario has been created using MATSim's evacuation contribution. Technical details about the evacuation contribution are given in Chapter 41.

Wilhelmsburg was severely flooded in 1962. Since then, many structural and operational improvements have been implemented. Back then, the housing situation was rather bad, many people lived in provisional housing due to destruction in World War II. Additionally to the by far more stable buildings, precautions for flooding have been taken and the walls have been heightened. Evacuation is nevertheless necessary under certain circumstances. The relocation of one of the major roads in Wilhelmsburg, the B75, will also influence the evacuation traffic, since it is one of the major north-south arterial roads. In this case study, the consequences of this relocation on the evacuation of Hamburg-Wilhelmsburg is investigated.

71.1 Brief Description

The scenario investigated here is the relocation of highway B75 in Wilhelmsburg. Two cases are investigated, as summarized in the following table. The investigation highlights differences in the evacuation traffic for both variants of the B75 trail. As seen in Figure 71.1, the new trail "B75 new" is located generally next to the existing railway track. In the south, the new variant is connected to the existing highway at the junction "Hamburg Wilhelmsburg Süd" (just north of the bridge across the river Elbe); in the north, it is connected to the existing highway just before the junction

1	Current location of B75 with restricted directional choice
2	New location of B75 with restricted locational choice

Table 71.1: Scenarios.

How to cite this book chapter:
Klüpfel, H and Lämmel, G. 2016. Hamburg Wilhelmsburg. In: Horni, A, Nagel, K and Axhausen, K W. (eds.) *The Multi-Agent Transport Simulation MATSim*, Pp. 437–444. London: Ubiquity Press. DOI: http://dx.doi.org/10.5334/baw.71. License: CC-BY 4.0

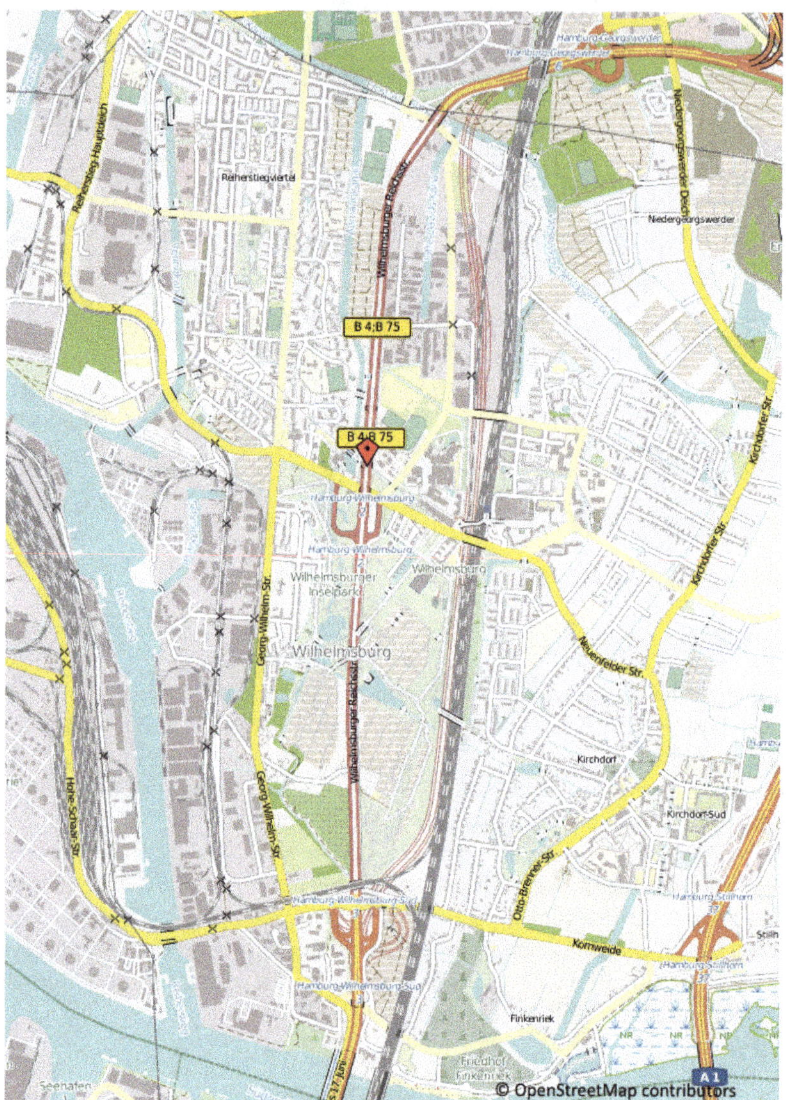

Figure 71.1: The current trail of highway B75 is shown in the center of the image. The new trail is east of it next to the railroad.

"Hamburg Georgswerder". The main differences between the two variants are the location of the access routes to highway B75 in the center of Wilhelmsburg.

71.2 Road Network

The MATSim road network is generated ("imported") from the Hamburg OSM file, downloaded from `http://www.geofabrik.de`. Fortunately, the OSM file already contains the new B75 highway track, marked by an attributed "open 2016". Therefore, the two networks for the "B75 old" and "B75 new" variants could be derived from the same OSM file. For the variant "B75 old", this file could be directly imported. For the variant "B75 new", the section of B75 to be relocated has been removed in a first step. In a second step, the new B75 track has been connected to the existing road network, i.e., the B75 north at junction "Georgswerder" in the north and junction "Hamburg Wilhelmsburg Süd" in the south.

(a) (b)

Figure 71.2: Comparison between network for the old and new track of the B75.

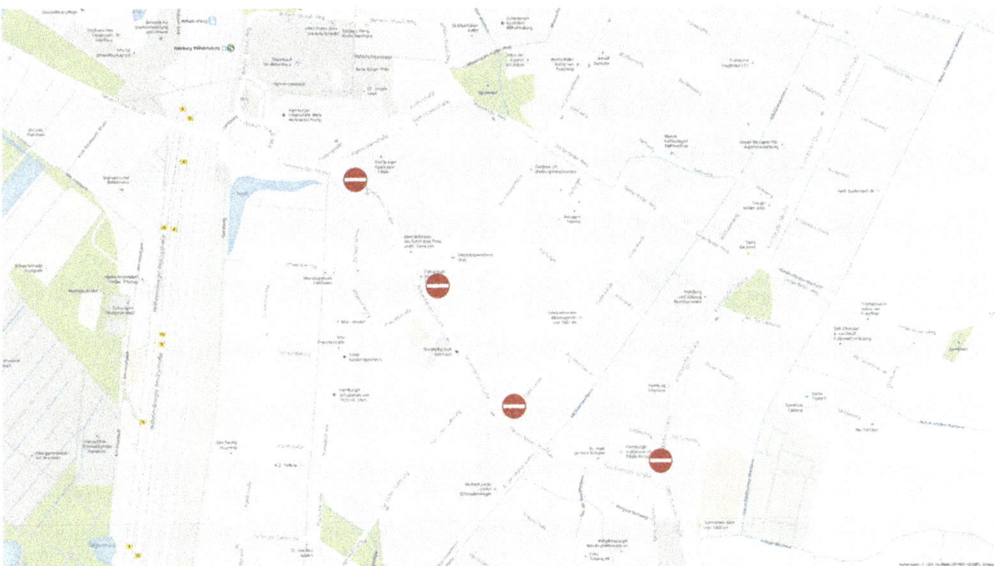

Figure 71.3: Roads closed during evacuation.

Additionally, the on and off-ramps to the B75 have been added. The two variants of the resulting road network, i.e., "B75 old" and "B75 new" are shown in Figure 71.2.

In an evacuation, some roads would be blocked to avoid intersecting and inbound traffic. The following streets were thus deleted in the OSM file:

- Neuenfelder Str.
- Im Schönenfelde
- Elsterweide
- Kirchdorfer Str.

An illustration is given in Figure 71.3.

71.3 Evacuation Scenario

The comparison of the two variants is based on overall evacuation time, clearing time of different cells (squares in the area that had to be evacuated) and the number of cars using the road network (utilization).

As described in Section 41.4, the input files for the network (OSM), the area (ESRI shp), and the population (ESRI shp), as well as the parameters for sample size and departure time distribution, have been specified and assessed via a GUI. The scenario XML file for the existing (or "old", in German "alt") track of highway B75 is shown in the following listing.

```
<?xml version="1.0" encoding="UTF-8" standalone="yes"?>
<grips_config
 xsi:schemaLocation="http://www.matsim.org/files/dtd
 http://matsim.org/files/dtd/grips_config_v0.1.xsd"
 xmlns:xsi="http://www.w3.org/2001/XMLSchema-instance">
    <networkFile>
        <inputFile>osm/hamburgB75_alt.osm</inputFile>
    </networkFile>
    <mainTrafficType>vehicular</mainTrafficType>
    <evacuationAreaFile>
        <inputFile>area/area.shp</inputFile>
    </evacuationAreaFile>
    <populationFile>
        <inputFile>population/population.shp</inputFile>
    </populationFile>
    <outputDir>
        <inputFile>matsim_output_B75alt</inputFile>
    </outputDir>
    <sampleSize>0.1</sampleSize>
    <departureTimeDistribution>
        <distribution>normal</distribution>
        <sigma>1800.0</sigma>
        <mu>1800.0</mu>
        <earliest>0.0</earliest>
        <latest>3600.0</latest>
    </departureTimeDistribution>
</grips_config>
```

71.3.1 Departure Time Distribution

The departure time distribution is specified in the file `scenario.xml`. The values were in seconds, i.e., a normal distribution with a mean value (mu) and a standard deviation (sigma) of 30 minutes in the range of zero (earliest) to one hour (latest) was chosen. More details about time distributions are discussed in Section 41.4. This distribution reflected certain assumptions made about evacuation procedure. The overall time frame, based on the warning time, is minimum 7 hours. The preparation phase is projected with three hours. Available time for the evacuation is three hours, with a one-hour buffer. The warning via radio will start at t=0 hours and local warning (e.g., by police cars, sirens, and via short messages) at t=1 hours; simulation reference point was set to t=3 hours. The overall time acceptance criterion for the simulation is the a required safe evacuation time (for simulation by car) of less than three hours (including reaction time). The reaction time set in the simulation could be interpreted as decision-making time after readying personal belongings. In short: ASET (Available Safe Evacuation Time) determined by flooding is 3 hours and the RSET (Required Safe Egress Time) is estimated by the simulation. The criterion for a successful evacuation is ASET > RSET.

71.3.2 Population Size

As explained previously, the population is not stored in the scenario file, but in the population shape file, possibly consisting of several polygons. The number of persons is stored as an attribute for each polygon. Here, one must assume that only part of the population would have to evacuate; for many, escape to higher ground might be sufficient. Detailed information on the different procedures can be found at `http://www.hamburg.de/sturmflut/3425646/sturmflut-download-1/` (in German).

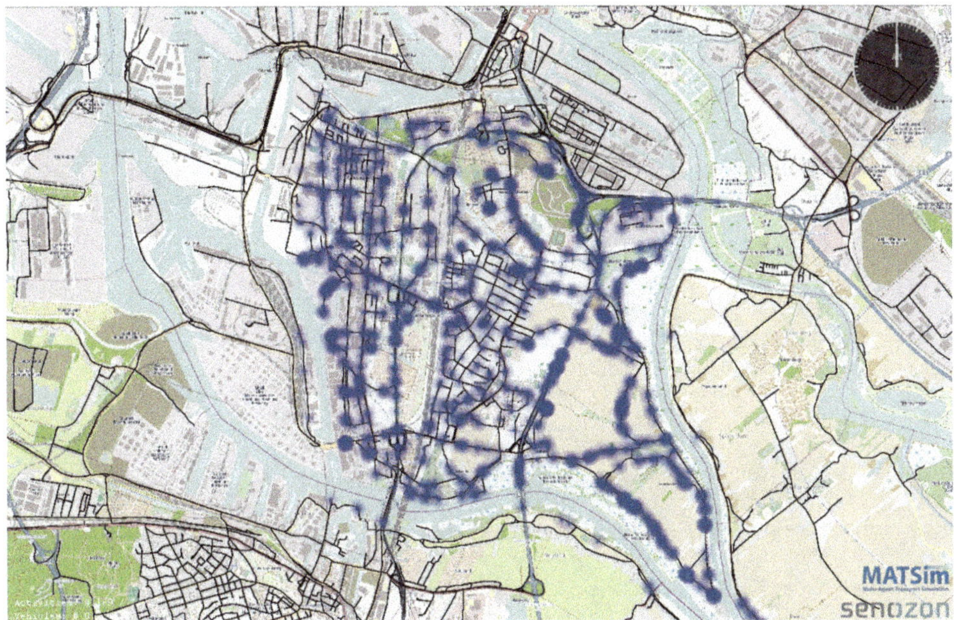

Figure 71.4: The initial distribution of the agents for the evacuation of Wilhelmsburg (for both cases, B75 new and old).

Each agent represented one evacuee traveling by car. To independently check the number of agents based on the simulation files, one could open the file `population.dbf` with a database or spreadsheet editor. Note that the number specified in the `population.shp` (resp. `population.dbf`) is multiplied with the `sampleSize` when converting the files to MATSim input, i.e., in this case, the `population.xml.gz` located in the output directory.

The population is initially distributed as shown in Figure 71.4. The algorithm that converted the area and population (i.e., `area.shp` and `population.shp`) is described in Section 41.2. It assigns agents to the edges of the network. In the case study, harbor areas are left out and agents are equally distributed to streets in the housing (and agricultural) areas of Wilhelmsburg (Figure 71.4). Of course, this could have been further refined by going to a block, or even house level and assigning the population according to detailed statistical housing data. This has not been undertaken for this simulation, for two reasons. First, many assumptions are made about behavior, initial location, and share of population that had to evacuate. Therefore, the level of detail seemed to be sufficient. Second, each agent represented a car driver, i.e., in the simulation, all cars registered in Wilhelmsburg left the area. Considering that inbound, as well as through traffic would be prohibited when flooding level exceeded a certain threshold, this is a "worst case" assumption resulting in a heavy traffic load. To summarize, the overall approach is justified to assess highway B75 relocation based on heavy traffic load with a reaction time span between 0 and 1 hour.

71.4 Simulation Results

The simulation results are summarized in Table 71.2. The 0th iteration is based on shortest distance only. This might have resulted in "strange" behavior, as illustrated in the following Figure 71.5 (south of the bridge across the Elbe river, near the junction "Großmoordamm"). The road network had a circular shape; it was cut out from the OSM road network according to the `area.shp`, which is, in this case, just a circle. Since all the agents are taking the shortest path in iteration 0, they headed to the nearest road out of the evacuation area. Technically, the boundary links in the network are connected to a super link when creating the MATSim network from the OSM file and the

Figure 71.5: Results for B75 old, iteration 0.

	B75 old	B75 new
Iteration	Time	(hh:mm)
00	04:45	05:00
10	01:52	01:58
20	01:42	01:46
30	01:40	01:42

Table 71.2: Results.

`area.shp`. This super-link is the destination in all evacuees' plans. A second factor that contributes to congestion in iteration 0 is a short cut via an on- and off-ramp of Autobahn A253 at "Großmoordamm". Capacity of the on- and off-ramp is 1 500 cars per hour, compared to 4 000 cars per hour on the highway. Thus, the short cut (which is shorter in distance, the reason agents chose it) was a bottleneck, resulting in artificial congestion in iteration 0. Therefore, the 0th iteration was unsuitable for assessing the overall evacuation time. As can be seen from Table 71.2 above, for both cases, from iteration 10 on, time presumably converges to some realistic value. This was also illustrated in Figure 71.6 where the situation at t=1:30 hours was shown for iteration 20.

In summary, relocation of highway B75 had no major influence on the overall evacuation time. The evacuation time of about two hours was also within the available safe egress time, as described in the previous section.

It would certainly have been possible to analyze the results further. The two screenshots above, for the situation in iteration 0 at t=3 hours and for iteration 20 at t=1.5 hours, were created with Senozon AG Via (the visualizer presented in Chapter 33). As a conclusion to this chapter and an illustration for the built-in capabilities of the evacuation contribution for analyzing simulation results, road utilizations of the two variants are shown in Figure 71.7.

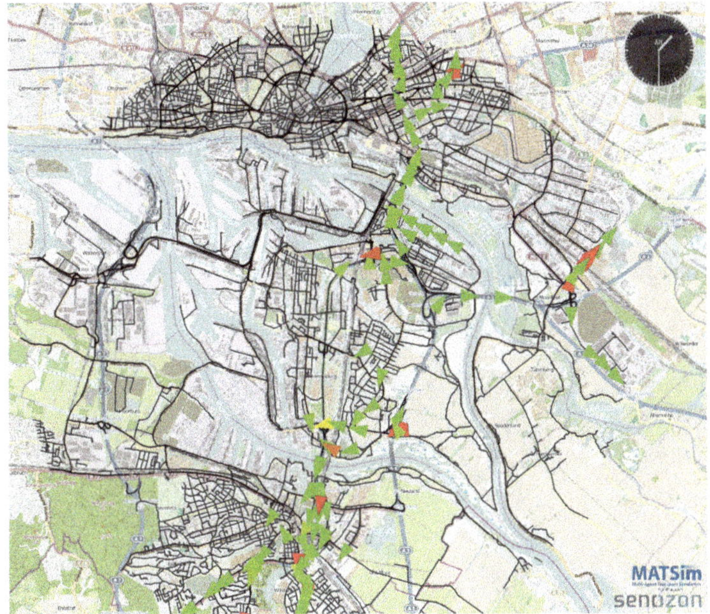

Figure 71.6: Results for B75 old, iteration 20.

(a) (b)

Figure 71.7: Comparison between network utilization for the old and new track of the B75.

CHAPTER 72

Joinville

Davi Guggisberg Bicudo and Gian Ricardo Berkenbrock

Joinville Prefeitura Municipal de Joinville (PMJ) (2015) is a mid-sized industrial city in the south of Brazil, with around 550 000 inhabitants. It has a large workforce, including commuters from neighboring cities and an intense industrial activity profile, meaning that companies work often in three shifts, causing peculiar traffic patterns. Many people also have 12-hour daily routines, encompassing work and higher education.

The Joinville traffic model was built as an initial step of a project to simulate the entire northeast region of Santa Catarina state, including air traffic, shipping, state highways and neighboring cities. The project aims to build a complete data base of people and freight movement in the region. The first version of the urban Joinville model is now complete, produced as a graduate thesis at the Federal University of Santa Catarina (UFSC) `http://ufsc.br`, Transportation and Logistics Engineering course `http://transporteslogistica.joinville.ufsc.br`.[1]

The scenario population was generated with data from the 2010 Brazilian census combined with demographic information from the city's travel survey; travel demand was generated from the same survey. Both were designed to fit into the MATSim, using Tutorial classes (with some adaptations).

The network was produced with vector data provided by the local Urban Sustainable Planning Institute of Joinville (IPPUJ) `https://ippuj.joinville.sc.gov.br`. The data came as a shapefile, with numerous connectivity problems. We were able to fix them using scripts in Python with the `NetworkX` module (Hagberg et al., 2008). Information was transformed from vector data into a graph, addressing issues with the help of QGIS and finally writing as the MATSim XML network format. The facilities were produced from land-use data provided by the city government.

For now, the model runs only with cars, using a full sample of the population. From the available data, we inferred 135 652 agents traveling by car; the rest were removed from the simulation. Figure 72.1 shows a screenshot of the Events using Via.

[1] The authors would like to thank their sponsors Federal University of Santa Catarina (UFSC) and Urban Sustainable Planning Institute of Joinville (IPPUJ).

How to cite this book chapter:
Guggisberg Bicudo, D and Berkenbrock, G R. 2016. Joinville. In: Horni, A, Nagel, K and Axhausen, K W. (eds.) *The Multi-Agent Transport Simulation MATSim*, Pp. 445–446. London: Ubiquity Press. DOI: http://dx.doi.org/10.5334/baw.72.

Figure 72.2 shows the comparison between simulated and count data for 20 links in the morning peak from 7 to 8 am. The count data available for comparison is still sparse and could not be used as effectively as we hoped; we know that calibration is needed for the next model versions. The good news is that the local authorities are installing more than a hundred counting stations throughout the city within the next couple of months and a new travel survey will be conducted this year.

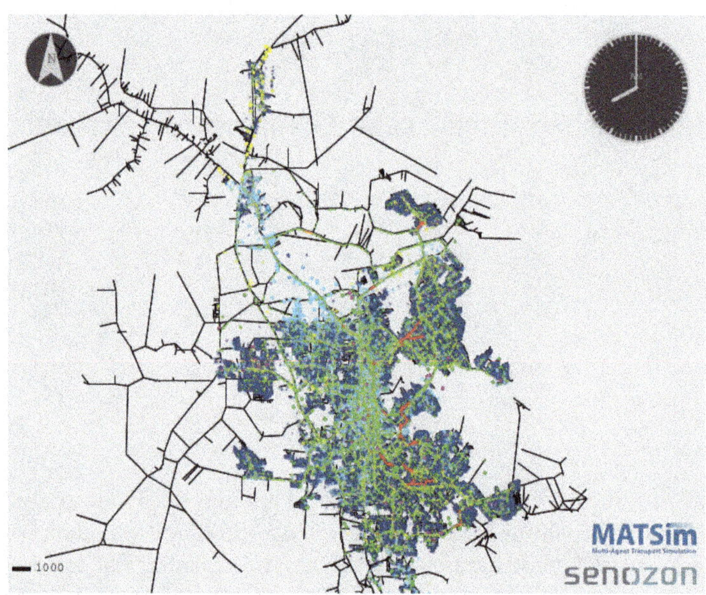

Figure 72.1: Screenshot of the simulation using Via.

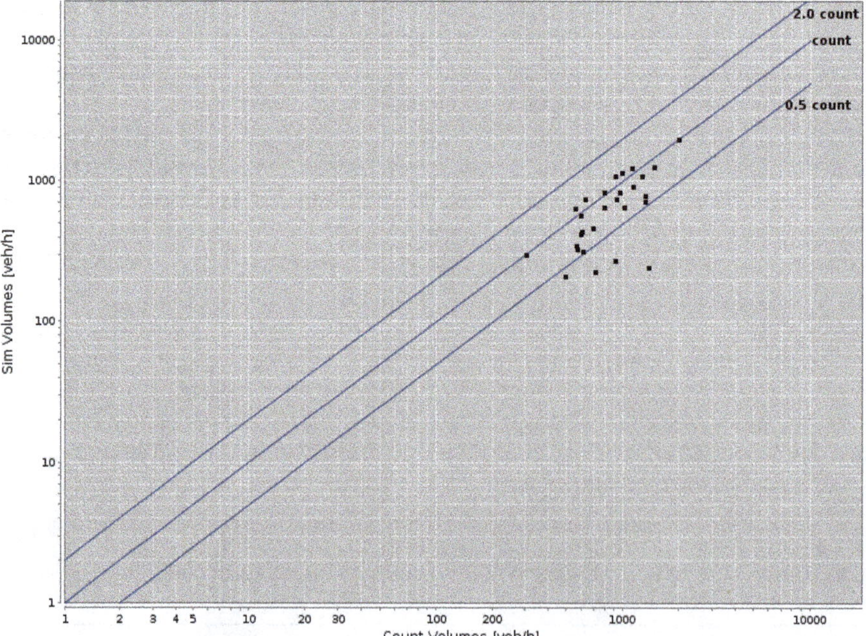

Figure 72.2: Count comparisons for the morning peak at 7-8 am.

CHAPTER 73

London

Joan Serras, Melanie Bosredon, Vassilis Zachariadis, Camilo Vargas-Ruiz, Thibaut Dubernet and Mike Batty

The building of a travel demand model for London started to take shape under the EUNOIA Project.[1] The core decisions around the model design were taken after two meetings with TfL (Transport for London), which was part of the Advisory Board in the project. In that respect, the main suggestion by TfL was the adoption of an activity-based approach.

The main traits from the current implementation of the London model are listed next:

- Our baseline year is 2010.
- The geographic extent of the case study area is contained within the M25 and includes around 9,4 million inhabitants (Census 2011).
- The types of activity included in the model are: home, work, shop, education, leisure and other.
- Four travel modes have been included: walk, cycle, car and public transport. The public transport mode includes buses, underground, rail, the Docklands Light Railway and the London Overground.
- The zones of analysis for the London model are the English Census 2011 Wards which we will refer to as wards from now on. Our case study is composed of 850 wards.

73.1 Supply

The assembly of the supply for our model includes the definition of the following three components:

- road network,
- public transport services, and
- land-use configuration

[1] see http://eunoia-project.eu

How to cite this book chapter:
Serras, J, Bosredon, M, Zachariadis, V, Vargas-Ruiz, C, Dubernet, T and Batty, M. 2016. London. In: Horni, A, Nagel, K and Axhausen, K W. (eds.) *The Multi-Agent Transport Simulation MATSim*, Pp. 447–450. London: Ubiquity Press. DOI: http://dx.doi.org/10.5334/baw.73.

The data used to build the road network is the Integrated Transport Network from the Ordnance Survey. The source network, which is defined at the navigation level, has been processed to remove some of the detail included. Decisions on the capacity of each road link has been based on the guidelines proposed by COBA Manual (2002)(Vol. 13),[2] by the UK's Department for Transport. This implies the usage of each road link's road type (Motorway or A road among others) and road nature (single carriageway, dual carriageway or slip road among others) to set the road capacity for each road link.

All the public transport services operating within the case study region have been obtained from timetable data held by the National Public Transport Data Repository from 2009[3]. This dataset, includes a very detailed account of all the services operating in the UK.

Finally, the land-use configuration for the London model has been produced using the Ordnance Survey `AddressBase` layer which keeps address records for all the United Kingdom with a definition of land-use for each one of them. We have processed the detailed spatial information in order to assign each address point to the nearest road link in the network; this means that after this process, each link in our network will contain a number of addresses which include the land-use associated to it. We have also mapped the wide categorization associated to each address point to the activity types from the model: home, work, shop, education, leisure and other.

73.2 Demand

In order to define the travel demand associated to London, we have followed the methodology adopted in TRANSIMS[4]. In this respect, we first generated a synthetic population representative of the case study area and then, we assigned the sequence of activities to each synthetic individual.

We created our synthetic population using a simulated annealing technique based on Metropolis et al. (1953). We have used the following two datasets: Census 2011 data for each of the 850 wards in London and the HSAR (Household Sample of Anonymised Records) for England in 2001. This technique is based on the selection of survey households from the HSAR which best match the overall socio-demographics from the Census 2011 for each of the 850 wards in London. The output of this technique includes a number of synthetic households associated to each ward and, correspondingly, the synthetic individuals which cohabitate within the household with very detailed socio-demographic information.

The assignment of each synthetic household to our network has been achieved using a probabilistic distribution based on the use of home-only activity locations within each ward.

The assignment of skeletal activity patterns for each synthetic individual has been executed using Classification and Regression Tree Algorithms much like in Speckman et al. (1998). More specifically, the multivariate regression tree algorithm. This technique aims to produce clusters of survey households whose activity patterns are similar through the use of socio-demographic data. Once the decision tree is built, it is used to assign each synthetic household to a given survey household through socio-demographic similarities between the two. In this case study, we have used the LTDS (London Travel Demand Survey) 2010/11 to generate the tree.

After assigning the skeletal activity patterns to each synthetic individual, the next step consists in assigning a location to each activity. In order to do this, we have used a multinomial logit choice model. This technique allows each synthetic individual to evaluate the benefit of performing a specific activity at a particular destination as a composite value based on objective metrics associated with this destination (e.g., number of relevant addresses), objective metrics associated with

---

[2] retrieved from `https://www.gov.uk/government/publications/coba-11-user-manual`

[3] see `http://data.gov.uk/dataset/nptdr`

[4] We used v3.1, corresponding to that developed in Los Alamos National Laboratory.

traveling from origin to destination (e.g., travel time) and subjective components following a probability distribution. The area units being considered in London have been the wards again. And, in this respect, the attractiveness of each ward has been quantified by the number of addresses for each activity type, and the accessibility throughout the region as the travel time across all wards using the crow-fly distances and the average speed for each travel mode in the case study. The calibration for each travel mode and activity type pair has been performed, and those parameters have been used to calculate the new activity locations for each synthetic agent.

73.3 Calibration and Validation

In terms of the calibration, the multinomial logit choice model applied described in the previous section could also be included here. On top of this, activity-related time values have been set in MATSim's configuration file using typical duration values observed from the LTDS dataset. Finally, the parameter values set by default in MATSim taking have also been adopted here. This is a limitation as the modal split currently in place is the one provided by the MATSim corresponding module. The related parameters should be first adjusted so that the observed modal share is similar across modes to the observed modal shares.

In terms of the validation, traffic counts from around 600 sensors have been made available to us for London. We hold values for the AM peak (8-9 am), the inter-peak (9 am-5 pm) and the PM peak (5-6 pm). The former and the latter are hourly counts; the inter-peak is an average value. Those counts are organized into so-called cordons (3) and screenlines (3). Comparisons are currently in place and we are still not in a position to evaluate in detail how the model validates to the observed data.

73.4 More Information

More detailed information on the building of the model including some results from each module previously described can be found at: `http://eunoia-project.eu/publications/` (Report on Case Study 1: London).

Figure 73.1: Snapshot of the road network for London's case study colored by road type (left) and map showing number of bus trips per road segment in London using timetable data (right).

Figure 73.2: Visual estimate of activities performed in London at 9 am using the Via software.

CHAPTER 74

Nelson Mandela Bay

Johan W. Joubert

The Nelson Mandela Bay metropolitan area is in the Eastern Cape province of South Africa and includes the cities of Port Elizabeth and Uitenhage, with a population of approximately 1.2 million inhabitants.

The issue of complexity drove the development of a scenario for this region. We needed an area where we could experiment with various modules and elements offered by MATSim, but one less complex than the mega-city region of Gauteng. The Nelson Mandela Bay case was attractive; it still had a substantial population, only one official bus operator (Algoa Bus Company) and one passenger rail operator (Metrorail). It also displayed the characteristic apartheid urban form of South African cities and towns, where many low-income commuters lived on the outskirts of spatially sprawled cities.

The population was, initially, generated from the 2001 census: later revised and updated to the 2011 census data. Travel demand was inferred from the 2006 travel diary conducted in the metro. The process of synthetic population generation is described in detail on the MATSim `https://matsim.atlassian.net/wiki/display/MATPUB/South+Africa` Confluence site. The population was generated as entire households, using MLIPF (Multi-Level Iterative Proportional Fitting) as published by Müller and Axhausen (2012). Households were also assigned to buildings, based on census description.

This was the first South African scenario to include private cars, freight and detailed public transport. The unique minibus taxis, a form of paratransit in South Africa and many developing countries, were incorporated in the Nelson Mandela Bay area and reported in Röder (2013) and Neumann et al. (2015).

How to cite this book chapter:

Joubert, J W. 2016. Nelson Mandela Bay. In: Horni, A, Nagel, K and Axhausen, K W. (eds.) *The Multi-Agent Transport Simulation MATSim*, Pp. 451–452. London: Ubiquity Press. DOI: http://dx.doi.org/10.5334/baw.74. License: CC-BY 4.0

CHAPTER 75

New York City

Christoph Dobler

The MATSim New York model was an example of an agent-based model based on a given activity-based demand generation process outcome: in this case, the NYBPM (New York Best Practice Model) of Parson Brinkerhoff (Vovsha et al., 2002; Parsons Brinckerhoff, 2005, 2009). It produced persons with individual activity chains; MATSim was chosen as the simulation-based alternative to conventional assignment processes.

Activity locations were selected on zonal level (3 824 zones), timings (i.e., start time and duration) were chosen using given distributions. As part of the conversion process to MATSim, locations were distributed within the zones, according to land use and buildings. For the route assignment, transport modes were converted into those supported by MATSim. The resulting population contained 5.3 million persons (25 % sample).

A multimodal network was created, containing car and public transport links, for the MATSim model. Car links were derived from the aggregated model network data, including capacity, number of lanes and speed limits. For the public transport network, a shape file containing every lines' routes was available. After converting and cleaning the data, the final multimodal network contained 498 000 nodes and 541 000 links. Based on further public transport-related data, a full schedule was created, including different public transport modes (bus, train, etc.).

An example for final model outcomes was shown in Figure 75.1 and Figure 75.2, depicting the car share of all performed trips within a region. Red indicated a high share, blue a low. In Figure 75.1, trips were aggregated on zonal level. In Figure 75.2, the MATSim model high resolution is shown; there, the trips were aggregated using hexagons with a side length of 500 meters instead of a zonal level.

Finally, Figure 75.3 shows traffic flows in Lower Manhattan. Cars were represented by rectangulars, public transport vehicles by arrows. Further model outcomes were presented by Balmer (2014). An online movie can be found at `http://senozon.com/news/2014-05/z%C3%BCrich-meets-new-york`

How to cite this book chapter:

Dobler, C. 2016. New York City. In: Horni, A, Nagel, K and Axhausen, K W. (eds.) *The Multi-Agent Transport Simulation MATSim*, Pp. 453–456. London: Ubiquity Press. DOI: http://dx.doi.org/10.5334/baw.75.

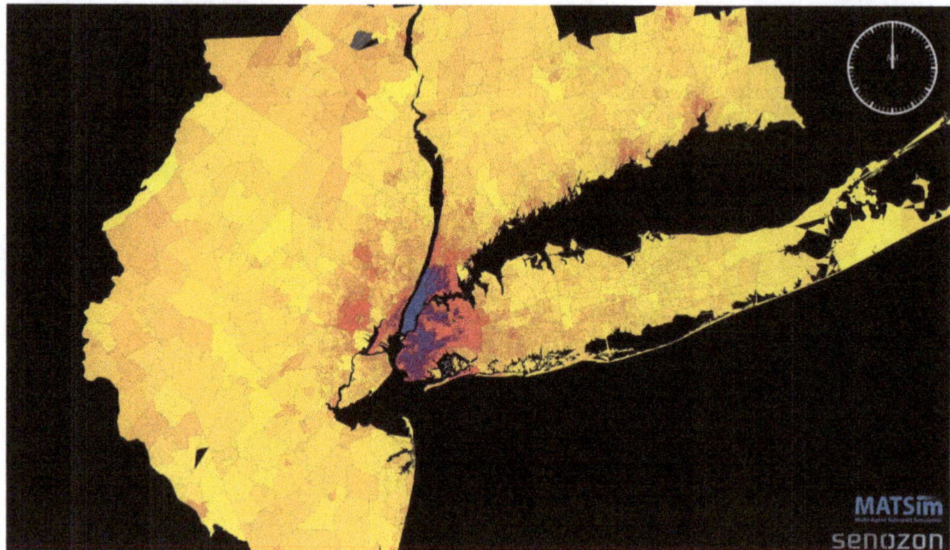

Figure 75.1: Car share (entire modeled area).

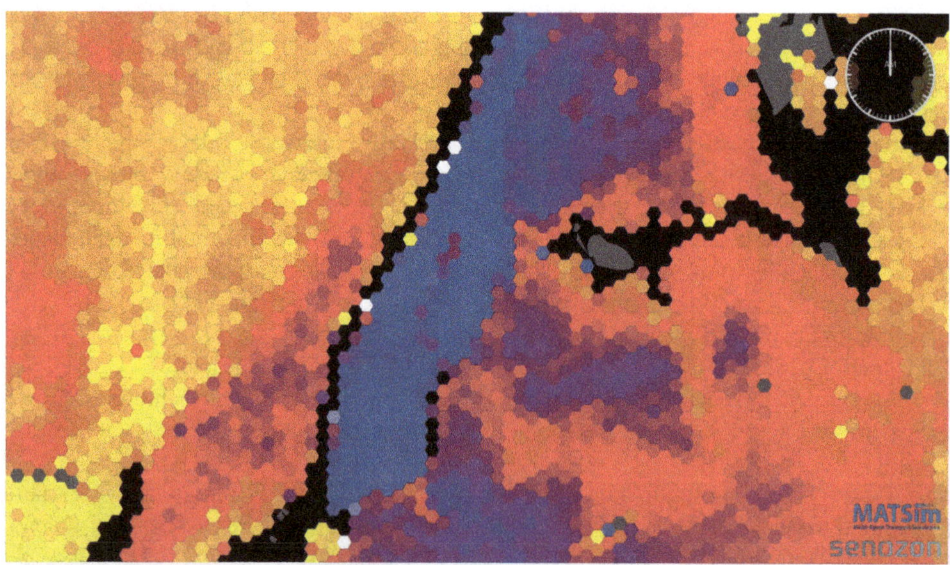

Figure 75.2: Car share (Manhattan).

Figure 75.3: Traffic flows in Lower Manhattan.

CHAPTER 76

Padang

Gregor Lämmel

The Padang scenario demonstrates the MATSim application to large-scale evacuation problems. The scenario has been created as part of the third party funded project "Last-Mile". Taubenböck et al. (2013) give a comprehensive overview. Padang is located on the west coast of Sumatra Island, Indonesia. In 2014, the city had a population of about 1 000 000 people. Because of its problematic location on the coast in a so-called "seismically locked" area (McCloskey et al., 2010), Padang is prone to earthquakes and subsequent tsunamis. In the "Last-Mile" project, a realistic tsunami scenario, triggered by an earthquake about 300 km off the coast, was identified (Goseberg and Schlurmann, 2009). The assumed tsunami would leave about 30 minutes for the evacuation. The flooding would reach as far as three kilometers inland, thus threatening up to 330 000 lives. Lämmel (2011) developed a MATSim scenario representing the city with its affected population. One unusual aspect of the Padang situation is the expected universal evacuation by foot; simulating pedestrians with MATSim was a novelty when this project started. The standard simulation model (see, e.g., Section 1.3) was thus adapted to deal with pedestrians. Details are discussed by Lämmel et al. (2009). Another important variation, contrary to most standard transport scenarios, is that network links would flood once the tsunami reached them. Thus, accessibility—flooded or not flooded—of the network links is time-dependent, which is modeled by a time-dependent network (Lämmel et al., 2010). In the time-dependent network concept, link attributes—like *freespeed*—can be changed, while the simulation is running, by precomputed network change events. For the Padang scenario, the network change events have been extracted from microscopic flooding simulation data.

Key Padang scenario facts:

- The network consists of about 6 000 nodes and 17 000 links.
- Synthetic populations for morning, afternoon, and night have been created, containing up to 330 000 agents.

How to cite this book chapter:
Lämmel, G. 2016. Padang. In: Horni, A, Nagel, K and Axhausen, K W. (eds.) *The Multi-Agent Transport Simulation MATSim*, Pp. 457–458. London: Ubiquity Press. DOI: http://dx.doi.org/10.5334/baw.76.

Figure 76.1: Evacuation time analysis for downtown Padang. Numbers showing average evacuation time in minutes, which are also indicated by the colors green, yellow, red.

- The flooding is modeled by a set of 109 network change events (one per minute), affecting 7 609 links.
- A set of 42 shelter buildings, which could be used for vertical evacuation, is also part of the scenario.

Based on the Padang scenario, various evacuation strategies have been investigated:

- A seemingly obvious evacuation strategy is the shortest path solution, where everyone is on the shortest path. This solution, however, ignores possible congestion and lead to unfeasible results.
- Shorter evacuation times are achieved with a Nash equilibrium approach, where everyone tries to find an optimal evacuation route through iterative learning (Lämmel et al., 2009).
- While the Nash equilibrium reduces individual evacuation time, total evacuation time might not be minimal. The marginal social cost-based simulation approach tries to minimize the total evacuation time (Lämmel and Flötteröd, 2009; Dressler et al., 2011).
- These three basic evacuation approaches are investigated in combination with flooding (Lämmel et al., 2010; Lämmel, 2011).
- Further, an evacuation strategy to reduce the exposure to risk has been developed by Lämmel et al. (2011).
- And finally, Flötteröd and Lämmel (2010) propose a method to integrate shelter buildings, which are evacuation sinks (i.e., safe places) with limited capacity, into the simulation.

CHAPTER 77

Patna

Amit Agarwal

Patna is a medium-size city in eastern India. As in other developing nations, traffic conditions are heterogeneous, composed of: a large number of bikes (37 %, including 4 % cycle rickshaws) and motorbikes (14 %). When this scenario was composed, public transport accounted for 18 % and walk for 29 %; only 2 % of all trips were made by car. Therefore, the MATSim queue simulation was modified to simulate travel demand under mixed traffic conditions (Agarwal et al., 2015b).

A detailed Patna scenario description can be found in Agarwal et al. (2013). The scenario was created using household survey data from a comprehensive Patna mobility plan (TRIPP et al., 2009), using the area within the Patna Municipal Corporation. The scenario consisted of 72 zones, with a population of about 1.57 million (year 2008). MATSim demand was generated using trip diaries, with car, motorbike and bike used as main congested modes (Figure 77.1). PCU (Passenger Car Unit) factors for different vehicle types were derived using effective area occupied by vehicles. The effective area occupied by a vehicle is calculated, and the ratio of area occupied by this vehicle to the area occupied by a passenger car is taken as PCU factor for the respective vehicle. To allow overtaking of slower vehicles (bike), by faster vehicles (car and motorbike), pre-existing, state-of-the-art FIFO queue simulation was overridden, using earliest link exit time as shown in Figure 77.2. Traffic behavior in modified queue simulation was then analyzed by plotting fundamental diagrams and space time trajectories for car, motorbike and bike (Agarwal et al., 2015b).

To address some special factors of Patna's travel time distributions, MATSim utility function was calibrated so that a mode share from real world data was replicated in the model, performed by allowing agents to switch modes. The model was validated using traffic count data and modal travel time distributions. The model's main shortcoming seemed to be overly short average travel times for motorbikes. Although no specific experiment was performed to analyze computational performance, no noticeable loss of performance was found during simulations. Thus, the model seems to be useful for many areas where mixed traffic conditions predominate.

How to cite this book chapter:

Agarwal, A. 2016. Patna. In: Horni, A, Nagel, K and Axhausen, K W. (eds.) *The Multi-Agent Transport Simulation MATSim*, Pp. 459–460. London: Ubiquity Press. DOI: http://dx.doi.org/10.5334/baw.77.

Figure 77.1: Patna: Various vehicles on network, car in red, motorbike in blue and bike in green.

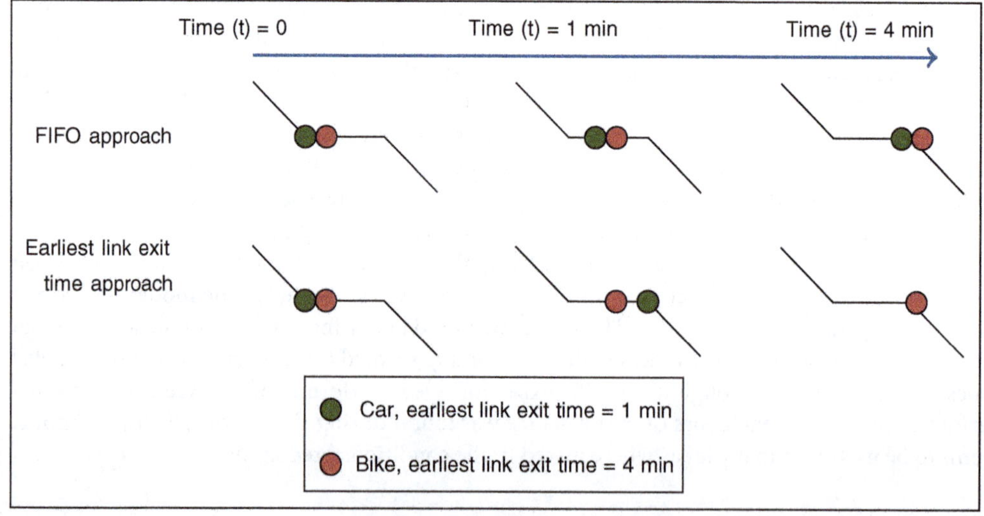

Figure 77.2: Patna: FIFO approach and passing of bicycle by car on a link (not to scale).

CHAPTER 78

The Philippines: Agent-Based Transport Simulation Model for Disaster Response Vehicles

Elvira B. Yaneza

This study's primary aim was adapting an agent-based traffic simulation model to assist planning agencies in determining road traffic routes for DRVs (Disaster Response Vehicles) in crises or disasters. After the initial disaster event period, road network management is crucial for disaster response operations, which must cope with travel demand increase. Depending on level of road damage, sections of the the road network may close. The degraded DRVs road traffic routes will result in longer travel times.

The model was developed using an agent-based simulation modeling paradigm implemented through MATSim. Road traffic routes were generated using Dijkstra's shortest path algorithm. MATSim output files stored each agent's routes, which represented traffic routes for DRVs; here, each route's calculated travel time was equivalent to each agent's running time (in actual motion, while using shortest paths from source to destination).

78.1 Literature Review

Road traffic routing studies generally use different modeling approaches and shortest path algorithms. In studies using modeling, Lefebvre and Balmer (2007) used MATSim for large-scale agent-based transport simulation, also investigating variations of Dijkstra's algorithm and A*-algorithm. Sumalee and Kurauchi (2006) used the Monte-Carlo simulation approach to approximate network capacity reliability, then evaluated traffic regulation policy performance, using the Kobe city (Japan) road network. Teknomo (2008) multi-agent simulation modeling approach considered route probability as a direct simulation output, rather than input, to the network. Sanders and Schultes (2017) outlined algorithms with faster run times than Dijkstra's algorithm for transportation network route planning. Their study focused on successful speedup techniques in static

How to cite this book chapter:
Yaneza, E B. 2016. The Philippines: Agent-Based Transport Simulation Model for Disaster Response Vehicles. In: Horni, A, Nagel, K and Axhausen, K W. (eds.), *The Multi-Agent Transport Simulation MATSim*, Pp. 461–468. London: Ubiquity Press. DOI: http://dx.doi.org/10.5334/baw.78.

road networks with fixed edge cost. Elalouf (2012) model incorporated joint analysis of expected route time and its variance, using dynamic-programming, shortest path algorithm as a basis for a fully polynomial time approximation scheme.

78.2 Design Details and Specifications

Element 1: Study Area During Tropical Storm Washi (Sendong), areas most affected areas were those near the Cagayan de Oro river (Ramos, 2011). Landslides near river banks, flash floods, as well as the overflowing river and its tributaries, caused some barangays (barrios)—already damaged by Tropical Depression Shanshan (Crising)—to be swept away (Del Rosario, 2011). The five most affected major bridges cross along the Cagayan de Oro River, connecting its two main areas, District 1 (west) and District 2 (east), in Misamis Oriental province (see Figure 78.1). The designated road network coverage has a total area of approximately 73.2 square kilometers, including the riverside (see Figure 78.2).

Element 2: Road Network and Facilities The model involved three main entities: road network, facilities and population and is described by two variables: nodes and links. It used graphical representation and had 3 847 nodes and 9 630 directed links (see Figure 78.3). A specific stretch of street consisted of nodes and links, representing intersections and street sections, respectively. MATSim handles only one-way links; in this model, one-way attribute had a default value of 1 and modes attribute were assigned only as car. Facilities were represented by their geographical coordinate locations in the network, which involved 21 entities from the following agencies: 10 hospitals with ambulance services, 3 fire stations, 8 police stations and 2 evacuation centers. Facilities were mapped on nearest road network links.

Element 3: Population and Demand Generation The population was classified into different types of DRVs, representing major agents in the traffic simulation model: ambulances, fire trucks and police cars. The hospitals, fire stations and police stations were assigned as agents' origins,

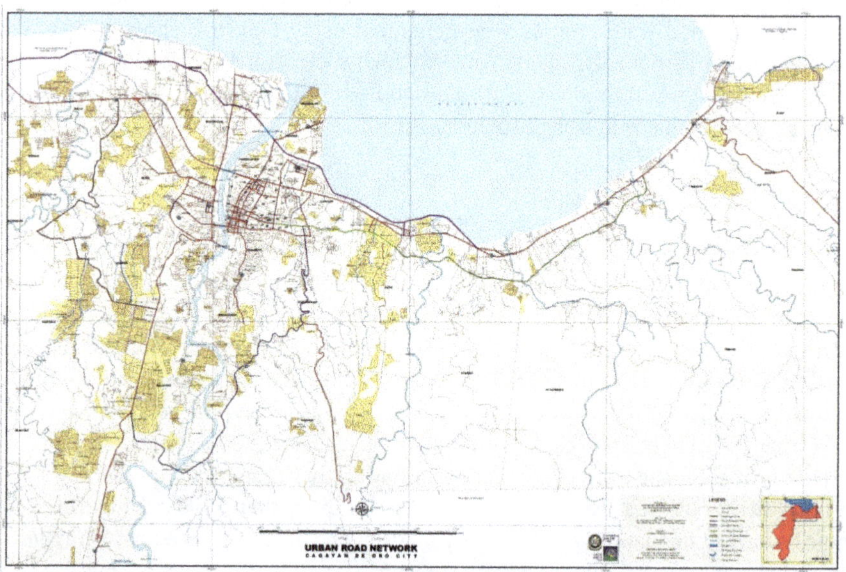

Figure 78.1: Cagayan de Oro City, Philippines urban road network.
Source: GIS City Planning Office, 2012

Figure 78.2: Spatial coverage of road network and locations of facilities in the network: it has 73.2 square kilometers including land and surrounding river and coastal areas. The facilities are mapped based on its actual geographical x and y coordinates in the road network. There are 23 facilities located in its nearest link in the network. These are: 10 hospitals, three fire stations, eight police stations and two evacuation centers.

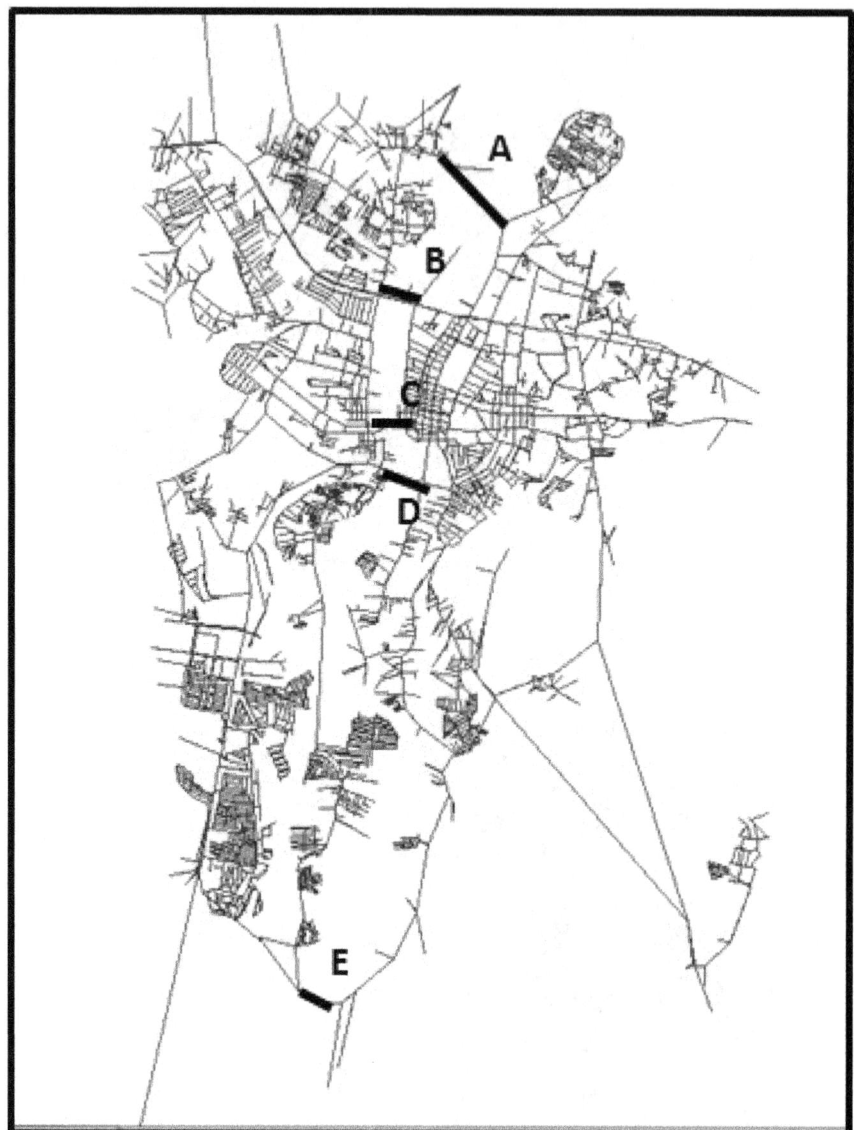

Figure 78.3: Nodes and links representation: Road Network has 3 837 nodes representing road intersections and 9 630 links representing the streets. It includes five major bridges along Cagayan River: (A) Kauswagan-Puntod Bridge, (B) Maharlika Bridge (formerly known as Marcos Bridge), (C) Gov. Ysalina Bridge (formerly known as Carmen Bridge), (D) Kagay-an Bridge (Rotunda Bridge) and (E) Emmanuel Pelaez Bridge.

where vehicles start and end their activities; evacuation centers were assigned as agent destinations. Population was characterized by four variables: person, plan, act and leg. The leg variable used a mode defining vehicle type, assigned as car. The model advanced by performing traffic routing activities. Each traffic routing activity, seven events, was processed in the following sequence: end activity event, agent departure event, wait to link event, enter link event, leave link event, agent arrival event and start activity event. The end activity event prompted the agent to depart from the origin facility and begin again in the same flow of events.

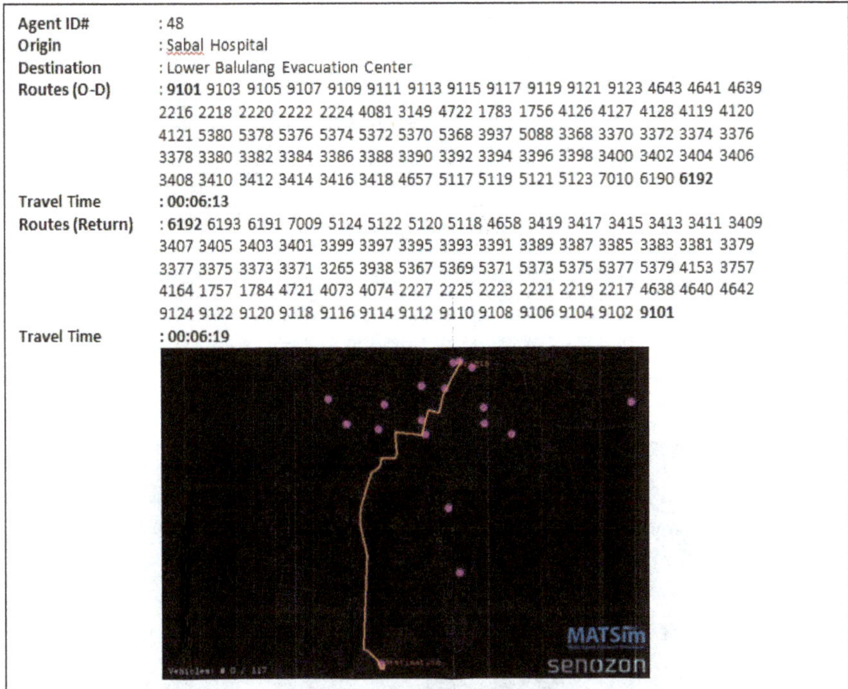

Figure 78.4: Screenshot of SCENARIO 1 (without bridge closures) using agent ID#4. DRVs trip starting from the Sabal Hospital (Origin) passing Carmen Bridge (Gov. Ysalina Bridge) going to Balulang Evacuation Center dropping point (destination) then back to its origin.

78.3 Model Scenarios

The simulation model was applied to the network of Cagayan de Oro City in Philippines. Two scenarios were assumed.

Scenario 1: No Bridge Closures The scenario was based on disaster response operations right after the disaster occurred; operations took place in Cagayan de Oro City. The scenario had two evacuation centers identified, (1) Balulang Elementary School Evacuation Area, located at the west side of Cagayan de Oro and (2) Burgos Barangay Hall Area, located on the east side of the city. The road network had 21 facilities as agents' origins, with 3 to 4 DRVs in each, dividing the network into 2 different evacuation centers. A total of 67 DRVs joined operations over time, as well as 50 additional vehicles from private institutions, traveling on their own rescue operations with different origins and destinations. No road obstructions were considered; traffic could access all five bridges defined in the network (see Figure 78.4). During the simulation run, DRVs were expected to cross the nearest bridge on their trips to destinations or evacuation areas: thus, using only shortest time traveled routes.

Scenario 2: With Bridge Closures In this scenario, road obstructions were represented as bridge closures in the network. The link IDs of bridges expected to close were required in data needed to run the java class for road closure generation. In the experiment performed, the link IDs for three bridges were entered; Carmen Bridge, Rotunda Bridge and Marcos Bridge. The same two evacuation areas and fifty additional vehicles were considered in the experiment and this time, only three bridges constituted road obstructions. The DRVs and other vehicles were expected to cross only the two remaining bridges (not included in the road closure generation): Taguanao Bridge and

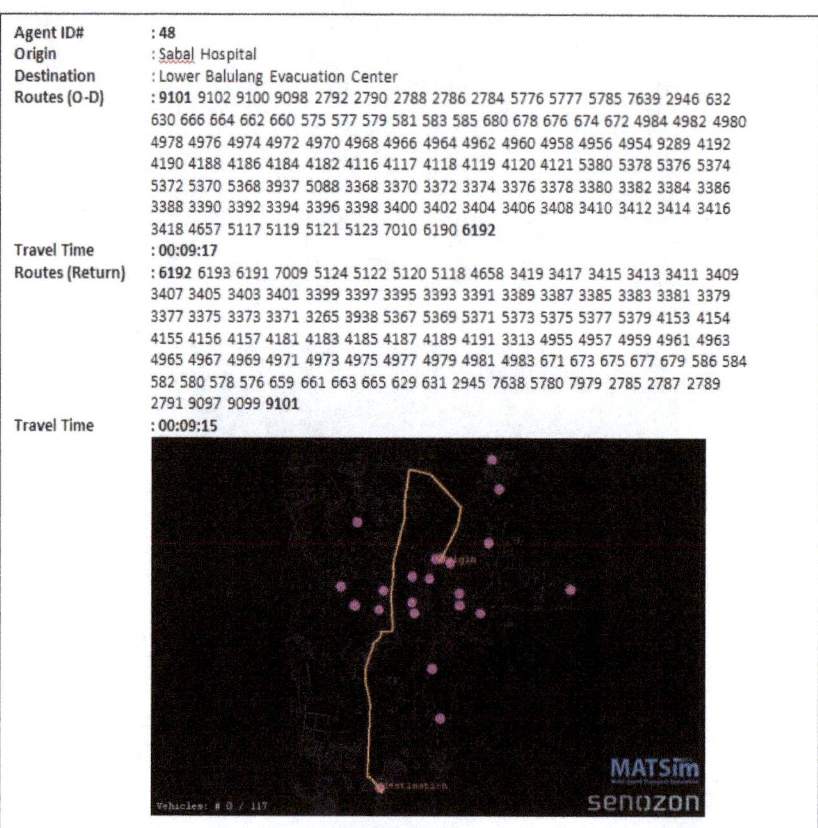

Agent ID#	**: 48**
Origin	: Sabal Hospital
Destination	: Lower Balulang Evacuation Center
Routes (O-D)	: **9101** 9102 9100 9098 2792 2790 2788 2786 2784 5776 5777 5785 7639 2946 632 630 666 664 662 660 575 577 579 581 583 585 680 678 676 674 672 4984 4982 4980 4978 4976 4974 4972 4970 4968 4966 4964 4962 4960 4958 4956 4954 9289 4192 4190 4188 4186 4184 4182 4116 4117 4118 4119 4120 4121 5380 5378 5376 5374 5372 5370 5368 3937 5088 3368 3370 3372 3374 3376 3378 3380 3382 3384 3386 3388 3390 3392 3394 3396 3398 3400 3402 3404 3406 3408 3410 3412 3414 3416 3418 4657 5117 5119 5121 5123 7010 6190 **6192**
Travel Time	**: 00:09:17**
Routes (Return)	: **6192** 6193 6191 7009 5124 5122 5120 5118 4658 3419 3417 3415 3413 3411 3409 3407 3405 3403 3401 3399 3397 3395 3393 3391 3389 3387 3385 3383 3381 3379 3377 3375 3373 3371 3265 3938 5367 5369 5371 5373 5375 5377 5379 4153 4154 4155 4156 4157 4181 4183 4185 4187 4189 4191 3313 4955 4957 4959 4961 4963 4965 4967 4969 4971 4973 4975 4977 4979 4981 4983 671 673 675 677 679 586 584 582 580 578 576 659 661 663 665 629 631 2945 7638 5780 7979 2785 2787 2789 2791 9097 9099 **9101**
Travel Time	**: 00:09:15**

Figure 78.5: Screenshot of SCENARIO 2 (with bridge closures) using agent ID#48. DRVs trip starting from the Sabal Hospital (Origin) passing Kauswagan-Puntod Bridge going to Balulang Evacuation Center dropping point (destination) then back to its origin.

Kauswagan-Puntod Bridge. Expected vehicle flow occurred, as seen during visualization output; see Figure 78.5.

78.4 Validation

Face Validation from Field Experts The goal was to verify and validate whether the simulation model reasonably represented the real-world system and its conformance to design and operational behavior specifications. Four domain experts were invited from the fields of: traffic engineering, computing, planning and management for face validation. Two evaluators were invited from the academy; one was a transportation engineering and built-environment specialist, the other a computer scientist. The other two evaluators were from local government units: one handled management and administration as a technical supervisor from the Road and Traffic Administration Office and the second was a planning coordinator with the Cagayan de Oro City Planning Office. Whether accepting or rejecting, the field experts evaluated the simulation model based on their areas of expertise. Generally, the four evaluators verified and accepted the simulation model design specifications, as well as validating and accepting its operational behavior.

Travel Time Validation Using Test Car Technique and Simulation Model Results When the plans file was scrutinized, from both scenarios, calculated travel time resulting from the simulation was actually equal to the running time when the vehicle was in motion. Running time was computed as equal to the difference between travel time and stopped time delay. Actual measurement of travel time and delay, using test car technique (Sigua, 2008) and travel time, using the simulation model, were compared. Delay time was the time lost by traffic due to traffic friction, traffic control devices and geometric designs. The actual running time computed was only 36 % of actual total travel time measured, due to of travel time delay. The difference between actual running time computed and running time from the simulation model was mostly caused by vehicle speed ranges.

78.5 Achieved Results

Scenario 1: No Bridge Closures Based on the generated events file, there were 667 directed links used by agents representing the DRVs, about 6.9 % of the total 9 630 directed links in the network. The events file stored all activities of 117 agents, 67 agents represented the DRVs and 50 agents represented the other vehicles. Finally, when no bridge obstruction occurred, the DRVs coming from 86 % of the entities crossed the Carmen Bridge. For faster road traffic access, it was suggested that the Carmen Bridge be restricted to DRVs during disaster response, together with the 667 directed links.

Scenario 2: With Bridge Closures Results showed that there were 841 directed links used by agents representing the DRVs, about 8.7 % of the total 9 630 directed links in the network. Note that three bridges (i.e., Marcos Bridge, Carmen Bridge and Rotunda Bridge) were considered for road closures. DRVs originated from 90 % of entities who crossed Kauswagan-Puntod Bridge. It was thus suggested that this bridge, and the 841 directed links, would be in the running when restricting routes for exclusive use of DRVs.

78.6 Conclusions

This study showed that the simulation model reasonably represented of the real-world system, as verified and validated by the four field domain experts and results confirmed the exclusive traffic routing system through the shortest path routes generated by Dijkstra's algorithm. The results were useful tools for traffic management decision-makers when determining traffic routes for exclusive use of disaster response vehicles.

Acknowledgements The author wishes to acknowledge the guidance and information received from the developers of MATSim, Prof. Dr. Kai Nagel of VSP at the ILS, in Berlin, Germany and Dr. Marcel Rieser of Senozon AG in Switzerland. The author would also like to thank Engr. Gerardo S. Doroja for several discussions we had when implementing this model.

CHAPTER 79

Poznan

Michal Maciejewski and Waldemar Walerjanczyk

At the time of the initial scenario, Poznan (population of over 550 000), was the fifth largest city in Poland; together with the neighboring suburban area, it made up an agglomeration inhabited by nearly one million people. The MATSim scenario development for the Poznan agglomeration began in 2012, and the model has been continuously extended and improved. Currently, it is a 24-hour microscopic model of private transport, with a goal of creating a 24-hour, multi-agent activity-based simulation of the Poznan agglomeration, combining both private and public transport.

The road network model was extracted from OSM and included all roads and link roads (such as entrances or exits from motorways). The final result was a high-detail road network model consisting of 17 026 nodes and 40 129 links. This model was calibrated to determine traffic flow parameters for links (e.g., flow capacity, storage capacity, free-flow speed) for each of the 13 modeled road classes (Piatkowski and Maciejewski, 2012).

The travel demand model was derived from the official trip-based 4-stage model used by the Poznan city planning department; this model dates back to 2000, but has been frequently updated since then. Since the official model was originally designed for morning and afternoon peak hours, it had to be extended to describe travel demand throughout the day, hour after hour. As a result, demand for private transport is represented by 24 sets of hourly O-D matrices, each set consisting of nine different matrices, one for each of nine travel motivations, namely home → work/education/shopping/other, work/education/shopping/other → home, and not related to home. This adds up to 216 O-D matrices (Piatkowski et al., 2014; Maciejewski et al., 2014).

The official model divided the agglomeration into less than 400 zones, insufficient for activity locations to be accurately modeled at the microscopic level. To increase accuracy, OSM land use data was used. Six types of land use—residential, industrial, green, commercial, schools and unclassified—were used to subdivide zones into homogenous subzones. As a result, home activities were located in residential subzones, education activities at schools, shopping in residential or commercial subzones, and so on. Figure 79.1 illustrates the distribution of *home* locations when land use was taken into account Piatkowski and Maciejewski (2013).

How to cite this book chapter:

Maciejewski, M and Walerjanczyk, W. 2016. Poznan. In: Horni, A, Nagel, K and Axhausen, K W. (eds.) *The Multi-Agent Transport Simulation MATSim*, Pp. 469–472. London: Ubiquity Press. DOI: http://dx.doi.org/10.5334/baw.79. License: CC-BY 4.0

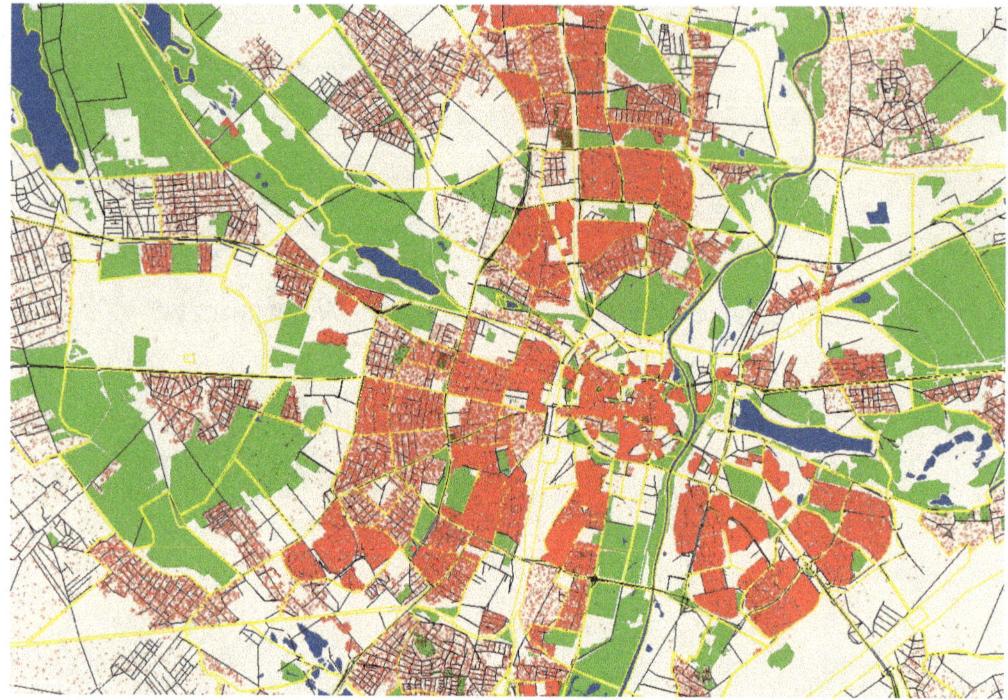

Figure 79.1: Distribution of home activities based on land use.

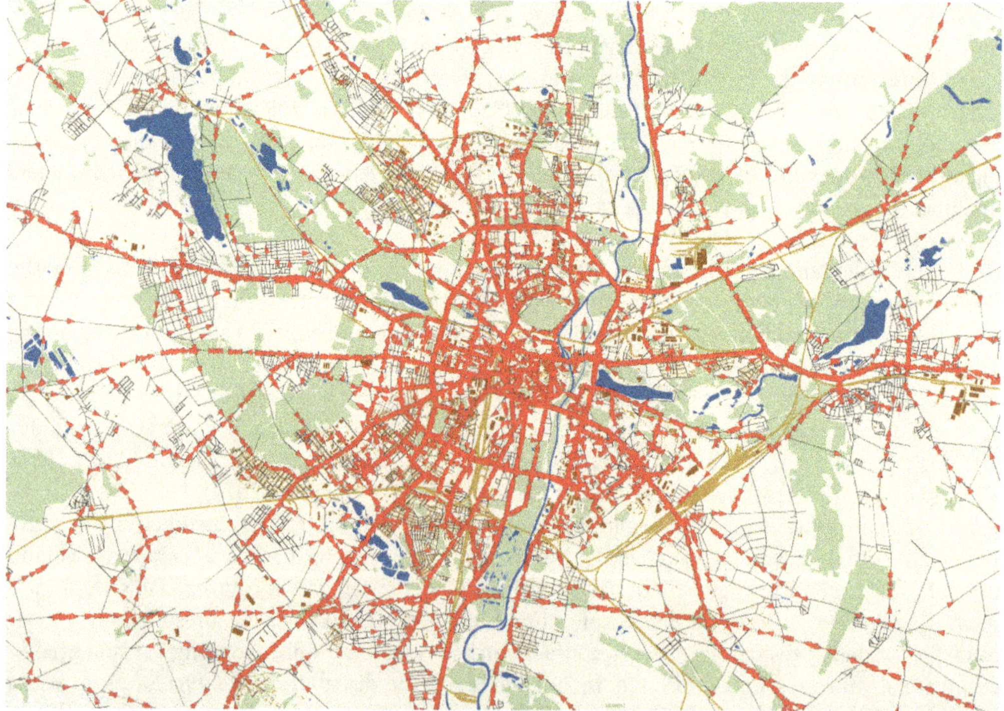

Figure 79.2: Road traffic in the Poznan agglomeration at 7 am.

Having calculated the O-D matrices for private transport and subdivided the area into homogenous subzones, the next step was to generate agents population. In the first attempt, it was assumed that each agent performed only one trip, so the number of agents equaled the demand represented by the O-D matrices, which was almost 840 000. Departure times were randomly distributed (uniform distribution) over each hour, and therefore, the only decision made by each agent during the replanning phase concerned the route choice for the preselected pair of locations. The whole simulation consisted of 120 iterations, yet it usually takes about 60 iterations to achieve a relaxed state. Figure 79.2 shows the state of traffic at 7 am.

Currently, the model is being updated according to a comprehensive travel study carried out in 2014. At the same time, the public transport system is being added, allowing for simulation of both private and public transport. The Poznan model has been used for simulation of real-time electric taxi dispatching, done through the DVRP contribution (see Chapter 23).

CHAPTER 80

Quito Metropolitan District

Rolando Armas and Hernán Aguirre

DMQ (Quito Metropolitan District, Ecuador) has grown rapidly in recent years, with increasing traffic congestion, gas emissions, pollution and energy use. Our research integrated evolutionary computation, traffic simulation, emission models and data mining tools to gain a better understanding of DMQ's complex mobility and transportation system and propose sustainable solutions.

As a first case study (Armas et al., 2014), we implemented a mobility scenario to optimize traffic lights under congested conditions. We focused on the DMQ's business district, an area covering 7x3 square kilometers, as shown in Figure 80.1. The area included only the primary and secondary pathways, where free speeds ranged from 30 to 80 kilometers per hour. The network had approximately 1 000 links and was derived from Geofabrik and OSM. 20 000 agents were simulated, each with a mobility plan consisting of three main trips: (1) home to work, (2) work to leisure and (3) leisure to home (see Figure 80.2). The plans were designed so that all agents moved first from south to north, completely crossing the geographical area of study. In their second trip, the agents moved from north to the central zone of the area under study and in their last trip, from the central zone to the south. Eleven signal lights were located on a main two-way street with flows in south-north and north-south directions (see Figure 80.1).

The evolutionary algorithm (the SOP (Signal Optimizer)) together with MATSim found optimal signal settings of the DMQ scenario, minimizing average travel time. First, we ran MATSim for 500 iterations, to ensure it reached a user equilibrium state without setting any traffic signals. After that, the SOP evolved a candidate solution population for a number of generations. Each solution represented a configuration of signals (signal control) for the transportation system. At each generation, the SOP called MATSim for each candidate solution to evaluate it. MATSim started from the equilibrium state, setting its signals controls with the tentative solution provided by the SOP and ran one additional iteration. This iteration's output was used to calculate travel time, which converted and feed back to the SOP as the fitness value. Figure 80.2 illustrates the

How to cite this book chapter:
Armas, R and Aguirre, H. 2016. Quito Metropolitan District. In: Horni, A, Nagel, K and Axhausen, K W. (eds.) *The Multi-Agent Transport Simulation MATSim*, Pp. 473–476. London: Ubiquity Press. DOI: http://dx.doi.org/10.5334/baw.80.

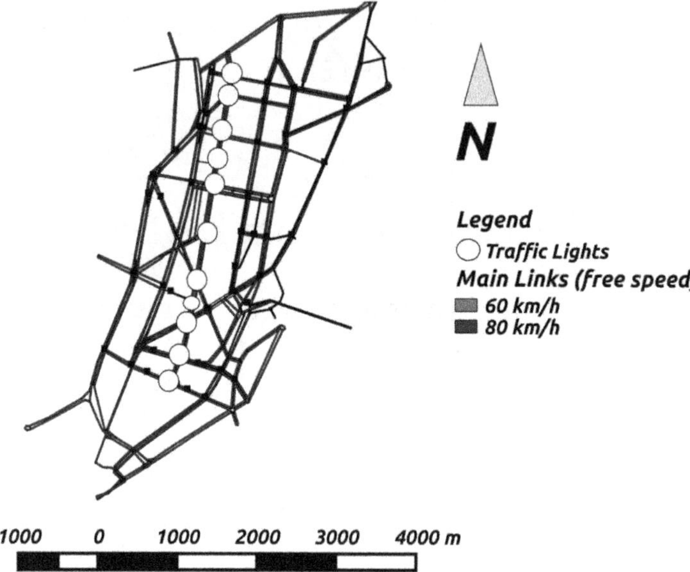

Figure 80.1: Study area.

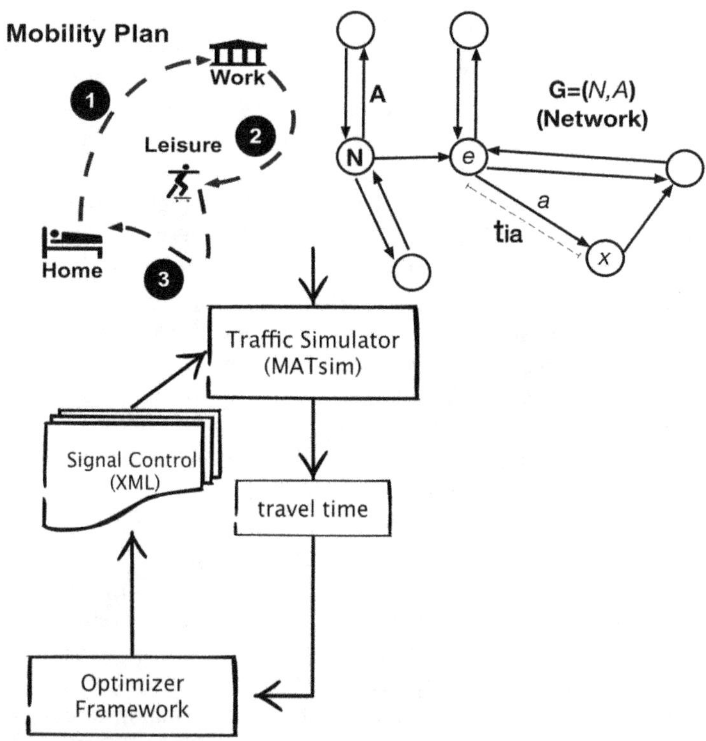

Figure 80.2: Optimization system.

interaction of MATSim and the SOP. The first case study (Armas et al., 2014) provided valuable insights into optimal traffic light setting in the business district of DMQ under congested conditions. This allowed us to validate problem representation used in the SOP and effectiveness of the mutation and recombination operators implemented to search solutions.

Currently, we are scaling up the number of traffic signals to be optimized and testing other mobility scenarios in the same area of study. Our next step is to incorporate an emissions model and use multi-objective evolutionary algorithms (Aguirre et al., 2013) to evolve optimal transportation and mobility system designs of the DMQ, satisfying multiple criteria for sustainability. These criteria include transportation and mobility policies, accessibility, reduction of emissions, reduction of energy use, as well as social and economic benefit.

CHAPTER 81

Rotterdam: Revenue Management in Public Transportation with Smart-Card Data Enabled Agent-Based Simulations

Paul Bouman and Milan Lovric

In Lovric et al. (2013) and Bouman et al. (2012), we proposed two scenarios for studying public transportation revenue management via time-based pricing strategies, like peak markups and off-peak discounts, currently being used by various transit agencies. To evaluate this approach, we developed agent-based simulations using MATSim and a transportation demand generated from smart-card data collected in a Dutch urban area. In the first scenario, we simulated only a metro network, while in the second scenario we considered a multimodal network, consisting of metro, tram and bus.[1]

In Lovric et al. (2013), we designed and implemented a decision support system for sustainable revenue management to evaluate the impact of various revenue management strategies on economic, social, and environmental performance. Figure 81.1 shows the decision support system structure built on top of the MATSim framework. Smart card transactions (individual check-in and check-out transactions made at stations' entrance and exit gates) were used to reconstruct individual passenger's daily tours. These were inputted into MATSim as initial demand; information about the transit network and vehicle schedule was extracted from the OSM data and the public transit operator's web site, respectively. Revenue management experiments were then

[1] This research was conducted at Rotterdam School of Management and supported by Netherlands Organisation for Scientific Research (NWO) Complexity Grant No. 645.000.001 awarded to Dr. Ting Li and Prof.mr.dr. Peter Vervest from Rotterdam School of Management. It was presented at MATSim User Meetings in 2011 and 2012, INFORMS International 2012 Beijing, the 7th Workshop on Agents in Traffic and Transportation at AAMAS 2012 Valencia, Erasmus University Rotterdam, Berlin Institute of Technology, Tsinghua University and Beijing Jiaotong University.

How to cite this book chapter:

Bouman, P and Lovric, M. 2016. Rotterdam: Revenue Management in Public Transportation with Smart-Card Data Enabled Agent-Based Simulations. In: Horni, A, Nagel, K and Axhausen, K W. (eds.) *The Multi-Agent Transport Simulation MATSim*, Pp. 477–480. London: Ubiquity Press. DOI: http://dx.doi.org/10.5334/baw.81.

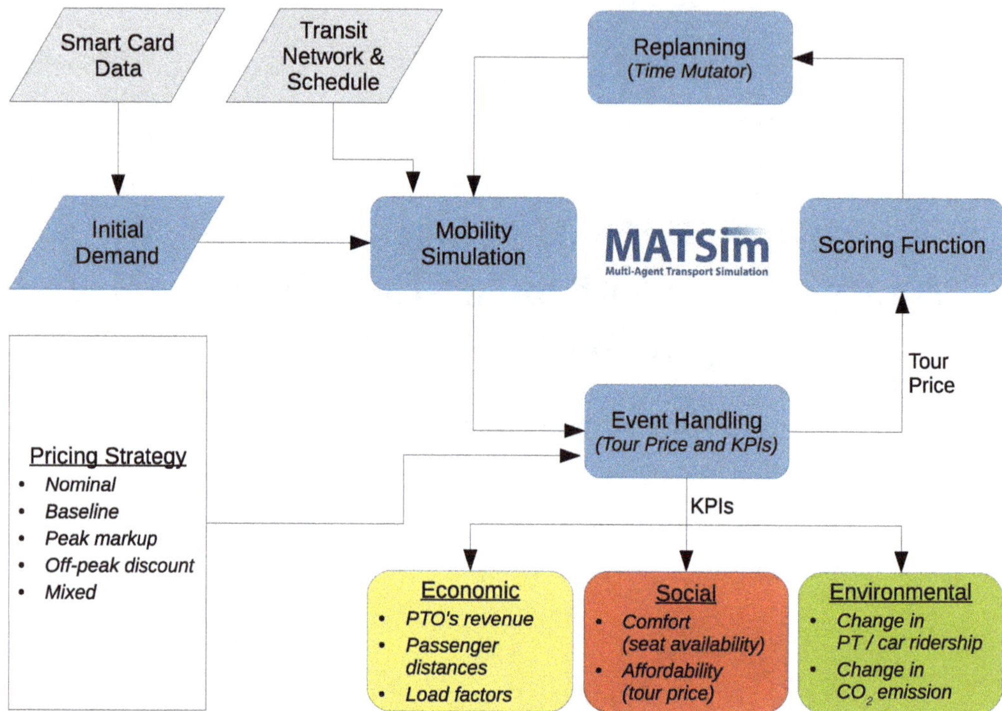

Figure 81.1: Rotterdam scenario: Decision support system for sustainable revenue management in public transportation.

conducted by applying various time-based pricing strategies defined as percentage-wise discounts or markups (applied on top of the nominal price) during specific periods of the day.

MATSim's loop (Section 1.2) was adapted for studying time-based pricing. First, an event handler was created to calculate whole daily tour travel fare for each individual (this was implemented using the real-life pricing scheme: a fixed fee applied at check-in, plus a variable distance-based fee applied at check-out). Second, we adapted the original Charypar-Nagel scoring function (Charypar and Nagel, 2005), by adding travel fare disutility. The existing MATSim's time mutator was used as a replanning strategy, allowing passengers to discover more affordable travel times when pricing strategies were enforced (however, a trade-off was introduced by applying a penalty for arriving outside the expected arrival window, based on the observed smart card data check-out times).

To capture revenue management impact on the three sustainability dimensions, we added event handlers to produce a number of relevant KPIs (Key Performance Indicators). The economic performance was measured by PTO (Public Transit Operator), passenger kilometers revenue and vehicle load factors. Social performance was measured by seat availability (a proxy for passenger comfort), calculated from vehicle loadings after the mobility simulation. We also looked at average tour price as the measure of public transportation affordability. Impact of a pricing policy on the environment was expressed as the change in carbon footprint occurring through a demand shift between public transportation and private cars (calculated from average tour price change and demand elasticity).

Our results showed that, by using a smart-card enabled decision support system and taking a customer-centric view, PTOs can better explore feasible solutions in a broader policy-making context that includes three dimensions of people-profit-planet sustainability. We validated our approach by comparing the simulation-generated travel fares in the nominal scenario with actual fares recorded in the smart card transactional database (see Lovric et al., 2013).

To further study smart card data opportunities in demand generation, Bouman et al. (2012), we introduced a pattern-based demand generation method using three different modalities' (metro, tram and bus) smart card transactions in a Netherlands urban area as input. In addition to using single day observations to generate activity-based demand, daily commuting patterns detected from longitudinal observations for a single smart card were generated. In this study, generated demands were utilized to analyze time-shifting behavior under two different revenue management policies: a plain tariff (with a fixed price per journey and a price per unit of traveled distance) and an off-peak discount. The experiment was repeated for different levels of pattern-based demand, where the varied parameter was the number of observed samples required for a smart card to be included.

In generated demand, agents not generated using pattern-based demand had to replicate their observed tour or trip within 15 minute windows of observed arrival and departure times. Pattern-based agents had time windows dependent on observed standard deviations in passengers' actual commuting travel patterns, which were used as a proxy for their time flexibility. This aspect of demand modeling was more detailed than Lovric et al. (2013), where agents were assumed to be homogeneous about their time flexibility. This flexibility was exploited by the time shift mutator, made available in MATSim as one of the replanning strategies. In future work, improvements in scoring function and use of more sophisticated pattern-based demand generation approach must be considered to create more realistic scenarios for a study of time-shifting behavior under revenue management policies.

CHAPTER 82

Samara

Oleg Saprykin, Olga Saprykina and Tatyana Mikheeva

82.1 Study Area

Samara is a major Russian city, regional capital of the Samara region, situated on the left bank of the Volga River between the mouths of the Samara and Sok rivers. The area is 466 square kilometers, made up of nine administrative districts, with a city population of 1 172 348 people (year 2014). There are more than 2.7 million people living within the city agglomeration (GKS, 2010).

Personal and public transportation are developed to varying degrees in Samara city. Automobilization of the population is 286 vehicles per 1 000 people (year 2014) (Gradoteka, 2015). Public transport consists of trams, buses, trolleybuses and subway. Transit of freight through the city is prohibited.

Samara is a major economic, transport, scientific, educational and cultural center. However, despite this, the city's street and road network is insufficiently developed, leading to the following problems.

- The street and road network has only two highways, which are connected by narrow streets; there are no transverse highways, resulting in traffic congestion. According to research from the Yandex company, Samara city was in fourth place for number of traffic jams in Russian cities (Yandex Company, 2013).
- Lack of sufficient parking areas leads to parking along city roads, creating additional traffic congestion.
- Active construction development in 2000, characterized by absence of an overall city building strategy, led to obvious violations in transport planning and significantly degraded transport infrastructure quality.
- Samara is located opposite the Samarskay Luka National Park, a region of unique natural beauty, but a destination for a huge number of summer weekend recreational trips, leading to uneven traffic flow distribution in the region.

How to cite this book chapter:
Saprykin, O, Saprykina, O and Mikheeva, T. 2016. Samara. In: Horni, A, Nagel, K and Axhausen, K W. (eds.) *The Multi-Agent Transport Simulation MATSim*, Pp. 481–484. London: Ubiquity Press. DOI: http://dx.doi.org/10.5334/baw.82.

In addition to these problems, Samara city is currently attempting to move toward more sustainable development, which raises new challenges:

- rapid growth of residential development within the city boundaries requires transport infrastructure modification,
- formation of new neighborhoods and new cottage villages within the urban agglomeration involves the construction of new roads, bridges and interchanges,
- hosting the FIFA-2018 World Cup requires traffic management organization in the downtown, stadium and festival/fans areas.

These issues and developments require street and road network modernization, impossible without traffic flow simulation modeling to support the projects.

82.2 Transport Demand

Population residence coordinates were taken from anonymized city population spatial distribution information provided by the National Population Census 2010 (GKS, 2010). Place of employment coordinates about Samara region companies and organizations were based on data from address directories.

Statistic package R was used for O-D matrices calculation; initial data relied on collected information about population distribution and employment locations. The estimation of O-D matrices was performed by the entropy model using the Shelehovsky-Shtskiy balance method (Nurminski et al., 2014; Autodor, 2013; Shvetsov, 2003). This approach is applicable for estimation of the O-D matrices values in case worker, business or recreation trips for private vehicles or freight transport. The O-D matrix was then obtained, which showed number of agents moving from one transport zone to another.

Activity chains were calculated for define path of each MATSim agent. Activity chains calculation was performed by a custom-developed method, using the author's algorithm described in Saprykina and T. Mikheeva (2012). Activity chains calculation uses O-D matrices as source data and resulting data was kept in the plans file and used in MATSim.

82.3 Transport Supply

As shown in Figure 82.1, the road network was extracted from OSM and saved to the MATSim network format, using the `NetworkEditor` module presented in Chapter 10. Detailed verification of the obtained network model revealed that some roads have incorrect number of lanes, requiring the writing of a utility that semi-automatically allowed for adjustment of the street and road network model according to the actual transportation planning scheme. Minor model inaccuracies were corrected manually in the `NetworkEditor`. The final network model consists of 4 365 nodes and 11 178 links.

The network model should contain transport infrastructure elements for adequate transportation planning reflection. The model takes into account certain traffic signs: speed limit, traffic lanes, movement on the interceptions and "no entry". Addition of traffic lights to the model is under development now. At this point, traffic light regulation schemes at specific intersections have been developed; work on their integration into the general city model is underway.

Transport simulation was performed only for private vehicles; Inclusion of public transport to the model is in process. Bicycle paths are still poorly developed in the city; therefore, their simulation is ow-priority.

Figure 82.1: The transport network extraction process.

82.4 Calibration and Validation

For calibration purposes, information about traffic flows at all intersections of Krasnoglinskoye highway and Voljskoe highway, as well as the intersections of central (historical) part of the city, was used. Traffic flow intensity data was received for the period from 19th to 24th of May 2009. Source data required pre-processing, which consisted of vehicle number alignment according to their type and calculation of total intensity in the target area. Maximum intensity requirement was utilized because intensity measurements were performed during “rush hours” from 8 am to 11 am and from 4 pm to 7 pm (Mikheeva, 2008).

For transport infrastructure mode validation and verification of its accuracy vs. real traffic conditions in the city, the following steps were completed:

- traffic flow parameters field measurements,
- data gathered from different traffic Web-services (Yandex Maps, Google Maps, etc.),
- comparative analysis of results obtained from the simulation, field explorations and Web-services (Saprykina and Saprykin, 2014).

82.5 Intelligent Traffic Analysis

The simulation results analysis is especially valuable to solve the relevant problems. With MATSim’s tools Senozon Via (Chapter 33) and OTFVis (Chapter 34), visual analysis of the model can be performed. However, a deeper understanding of the model can be achieved by applying data mining tools to simulation results to identify hidden patterns and correlations, supplying more information to address applied problems.

At this point, the simulation output folder contains files with events and actual plans, containing all actions performed by agents. For loading the data to the mathematical package R, they were converted into `.csv` format through specially designed utility and MS Excel applications. Transport infrastructure information from external sources was also imported to R as a table, containing the coordinates and types of the object. This made it possible to process the MATSim output using all power of the programming language R.

The search for hidden patterns was performed using the `NeuralNet` package installed in R. One of the goals was finding dependencies of tension at transport flows’ gravity points from transport infrastructure spatio-temporal parameters. To solve this problem, a feed-forward neural network was used, trained by resilient back-propagation with weight backtracking algorithm. Source data was split into training and test sets in a 70/30 ratio. Verification was carried out by the regularity criterion (Mikheeva et al., 2012).

The study produced a trained neural network, able to predict gravity points' tension during changing transport infrastructure parameters. This eliminates the need to restart the simulation to test the hypotheses for city transport infrastructure changes, allowing an overview of changes on the fly. Figure 82.2 shows how the trained neural network displays the tension calculation process at the intersection.

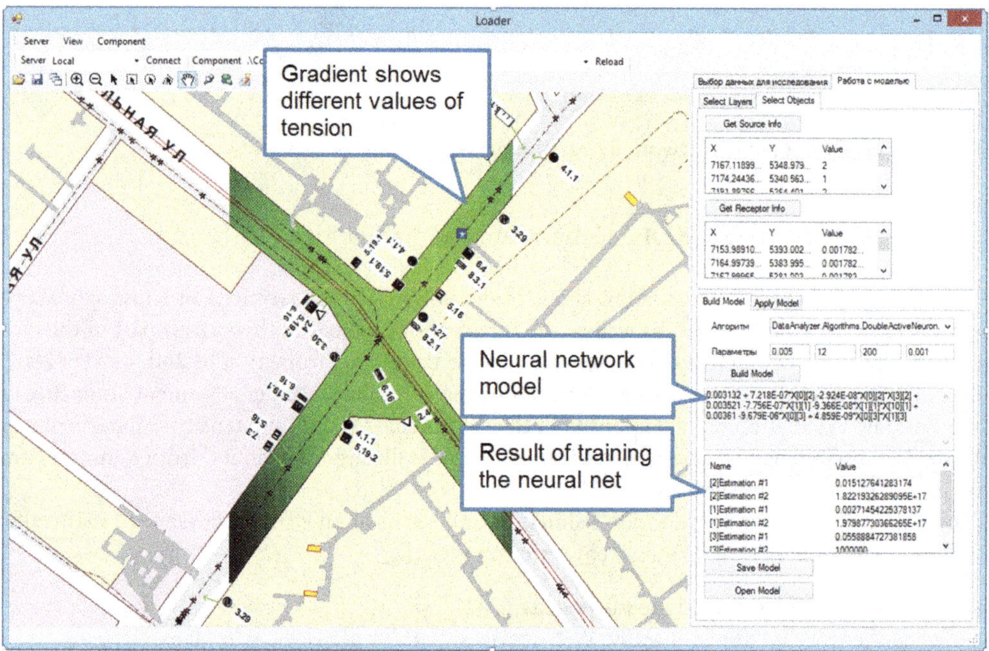

Figure 82.2: The tension calculation process by the trained neural network.

CHAPTER 83

San Francisco Bay Area: The SmartBay Project - Connected Mobility

Alexei Pozdnoukhov, Andrew Campbell, Sidney Feygin, Mogeng Yin and Sudatta Mohanty

83.1 Introduction

Novel mobility-as-a-service paradigm, enabled by ICT and mobile computing, is changing the transportation landscape faster than traditional data sources, such as travel surveys, are able to reflect. The development of on-demand transportation, the rising popularity of car- and ride-sharing services and the growing tendencies towards multi-modality pose new challenges for supply side modeling. This is particularly true in the San Francisco Bay Area (California, USA) as the influx of people and businesses to the city, volatility of job markets, evolving demographics and internal migration further increase the variability of mobility patterns evolution. It is more important than ever to be able to measure, realistically model and forecast travel demand in near real-time. The baseline scenario of the SmartBay project spans the nine counties in the area and is designed to extend the state-of-the-art in activity-based simulations in two respects. First, the SmartBay's demand model is based on the anonymized cellular network infrastructure data stream. Second, agents' population is connected to a social network and their scoring functions are tailored to study the implications social influence exerts, particularly in mode and secondary destinations locations choice.

83.2 The Study Area and Networks

The baseline SmartBay simulation implements a typical working day scenario within the nine San Francisco Bay area counties. As of 2015, total area population is 7.5 M people, with an estimated 3.4 M commuters, of whom 350 K use public transport as their only commute mode. Driving is the major mode for home to work trips, with 75 % of trips made by a driver alone. While average commute duration is estimated to be 28 minutes, severe congestion at peak hours is widespread.

How to cite this book chapter:
Pozdnoukhov, A, Campbell, A, Feygin, S, Yin, M and Mohanty, S. 2016. San Francisco Bay Area: The SmartBay Project - Connected Mobility. In: Horni, A, Nagel, K and Axhausen, K W. (eds.) *The Multi-Agent Transport Simulation MATSim*, Pp. 485–490. London: Ubiquity Press. DOI: http://dx.doi.org/10.5334/baw.83.

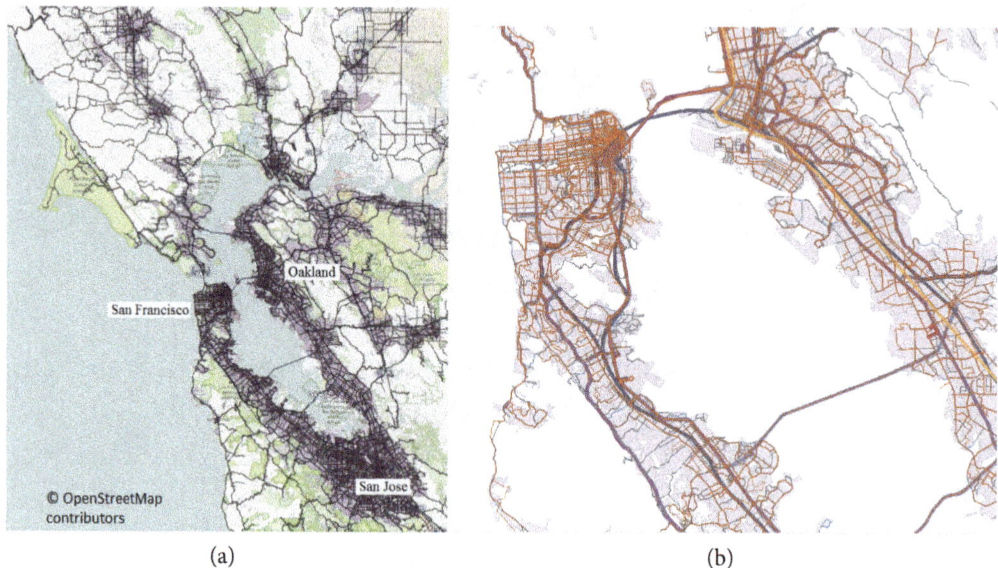

Figure 83.1: The geographical extent of the SmartBay simulation (left) and a close-up view on the multimodal network spanning San Francisco-Oakland area (right).

The road network used in the scenario consists of a total of 96 000 links, with a mix of freeways, state routes, all major arterial and countryside roads. Road network geometries were extracted from the OSM data: then verified and augmented with the speed limits, capacities and number of lanes. The network was extended with all major public transit lines available through GTFS, provided by the respective agencies. There are 9 major bus agencies, several minor bus line operators, a light rail system, and commuter trains. The major rapid rail carrier is a Bay Area Rapid Transit system that serves 400 K daily trips over four inter-connected lines. GTFS includes schedules and capacities of transit vehicles.

83.3 Population and Demand Generation

There are 1454 TAZs in the area developed by the MTC (Metropolitan Transportation Commission), used as origin and destination units of a demand model developed and supported by the MTC, as well as for population and workplace projections made on a regular basis for different time horizons. The MTC model adopts the activity-based approach, with a tour-trip hierarchy of mandatory (home, work, school trips) and secondary trips, with the respective mode choices, composition of tours and departure times governed by a rich set of discrete choice models calibrated from recent California Household Travel Survey data (CHTS, 2010-2012) and inherited from other California agencies' relevant studies.

SmartBay scenarios use the anonymized cell phone data logs to adjust MTC demand models. Cell phone data are routinely collected and managed by AT&T Inc., the second largest nationwide telecom operator in the United States with 120 M users nationwide (which translates to a sample size of more than 1 M commuters in the SF Bay Area). Data used for mobility modeling originates from anonymized CDRs, recorded at the spatial resolution of the deployed cell phone towers (or antennas) and is usually available with a time latency of several minutes. Historical CDRs analysis allows detection of important places for each user based on frequency of calls, texts or data packets sent through a given cell tower (Isaacman et al., 2011; Becker et al., 2013). This approach is most robust in identifying primary locations of frequent and recurrent visits, such as home, work or school. The data is stored and processed internally at secure AT&T servers. A rescaling procedure, based on area-to-point pycnophylactic interpolation (Kaiser and Pozdnouhkov, 2013) and a variant of iterative proportional fitting was used to project aggregates from cell tower level to areal units defined by the TAZs. Population census data were used to estimate correction coefficients and adjust cell phone user counts for the total population. This adjustment resulted in an up-to-date and more accurate representation of mandatory trip O-D flows related. When compared with the MTC demand models, notable discrepancies detected include new urban developments, as well as major shifts in employment re-distribution due to the fast IT sector evolution in Silicon Valley.

83.4 Work Commute Model Evaluation

MATSim instance was deployed on AT&T servers to simulate the home-to-work commute scenario for a typical weekday. Scenario runs with 15 % to 30 % commuting population sample were evaluated (550 K to 1.1 M agents). Driving and public transit were set as the only modes; mode share at the beginning of the mode re-planning in MATSim was set according to MTC findings from CHTS. Resulting link volumes were validated based on hourly traffic counts collected by California Department of PEMS (Transportation Performance Management System) inductive loop detectors, deployed on all major freeways. Sample count histograms are presented in Figure 83.2. The model met the Federal Highway Authorities accuracy specifications.

83.5 Extensions and Work in Progress

Main extensions developed in the SmartBay project are related to simulating a population explicitly connected to a social network; current work is directed toward two domains. First, an extension of location choice is approached with machine learning tools that model social influences in destination choices for secondary activities and the second extension introduces social connections to scoring functions and aims to capture peer pressure effects in mode choices.

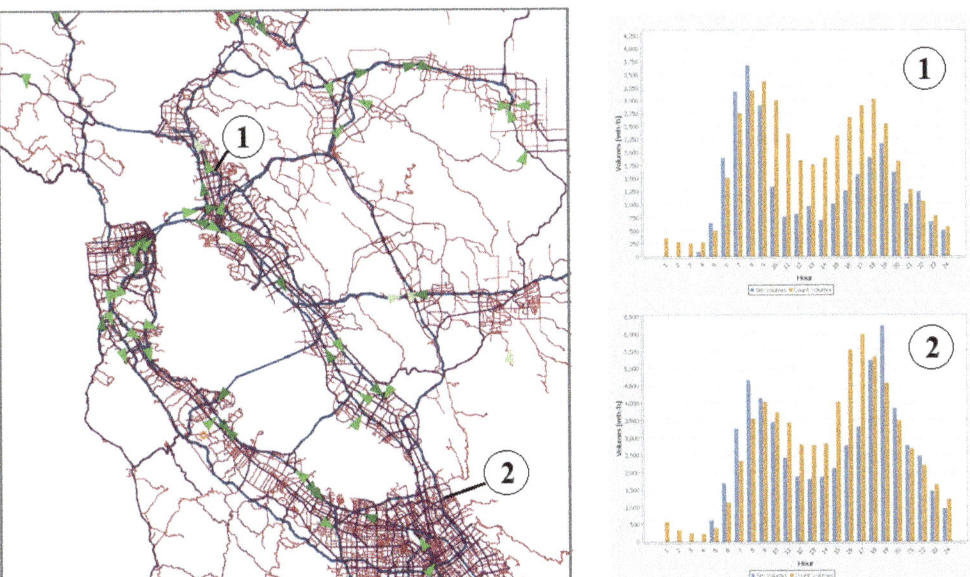

Figure 83.2: A sample of the simulated vehicles and the examples of the observed (light/orange) and simulated (dark/blue) counts at two particular validation locations. Secondary trips, mainly occurring at midday, were not included in this scenario.

Social Influence in Destination Choice There is evidence that population social network geography in an area is a strong predictor of destination choice for secondary trips. This is valid both for trips directly related to social activities, as well as when destination choice was conditioned by recommendations received from peers in the past. As such, this provides a way to use machine learning-based approaches for predicting destination choice from historical data and social ties. This approach requires building a social connections model for the virtual agent population, i.e., defining a weighted graph with edges P_{ij} for each pair of agents i and j. Our preliminary work is based on the model proposed in McGrath and Pozdnoukhov (2014) and is applied at the home level TAZs, instead of an individual. This approach requires a seed network to be derived from the cell phone CDRs, with the weights P_{ij} emphasizing recurrent reciprocal calls, as evidence of a social tie between i and j. The seed network is then removed from the model, resulting in a connected virtual population with similar network statistics that replicates the geographical community's real social network structure in the area.

SmartBay currently adopts the MTC secondary activities classification that includes eight categories for non-mandatory trips. There are 120 K venues derived from the `Factual.com` API, introduced to the simulation as destinations for secondary trips. Hierarchical spatial clustering was applied to the venues set to reduce the number of venues to 1 200. This approach is justified both by the need to reduce computational expenses in the re-planning stage, as well as evidence of spatial hierarchies in human spatial cognition and decision making. A spatial choice model for the secondary home- and work-based trips is calibrated from the CDRs, using the McArdle et al. (2014) approach. A key parameter set in this model is the attractiveness of agent venues, which is assumed to be proportional to the number of peers who also visit the venue. A thorough experimental validation of the full-scale scenario, with secondary trips, is computationally expensive and is ongoing.

Social Influence in Mode Choice The following extension to the conventional Charypar-Nagel scoring function is considered:

$$U_i = U_i^{CN} - \gamma \sum_{j=1}^{N} P_{ij} \left\| a_i - a_i^0 \right\| + \theta \sum_{j=1}^{N} P_{ij} \left\| a_i - a_j \right\|$$

Here, an agent specification is extended with an attribute vector a_i, describing an agent's profile as it relates to membership in a particular group (such as drivers or transit users). We define attribute components as continuous within [0, 1] interval, corresponding to an agent's tendency to drive or take transit as his/her primary commute mode. This attribute value is also used to define the probability of the current plan's primary mode choice to be selected for mutation in the evolutionary optimization re-planning step. U_i^{CN} represents the Charypar-Nagel score of the daily plan, augmented with two terms. The first term describes peer pressure effect toward a pre-specified "socially-responsible" choice a_i^0. The second term describes an agent's tendency to behave similarly to his/her immediate peers in regard to choice attributes. As these two effects appear only with evidence of a social tie, both terms include a summation over the agent peers, with connection strength P_{ij} defined as described in the previous subsection. The resulting mode choice sensitivity to parameter values γ and θ is determined through currently ongoing computational experimentation.

83.6 Conclusions and Acknowledgments

An increasing pace of urbanization severely tests city infrastructure systems. The transportation field is responding to these global challenges by evolving at an ever-increasing pace. More flexible and powerful tools are required to support decision making in planning, operations, and policy regulation applied to emerging mobility technologies. SmartBay project has developed a MATSim-based platform capable of ingesting demand models based on big data and extending the utility functions specifications to study social influence on mobility behaviors. It also incorporates semi-parametric machine learning models applied to destination location choice predictions for socially-related secondary trips. With encouraging results obtained in baseline scenario simulations, these advanced developments are currently ongoing.

The authors acknowledge the contributions from our collaborators at AT&T Research: Dr. J.-F. Paiement, Dr. J. Pang, Dr. A. Skudlark, Dr. C. Volinsky. Funding support from State of California Department of Transportation (CalTrans) through UCCONNECT faculty research grant program, agreement 65A0529, is also acknowledged.

CHAPTER 84

Santiago de Chile

Benjamin Kickhöfer and Alejandro Tirachini

84.1 Introduction

This section describes the creation process of the freely available MATSim scenario of Santiago de Chile. The first version of a calibrated scenario is available online[1] and is documented in Kickhöfer et al. (2016). For the scenario setup, three open data sources are used: (i) car network information from OSM (OpenStreetMap), (ii) PT (Public Transport) supply data from GTFS (General Transit Feed Specification), and (iii) travel diaries from Santiago's 2012 Origin-Destination Survey.

Multiple interventions in Santiago's transport system in the past 20 years make this city an interesting case study for the analysis of alternative transport policies. Santiago has a Metro (subway) network of five lines over 104 kilometers, with two new lines to be launched in 2017 and 2018, adding 37 kilometers to the network. In the city, there is a full-scale integrated public transport system launched in February 2007—the Transantiago system (Muñoz et al., 2014), which has fare integration between all urban buses and the Metro through the use of a single prepaid (smartcard) payment method. There also exists a network of 200 kilometers of tolled urban highways. In winter, the air pollution problem is tackled, in part, by introducing plate-number based car driving bans on the most polluted days. All these elements make Santiago an appealing case study for the application of a metropolitan-scale transport and activity simulator.

---

[1] See `https://svn.vsp.tu-berlin.de/repos/public-svn/matsim/scenarios/countries/cl/santiago/` or search from `http://matsim.org/datasets`.

How to cite this book chapter:
Kickhöfer, B and Tirachini, A. 2016. Santiago de Chile. In: Horni, A, Nagel, K and Axhausen, K W. (eds.) *The Multi-Agent Transport Simulation MATSim*, Pp. 491–494. London: Ubiquity Press. DOI: http://dx.doi.org/10.5334/baw.84.

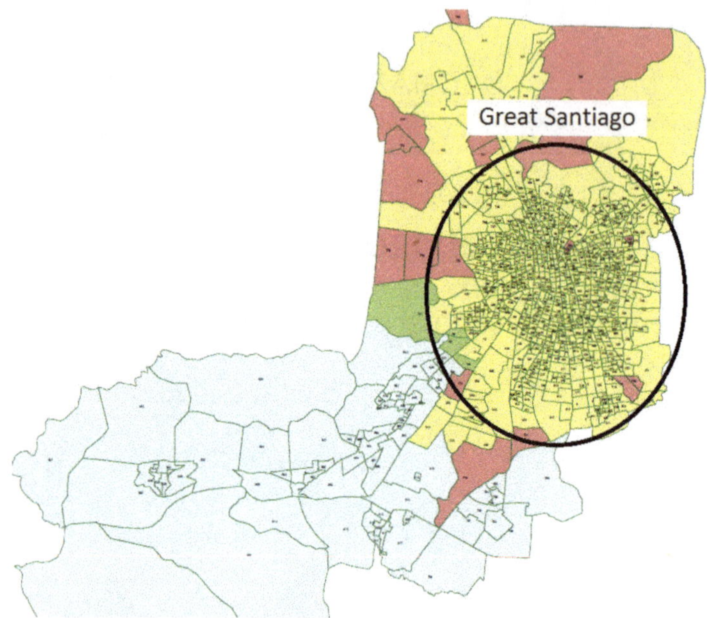

Figure 84.1: 2012 ODS study area and zones, adapted from SECTRA (2014).

84.2 Data

84.2.1 The 2012 origin-destination survey

The travel demand and activity patterns of the MATSim Santiago scenario are based on the travel and activity data collected in the 2012 Origin-Destination Survey (ODS), whose database and results were released to the public in March 2015.[2] The surveyed area encompasses 45 *comunas* (municipalities) of the Santiago Metropolitan Region, with an estimated population of 6.65 million people. The survey goes beyond the Great Santiago Area to include the neighboring municipalities of Colina, Lampa, Pirque, Calera de Tango and Melipilla. The total area has 2 million households with an average of 3.24 persons per household. The sample size is 18 000 randomly chosen households along 866 zones that were defined for the survey. Figure 84.1 shows a map of the survey area and zones. The Great Santiago Area is highlighted by an ellipse, in which 91 % of the population is concentrated.

It is estimated that, on a normal working day, there are 18.5 million trips, from which 38.5 % are by non-motorized means (walking and cycling). Around 25 % of the total trips are made using the Transantiago public transport system, out of which 52.4 % are bus-only trips, 22.2 % are metro-only trips and 25.4 % are combined bus-metro trips. Car travel has a modal share of approximately 26 % of the total trips.

In total, 60 054 individuals were interviewed in the 2012 ODS, with a total of 113 591 trips. Omitting all individuals that do not have two activities plus one connecting trip reduces the sample size to 42 459 synthetic agents (70.7 % of all interviewees). Therefore, considering the population of the whole metropolitan area of the sample (6.65 million), the MATSim synthetic population represents

[2] The survey form, reports and full database are available at the website of Chile's Transport Planning Office (SECTRA), http://www.sectra.gob.cl/biblioteca/detalle1.asp?mfn=3253, accessed 16 August 2015.

approximately a 0.65 % sample, with agents performing activities of the following types: home, at work, work-related, education, health-related, visit someone, shopping, leisure and other.

84.2.2 *Road network and public transport supply*

The source data for the MATSim Santiago road network is taken from OSM. The source data for the Transantiago PT routes and departure times/service frequencies at the stops over time-of-day is a GTFS file[3], published and continuously updated by Santiago's Metropolitan Public Transport Authority (Directorio de Transporte Público Metropolitano, DTPM). The GTFS file includes all bus and Metro services.

From the MATSim transit schedule, a pseudo transit network is created along with the transit vehicles. This transit network connects—for each transit line—the stops directly to each other. It is not connected to the car network, and only follows the car network's geometry where the resolution of transit stops is high (i.e., where a transit line has a stop at every corner). In consequence, cars and buses run in separate networks; as a result it is currently not possible to analyze, for example, cross-congestion effects between modes. Nonetheless, current congestion patterns of PT are exogenously included, since bus travel times are set to be larger in peak periods, calibrated using historical data from buses that are equipped with GPS devices.

84.3 Setting up the Open Scenario

84.3.1 *Scenario specifications*

By converting the input data into MATSim format, several files are generated to run the simulation. Since there are no data restrictions, these files are provided as an open scenario.[4] The code for obtaining this data from the input data is also publicly available.[5] Behavioral parameters are taken from a study by Munizaga et al. (2008) and converted into MATSim parameters (Kickhöfer et al., 2016). When performing mode choice, in the present version of the model, agents are only allowed to switch between the transport modes car, PT and walk. Trips performed by any other mode (bike, colectivo, other, ride, taxi, train) remain fixed but can be included in the choice set in future versions. PT captive users are taken into account since agents are only allowed to use a car if they have access to a car according to the survey data. Otherwise their only options are PT and walk. The attributes of the three different modes considered in the present study are travel time (car, PT, walk) and monetary costs (car, PT). Travel time for car trips is a direct output of the simulation where vehicles interact on the road network. Hence, the car travel time also includes road congestion. Travel times for PT results from the GTFS data (station-to-station travel times including transfer time) plus access and egress times done by the walk mode. Hence, the PT travel times do only partly include road congestion, i.e., as long as it is approximated correctly by the schedule, which uses longer travel times in peak periods. Travel times for walk are approximated by teleporting agents between their activities q and $q+1$ with a travel time of $t_{trav,q} = \frac{1.3 \cdot d_{trav,q}}{4.0\ km/h}$, where $d_{trav,q}$ is the beeline distance between the two activities.

Travel times for all other transport modes are approximated by congested car travel times (for colectivo, other, ride, taxi) or by teleportation similar to the walk mode (bike, train) with different

---

[3] See `http://datos.gob.cl/dataset/1587`, accessed 13 August 2015.

[4] See `https://svn.vsp.tu-berlin.de/repos/public-svn/matsim/scenarios/countries/cl/santiago/` or search from `http://matsim.org/datasets`.

[5] Currently, see `https://github.com/matsim-org/matsim/tree/master/playgrounds/santiago/src/main/java/playground/santiago`.

teleportation speeds (10.0 and 50.0 km/h, respectively). Monetary costs are also approximated. However, as long as switching from/to these modes is not allowed (see next paragraph), this essentially has no effect on simulation results.

84.3.2 Calibration/validation

The Alternative Specific Constants of the different modes are determined in the calibration process. The procedure to run the first simulation with 200 iterations, together with the calibration of the constants is explained in Kickhöfer et al. (2016).

Another standard verification of MATSim simulation output is the comparison of traffic flows to data from real-world counting stations. 49 counting stations are available within the Santiago greater area, 40 on major roads, 9 on (parallel) local roads. The counts data is recorded in July 2011. After cleaning the data, 36 counting stations remain with data from 6:00 am to 11:30 pm in 15 minutes time bins. MATSim traffic output versus observed traffic is analyzed in Kickhöfer et al. (2016), which indicates the need for further calibration efforts once the population is expanded to a 10 % or 100 % sample.

84.4 Conclusion and Outlook

This section summarized a MATSim scenario set up from input data that is open and publicly available. This makes the scenario an interesting tool for transparent decision making of public administrations, for advancing transport modeling and policy research as well as for stimulating innovation activity of the private sector. Possible applications include the (economic) evaluation of planned transport policies and projects and the development of business ideas based on the simulated mobility of individuals in Santiago. A number of future model improvements to be implemented in the scenario are provided in Kickhöfer et al. (2016). A non-exhaustive list of potential research problems to be analyzed with the MATSim Santiago scenario is the following:

- the effects of road pricing strategies on travel times, traffic volumes, public transport and demand for non-motorized mobility, air pollution, noise levels, etc.,
- the introduction of alternative interventions such as (full or partial) pedestrianization of the city center, speed limitations, roads with exclusive right-of-way for public transport, plate-number based car driving restrictions, parking restrictions, road closures and road openings, restrictions on truck traffic, new cycleways and new Metro lines, and
- the extraction of accessibility measures to study the land use impacts of transport interventions.

CHAPTER 85

Seattle Region

Kai Nagel

A MATSim model of the Seattle region—more precisely the PSRC (Puget Sound Regional Council) area—was developed during K. Nagel's sabbatical stay with the UrbanSim team in Seattle in 2008. The model resulted from a prototypical integration of the UrbanSim software (e.g., Waddell et al., 2003) with MATSim.

The base was an existing PSRC UrbanSim model, which used an established EMME/2 model as a travel model. The investigation centered around how difficult it would be to replace the EMME/2 model with MATSim.

The network was taken, by conversion, from the existing EMME network, resulting in 15 478 links and 5 025 nodes with attributes length, free speed, and capacity.

Demand was generated as output from UrbanSim. Evidently, the UrbanSim simulation already contained a full synthetic population. The UrbanSim model was also set up with workplace choice, so that each synthetic person with "working" status had a workplace assigned. Since that version of UrbanSim worked on the parcel level, this meant that home-to-work trips could be extracted directly, with coordinates, from the model. As so often for initial MATSim studies, this home-to-work demand was then extended to home-work-home plans.

The configuration used standard MATSim scoring parameters: a 7 am workplace opening time and latest work start time of 9 am. The iterations were run with re-routing and time mutation enabled until convergence. Since this was an exercise in rapid prototyping, only a 1 % sample of the full synthetic population was used; road network flow and storage capacities were scaled down accordingly. Figure 85.1(a) shows a result. Figure 85.1(b) shows households most affected by a hypothetical closure of the Alaskan Way viaduct, which traverses the Seattle downtown area on the waterfront side to the west.

How to cite this book chapter:

Nagel, K. 2016. Seattle Region. In: Horni, A, Nagel, K and Axhausen, K W. (eds.) *The Multi-Agent Transport Simulation MATSim*, Pp. 495–496. London: Ubiquity Press. DOI: http://dx.doi.org/10.5334/baw.85.

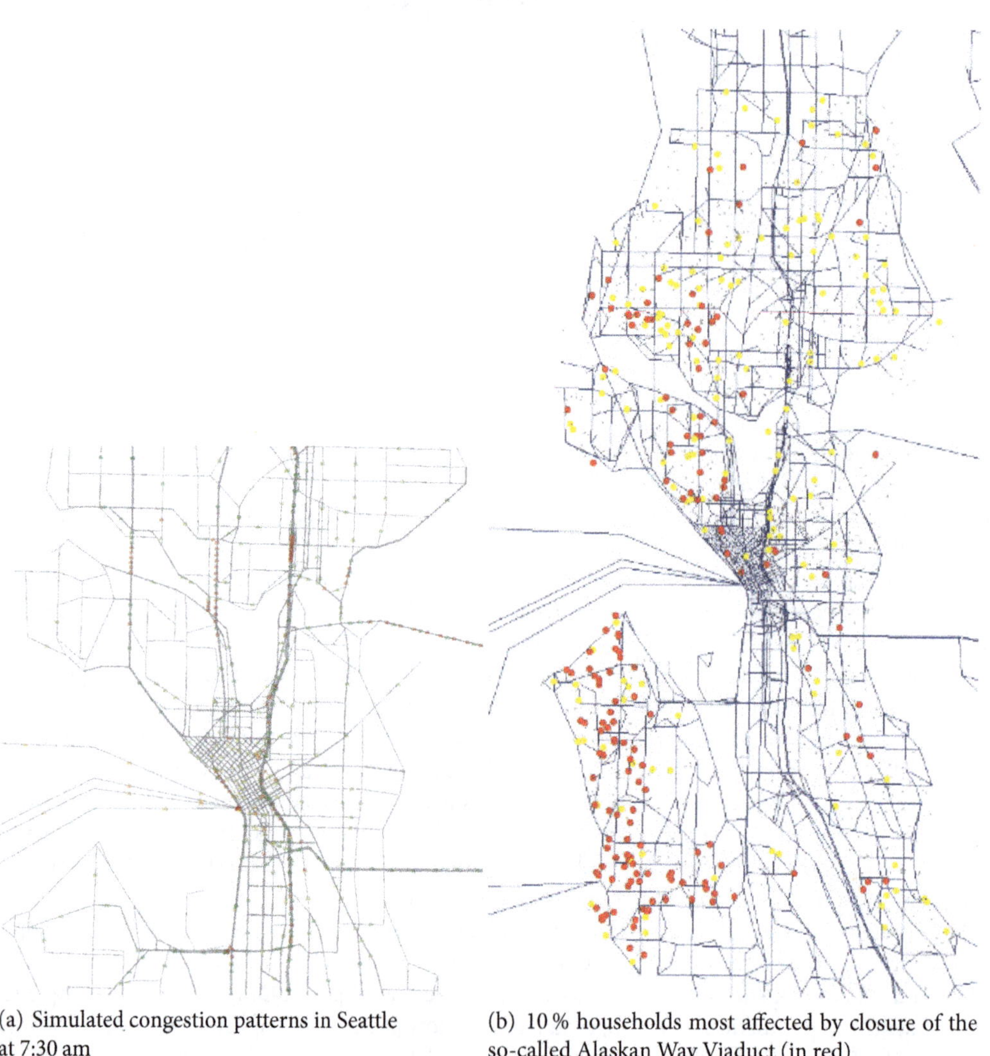

(a) Simulated congestion patterns in Seattle at 7:30 am

(b) 10 % households most affected by closure of the so-called Alaskan Way Viaduct (in red)

Figure 85.1: Seattle region scenario.

CHAPTER 86

Seoul

Seungjae Lee and Atizaz Ali

The MATSim model of SMA was developed in 2012, as a result of long-term research collaboration between the University of Seoul (Prof. Seungjae Lee) & ETH Zürich (Prof. Kay W. Axhausen). The model was updated yearly and demand was generated based on 2012 HHTSD. Demand statistics (input) are summarized as follows.

Study area was the SMA (Gyeonggi-do province, with emphasis on the Seoul Metro, comprised of 25 main administrative districts). A population synthesizer was developed to generate the MATSim input demand, based on HHTSD 2012. Total population of SMA was 21.5 million; therefore, a 10 % sample was generated and simulated (2.15 million agents). A detailed nodes and links network was generated, capturing all details (16 384 nodes and 32 768 links) for railways, highways, arterials, pedestrians, expressways and bus-only lanes. EMME/2 network was converted to MATSim format. The 2012 Korean Transport Database was utilized to generate transit schedules and vehicle definitions, according to bus types, railway and metro lines. Total number of routes was 1 317 (contained regional buses, inter-city buses, feeder line buses and metro lines, etc.). In collaboration with Senozon AG, a more realistic door-door demand was generated in Seoul City in July, 2014. Data source was the Korean GIS department.

In Seoul, MATSim has been widely used for various research purposes to aid policy evaluation Kim et al. (e.g., 2012); Lee and Ali (e.g., 2014).

A master's thesis on transit demand generation and calibration using smart card data in SMA is currently underway by this chapter's second author, sequenced as follows. A video is available from the authors on request:

- data mining (trimming off non-useful data),
- converting disaggregate transactions (O-D) to individual trips and trip segments based on user ID,
- activities inference and assignment in SPSS (Statistical Package for the Social Sciences) database,

How to cite this book chapter:

Lee, A and Ali, A. 2016. Seoul. In: Horni, A, Nagel, K and Axhausen, K W. (eds.) *The Multi-Agent Transport Simulation MATSim*, Pp. 497–500. London: Ubiquity Press. DOI: http://dx.doi.org/10.5334/baw.86.

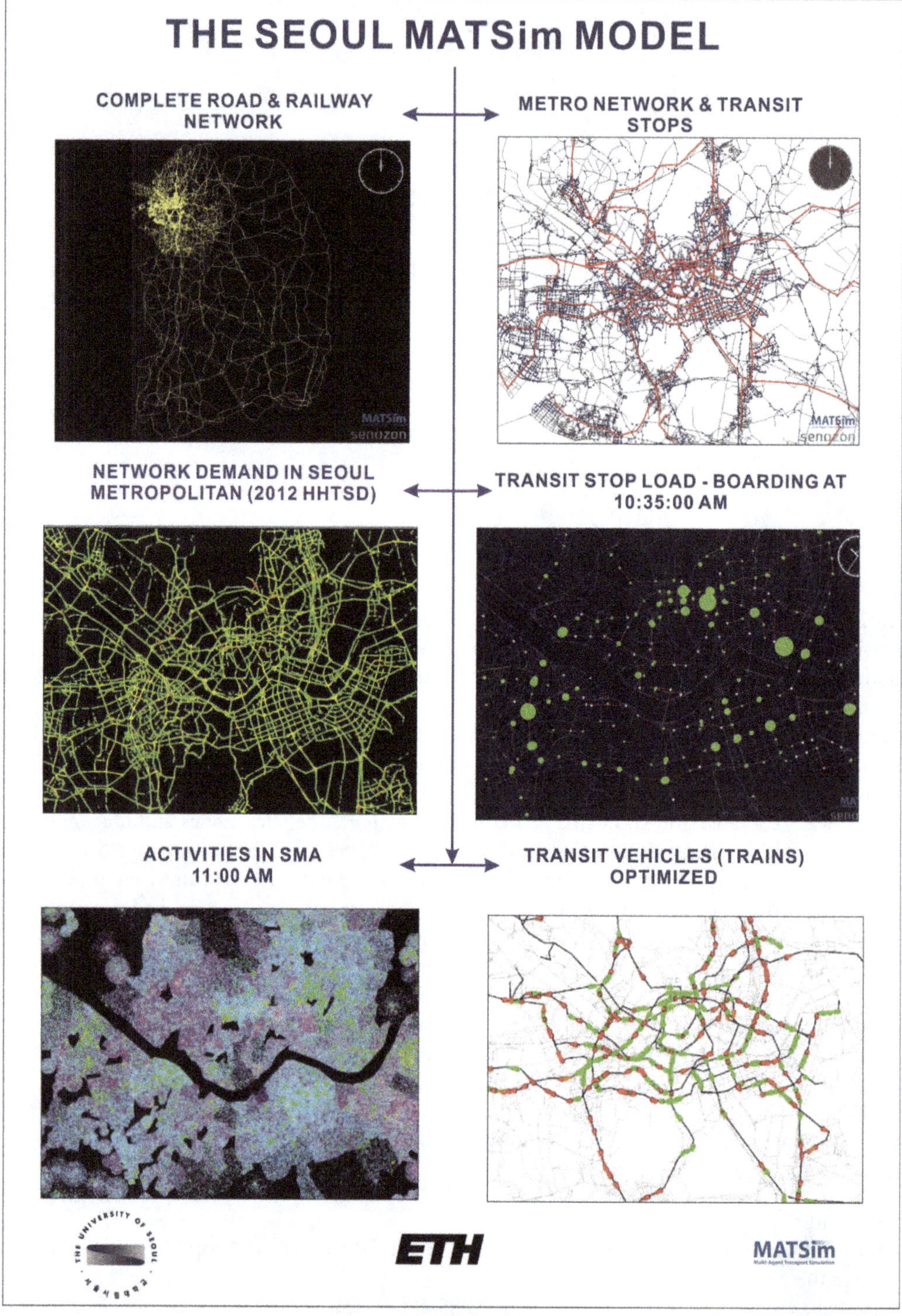

Figure 86.1: Seoul scenario.

- generating transit demand (MATSim input format),
- updated transit network & schedule for running the simulation, and
- model calibration (in process).

MATSim tutorials were also presented during the fall semester 2014 to help Department of Transportation Engineering undergrad and grad students gain a thorough working knowledge of MATSim.

CHAPTER 87

Shanghai

Lun Zhang

Shanghai, with a population of about 20 million and 6 073 square kilometers land area, is the biggest metropolis in China. To fully integrate activity-based demand modeling and further public transport models, the full implementation of MATSim for Shanghai was built to forecast precise traffic demand on network, as well as scientific policy evaluation. The scenario contained 200 000 synthetic persons, simulated on a network with 50 000 links. Shanghai scenario key features are as follows.

A 1 % sample of the actual population, about 0.2 million agents, was used. To generate the population individual with personal attributes, the Monte Carlo method was used to disaggregate available census data from the 4th Travel Survey of Residents.

Demand generation was based on 24 hour O-D matrices generated from the GPS data and synthetic population; the O-D were then disaggregated into individual trips. The activity-based modeling was used to generate initial population plans in five steps: activity chain choice, duration choice, mode choice, destination choice and route choice, where the MNL model was used to estimate and serialize choices of agents. During the simulation, activity replanning were introduced to discern better travel plans; while scoring for a plan was modeled using a utility-based approach.

The Shanghai street network was extracted from the overall OSM network and then merged with the Shanghai expressway network. Road attributes, such as number of lanes per direction, or flow capacities, were set through road classification specification. To simplify the original network, optimization rules were designed to remove unneeded information that increased computational burden.

All facilities from O-D pairs were classified into particular zones using their geographical coordinates. Three main facilities types, home, work/education and leisure, are used. Origin and destination facilities' names were obtained via reverse geocoding; these facilities are classified by their names. The unit resolution of facilities was the hectare, in which facilities and types are randomly created according to their coordinates.

How to cite this book chapter:

Zhang, L. 2016. Shanghai. In: Horni, A, Nagel, K and Axhausen, K W. (eds.) *The Multi-Agent Transport Simulation MATSim*, Pp. 501–502. London: Ubiquity Press. DOI: http://dx.doi.org/10.5334/baw.87.

Simulated modes were as follows. A public transport system, subways and buses, was integrated with both motor traffic and non-motorized traffic. Transit schedules were considered for public transport. A travel time based transport mode choice model—between car and public transport—was developed.

Activity replanning was used to optimize activity plans of agents; stable simulation system state was reached after 100 replanning procedures iterations. The effectiveness of the Shanghai MATSim transport simulation model was validated against observed counts from vehicle detectors and mode split from travel surveys. Extensive simulation results indicate that most traffic simulation volumes matched quite well with observed counts, which demonstrated MATSim's potential for large-scale dynamic transport simulation. It provides researchers and policy makers with a useful tool to evaluate traffic policies.

Specific algorithms integrating new data in Shanghai with MATSim inputs, such as synthetic population, facilities and network, were separately designed according to data characteristics. To see more detailed work about the Shanghai scenario, please see Zhang et al. (2014).

CHAPTER 88

Sochi

Marcel Rieser

Major sport events usually attract huge crowds of spectators, as well as media reporters, necessitating numerous official helpers in various locations to guide and support attendants; naturally, all athletes must also be at the right place at the right time. For large, international contests like Olympic games or soccer championships, accommodations are rarely close to the event facilities, making it necessary to transport spectators, media, helpers and athletes efficiently over long distances. As such events typically run for multiple days, or even weeks, with ever-changing combinations of locations and times where actual competitions take place, substantial planning is required to ensure that all attendants and participants reach their event locations in time.

Masterconcept Consulting GmbH (Gesellschaft mit beschränkter Haftung), an Austrian consulting company, has positioned itself to provide high-level concepts for large sport events. To better serve its clients, it developed ITSOS (Intermodal Transport Simulation & Operation System), a GIS-based system to support its transport planners in the creation of mobility concepts for major events, as well as regional planning. When simulating the planned events, ITSOS depends heavily on MATSim to verify that special infrastructure at major events can handle transport within required time frames, to and from specific event locations.

Senozon AG was responsible for integrating MATSim with ITSOS and adding ITSOS-specific functionality to MATSim. Together, they created a test scenario depicting the 2014 Olympic winter games in Sochi.

88.1 System Overview

ITSOS used ArchGIS for storing and editing infrastructure data, like road and train networks and event facilities. A custom plug-in also provided a graphical user interface inside ArchGIS to specify transit routes and schedules, vehicle types and their assignments to lines and departures, as well as methods describing expected travel demand. Transport planners could create and manage scenarios and scenario variants directly from the custom user interface available inside ArchGIS.

How to cite this book chapter:
Rieser, M. 2016. Sochi. In: Horni, A, Nagel, K and Axhausen, K W. (eds.) *The Multi-Agent Transport Simulation MATSim*, Pp. 503–506. London: Ubiquity Press. DOI: http://dx.doi.org/10.5334/baw.88.

After successful modeling of a scenario in ArchGIS, a planner could export the network and transit schedule in MATSim's XML format directly from ITSOS to a local directory. The travel demand information, consisting of activity-chains, with zone- or facility-references and number of persons having such a chain, was exported as tabular information. A special program created a MATSim population file from this tabular data, along with a default config file.

The user could then start the MATSim simulation, using a simple bat-file on Windows. After the simulation ended, events were preprocessed and imported into a database, from which they could be queried and used within ArchGIS for analysis and visualization purposes.

88.2 Extensions to MATSim

The various groups at major sport events require different handling; in addition to athletes, there are media reporters, officials, helpers, caterers, and, of course, many spectators. Persons from different groups attending the same event will have different requirements about when to be at the event location, what entrance to use for the event location and the kind of transport necessary to reach the location. For this reason, supporting sub-populations for replanning and scoring was an important issue. Different transit offerings were also defined for different agent groups, because spectator mass transport must usually be separate from athlete and official transport.

To facilitate transport planners' work, transit lines in ITSOS were defined with adaptive schedules; given a base headway, additional departures were scheduled between iterations, if high occupancy was expected to occur on a line during specific hours. This adjustment was based on a rule set that ensured a minimum duration for the shorter headway, as well as a minimum duration for the base headway between the shorter headways. Figure 88.1 shows the graphical schedule of an adaptive transit line after 80 iterations.

In addition to private car traffic and schedule-based public transport, athletes, media and officials also use special transportation offerings: shuttle buses, or even limousine services operating on demand, between only two or more fixed locations. Termed "transit on demand" in ITSOS, transit lines with stops along a route were defined, but without scheduled departures. Instead, a within-day-like operator was implemented, scheduling vehicles whenever someone from an agent group wished to depart. The rule-based operator had additional constraints, like minimum occupancy of on-demand vehicles before departure (to prevent every on-demand vehicle transporting only

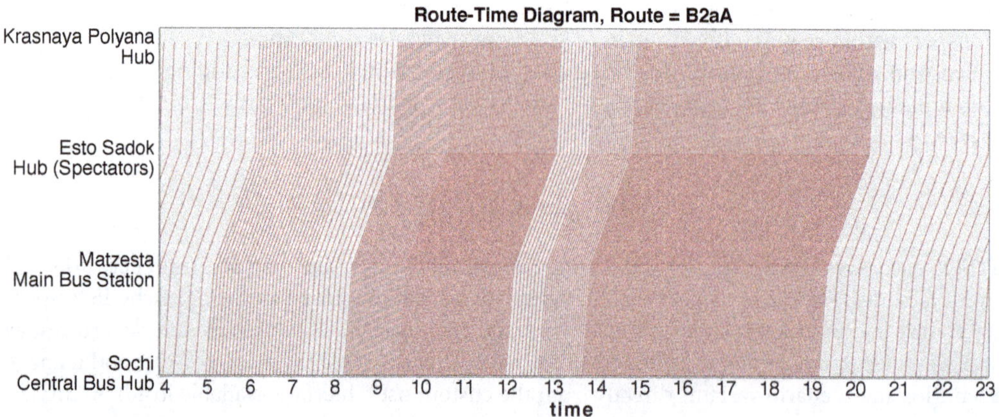

Figure 88.1: Bus schedule with automatically adapted headways based on simulated demand for bus line from Sochi (Central Bus Hub) to Krasnaya Polyana (Hub).

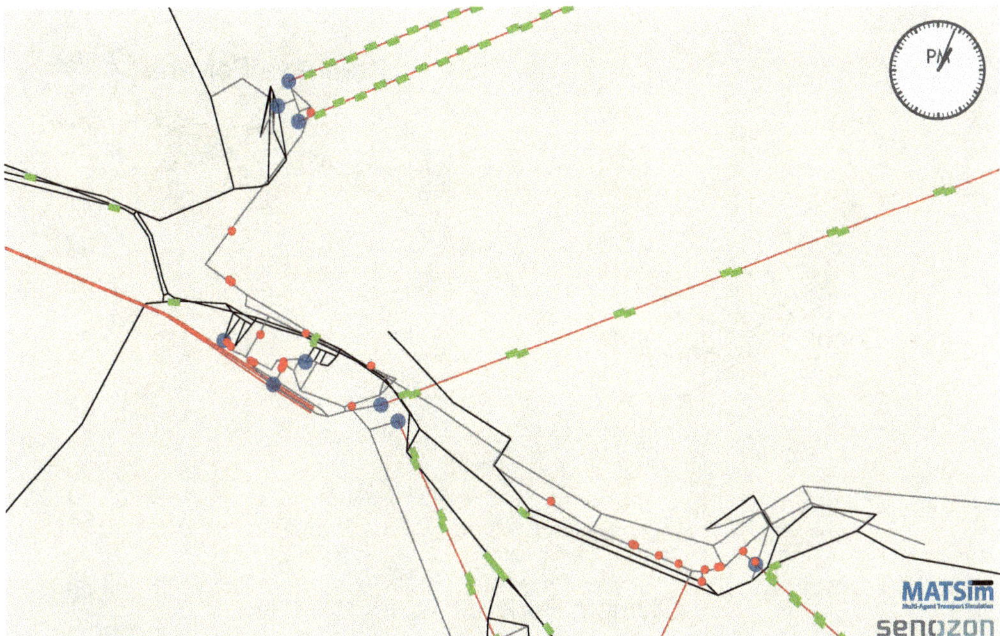

Figure 88.2: Simulated pedestrians (red circles) at Krasnaya Polyana hub. Transit vehicles (incl. cable cars) shown as green boxes, transit stops as blue circles.

a single agent), as well as a maximum waiting time before departure for such vehicles (to prevent agents in remote locations having to wait forever).

At sport events, large number of spectators have to share both common entrances to event facilities and common access paths to those facilities. This made it necessary to simulate more detailed pedestrian flows (in certain places) than just the default teleportation approach typically used by MATSim. For Olympic games, this was even more crucial because, in several locations, security checks created additional bottlenecks. This requirement was solved by implementing a special router for the walk mode, along with a custom departure handler. The router tried to find a path on the network for walk legs, assessing distance from the closest walk link to/from a facility to decide if the link functions as an access to the facility or not. If no nearby link was found, or no route found between two access links, an empty route was stored in the leg. The departure handler checked whether the route was empty or not, either teleporting the agent or putting it on a walk link in the network. Walk links are regular queue-based network links with capacity and free-speed set, according to the simplified physics of directed pedestrian flows. This approach readily allowed modeling of security screening gates' bottleneck effects and considered essential walk path locations where necessary. These were modeled, omitting them on non-critical routes. Figure 88.2 shows an example of simulated pedestrian movements at Krasnaya Polyana, the mountain area near Sochi where numerous events took place.

88.3 Simulation of Sochi

To test ITSOS applicability for major events transportation planning, a model of the 2014 Olympic winter games in Sochi (Russia) was built. Data was either collected either by Masterconcept employees or cooperating companies, or received from Russian governmental institutions.

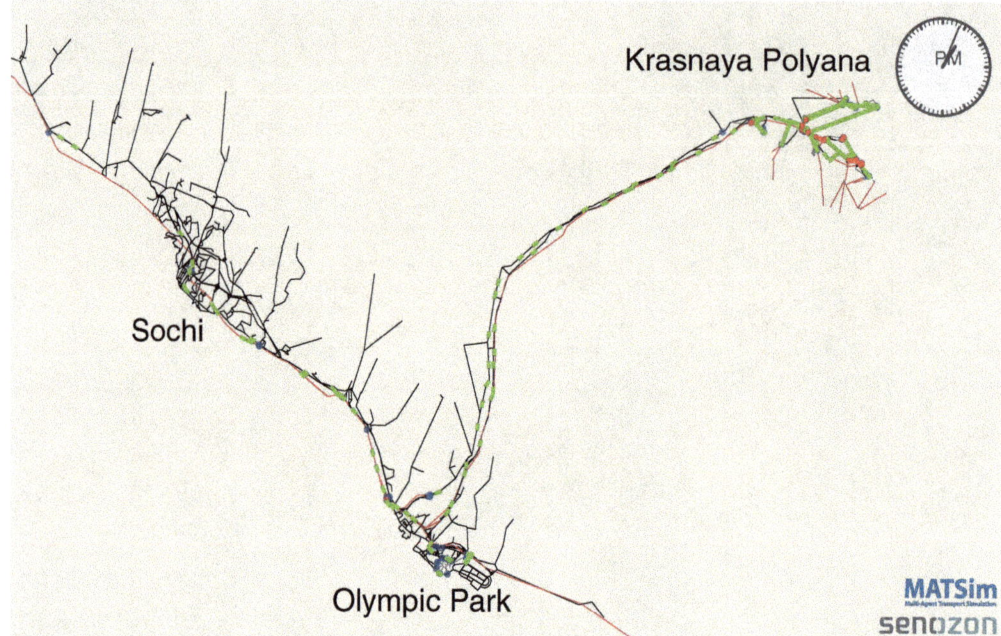

Figure 88.3: Overview of the Sochi model.

Road and train networks were modeled in ArchGIS, using the ITSOS extensions. The transit schedule included 55 transit lines, a mix of bus lines, train lines, and cable cars, going up into the mountain areas. 24 of those lines were defined to be adaptive, 19 lines operated on-demand as shuttle services.

Travel demand was defined for each day of the games, based on the actual schedules, making assumptions about how many spectators would visit each different competition during the day. While size of event facilities can be used as a upper limit for number of spectators, substantial experience and knowledge from Masterconcept was used to define actual numbers of people expected at each event.

Events often start and end at different times of day, because many event locations share, at least partially, a common route to reach them; it was important to simulate whether the transport services offered could cope with the combined travel demand generated by multiple, separate events.

A typical simulation run of Sochi included about 150 000 agents. To speed up simulations, parallel events handling and parallel qsim was used. The simulation generated around 15 million events per iteration. Figure 88.3 shows a screenshot of the Sochi scenario, visualized in Via.

88.4 Outlook

In addition to the test case of the 2014 Olympic winter games in Sochi, ITSOS/MATSim was also used to simulate traffic in St. Johann (Pongau, Austria), with particular emphasis on pupils, who often must take a combination of buses and trains to get to school.

A new company, Masterconcept Mobility GmbH, was split off from Masterconcept Consulting GmbH in 2014; this new firm offers major event transportation planning services, as well as regional planning services based on the combination of ITSOS and MATSim.

CHAPTER 89

Stockholm

Joschka Bischoff

The Stockholm scenario was created as a student project at TU Berlin in summer, 2014. Because several groups worked on the project, the common base was a census data synthetic population, an OSM-based network and counts data.

The network was taken from OSM 2013 data. Within the city, all roads were used; in outlying regions, only mayor roads were included in the network. Demand consisted of home-work-home-plans only. The population sample size was—depending on the student group—between 1 and 5 %. Agents used car and (pseudo) public transit.

Count data for the morning peak along a mayor road, the E4, was used to calibrate the scenario. This calibration was handled differently by the groups; some just added traffic, others tried to imitate the Stockholm toll. Further scenario documentation is available in German.

How to cite this book chapter:

Bischoff, J. 2016. Stockholm. In: Horni, A, Nagel, K and Axhausen, K W. (eds.) *The Multi-Agent Transport Simulation MATSim*, Pp. 507–508. London: Ubiquity Press. DOI: http://dx.doi.org/10.5334/baw.89.

CHAPTER 90

Tampa, Florida: High-Resolution Simulation of Urban Travel and Network Performance for Estimating Mobile Source Emissions

Sashikanth Gurram, Abdul R. Pinjari and Amy L. Stuart

90.1 Introduction

Mobile sources are significant contributors to ambient traffic-related air pollution associated with adverse health impacts in urban areas. Thus, it is important to accurately characterize mobile source emissions and population exposure to those emissions; this requires a high-resolution simulation of urban travel. In this study, using activity-based travel demand modeling and MATSim-based dynamic traffic assignment modeling, we demonstrate a large-scale, high-resolution simulation of resident population travel activity and highway network performance in Tampa, Florida. Such high resolution simulation outcomes are useful in estimating mobile source emissions and human exposure to those emissions.

90.2 Study Area

Hillsborough County, a large section of the Tampa Bay region in Florida, is our study area. The county's geographic context is presented in Figure 90.1. The freeway road I-275, acts as a major commuter corridor connecting the area north of Tampa to the central business district to the south. The freeway roads I-75 and I-4 run north-south and east-west, respectively, and serve as major highways for intra-city, inter-city and inter-state travel. The county has a diverse mix of air pollution sources and population demographics, few public transportation options, an unsatisfactory air quality record and a sprawling urban form. These make it an interesting test case from an air pollution perspective.

How to cite this book chapter:

Gurram, S, Pinjari, A R and Stuart, A L. 2016. Tampa, Florida: High-Resolution Simulation of Urban Travel and Network Performance for Estimating Mobile Source Emissions. In: Horni, A, Nagel, K and Axhausen, K W. (eds.) *The Multi-Agent Transport Simulation MATSim*, Pp. 509–514. London: Ubiquity Press. DOI: http://dx.doi.org/10.5334/baw.90.

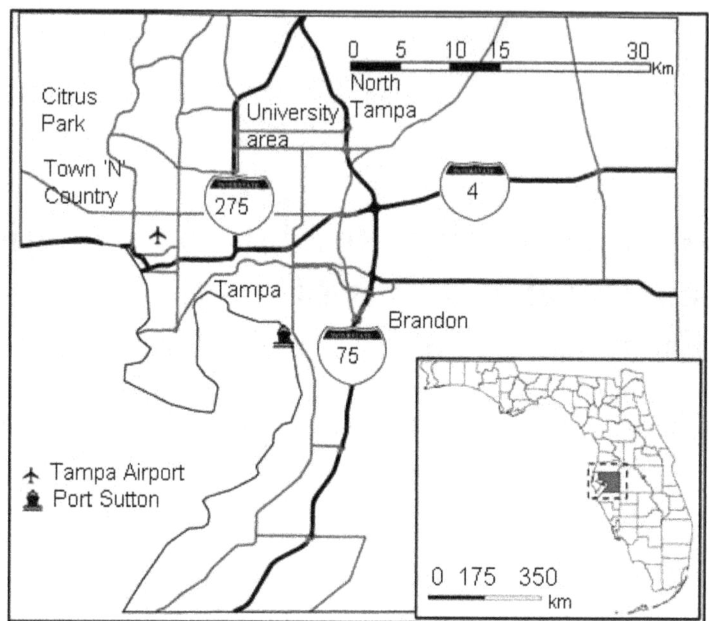

Figure 90.1: The study area of Hillsborough County, Florida.
Source: Gurram et al. (2015)

90.3 Modeling Framework

Figure 90.2 depicts the modeling framework used to simulate urban population activity and transportation network performance for the study region. An activity-based model (ABM) of travel demand (DaySim) is coupled with MATSim, here applied as a dynamic traffic assignment (DTA) model. The DaySim framework was originally developed for the Sacramento region (Bradley et al., 2010) and calibrated for the Tampa Bay region, using local household travel data (Gliebe et al., 2014). An appealing feature of DaySim is its use of fine, parcel-level representation of space, which leads to high spatial resolution in the simulated activity locations. Similar to other ABMs (Activity-Based Models), inputs to DaySim include detailed population demographics, land-use patterns and transportation system characteristics in the study region. The demographic inputs come from a population synthesizer called PopGen (Pendyala et al., 2011) that generates a synthetic population of individuals and households to match aggregate-level distributions of both household- and person-level characteristics from the U.S. Census. Demographic variables not controlled in PopGen (e.g., household car ownership and individual employment characteristics), were estimated using econometric models based on local data. Taking all the above as inputs, DaySim simulates the daily activity and travel patterns of all residents in the study region, including the timing, duration and location of activities and the mode of travel between different activity locations. We ran the model on an eight-core Windows machine with a 2.8 GHz Intel Xeon processor and 24 GB RAM. The run time was approximately 5 hours for the entire Tampa Bay population of about 3 million individuals.

DaySim does not simulate travel route information between different activity locations. However, information on travel routes and network performance (i.e., link speeds and volumes) is essential for estimating emissions and human exposure to those emissions. Therefore, MATSim was used to simulate travel routes and network performance (Balmer et al., 2008). To do so, outputs from the Tampa ABM were processed using SPSS and Java programming to provide the initial set of plans for

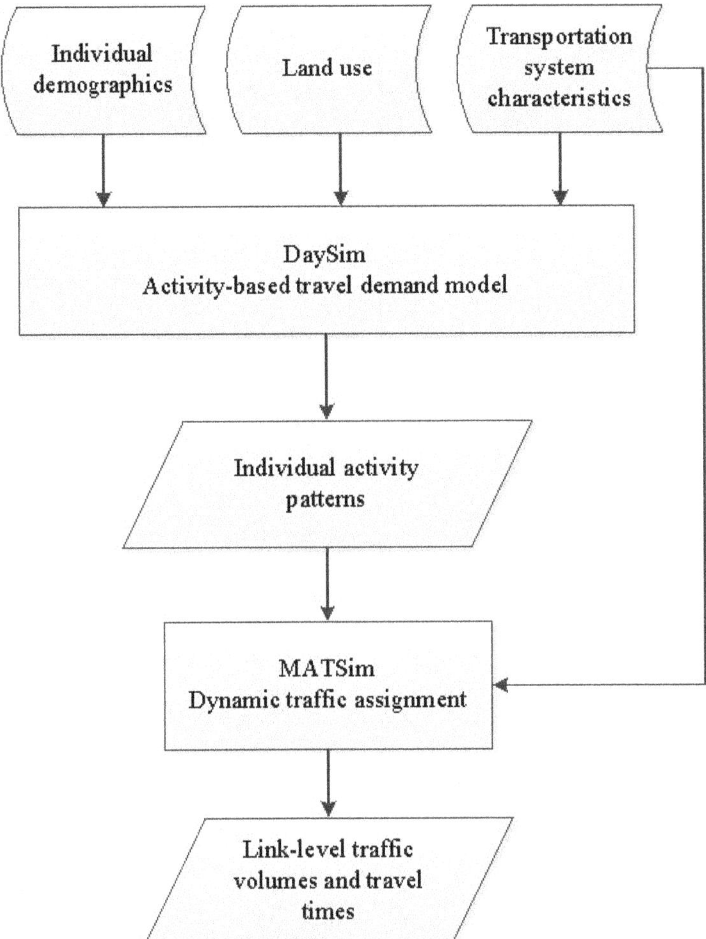

Figure 90.2: The transportation modeling framework.

MATSim. Similarly, the ArcGIS road network file for the Tampa Bay area was processed to create network inputs for MATSim. Since most travel in Tampa is by automobile (with close to 90 % mode share), only these trips were simulated in MATSim. It is worth noting, however, that a large number of automobile trips were simulated. Specifically, 9.7 million trips made by approximately 2.3 million residents of the study region during a 24 hour period were simulated. The simulation was run for 300 iterations, with the storage capacity factor for the links set to 3. Additionally, maximum plan memory size for each agent was set to 3. The `BestScore` and `ReRoute` replanning modules were used with a probability of 0.9 and 0.1, respectively. To undertake this large-scale and computationally intensive simulation, 48 parallel processors each with 25 GB of RAM from a university research computing cluster setup were utilized, requiring 5.2 days total run time. Link-level outputs from the simulation, including hourly traffic volumes and travel times, were written to a `linkstats` file; trip-level route information was written to a `plans` file.

90.4 Results

Diurnal patterns of link-level passenger car volumes and travel speeds for Hillsborough County are presented in Figure 90.3 (in the form of bi-hourly averages). As expected, traffic volumes, shown in

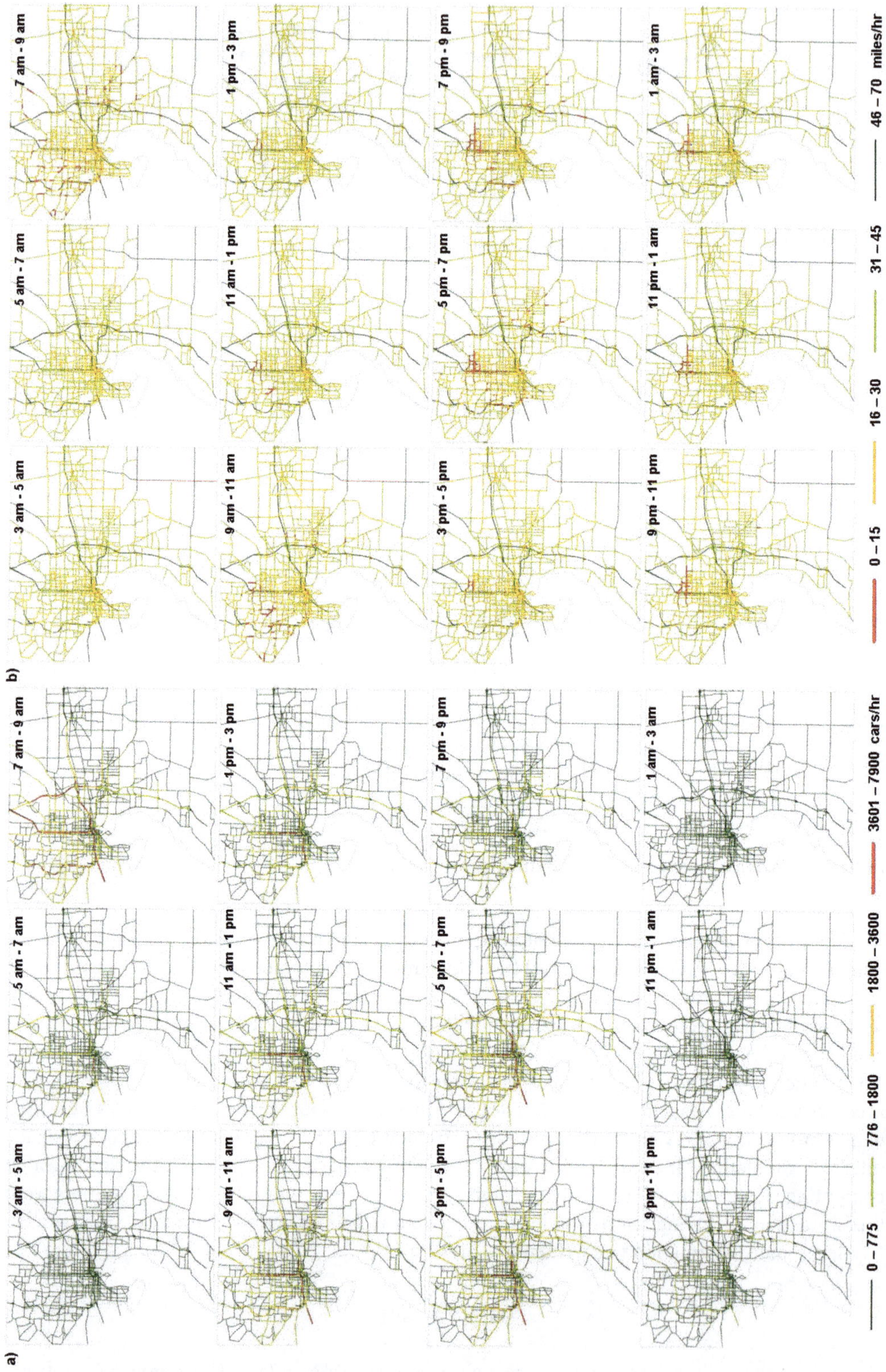

Figure 90.3: Simulated bi-hourly varying a) passenger car volumes and b) travel speeds (mph) for Hillsborough County on a typical weekday.

Figure 90.3 a), are higher during the morning (7 to 9 am) and the evening (4 to 7 pm) peak hours than the rest of the day. Additionally, traffic volumes during evening peak hours are higher than volumes during morning peak hours, perhaps partly because the evening commute has a higher propensity for trip chaining compared to the morning commute (Chu, 2003). Travel speeds shown in Figure 90.3 b) correspond to the diurnal pattern of traffic volumes, with lower speeds during the morning and evening peak hours.

Spatially, higher volumes are observed along major freeway corridors—I-75, I-275, and I-4 as expected. High traffic volumes are also observed along the road network near suburban locations, including Brandon, Citrus Park and Town 'N' Country. Accordingly, travel speeds are lower in these suburban locations along with the North Tampa area, University area and a few sections of the freeway corridors.

The root mean squared error between the estimated traffic volumes and observed traffic volumes at eight different traffic counting stations is 0.41. Further, the error between estimated and observed traffic flows for inter-city roads was higher than those for intra-city roads, presumably because the current model system does not consider long-distance (or inter-city) travel, visitors' travel and freight movement in detail. Nevertheless, the high temporal and spatial resolution of the population activity (including individuals' travel routes) and network performance (i.e., link volumes and speeds) simulated using the model system is promising for future detailed estimation of traffic pollutant emissions and human exposures to those emissions.

90.5 Future Work

The next steps of this study include addition of inter-city, visitor and freight travel to the model system. Utilizing the fine-resolution, link-level traffic volume and speed outputs from MATSim, EPA's MOVES software is being used to estimate mobile source emissions. Mobile source emissions can be combined with other sources of emission and meteorological data, using a pollutant dispersion model, to estimate diurnal cycles of hourly varying pollutant concentrations. The resulting pollutant concentrations will be combined with the diurnal locations of individuals (obtained from the ABM and MATSim) to estimate individual-level exposure to traffic-related pollutants, such as nitrogen oxides. Such individual-level exposure measures will be utilized to estimate demographic group-level exposures for assessment of inequality in exposure to traffic-related air pollution, as we have done previously using travel survey data (Gurram et al., 2015). The model system described above will be used to obtain estimates of population exposure, for alternative scenarios of urban land-use design and transport policies.

90.6 Conclusion

In this study, we simulated urban travel using activity-based travel demand modeling and dynamic traffic assignment, to obtain network performance measures, including link-level traffic volumes and speeds, at a high spatial and temporal resolution for Hillsborough County in Florida. As expected, simulated traffic volumes are higher and travel speeds lower during morning and evening peak hours. Spatially, higher volumes and lower speeds are observed along the freeway corridors and suburban locations than other locations. Model performance (vis-à-vis observed traffic patterns) is better for inter-city roads than intra-city roads, highlighting the need for better modeling of long distance passenger travel and freight movement. When the ABM-DTA framework built in this study is expanded to consider mobile source emissions and pollutant dispersion, the resulting transportation and air pollution modeling system will be useful for understanding interactions between urban transportation design, air pollution and population exposure to pollution.

CHAPTER 91

Tel Aviv

Christoph Dobler

The initial Tel Aviv MATSim scenario (Bekhor et al., 2011) was recently extended by adding destination choice to the MATSim iterations (Dobler et al., 2014).

The modeled area was divided into 1 219 TAZ (Figure 91.1(a)); geometry was provided as a ESRI shape file (ESRI, 1998). Zonal attributes contained information on the population living in the zone, as well as types of activities that can be performed.

The population was created using population generator outcomes from the Tel Aviv activity-based model, containing socio-demographic attributes and daily schedules with up to six activities. This kept computational effort manageable; a 10 % population sample was simulated. Additional data was provided for external trips; for each of the three types (car, truck, commercial), O-D matrices for three different time periods were available.

Network input data was taken from the EMME/2 model (see INRO, 2015), also used by the Assignment Unit of the existing Tel Aviv Model. Conversion process details can be found in Gao et al. (2010). Turning restrictions were handled by adapting the network structure, resulting in a network containing 9 474 nodes and 18 570 links (Figure 91.1(b)). Some major road capacities were obviously too low (e.g., noticeably lower than traffic counts indicated) and were corrected manually.

The Tel Aviv scenario contained road pricing for two arterial highways; count data for validation was available for three arterial roads.

How to cite this book chapter:

Dobler, C. 2016. Tel Aviv. In: Horni, A, Nagel, K and Axhausen, K W. (eds.) *The Multi-Agent Transport Simulation MATSim*, Pp. 515–516. London: Ubiquity Press. DOI: http://dx.doi.org/10.5334/baw.91.

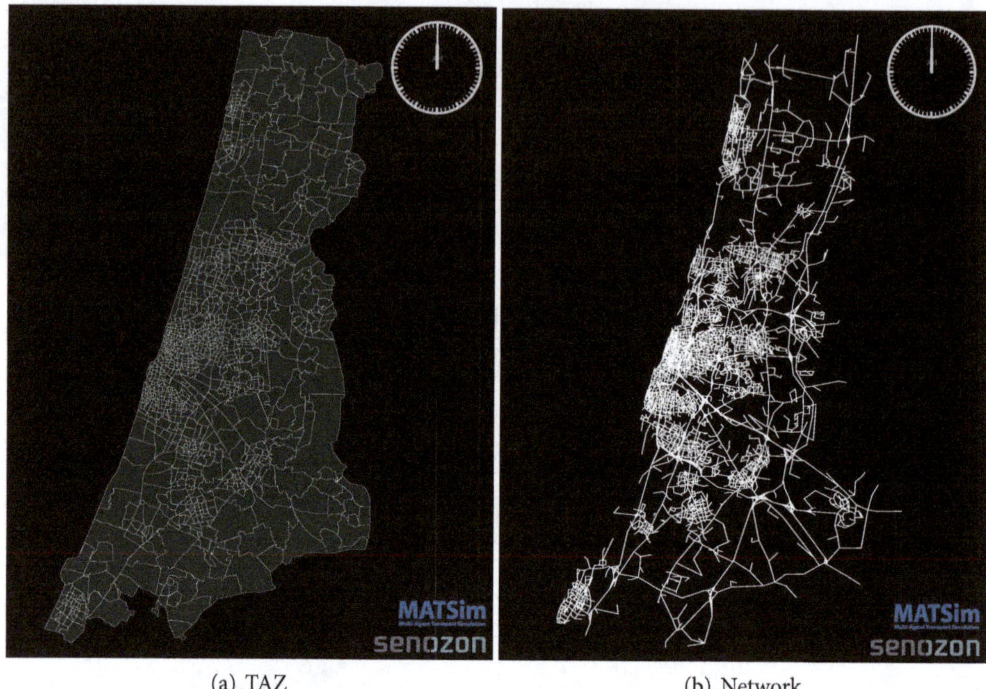

(a) TAZ (b) Network

Figure 91.1: Tel Aviv scenario.

CHAPTER 92

Tokyo: Simulating Hyperpath-Based Vehicle Navigations and its Impact on Travel Time Reliability

Daisuke Fukuda, Jiangshan Ma, Kaoru Yamada and Norihito Shinkai

92.1 Introduction

Most standard commercial vehicle navigation systems usually rely on fixed travel times as link weights; sophisticated algorithms deal mainly with stochastic travel time. Reliable routing incorporating such travel time variability could provide extra benefits to drivers. However, implementations of many reliable routing algorithms might become impractical, mostly due to heavy computational loads. The hyperpath-based navigation demonstrated in this chapter would consider only lower and upper bounds travel times for each link as inputs and produce a set of potentially optimal links with recommended link choice possibilities.

The basic concept of hyperpath is: "Don't put all your eggs in one basket in an uncertain environment". In literal terms, actual routes are more widely distributed as congestion increases. Thus, delay risk due to induced congestion would be reduced and the network burden—congested links—would be lightened (Figure 92.1). Based on the idea of "Optimal strategy", widely employed in frequency-based transit assignment (see Spiess and Florian, 1989), Bell (2009) proposed the shortest hyperpath search algorithm called "Hyperstar". Algorithm variations under various conditions have been further developed in Bell et al. (2012) and Ma et al. (2013) for risk-averse vehicle navigation.

Hyperpath-based navigation can be beneficial in at least three ways:

1. The concept of hyperpath could benefit drivers by helping reduce travel time unreliability; it provides an 'adaptive choice opportunity' to potentially avoid stops at intersections, or long delays on links. For example, other than typical navigation systems, the turn notification

How to cite this book chapter:
Fukuda, D, Ma, J, Yamada, K and Shinkai, N. 2016. Tokyo: Simulating Hyperpath-Based Vehicle Navigations and its Impact on Travel Time Reliability. In: Horni, A, Nagel, K and Axhausen, K W. (eds.) *The Multi-Agent Transport Simulation MATSim*, Pp. 517–522. London: Ubiquity Press. DOI: http://dx.doi.org/10.5334/baw.92.

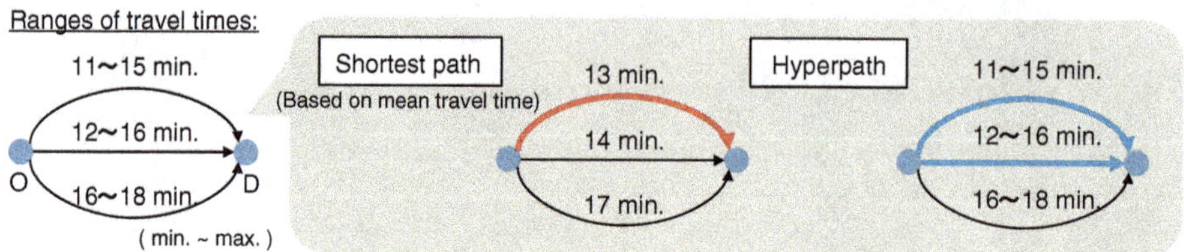

Figure 92.1: Concept of hyperpath under travel time uncertainty (1 O-D - 3 routes example).

received by drivers before entering intersections could be "go straight or turn right". In this case, drivers may decide to turn right when encountering a red light for going straight. Even if the final experienced travel time is slightly longer than the non-adaptive drive, driving experience could be better (say, eight minutes driving plus two minutes waiting versus ten minutes non-stop driving).

2. For drivers, hyperpath also could cater more to individual tastes without pre-defining drivers' actual route choice preferences. Comparatively, shortest path (SP), or multiple shortest paths with different criterion, would require modelers' definition of "shortest". In the long run, the hyperpath model has the potential to evolve with reinforcement learning technologies and provide more customized adaptive route guidance.
3. Existing commercial navigation systems seldom take their effect on networks into consideration; sometimes congestion is actually produced by navigation systems. Thus, DTA, along with route guidance, might be still be mostly academic or hypothetical. Classical DTA are largely based on time-dependent K-shortest paths and aim to analyze equilibrium conditions as ideal states. For example, Dynamic User Equilibrium defines the equilibrium where drivers cannot change their trip plans to reduce actual experienced travel time. However, experienced travel time can never be known beforehand, since real-life transportation is much more complicated than laboratory DTA settings. Hyperpath-based route choice does not search for equilibrium, but it might be equilibrium-like to some extent, as it is strategically reactive to delay changes.

The hyperpath-based route recommendation could thus reduce overall network congestion, because it recommends a potential optimal set of paths instead of the shortest (single) path and leads to appropriate dispersion of traffic. However, its impact on the entire traffic networks has not been well analyzed. In this chapter, we demonstrate—using MATSim—how hyperpath-based vehicle navigation market penetration would affect overall network performance. Though development of real-time traffic information for navigating vehicles has benefited drivers, to some extent, market diffusion of these technologies may not lead to the reduction of traffic congestion, mainly due to the concentration of traffic into particular paths or links in the traffic networks. Certain unexpected phenomena, such as "Hunting (e.g., Oguchi et al., 2003)", might occur. We changed the ratio of vehicles with risk-averse route guidance, conducted traffic simulation and then checked traffic performance.

92.2 A Small-Sized Network Case

MATSim was utilized as the simulation tool. In the early stage, we conducted such simulations on the Sioux Falls network (see also Chapter 59) with synthetic O-D demands (see (Yamada et al., 2013) for details). Hyperpath algorithm was initially written as an external route planning module in Python and the market share can be configured by setting the "ModuleProbability" item. Figure 92.2 illustrates the configuration sample for hyperpath with 20 % market share. Figure 92.3

```
<param name="ModuleProbability_1" value="0.2" />
<param name="Module_1" value="ExternalModule" />
<param name="ModuleExePath_1" value="python &INBASE;/external_parallel_v5.py" />
<param name="ModuleProbability_2" value="0.8" />
<param name="Module_2" value="ReRoute" />
```

Figure 92.2: Setting for the case that 20 % of vehicles follow hyperpath-based vehicle navigation.

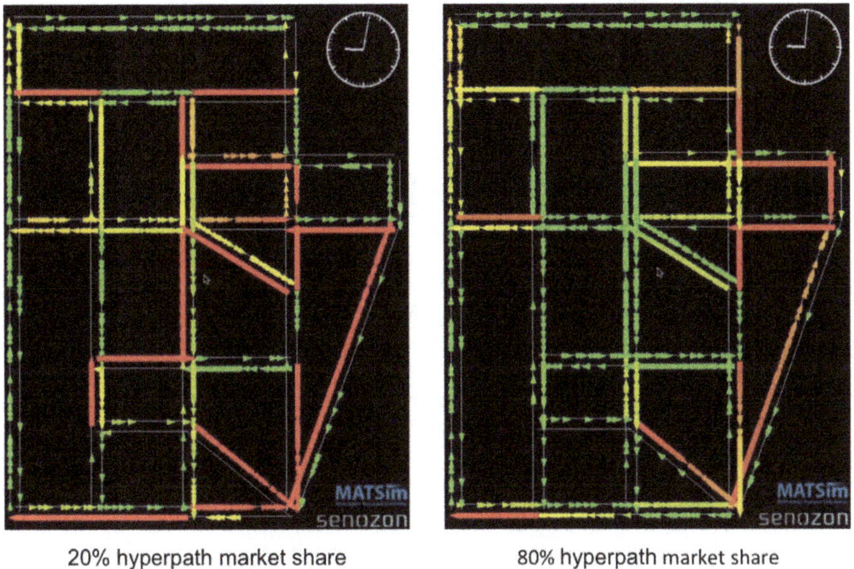

Figure 92.3: Link travel speeds for different levels of market penetrations.

also shows the network state (in terms of link speed) improvement with increase in the market share of hyperpath from 20 % to 80 %.

92.3 Simulation in Tokyo's Arterial Road Network

Based on early-stage experiments on the Sioux Falls network, we were interested in a similar simulation in Tokyo's large-scale arterial network with actual traffic data.

92.3.1 Network and Travel Demand

The arterial road network, including the whole Tokyo Metropolitan Area (Figure 92.4), was prepared from Digital Road Map version 2011 (DRM2403) in a radius of about 70–80 kilometer from downtown. The traffic network consisted of 444 220 nodes and 177 971 links after being cleaned using the "networkcleaner" API in MATSim. Capacity and free flow speed of each link were set up considering road hierarchy information, type of links and their corresponding speed limits.

We analyzed car traffic during morning rush hours and the O-D table was subtracted from a large-scale travel survey (Person Trip Survey 2007) to create agents' plans. The total number of the O-D pairs was 17 186 and there were about 2 307 000 vehicular trips during the target time period within the whole area. From the data, 219 642 agents (approximately 10 % sampling rate) were randomly created and each agent had only one activity, commuting from his/her home to the

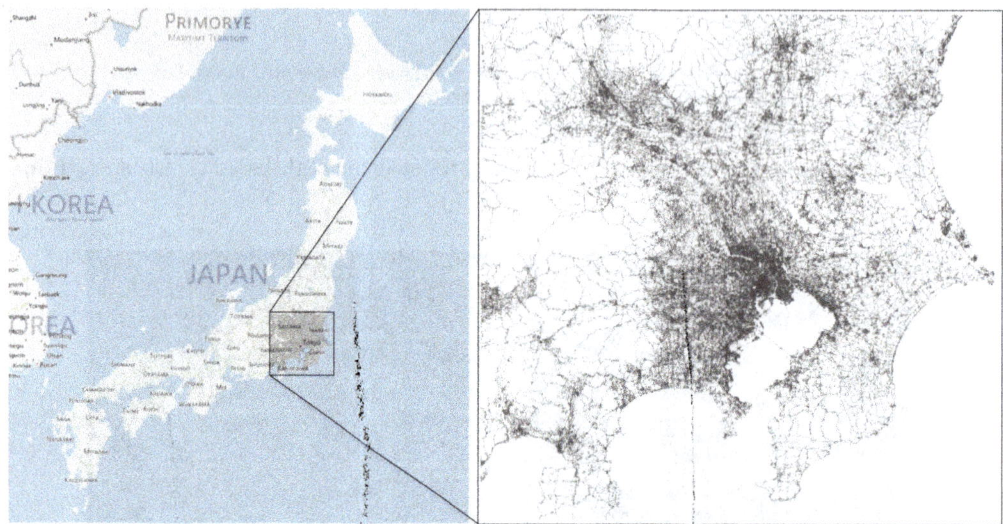

Figure 92.4: Arterial road network in Tokyo Metropolitan Area.

workplace. For drivers' departure time from home, a normal distribution with the mean of 7 am and the standard deviation of 1 hour was assumed.

92.3.2 *Setup of Day-to-Day Simulation Experiments*

Although the Logit-based route planning module had already been employed as one of the MATSim routing strategies, it may have had less supportive route guidance explanations. We thus focused on the combination of "re-route" and "best-score" planning modules, which meant that some drivers adjusted their daily travel plan according to yesterday's experience, while the others simply chose the most positive route from their past choices.

To get travel time data for creating hyperpaths, simulation runs for 30 iterations (i.e., 30 days) were firstly performed with no HP-based drivers (i.e., 100 % of SP-based drivers) to obtain travel time distribution. Then, maximum delays in each during these 30 days were computed and used in the following main simulation, with the market diffusion of HP-based navigated drivers. Figure 92.5 illustrates a simulation with MATSim for the downtown Tokyo.

92.3.3 *Results*

We conducted five different cases of traffic simulation by changing the shares of HP-based drivers from 0% to 80% by 20%. The simulation runs were conducted for 30 days for each case to evaluate network-level travel-time savings, as well as reliability.

Average travel time per unit length of all agents in each one day was plotted for different cases in Figure 92.6. Since the traffic network in Tokyo is quite large and drivers' trip lengths are diverse, we plotted the average travel time per unit length (shortly ATTPUL) for a fair comparison. Apparently, there were high levels and large fluctuations in ATTPUL when there were no HP-based drivers (HP %, that is SP 100 %). But it is obvious that, as the market diffusion rate of HP-based drivers increased, fluctuation and levels of ATTPUL would be significantly reduced.

Table 92.1 summarizes the result of Figure 92.6 by computing the ATTPUL ($\bar{t}_{unit}$) average, as well as the ATTPUL (σ_{unit}) standard deviation over the 30 days. It is clear from this table that

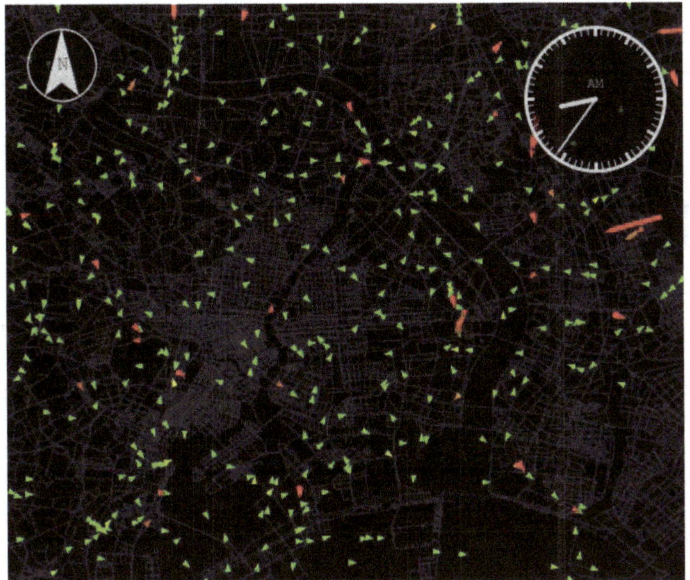

Figure 92.5: Snapshot of the hyperpath-based traffic simulation in Tokyo.

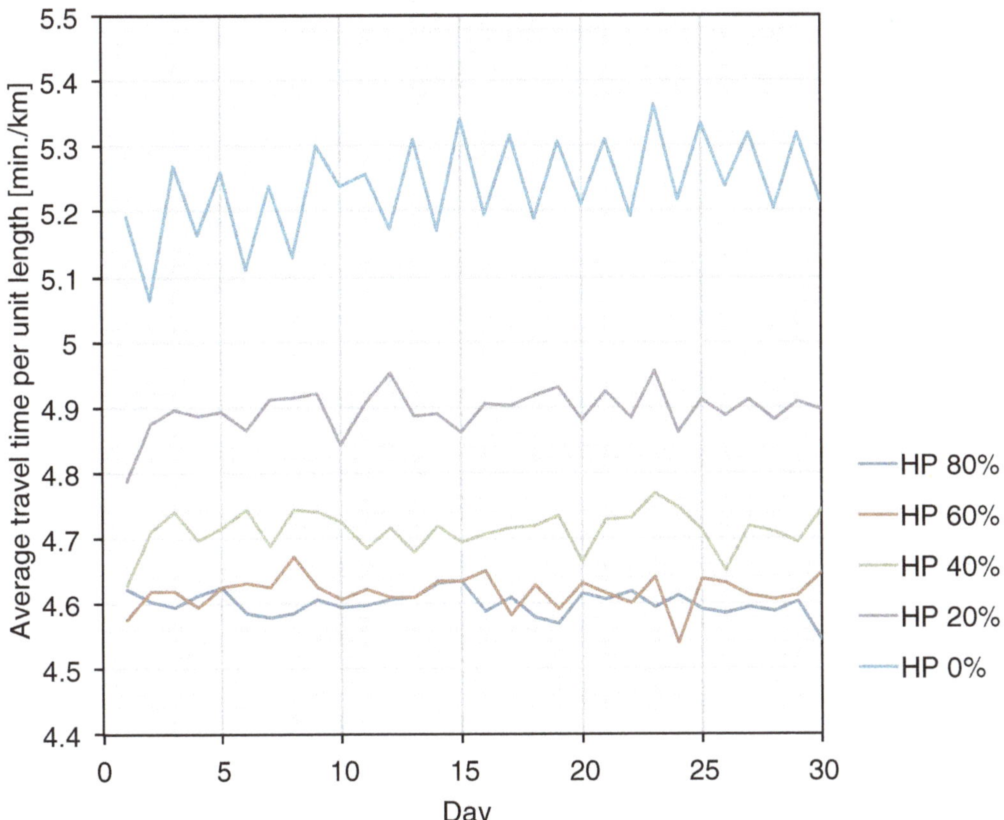

Figure 92.6: Average travel time per unit length for all vehicles.

Case	$\bar{t}_{unit}$ [min./km]	σ_{unit} [min./km]
HP 80%	4.60	0.02
HP 60%	4.62	0.02
HP 40%	4.71	0.03
HP 20%	4.90	0.03
HP 0%	5.24	0.07

Table 92.1: Summary of the network perfomance.

both average and standard deviation of travel time tended to decrease when the mixing ratio of HP-based route guidance was increased. This result indicates that HP-based route guidance would be superior in terms of of travel time reduction and reliability given day-to-day patterns of average travel time for heavy Tokyo traffic.

92.4 Validation of Hyperpath-Based Navigation

A field experiment was conducted to verify the benefits of travel time reliability improvement for drivers by Ito et al. (2015) in Tokyo. Drivers equipped with the time-dependent shortest path and those equipped with the time-dependent hyperpath navigation systems were compared. Both navigation systems use the same historical travel time sourced from probe vehicle data. Based on results collected from two weeks of experimental driving by different drivers, the hyperpath produced a significantly better result, especially when the network was congested.

CHAPTER 93

Toronto

Adam Weiss, Peter Kucireck and Khandker Nurul Habib

93.1 Study Area

The GTHA (Greater Toronto and Hamilton Area) is located northwest of Lake Ontario, in the province of Ontario, forming Canada's largest urban region. The GTHA's current population is over 6.5 million, with projected growth to approximately 8.6 million by 2031.

93.2 Population, Demand Generation and Activity Locations

The TTS (Transportation Tomorrow Survey) was the basis for travel demand used for the multimodal assignment simulation. TTS was a retrospective telephone survey, conducted in the GTHA every five years. The TTS sampled just over 5 % of GTHA households; the survey collected household socioeconomic and geographical data, characteristics of each household member and a full 24 hours travel diary for each household member. Current MATSim models use the TTS travel diary records to generate the plans file. Integration of the TASHA activity based model, developed for the glsgtha, was also investigated. Irrespective of the demand data source, both sources provided the traffic zone location for all activities. The Toronto implementation then randomly distributed activities around the traffic zone, which resulted in unique x-y coordinates for each activity. Within the current MATSim implementation in Toronto, no MATSim facilities development has been attempted.

93.3 Network Development and Simulated Modes

The GTHA MATSim implementation used a pre-existing planning level network for static user equilibrium assignment,employing the EMME traffic assignment software. This network was converted to a MATSim network, using a conversion tool found in the MATSim Toronto playground.

How to cite this book chapter:

Weiss, A, Kucireck, P and Habib, K N. 2016. Toronto. In: Horni, A, Nagel, K and Axhausen, K W. (eds.) *The Multi-Agent Transport Simulation MATSim*, Pp. 523–524. London: Ubiquity Press. DOI: http://dx.doi.org/10.5334/baw.93.

More recently, this network was merged with GTFS data for five of the eight major regional transit agencies to allow for multimodal demand assignment.

93.4 Calibration, Validation, Results

The Toronto MATSim implementation was compared to more conventional, large-scale assignment models with varying success. The work of Gao et al. (2010) found that travel time, travel distance, link flows and speeds were reasonably comparable, in fact more plausible, than those achieved through the EMME assignment. Conversely, work on transit assignment, first done by Kucirek (2012) and then by Weiss et al. (2014), found limitations associated with predicting line boarding figures; these were based on different transit technologies and agencies and utilized different fare structures, suggesting that further work to calibrate the multimodal assignment model is required. These issues are exacerbated by the current implementation's inability to distinguish between in-vehicle dwell times and out-of-vehicle wait times; these should ideally be weighted differently, particularly given the climate and predominance of outdoor bus stops in the region.

CHAPTER 94

Trondheim

Stefan Flügel, Julia Kern and Frederik Bockemühl

The Institute of Transport Economics (TØI), in cooperation with Julia Kern from TU-Berlin and Frederik Bockemühl from Hasselts University, built a first prototype model for the region of Trondheim (Norway) (Flügel and Kern, 2014).

The road network data was imported from a publicly accessible data base (Elveg). Figure 94.1 illustrates the network. Most required link information could be directly inferred from the data base. The lane capacity (vehicles per hour) was assumed to be a flat 1 800 per lane. Existing toll stations, with their current toll structures, were coded manually in the network file. The public transport, walk and cycle networks had not been implemented at this time. Agents using one of these modes were teleported; travel times were calculated with predefined speeds per transport mode. Initial demand was derived from the National Travel Survey (NTS 2009) travel diaries. 4 453 respondents were simply scaled up to 191 676 agents; activity locations and departure times were slightly randomized to avoid clusters. This model differentiated only between work and "other" activities. Desirable working hours were specified as eight hours; demand consisted only of private cars (no trucks).

Standard utility functions were applied, but in the calibration process, default values for travel time disutility in different transport modes were adjusted so that the model would reproduce observed market shares. The simulated traffic fit (in the reference scenario) against real-world counts was deemed satisfactory for a first implementation (Bockemühl, 2014).

Standard behavioral modules in MATSim were included in the Trondheim model. Agent could react to policy measures through three choice dimensions: changing route, changing transport mode and changing departure time. To test whether MATSim predicted reasonable behavioral changes, a small case study was performed. Additional tolls on streets (bridges and tunnels) to Trondheim city center were coded in the network and three congestion price structure were tested. Figure 94.2 illustrates the effects on the simulated cars entering and leaving Trondheim city center. Compared to the reference scenario without tolls, total number of cars was reduced in

How to cite this book chapter:

Flügel, S, Kern, J and Bockemühl, F. 2016. Trondheim. In: Horni, A, Nagel, K and Axhausen, K W. (eds.) *The Multi-Agent Transport Simulation MATSim*, Pp. 525–526. London: Ubiquity Press. DOI: http://dx.doi.org/10.5334/baw.94. License: CC-BY 4.0

all toll scenarios. Some agents changed transport modes; others, who would have driven through Trondheim center, changed their route. Comparing the three different congestion-pricing structures, it was also evident that agents changed departure time. The difference between the 15 NOKs flat scenario and the 10/20 NOKs scenario was small; the effect in the 50 NOKs rush scenario was substantial. Actually, in this scenario, traffic was heavier before 3 pm and after 5 pm implying that many agents changed departure time to avoid high congestion pricing.

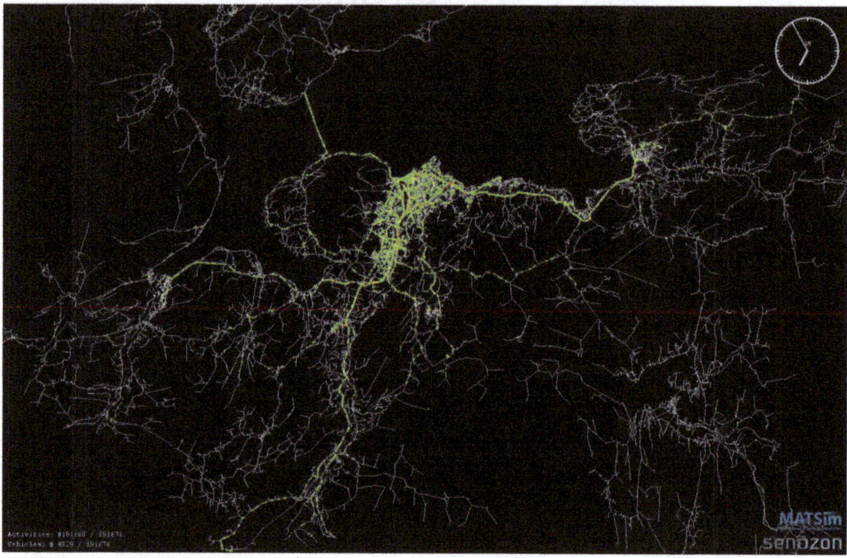

Figure 94.1: Network and simulated traffic in Trondheim and surroundings for 6:55 am (source Flügel et al., 2014) (visualized with Via).

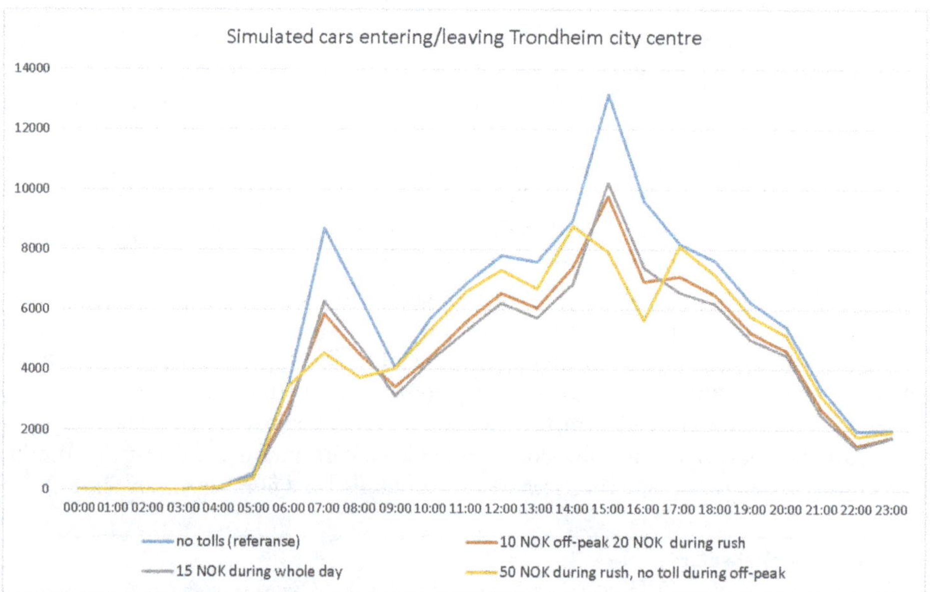

Figure 94.2: Cars entering/leaving Trondheim city center in reference scenario and three congestion pricing scenarios (source Bockemühl, 2014).

CHAPTER 95

Yarrawonga and Mulwala: Demand-Responsive Transportation in Regional Victoria, Australia

Nicole Ronald

In November 2013, Public Transport Victoria implemented a service called Flexiride in twin regional Victoria towns, consisting of an on-demand public transport service using taxis. This service replaced an existing fixed-route bus service, which was poorly patronized.

This scenario was designed to investigate operational performance change between two different DRT schemes: Flexiride and a completely ad-hoc scheme. More details can be found in (Ronald et al., 2015). This work was a first step in developing a decision-support tool to evaluate different DRT schemes, particularly when integrated with other transport modes.

The scenario was part of a larger project exploring the viability of mobility-on-demand, focusing on ridesharing and DRT services (Ronald, 2014).

The scenario covered twin towns on the border of Victoria and New South Wales, Australia, separated by the Murray River. Yarrawonga (Victoria) has a population of 7 057 and an area of 95.0 square kilometers, while Mulwala (New South Wales) has a population of 1 904 and an area of 18.6 square kilometers.

The Flexiride scheme delivered six services on weekdays and three services on Saturday, leaving Yarrawonga center (Orr St) at fixed times. The local taxi operator was paid a holding fee by Public Transport Victoria to have a taxi available at Orr St at the nominated time. The taxi returned to normal service when there were no bookings or passengers waiting.

Passengers could ride either by starting their trip at Orr St, or by phone booking, at least 10 minutes before a scheduled departure from Orr St. Existing bus stops were used as pickup and drop-off points.

Flexiride drivers recorded pickup and drop-off locations for each service. Using this data, probabilities of trips occurring between two zones were developed, using the process in Deflorio (2011). A continuous departure time distribution was derived from evenly spreading demand for particular services to either side of that service.

How to cite this book chapter:

Ronald, N. 2016. Yarrawonga and Mulwala: Demand-Responsive Transportation in Regional Victoria, Australia. In: Horni, A, Nagel, K and Axhausen, K W. (eds.) *The Multi-Agent Transport Simulation MATSim*, Pp. 527–528. London: Ubiquity Press. DOI: http://dx.doi.org/10.5334/baw.95. License:

The network was extracted from OSM. Some bus stops were removed if they were assigned to the same link in MATSim, e.g., stops on the same road between intersections.

Only passengers for the demand-responsive service were included. However, the use of MATSim for this initial model means that other modes could be added in later versions.

This was an exploratory simulation that demonstrated how DRT could be modeled for exploring viability and comparison of different schemes.

Using MATSim, experimentation with varying demands, two different scheduling algorithms and an altered Flexiride service, with more services, were carried out. Outcomes like drive time, vehicle-kilometers traveled and passenger wait time could be measured.

Results showed that the two schemes performed differently for operators and passengers. Optimization schemes had little effect in low demand situations, while seating requirements showed more variability in the ad-hoc scheme, as demand increased. Future work involves estimating both schemes' costs for further comparison.

This work was supported by a grant from the Australian Research Council (LP120200130). We are also grateful to Michal Maciejewski for his assistance with the DVRP contribution (see Chapter 23).

CHAPTER 96

Yokohama: MATSim Application for Resilient Urban Design

Yoshiki Yamagata, Hajime Seya and Daisuke Murakami

96.1 Introduction

In Yamagata and Seya (2015), we proposed the concept of a resilient local electricity-sharing system as a complement, or alternative, to a FIT (feed-in tariff) to achieve CO_2-neutral transportation in cities. In our proposed system, electricity generated from widely introduced solar PVs (Photovoltaic Panels) is stored in cars "not in use" in a city. In Japan, almost half the central Tokyo metropolitan area cars are used only on weekends and thus are kept parked weekdays. These cars could represent a huge new storage potential if they were replaced by EVs; that is, they could be used as storage batteries in a V2C (Vehicle to Community) system.

This study analyzed the potential of EVs as storage batteries in emergency cases. Specifically, we focused on the following three questions:

1. How much residential demand can be met (in each 24 hour) by electricity from just PVs, which are installed on the roofs of all detached houses in the study area?
2. How many EVs are needed to store all surplus electricity (PV supply minus demand)?
3. How does EVs driving change the load curve and how can mass-adopted PVs fulfill total demand?

To answer our second and third questions, we needed to know (a) the number of cars parked at home during each hour (that is, the time each car arrived at home after use) and (b) the amount of battery charge consumed by each driver during his/her daily trips (that is, trip duration). For this simulation, we used MATSim. In this chapter, we briefly introduce our MATSim application for a local electricity-sharing system in Yokohama city, based on Yamagata and Seya (2013); Yamagata et al. (2014, 2015).

How to cite this book chapter:

Yamagata, Y, Seya, H and Murakami, D. 2016. Yokohama: MATSim Application for Resilient Urban Design. In: Horni, A, Nagel, K and Axhausen, K W. (eds.) *The Multi-Agent Transport Simulation MATSim*, Pp. 529–532. London: Ubiquity Press. DOI: http://dx.doi.org/10.5334/baw.96.

96.2 Results

We assumed that PV was installed on the roof of each detached house in Yokohama city. Then, we calculated the amount of electricity supplied each hour throughout the whole day by employing simple intensity method. The O-D trip data used are from the Fourth Person Trip Survey in Tokyo Metropolitan Area, implemented in 1998. The data are available through the People Flow Project (`http://pflow.csis.u-tokyo.ac.jp`) on request (application) and include the O-D trips by traffic mode, time of day, purpose, etc. for each micro district, called *cho-cho-moku*. The Person Trip survey is a national survey that focuses on people's travel behavior during a given few days of each month, from October to December. Because the number of cars in Yokohama for each cho-cho-moku was unknown, the city-level value was allocated to the cho-cho-moku (areal weighting) and adjusted for the size of the population. The road-network information was taken from the National Digital Road Map Database and included sufficient data on road capacity, width classification, link length, number of lanes and travel speed to perform traffic simulations in MATSim. MATSim requires a daily "plan file" for each agent (car driver); we prepared these files by using the Fourth Person Trip Survey, which captured the daily movements of 722 000 people. Because the Fourth Person Trip Survey sampled approximately 2 % of the population of the Tokyo metropolitan area, the plan file was replicated according to the intensity factor provided by the People Flow Project, resulting in 505 335 agents. From the MATSim simulation, we had obtained each agent's trip duration and arrival time.

Considering load curve changes due to the EVs driving, we then asked if massively adopted PVs would be enough to satisfy total energy demand in Yokohama. In Figure 96.1, the solid and dashed lines represent electricity surplus cumulative distribution, charged to or discharged from the batteries of EVs, not in use and used only for charging the EVs in use, during May and August (solid line, maximum; dashed line, average). The dotted line in the figure represents the scenario where electricity surplus was both charging EVs and satisfying households' typical electricity demand

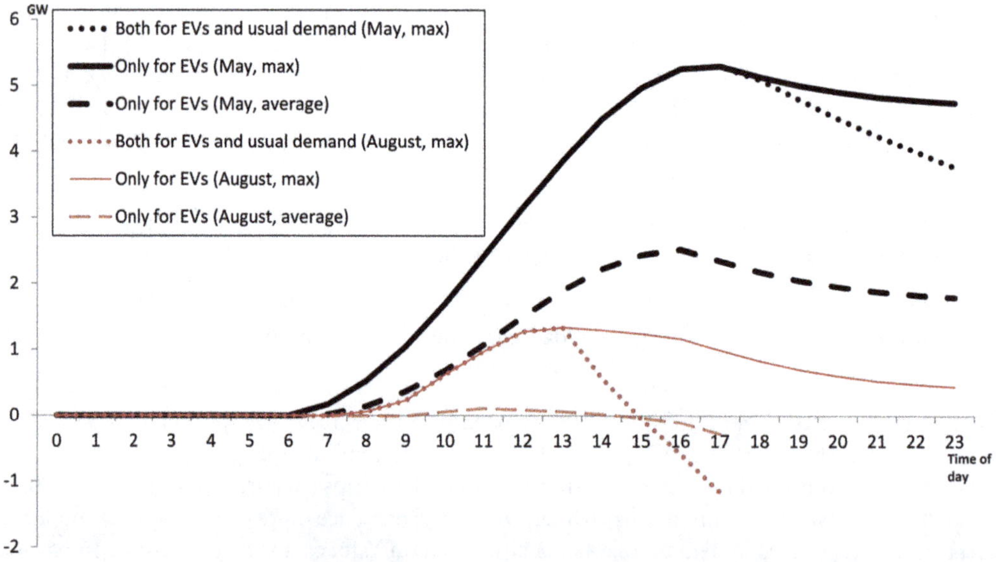

Figure 96.1: Cumulative distribution of electricity surplus charged to or discharged for electricity demand (y axis denotes the cumulative distribution of electricity surplus).

Source: Reproduced by permission of the Institution of Engineering & Technology: published in Yamagata and Seya (2015, Figure 6)

under maximal/average solar irradiance. However, in August (high demand, high PV supply), the electricity surplus was sufficient for charging EVs, but not enough to meet the households' huge electricity demand due to evening use of air-conditioning.

To meet household electricity demand, PV electricity needs be efficiently stored in EVs and locally shared. For example, if a high-affordability zone (storage capacity is greater than electricity surplus) is adjacent to a low-affordability zone (storage capacity is smaller than electricity surplus), then the share of their EV capacity increases the ratio of stored PV electricity. Because storage affordability (storage capacity minus electricity surplus) is significantly different regionally (see Figure 96.2), clustering of community-based local sharing must be carefully designed. In this study, we attempted to optimize community clusters using several different algorithms. Firstly, the number of clusters was assumed 18 to be the same as the number of Yokohama city wards. Then, cluster optimization was performed by minimizing (the sum of storage affordability in the 18 clusters) plus k (minimum circularity in these clusters), where k was the weight for the circularity. The first term balanced storage capacity and electricity surplus to increases the rate of stored PV electricity; the second term decreased inter-point distance within each cluster, as well as electricity sharing (transmission) cost. The minimization was conducted in every month through a simulated annealing algorithm to find optimal spatially clustered communities.

Figure 96.3 shows four-month clustering results; all clusters indicate positive storage affordability in April, May, June, July, September, and October. In other words, PV electricity covers whole household electricity demands, if EV capacities are shared with these optimized clusters.

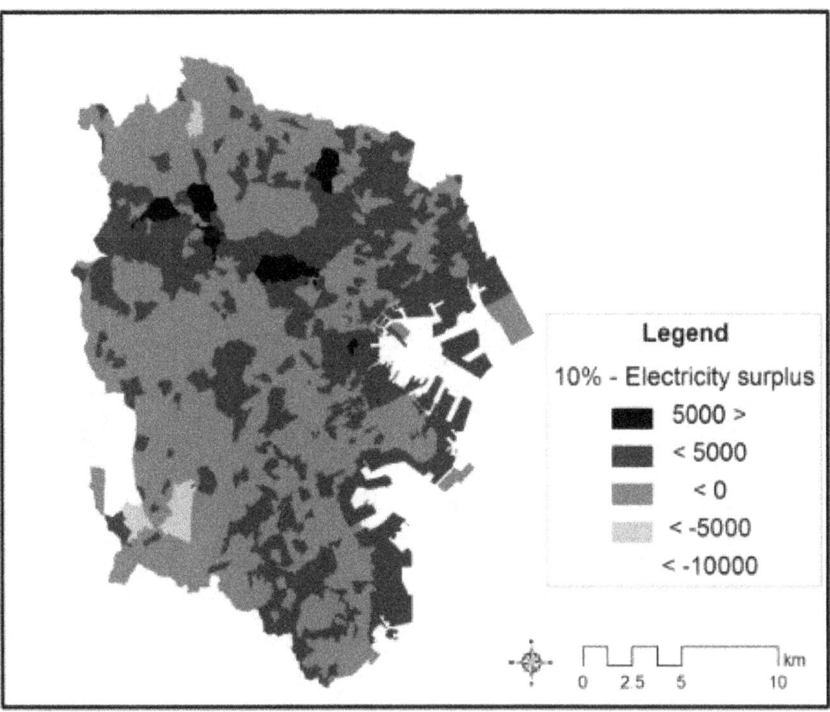

Figure 96.2: Storage affordability: Storage capacity minus electricity surplus in kWh/day (10 % of EVs not in use being used as battery).

Source: Reproduced by permission of the Institution of Engineering & Technology: published in Yamagata and Seya (2015, Figure 10.a)

Figure 96.3: Monthly clustering results.

In summary, we applied MATSim to analyze the potential of EVs in a V2C system and found that EVs can cover typical household electricity demands in some months and the cover ratio can be increased by community clustering for local electricity sharing. In the future study, we plan to use MATSim to simulate mobility behavior for electricity sharing community scenarios and extend our clustering analysis utilizing simulated behavior. Finally, development of community level mobility sharing service would be a very important topic to integrate MATSim simulations with our land use and transportation scenarios, such as compact and dispersion scenarios (see Yamagata et al., 2013).

CHAPTER 97

Research Avenues

Kai Nagel, Kay W. Axhausen, Benjamin Kickhöfer and Andreas Horni

The on-going work documented and interest expressed in the various scenarios proves that this system has not at all exhausted its possibilities as a platform for research both on and with it. This chapter highlights chances for further discussion and action.

97.1 MATSim and Agents

97.1.1 Complex Adaptive Systems

The core MATSim architecture, where agents learn utilities for plans, was originally derived from the field of Complex Adaptive Systems (CAS; e.g., Axelrod, 1984; Holland, 1992; Hraber et al., 1994; Palmer et al., 1994) (also see Section 46.1 of this book). Arthur (1994) addresses a coordination problem where agents receive a payoff only when less than 60 out of 100 go to an event. He addresses this by first generating a large number of heuristic predictors for the next round's attendance, such as "same as in last round" or "trend from last four rounds". He next gives each agent a randomly selected handful of these strategies, so that agents have different sets of predictors. Then, many rounds of the game are played, where the score of each predictor is updated based on its prediction quality, and agents act based on their currently best predictor. Simulations demonstrate that the approach leads to successful coordination, i.e., around 60 agents show up in every round. That approach, in turn, builds on work by Palmer et al. (1994), who simulate a stock market, Holland (1992), whose classifier systems have more structure than Arthur's model, but a similar model of performance learning, or Axelrod (1984), who investigates adaptive agents in the face of repeated non-cooperative games.

Arthur (1994) keeps each agent's predictors fixed after initialization. In contrast, Hraber et al. (1994) simulate an artificial ecosystem, where individual agent strategies are based on so-called genes, adapted over the rounds/iterations by genetic algorithms (Goldberg, 1989).

How to cite this book chapter:

Nagel, K, Axhausen, K W, Kickhöfer, B and Horni, A. 2016. Research Avenues. In: Horni, A, Nagel, K and Axhausen, K W. (eds.) *The Multi-Agent Transport Simulation MATSim*, Pp. 533–542. London: Ubiquity Press. DOI: http://dx.doi.org/10.5334/baw.97.

97.1.2 *Artificial Intelligence*

CAS focuses on many agents, agent interaction and emergence. Artificial Intelligence (AI) in contrast, concentrates on single agents. In AI terms, the original MATSim agents (those doing day-to-day learning) are very simple reinforcement learning agents (Russel and Norvig, 2010, Chapter 21.3). Since these MATSim agents have only one state (the initial/nightly state) and each action is simply a plan, the distinction between Q-learning and utility learning (as defined by Russel and Norvig, 2010) actually collapses; what remains is the temporal difference learning (again as defined by Russel and Norvig, 2010) scheme for the utility, which translates to the MATSim situation by updating the score/performance/utility value of each plan every time it is selected.

97.1.3 *Synthesis*

The original MATSim system thus took the focus on large systems, interaction, emergence, and strategy innovation from CAS, while the score updating comes from the AI field. In consequence, a clear path to move on is the inclusion of more modern AI aspects into the MATSim agents. Examples include:

- Extend MATSim to agents that can react immediately, rather than having to wait for the next iteration or round. In transport, this is sometimes called en-route or within-day replanning (e.g., Emmerink et al., 1995; Balijepalli et al., 2007; Axhausen, 1990). See Chapters 30 and 23 as well as Section 97.2.
- Improve the MATSim agents with respect to choice set generation. This may include both better creative capability for the agents to come up with innovative new strategies to handle their virtual lives, as well as consistency considerations between choice set generation and estimated choice models. See Sections 49.2 and 97.3.

97.2 Within-Day Replanning and the User Equilibrium

Within-day replanning, i.e., the ability of the agents to respond to the immediate context, is the standard mode of operation for simulation models. In the transport domain, note the traffic flow models as an example, where aspects such as acceleration/braking, or lane changing, are (obviously) computed reactively, while the simulation is running and not before the simulation starts (e.g. Wiedemann, 1974). Many traveler-oriented or agent-based models of travel demand adopt the same approach, cf. ORIENT (Sparmann and Leutzbach, 1980), ORIENT/RV (Axhausen and Herz, 1989), MobiTOPP (Schnittger and Zumkeller, 2004). For many aspects of Intelligent Transport Systems (ITS) systems, within-day replanning is indispensable (e.g., Hall, 1993; Emmerink et al., 1995; Dobler, 2013). None of these systems aim for equilibrium in the same way as MATSim, carried forward from TRANSIMS and originally inherited from static assignment.

One may argue that, if supplied with a learning approach, these within-day models should approach equilibrium after many iterations, as agents with a suitable memory structure would avoid plans that could put them at a disadvantage. This memory, which would need to be agent-specific and covering the very large set of choice options, makes the approach costly to implement. Importantly, the solutions may be different: when faced with a stochastic environment, an agent able to react within-day could be better off than an agent following a pre-computed plan. This is important: finding a plan with the highest expected score is not the same as finding a conditional strategy with the highest expected score.

Still, there are contexts where this immediate response ability can be used within MATSim to explore the choice set more effectively, especially if the choice alternatives are limited and within geographic reach. Waraich et al. (2013a) proposes within-trip replanning to find the best parking space near a destination. This localized search reduces the need for full iterations considerably and allows addition of behavioral detail at this point (here the type of parking), walking distance to final destination, and parking fee trade-off.

While within-day replanning can be used as described above within the framework equilibrium search, it can also be added to open up the MATSim framework to contexts where such an equilibrium is inappropriate. Dobler (2013) uses the MATSim calculated equilibrium as the starting point for his model of evacuations and the behavior of evacuees. He also explains that his approach finds a set of executed plans close to the MATSim equilibrium, but for much lower computing cost. While the benefits of an equilibrium solution in comparison with an approximation have been extensively discussed for aggregate assignment models, for MATSim the issue is whether these fast approximations could be used to speed up the overall equilibrium search; similar to starting a Frank-Wolfe search based on four or five incremental loadings of the network (Jourquin and Limbourg, 2006).

While aggregate assignment can identify the routes chosen as belonging to the equilibrium, research in the agent-based context is needed to see: a) if the approach is indeed faster and b) if the resulting set of plans is unbiased by the fast initialization.

97.3 Choice Set Generation

As described at several places in this book (e.g. Sections/Chapters 3.1, 4.5.1, 49.2, 27, 47, 49), the MATSim iterative process in its standard version modifies each agent's choice set (= each agent's set of plans) over the iterations. Clearly, an agent can only select a plan generated by this process. Thus, search space definition is important.

97.3.1 The Statistical Weight of Each Plan

Econometric research (e.g., Ben-Akiva and Lerman, 1985, Chapter 8 and 9) points out that it is not sufficient if certain alternatives are eventually discovered by the search process; rather, it is important that they are generated with probabilities consistent with the choice model. This, however, is at odds with the CAS approach, where solutions are generated rather arbitrarily. For example, Arthur (1994) "create[s] 'an alphabet soup' of predictors" that are "randomly ladle[d] out".[1] Research is needed to clarify when statistical properties of the choice set need to be tightly consistent with the choice model and when not.

As a result, in the MATSim context, it is important to look not only at plans generation/innovation (e.g., Section 4.5.1.1), but also at plans removal. The default MATSim approach is to remove the plan with the worst score. This is, however, problematic both from a CAS and an econometric perspective. From a CAS perspective, such an approach simply does not generate enough diversity, since similar scores rather often mean similar plans; thus, the approach has a tendency to remove the most different plan, typically leading to a set of plans that are all quite similar. From an econometric/discrete choice perspective (cf. Section 49.2), the combination of plans generation and plans removal needs to ensure that each plan's probability of being in the choice set corresponds to its weight used in the choice model estimation.

Section 49.2 discusses a version of the plans' generation/removal process, but makes rather strong assumptions about the capability to compute best-response plans. Here, let us instead consider a heuristic argument. Assume that plans i for a person n are created with a certain probability $p_{n,i}^{create}$; the person index n will be dropped in the following. Also assume that plans are removed with probability p_i^{remove}. The master equation for the probability q_i of plan i to be contained in the choice

[1] To be precise, Arthur uses strategies that eventually *generate* choices.

set becomes in leading order[2]

$$\frac{dq_i}{dt} = -q_i \cdot p_i^{remove} + p_i^{create} .$$

The steady state solution, obtained from $dq_i/dt = 0$, is given by

$$q_i = \frac{p_i^{create}}{p_i^{remove}} . \tag{97.1}$$

That is, quite obviously, if one wants to control the statistical distribution q_i within the MATSim process, one needs to look not only at plans generation, but also at plans removal.

MATSim's plans generation model can be approximately described by a creation probability of $p_i^{create} \sim \exp(\beta^{create} S_i)$, with relative large β^{create}, corresponding to an approximate best-response model.[3] At the same time, removal of the worst plan corresponds to $p_i^{remove} \sim \exp(-\beta^{remove} S_i)$ with a very large β^{remove}. Overall, thus

$$q_i \sim \exp((\beta^{create} + \beta^{remove}) S_i) .$$

Combining this with a choice model that selects with $\sim \exp(\beta^{choice} S_i)$ from the set of plans, i.e. `ChangeExpBeta` or `SelectExpBeta`, leads to

$$p_i \sim \exp(\beta^{choice} S_i) \cdot q_i \sim \exp((\beta^{choice} + \beta^{create} + \beta^{remove}) S_i) . \tag{97.2}$$

Let us again stress that this is not an exact statistical analysis of the MATSim dynamics, but instead an illustrative approximation to gain insight. From this approximation, it becomes clear that MATSim in its current form, because of the strong additional effects of plans generation, expressed through β^{create}, and plans removal, expressed through β^{remove}, strongly over-weighs plans with high scores. *It is thus important to include plans removal in all considerations,* since otherwise the very large β^{remove} in Equation (97.2), coming from always removing the worst plan, will dominate the statistical distribution.

97.3.2 Heterogeneity in Plans Removal

Clearly, removing not the worst plan, but instead according to some logit model with a smaller β^{remove} would improve the situation. In addition, to increase diversity and simultaneously correct for correlations between alternatives, one could use an (inverse) path-size logit (e.g., Frejinger and Bierlaire, 2007; Prato, 2009; Schüssler, 2010) model, i.e.,

$$p_i^{remove} \sim \exp(-\beta^{remove} S_i + \alpha \, PS_i)$$

where PS_i would be an index of similarity of plan i to all other plans in the plans set. As a result, plans very similar to other plans in the set would have a greater chance of being removed. The last of such similar plans would no longer be similar to any other plan, thus PS_i would be small; that plan would be less likely to be removed.

Such an approach is experimentally available as `PathSizeLogitSelectorForRemoval`. It possesses an ad-hoc similarity computation of one plan to all other plans in the set (Grether, 2014). Further investigations using this approach should be performed.

---

2 In higher order, one would have to correct for the possibility that a plan may appear more than once in the choice set.

3 The operator $\sim$ means "proportional to in leading order". It neglects, for example, the effect of the denominator in a logit model.

97.3.3 Heterogeneity in Plans Generation

No extended research has yet been undertaken to see whether MATSim could adopt the strategy of regularly introducing new random starting solutions to avoid local minima. One challenge: the generation of such random plans would result—in most cases—in nonsensical plans, which would need to be removed through computationally expensive iterations. See Feil (2010) for difficulties in constructing optimal alternative plans for the number and sequence of activities. One possibility is to initially allocate a small set of randomly chosen alternative day plans and measure their similarity throughout the simulation. There is no research on a replanning technique involving switching of the day plan activity order, which again would produce more dissimilar plans than currently possible.

Moyo Oliveros and Nagel (in press) and Nagel et al. (2014) report computational experiments where a randomized Pareto router is used to generate a different route every time it is called. The Pareto router randomly draws a trade-off between different utility function contributions, such as fare/toll, travel time, access/egress time, then computes an optimal route based on the resulting generalized cost. The randomized approach considerably reduces the requirement that the router be consistent with the scoring function. The randomized Pareto router generates a collection of possible routing solutions; each agent then can select one that best suits its own trade-off between monetary budget and time pressure. Heterogeneity is generated by each synthetic traveler having a different trade-off.

The approach of Horni (2013, also see Chapter 27 of this book) can also be seen in this sense: attaching a random error term to each location-person-pair means that two persons—at exactly the same home location with exactly the same activity pattern—will select different locations for their activities. So far, this describes heterogeneity between persons. However, the approach also generates more heterogeneity per person, since the destinations attractive to each synthetic person will be spatially more spread out than they might otherwise be.

97.3.4 Deliberate Search Strategies

The need for a strategies meta-search, as sketched by Arthur (1994), remains an open question. In the MATSim context, all decisions, based on explicit search for alternatives, can be studied to see how far apart choice set generation strategies of discrete choice modeling (which draw from the universal choice sets), are from explicit construction strategies. One idea would be to observe the second step to see what impact these would have on the results and the policy conclusions.

A good example is parking search (Waraich et al., 2012), for which multiple strategies have been documented and which explains the empirical observations (Shoup, 2005). In a discrete choice model context, distribution of parking preferences can mimic choice strategies, but the approach could not capture the context-specific strategy choice. In MATSim, this set of strategies could become the object of a meta-search to see which agents would retain which strategies and how these would be used by the agents. Empirical work could be conducted to see whether these sets and their distributions match travelers' practices.

In the same vein, one could look at the leisure destination choice, where different strategies can be observed, although they have not yet been subject of empirical study. If longer-term choices were added to the MATSim framework, residential and workplace choice could also be considered.

MATSim plans' convergence towards a single optimal structure can be seen as the absence of search strategies on the plan level. This overlaps strongly with the question of number choice and activities sequence, where these alternative plans are needed.

97.3.5 *Transients Versus the Notion of "Learning"*

As in other similar simulations, interpretation of the relaxation procedure (iterations) of MATSim is unclear. Sometimes, the relaxation process is ascribed a behavioral interpretation: for example, day-to-day learning, where the transition process, as well as the final equilibrium, has a meaning (Liu et al., 2006, p.128), (Nagel and Barrett, 1997, p.523). An opposite viewpoint exists, where the relaxation procedure is just a numerical method to compute the equilibrium state, or states, without a behavioral basis of the transitions. Although this interpretation ambiguity has not hampered development process so far—also because, in discrete choice modeling, the same ambiguity exists—it is obvious that future questions about adoption of behavioral versus statistical methods require MATSim interpretation.

97.4 Scoring/Utility Function and Choice

97.4.1 *Discussion of the Present Scoring Function Mathematical Form*

The current logarithmic MATSim activity scoring function,

$$S_{act,q} = \beta_{dur} \cdot t_{typ,q} \cdot \ln(t_{dur,q}/t_{0,q})$$

(cf. Equation (3.2), with $t_{0,q}$ as defined by Equation (3.7), is not suitable for modeling activity addition and dropping (Feil, 2010, p.127f). As already stated in Section 3.3.1, the problem is that, at the typical duration, i.e., at $t_{dur,q} = t_{typ,q}$, all activities generate the same score, independent of their actual duration; thus, it makes sense to first drop the longest activity, since that generates the least amount of utility *per time unit.* This is typically the home or work activity; dropping this first clearly is nonsensical.

The property that all activities have the same utility at their typical duration is obtained by computing the value of the parameter $t_{0,q}$ from the condition[4]

$$const \cdot \beta_{dur} \stackrel{!}{=} S_{act,q}\Big|_{t_{dur,q}=t_{typ,q}} = \beta_{dur} \cdot t_{typ,q} \cdot \ln(t_{typ,q}/t_{0,q}) \tag{97.3}$$

and therefore

$$t_{0,q} = t_{typ,q} \cdot \exp\left(-\frac{const}{t_{typ,q}}\right) \tag{97.4}$$

(cf. Equation (3.7) with $10h \to const$ and $prio \to 1$).

97.4.2 *Utility at Typical Duration Proportional to Typical Duration*

As an alternative, Equation (97.3) could be replaced by the requirement that all activities at their typical durations yield a score proportional to their typical duration, i.e.,

$$const \cdot \beta_{dur} \cdot t_{typ,q} \stackrel{!}{=} S_{act,q}\Big|_{t_{dur,q}=t_{typ,q}} = \beta_{dur} \cdot t_{typ,q} \cdot \ln(t_{typ,q}/\tilde{t}_{0,q}) \,, \tag{97.5}$$

leading to

$$\tilde{t}_{0,q} = t_{typ,q} \cdot \exp(-const) \,. \tag{97.6}$$

That is, replacing Equation (97.4) by Equation (97.6) in the MATSim scoring function would make, in first order, all activities equally likely to drop. Starting with MATSim release 0.8.x, there will be a config switch

---

4 The notation $S\Big|_{x=a}$ means that the expression S shall be evaluated at $x = a$.

```
<param name="typicalDurationScoreComputation" value="..." />
```

where `uniform` will mean the old behavior and `relative` the behavior suggested in this section. Consequences of this still need to be investigated.

97.4.3 S-Shaped Function

A new S-shaped function, proposed by Joh (2004), was tested by Feil (2010, p.129ff). It starts horizontally at zero duration, bends upwards with a positive second derivative and then changes curvature to the normal negative second derivative at longer durations. The function was motivated by the observation that utility functions with infinite (i.e., diverging) first derivative at duration zero lead to "doing a little bit of everything". This is also known from regular consumer theory, with activities replaced by goods. The S-shaped function avoids that problem, instead implying that activities below a certain duration should instead be dropped completely.

Estimates of the new function, based on the Swiss microcensus, were provided; this estimation, however, was difficult, which was attributed to the non-linearities of the function, and to the difficulty in generating sufficiently large choice sets. In addition, many daily activities and their durations were not chosen freely by the individual. Consequently, it is currently *not* advisable to replace the MATSim default scoring function with the Joh/Feil approach.

97.4.4 Heterogeneity of Alternatives and Challenges of Estimation

It is normal to differentiate between types of alternatives in the average; for example, trips by different modes, or with different purposes, are commonly assigned different time values. However, there are also large deviations from those averages between travelers. A possible approach to address this are so-called taste variations, i.e., to make some parameters of the utility function random, but fixed per agent; parameters of this randomness are made part of the choice model estimation. However, some of this apparent randomness may, in fact, be causal. For example, higher values of time for commuting than for leisure may be caused by the more crowded daily schedule on working days. Similarly, the strength of a preference for public transit may be caused by the walking distance to that transit stop serving the desired destination.

Simulation systems such as MATSim should be able to explicitly integrate alternatives' heterogeneity. Besides the aspects discussed in Section 97.4, it is desirable to know how the following aspects influence the scoring function:

- access/egress times to/from public transit,
- transfers between public transit lines,
- crowding in public transit vehicles,
- parking search,
- types of parking (on-street, guarded, sheltered, etc.), and
- personal or household income.

Clearly, this list is not complete.

For most of these aspects, initial studies within the MATSim context are available, see, e.g., Moyo Oliveros and Nagel (2012, in press) for access/egress times and transfers to PT (Public Transport), Bouman et al. (2013), Sun et al. (2014a) or Erath et al. (in preparation) for crowdedness, Waraich et al. (2013b) for parking search, or Kickhöfer et al. (2011) for income. In some cases, it is even possible to configure these elements through the standard config file, completely without Java programming. It is also quite clear that these issues were addressed outside the MATSim context.

A challenge, however, is that it is normally not possible to just collect and combine results from different studies, for the following reasons:

- It is not correct to take an estimated utility function and then change the list of attributes. For example, if walking access/egress to/from PT is not included in the estimation, then its effect may be partially be included in the alternative-specific constant, or in the population density (which may serve as a proxy for the density of PT access points). Just adding the effect of walking access/egress from some other study is thus incorrect.
- Even when MATSim is able *in principle* to add these elements, doing so in practice poses a considerable statistical challenge. For example, one may assume that households inside a zone self-select their precise residential location based on the PT accessibility of their regular destinations. In contrast, a typical MATSim initial demand generation process will first assign residential locations, then generate their destinations, e.g., their workplaces. Thus, persons who might reach their destinations easily by PT might have their MATSim residences far away from the relevant PT stop.

Therefore, it is necessary to estimate the scoring function with exactly those attributes available in the simulation with sufficient precision. Kickhöfer (2009) has, in consequence, re-estimated his scoring function based on data from Vrtic et al. (2008). For the same reasons, it is not possible to combine functions independently estimated for different choice dimensions. This is not even possible when they all contain monetary units. For example, assume that one has

$$\ldots + \beta_t \, t_{trav} + \beta_m \, \Delta m + \ldots$$

for mode choice, and

$$\ldots + \beta_r \, \rho + \tilde{\beta}_m \, \Delta m + \ldots$$

for parking, where t_{trav} is the travel time, ρ is congestion in a parking lot, and Δm is, in all cases, the change in the monetary budget, e.g., cost for gas, PT fare, or parking. Even then, it is not possible to say

$$\ldots + \beta_t \, t_{trav} + \beta_m \, \Delta m + \beta_r \, \frac{\beta_m}{\tilde{\beta}_m} \, \rho + \ldots ,$$

since that confuses the scale parameters of the two separate estimations.[5] If only travel time is available as common attribute, the situation deteriorates, since time valuation in MATSim is non-linear; thus, operating points for linearization need to be defined, or found by iterative procedures (Horni, 2013, p.75ff).

As a long-term perspective, one could also imagine estimating choice models directly inside MATSim, possibly taking hints from UrbanSim which has such an approach at its core. An early step in this direction within MATSim using Cadyts (see Chapter 32), is described by Flötteröd et al. (2012).

97.4.5 *Agent-Specific Preferences*

MATSim scenarios so far consider a relatively small set of agent attributes, essentially because of missing data suitable for deriving detailed large population attributes (Müller and Flötteröd, 2014). Some studies, however, used larger sets of attributes. Grether et al. (2010); Kickhöfer et al. (2011) estimated individual income-contingent utility functions. Horni and Axhausen (2012b,a)

---

5 Realistically, combining separate estimations via their conversion in monetary terms may be the best one can do in many situations.

incorporated agent-specific travel preferences and individual income-dependent marginal utilities of money; preference values, however, were assigned randomly. Because consideration of agent-specific preferences is one of the cornerstones of agent-based microsimulations, future work should exploit this avenue.

97.4.6 Frozen Randomness for Choice Dimensions Other Than Destination Choice

For destination choice, an iteration-stable random error term has been successfully applied to incorporate unobserved heterogeneity not included by the stochasticity of the co-evolutionary process (see Chapter 27). Other choice dimensions might also benefit from explicit agent-specific error terms. This could incorporate a mechanism to generate the error terms with the correct correlation structures.

More formally: The current MATSim choice process can be interpreted as maximizing, for each agent n,

$$U_{ni} = V_i + \tilde{V}_{ni} + \boldsymbol{\beta}^T \boldsymbol{\eta}_{\mathbf{ni}} + \tilde{\varepsilon}_{ni} \,, \tag{97.7}$$

where V_i is the systematic utility of alternative i, $\tilde{V}_{ni}$ is an agent-specific addition, $\boldsymbol{\beta}^T \boldsymbol{\eta}_{\mathbf{ni}}$ describes randomness inserted by the network loading model (see Equations (49.4) and (49.5)), and $\tilde{\varepsilon}_{ni}$ is remaining (unexplained) noise. Two challenges are:

- $\tilde{V}_{ni}$ denotes aspects often assumed as random in choice models, but fixed in typical MATSim runs. An example is walking distance to the next PT stop, which may have to be assumed as random in an estimation context based on travel analysis zones, but which is fixed in the context of a MATSim run.

  $\rightarrow$ To be consistent, a choice model and a MATSim implementation used together should use exactly the same disaggregated attributes.
- In most MATSim runs, the $\tilde{\varepsilon}_{ni}$ are either assumed as zero (`BestScore`), or are parameterized by the MATSim choice model (`ChangeExpBeta` or `SelectExpBeta`), which can be interpreted as that the $\tilde{\epsilon}_{ni}$ are re-drawn from the distribution every time a choice is made. This leads, for example, to purely random "logit" switchers between a base and a policy case (e.g., Grether et al., 2010).

  Moreover, the default plans removal (Sections 4.5.1.4 and 97.3.2) has a tendency to remove all alternatives except the best, effectively setting the $\tilde{\varepsilon}_{ni}$ to zero for *all* typical MATSim configurations when run for sufficiently many iterations.

  This is acceptable in situations where most of the noise can be assumed to be in the $\tilde{V}_{ni}$ and/or the $\boldsymbol{\beta}^T \boldsymbol{\eta}_{\mathbf{ni}}$ (and thus generated with hopefully plausible structure by the MATSim dynamics); this may be the case for the choice dimensions of route, mode, and time. It is clearly wrong for locations where $\tilde{\varepsilon}_{ni}$ subsumes preferences that are specific to each person-alternative-pair and that often cannot be included into the $\tilde{V}_{ni}$. For example, a person may have a strong preference for "swimming" in a situation where the data only knows about "leisure" facilities. In this situation, a possible approach is to generate random but "frozen" $\tilde{\varepsilon}_{ni}$, as described in Chapter 27.

  $\rightarrow$ One should thus evaluate how far, and how, a similar approach could be introduced for choice dimensions beyond destination choice.

97.4.7 Economic evaluation

As the above Section 97.4.6 already indicates, further work is desirable to better understand the connection between MATSim scores and utility from consumer theory. At face value, Equation (97.7) could be taken as each agent's utility. As also discussed in Section 51.2.5, problems arise when the $\tilde{\epsilon}_{ni}$ are not explicitly known for each person-alternative-pair *ni*.

In many past MATSim studies, their effect has therefore been parametrized by a logit choice model (with the use of `Change/SelectExpBeta`). In these cases, the logsum of all plans' scores of an agent is each agent's correct utility measure. See Section 51.2.5.1 for details.

In other MATSim studies, the $\tilde{\epsilon}_{ni}$ were effectively assumed to be zero (with the use of `BestScore`). In these cases, the highest of all plans' scores of an agent is each agent's correct utility measure. Because of the `BestScore` plan selection model, the plan with the highest score will at the same time also be the selected plan. See Section 51.2.5.2 for details.

For some of us, it seems attractive to move into the direction of working with frozen randomness, as discussed in Section 97.4.6. That approach would combine the advantages of the two approaches from above: It would inherit the parsimonious interpretation of the `BestScore` approach, where only the plan with highest score (= the selected plan) of each agent needs to be considered, and at the same time include the idea of random utility theory, and, hence, the effect of the ϵ_{ni} on individual choices.

Another avenue of research is to further push the understanding of the econometric and statistical properties of the MATSim choice modeling, cf. Section 51.2.5.5.

97.5 Double-Queue Mobsim

The standard MATSim mobsim QSim implements a single-queue model as described in Chapter 50. The associated FD (flow vs. density) is horizontal for medium densities, and falls to zero very steeply at very high densities. This is consistent with the fact that a vehicle leaving a link opens up its space already in the next time step; jam patterns thus have a backwards traveling speed of $L/1\,s$ (L is the length of respective link) rather than the conventional approx. 15 kilometers per hour (see also Charypar et al., 2009).

The JDEQSim (Section 4.3.2) and the deprecated DEQSim (Section 43.1) implement a double-queue model with backward traveling gaps. Recently, the QSim has also implemented a double-queue variant (Agarwal et al., 2015a), switched on by using a "holes" option in the config; it is, however, not yet thoroughly tested.

97.6 Choice Dimensions, in particular, Expenditure Division

As shown in Section 46.2.2.3 and pictured in Figure 46.1, the Zürich group targets a fuller scheduling model. In addition to standard choice dimensions (printed in red in the cited figure), numerous choices are subject to ongoing research. In particular, "expenditure division" is unexplored not only in MATSim, but in transport planning in general; studies have focused on single-travelers or household-based groups. The field's understanding of both expenditure patterns and allocation styles inside a household are poor, which is no surprise since relevant questions are missing in surveys. First tests for necessary survey works are currently in process and will lead to a better understanding of activity participation and time values that travelers bring to their decisions.

97.7 Considering Social Contacts

Apparently, social contacts, within households as well as within extended social networks, have a substantial influence on travel decisions, particular for social activities in leisure time (Kowald et al., 2009). An early social networks study, in context, but not based on MATSim, is by Marchal and Nagel (2005). Further work based on or, again, in context of MATSim was undertaken by Hackney (2009); Illenberger (2012); Illenberger et al. (2011); Kowald et al. (2009). The most recent work on joint trips is reported in Chapter 28. Despite this range of valuable work, future research is required on this topic, especially for leisure destination choice (Horni, 2013).

Acronyms

ABM Activity-Based Model. 510, 513
ABMS Agent-Based Modeling and Simulation. 203, 204
AI Artificial Intelligence. 24, 534
API Application Programming Interface. xxxi, xxxii, 9, 231, 232, 255, 267, 290, 294, 296, 297, 488, 544
ARC Australian Research Council. xxiii
ARTEMIS Assessment and Reliability of Transport Emission Models and Inventory Systems, a harmonized emission model in the EU, possibly a predecessor of HBEFA, version 3.1. 248
ASET Available Safe Evacuation Time. 440
ASTRA BundesAmt für STRAssen, French: Office fédéral des routes OFROU, Italian: Ufficio federale delle strade USTRA. xx

BABS Bundesamt für Bevölkerungsschutz, Switzerland. xxii
BASt Bundesanstalt für Straßenwesen. 434
BCA Benfit-Cost Analysis. 353, 360, 362
BDI Belief Desire Intention. xxiii, 201–210
BEV Battery Electric Vehicle. 95
BfS Bundesamt für Statistik - Federal Statistical Office. xxii
BMBF Bundesministerium für Bildung und Forschung/Federal Ministry of Education and Research. xxi, xxii
BVG Berliner Verkehrsbetriebe (started as Berliner Verkehrsaktien-Gesellschaft). xxii, 113, 369

CA Cellular Automaton. 403
Cadyts Calibration of Dynamic Traffic Simulations. See Chapter 32 and Flötteröd(accessed 2015). 213–215, 372, 433–435, 540
CART Committee on Advanced Road Technology. xxiv
CAS Complex Adaptive Systems. 3, 533–535
CCEM Competence Center Energy and Mobility. xxii
CDR Call Detail Record. 397, 486, 488
CEMDAP Comprehensive Econometric Microsimulator for Daily Activity-Travel Patterns (Bhat et al., 2008). 372

CO_2 Carbon Dioxide. 407
ComPuTr Complexity in Public Transport. xxiv
CONICYT Comisión Nacional de Investigación Científica y Tecnológica - National Commission for Scientific and Technological Research. xxiii
COOPERS Co-Operative Networks for Intelligent Road Safety. xxi
CREATE Campus for Excellence and Technological Enterprise. 379
CSV Comma-Separated Values. 257
CTI Commission for Technology and Innovation. 295
CUDA Compute Unified Device Architecture, a parallel computing platform and API by NVIDIA. 267, 303

DAAD Deutscher Akademischer Austauschdienst - German Academic Exchange Service. xxii
DB AG Deutsche Bahn AG. 431
DB ML AG DB Mobility Logistics AG. 431
DEQSim Discrete Event Queue Simulation. 6, 38, 267, 285, 542
DFG Deutsche Forschungsgemeinschaft. xx–xxii, 295
DMQ District Métropolitain de Quito, Quito Metropolitan District, Ecuador. 473–475
DRT Demand Responsive Transport. 152, 527, 528
DRVs Disaster Response Vehicles. 461, 462, 465–467
DTA Dynamic Traffic Assignment. 213, 315, 320, 326, 513, 518
DTD Document Type Description. 43, 61
DVRP Dynamic Vehicle Routing Problem. 146–148, 151, 152, 471, 528

EBPTR Events-Based Public Transport Router. 124, 127, 128, 130, 131
EMEF Enquesta de Mobilitat en dia Feiner, Barcelona's annual transport survey. 398
EMME Equilibre Multimodal Multimodal Equilibrium. See `http://www.inrosoftware.com/en/products/emme/`. 50, 62, 115, 495, 523, 524, 544
EMME/2 Version 2 of EMME. 115, 495, 497, 515
EMT Emission Modeling Tool developed by Hülsmann et al. (2011) and Kickhöfer et al. (2013), see Chapter 36. 247–251, 383
EMU Expected Maximum Utility. 358, 550
EPSG European Petroleum Survey Group. 13
ERA European Research Action - Country consortia. xxii, xxiv
ERD Entity Relationship Diagram. 256, 257
ERP Electronic Road Pricing. 381
ESRI Environmental Systems Research Institute. 224, 274, 276, 440, 515
ETH Eidgenössische Technische Hochschule. xx–xxii, 159, 219, 290, 309, 310, 313, 379, 397, 497
EU European Union. xx–xxiv, 360, 384, 543
EUNOIA Evolutive User-centric Networks for Intraurban Accessibility `http://www.eunoia-project.eu`. xxiii, xxiv, 397
EV Electric Vehicle. 92–95, 529–532

FCL Future Cities Laboratory. xxii, 265, 290, 379
FD Fundamental Diagram. 347, 348, 351, 542
FEATHERS Forecasting Evolutionary Activity-Travel of Households and their Environmental Repercussions (Arentze and Timmermans, 2006). 176, 372
FIFO First In, First Out. 79, 152, 459, 460
FIT feed-in tariff. 529
FONDECYT Fondo Nacional de Desarrollo Científico y Tecnológico. xxiii

GA Genetic Algorithm. 394

GB Gigabyte. 11, 128, 131, 132, 310, 510, 511
GDP Gross Domestic Product. 429
GFIP Gauteng Freeway Improvement Project. 429
GIS Geographic Information System. 13, 240, 254, 277, 280, 281, 386, 408, 497, 503
GmbH Gesellschaft mit beschränkter Haftung. 503, 506
GPL GNU General Public License. 161
GPLv2 GNU General Public License version 2.0. 290
GPLv3 GNU General Public License version 3.0. 213
GPS Global Positioning System. 13, 19, 115–118, 120, 155, 223, 501
GRIPS GIS-based Risk analysis, Information, and Planning System for the evacuation of areas. xxi
GTFS General Transit Feed Specification. xix, 116, 117, 120, 121, 486, 491, 493, 524
GTHA Greater Toronto and Hamilton Area. 523
GUI Graphical User Interface. 10, 21, 254, 257, 272, 274, 278, 281, 440

HAFAS HaCon Fahrplan-Auskunfts-System. 110
HBEFA Handbook on Emission Factors for Road Transport, version 3.1. See `http://www.hbefa.net`. 247–252, 543
HHTSD Household Travel Survey Data. 497
HITS Household Interview Travel Survey. 379, 380
HOV Highly Occupancy Vehicle. 407
HPCC High-Performance Computing Clusters. 37
HSAR Household Sample of Anonymised Records. 448

IATA International Air Transport Association. 422
ICT Information and Communications Technology. 145, 485
IDE Integrated Development Environment. 66, 295, 309
IfV Institut für Verkehrswesen/Institute for Transport Studies. 312, 313
ILS Institut für Land- und Seeverkehr – Institute for Land and Sea Transport Systems. 290, 467
INTS Irish National Travel Survey. 414
IPF Iterative Proportional Fitting. 373, 379
ITS Intelligent Transport Systems. 534
ITSOS Intermodal Transport Simulation & Operation System. 503–506
IVT Institut für Verkehrsplanung und Transportsysteme – Institute for Transport Planning and Systems. xxii, 159, 290, 377

JAR Java ARchive. 10, 35, 66, 100
Java SE Java Standard Edition. 9
JAXB Java Architecture for XML Binding. 292
JDBC Java Database Connectivity. 255
JDEQSim Java Discrete Event Queue Simulation. 6, 38, 90, 263, 267, 268, 285, 347, 349–351, 542
JOGL Java OpenGL. 231
JOSM Java Open Street Map Editor. 65, 66

KiD Kraftfahrzeugverkehr in Deutschland. 161
KPI Key Performance Indicator. 478
KTH Kungliga Tekniska Högskolan – Royal Institute of Technology. xxiv
KTI Kommission für Technologie und Innovation. xxii
KVMZH Kantonales Verkehrsmodell Zürich. 155, 375
KWM Kinematic Wave Model. 348, 350, 351

LANL Los Alamos National Laboratory. 308
LAU Local Administrative Unit. 400, 431
LTA Singapore Land Transport Authority. 380
LTDS London Travel Demand Survey. 448, 449
LUTI Land-Use and Transport Interaction. 403

MATSim Multi-Agent Transport Simulation. See `http://www.matsim.org`. xix–xxii, xxxi, xxxii, 3–7, 9–21, 23, 24, 26, 27, 29–32, 35–42, 47–51, 55–58, 61, 62, 65–67, 70, 77, 78, 83–93, 95, 97–100, 105–107, 109–113, 115, 121, 123, 124, 126–128, 130–132, 135, 136, 140, 142, 143, 146–151, 155–157, 159, 165–168, 170–172, 175, 177–179, 183, 188, 190–192, 196, 197, 200–210, 214, 215, 219–221, 223–227, 231, 237–245, 247, 249–255, 257, 259, 263, 264, 267, 268, 272–274, 278, 280, 281, 283–286, 289, 290, 292–301, 303, 304, 307–315, 320–322, 324, 326–332, 335, 337–344, 347–351, 353–356, 358–360, 362, 363, 367–369, 371–373, 375–377, 379, 381, 383, 385–387, 391–395, 399–402, 405, 408, 409, 411, 413, 414, 416, 417, 419–421, 423, 427–429, 431, 433, 434, 436–438, 441, 445, 449, 451, 453, 457, 459, 461, 462, 467, 469, 473, 474, 477–479, 482, 483, 487, 489, 491–495, 497, 499, 501–506, 509–511, 513, 515, 518–520, 523–525, 528–530, 532–542, 549, 553, 554
MAUP Modifiable Areal Unit Problem (Openshaw, 1984). 242
MB Megabyte. 11, 130, 131
MFD Macroscopic Fundamental Diagram. 387
MH Metropolis-Hastings. 342, 343
MiD Mobilität in Deutschland (MiD 2002, Follmer et al., 2004). 383, 432, 434
MIMOSA Modélisation Isentrope du transport Méso-échelle de l'Ozone Stratosphérique par Advection. 248
MINTE MItigating Negative Transport Externalities in industrialized and newly industrializing countries. xxi
MLIPF Multi-Level Iterative Proportional Fitting. 451
MNL Multinomial Logit Model. 41, 169, 286, 338, 354, 357, 363, 423, 501
mobsim Mobility simulation. Also, depending on the context and specific mobility simulation capabilities, called network loading, traffic flow simulation or synthetic reality. 4–6, 11, 12, 16, 17, 19, 21, 27, 31, 36, 37, 127, 130, 134–136, 170, 191, 192, 197, 202, 226, 227, 230–232, 263, 267, 268, 286, 292, 297, 298, 301, 303, 304, 309, 310, 312, 347, 350, 542, 551
MOVES Motor Vehicle Emission Simulator. See `http://www.epa.gov/otaq/models/moves/`. 248
MPI Message Passing Interface. 267, 310
MRE Mean Relative Error. 394
MRT Mass Rapid Transit, Singapore. 130, 381
MSA Method of Successive Averages. See, e.g., Liu et al. (2007). 31, 32, 41, 322, 324, 325, 338
MTC Metropolitan Transportation Commission. 486–488
MVI An OTFVis Movie File, not to be confused with the "Musical Video Interactive" file usually abbreviated mvi. 225–231

NFP Nationales Forschungsprogramm. xxii
NFPA National Fire Protection Association. 389
NHTS National Household Travel Survey. 429
NO_2 Nitrogen Dioxide. 384
NOK Norwegian Krone. 526
NPTS US Nationwide Personal Transportation Survey. 386
NRF Singaporean National Research Foundation. xx, xxii
NYBPM Nederlandse Organisatie voor Wetenschappelijk Onderzoek – Netherlands Organization for Scientific Research. xxiv

NYBPM New York Best Practice Model. 453

O-D Origin-Destination. 126, 224, 302, 316–319, 325, 371, 372, 386, 391, 405, 408, 409, 422–427, 429, 431–436, 469, 471, 482, 487, 497, 501, 515, 518, 519, 530, 552
OAG Official Airline Guide: `http://www.oag.com/`. 420–422
ODBC Open Database Connectivity. 254
OpenGL Open Graphics Library. 231, 232
OS Operating System. 226, 231
OSM OpenStreetMap (OpenStreetMap, 2015). xix, xxiv, 20, 58, 62, 65, 66, 220, 242, 245, 273, 274, 276, 281, 387, 391, 400, 401, 405, 408, 411, 413, 414, 429, 432, 438–441, 469, 473, 477, 482, 486, 491, 493, 501, 507, 528, 551
OTFVis On The Fly Visualizer. 19, 225–233, 483

PCU Passenger Car Unit. 459
PEMS Transportation Performance Management System. 487
PETRA PErsonal TRansport Advisor. xxiv
PhD Philosophiae Doctor - Doctor of Philosophy. 290, 308, 312
PHEM Passenger Car and Heavy-duty Emission Model. 248
PHEV Plugin Hybrid Electric Vehicle. 94
POI Point of Interest. 414
POWSCAR Place of Work, School or College Census of Anonymised Records. 414
PSim Pseudo-Simulation. 263–266, 268
PSRC Puget Sound Regional Council. 495
PT Public Transport. 133, 134, 357, 491, 493, 539–541
PTO Public Transit Operator. 478
PV Photovoltaic Panel. 529–531

QGIS Quantum GIS. 13, 50, 391, 445

RAM Random Access Memory. 11, 226, 268, 309, 510, 511
RMIT Royal Melbourne Institute of Technology. xxiii
RSET Required Safe Egress Time. 440

SANRAL South African National Roads Agency Limited. 429
SATURN Simulation and Assignment of Traffic to Urban Road Networks (SATURN, 2014). 405
SBPTR Schedule-Based Public Transport Router. 123, 128, 130, 131
Scala SCAlable LAnguage. See `http://www.scala-lang.org/`. 50
SCCER Swiss Competence Center for Energy Research. xxii
SEC Singapore-ETH Center for Global Environmental Sustainability. 379
SI Système International (d'Unités): International System (of Units). 56
SMA Seoul Metropolitan Area. 497
SNF Schweizerischer Nationalfonds. xx, xxii, 294
SOC State of Charge. 95
SOP Signal Optimizer. 473, 474
SPI Service Provider Interface. 298
SPSS Statistical Package for the Social Sciences. 497
SQL Structured Query Language. 254, 257
SrV System repräsentativer Verkehrsbefragungen (Ahrens et al., 2009). 372
SUE Stochastic User Equilibrium. 316, 317, 319–322, 324, 326
SUMO Simulation of Urban Mobility. See `http://www.dlr.de/ts/sumo/en/`. 50
SURPRICE Sustainable mobility through Road User Charging. xxii

TASHA Travel Activity Scheduler for Household Agents. 368, 523
TAZ Traffic Analysis Zone. 389, 431, 433, 486–488, 515, 516
TESF Transportation Energy Simulation Framework. 93–95
TfL Transport for London. 447
THELMA Technology-centered ELectric Mobility Assessment. xxii
TML Transport & Mobility Leuven. 377
ToPDAd Tool supported Policy Development for regional Adaptation. xxii, 376
TransCAD Transportation Computer Assisted Design. 397
TRANSIMS TRansportation ANalysis and SIMulation System. See `https://code.google.com/archive/p/transims/`. 3, 309, 310, 448, 534
TRENoP Transport REsearch with Novel Perspectives. xxiv
TRV Trafikverket - Swedish Transport Administration. xxiv
TSA Transport System Analysis. 399
TTS Transportation Tomorrow Survey. 523
TU Technische Universität. xx, xxi, 50, 290, 369, 507

UCL University College London. xxiv
UE User Equilibrium a.k.a. Wardrop's first principle. 316, 317, 319–322
UIB Universitat Autónoma de Barcelona. xxiii
UK United Kingdom. xxiv
URL Uniform Resource Locator. 10
UTC Temps Universel Coordonné - Coordinated Universal Time. 422
UTM Universal Transverse Mercator. 12, 13
UVEK Eidgenössisches Departement für Umwelt, Verkehr, Energie und Kommunikation. 374

V2C Vehicle to Community. 529, 532
V2G Vehicle-to-Grid. 94, 95
VGI Voluntary Geographic Information. 242
VISSIM Verkehr In Städten - SImulationsModell. See `http://www.ptv.de`. 248
VISUM Verkehr In Städten - UMlegung. See `http://www.ptv.de`. 50, 63, 106, 110, 115, 292, 369, 383
VPL VerkehrsPLanung. See `http://www.ivt.ethz.ch/vpl/`. 290
VRP Vehicle Routing Problem. 95, 146, 147
VSP VerkehrsSystemPlanung und Verkehrstelematik - The Transport Systems Planning and Transport Telematics group at TU Berlin. See `https://www.vsp.tu-berlin.de`. 290, 467
VTTS Value of Travel Time Savings. 26, 28, 354, 356, 360

WKT Well-Known Text. 13

XML Extensible Markup Language, see `http://www.w3.org/XML/`. 14, 57, 61, 70, 250, 254, 255, 257, 274, 275, 292, 298, 299, 309, 440, 445, 504

Glossary

Activity The central element of modern activity-based modeling (see below). 4, 221, 552

Activity location People perform activities at activity locations, which can be as small as one single building or large zones. In MATSim, activity locations are often further specified by using the facility object, which (in addition to others) define open times. xix, 14, 550

Activity-based Modern transport planning assumes that "*travel demand is derived from activity demand*" (Jones, 1979; Bowman, 2009a,b; Bhat and Koppelman, 2003; Ettema and Timmermans, 1997; Bowman and Ben-Akiva, 1996, 2001). People travel because they want to perform a certain activity, which is best captured by activity-based models with activities the central element of modeling. 4

Agent According to Wooldridge (2009, p. 21) an agent is "*a computer system situated in an environment, capable of autonomous action in this environment to meet its delegated objectives*". 307

Algorithm A set of operations to solve a specific problem. 7, 30

ArcGIS ESRI's geographic information system. 254, 503, 504, 506

C++ An object-oriented programming language with full control of memory management. 6, 285, 309, 310

C# An object-oriented .NET programming language. 310

Configuration file The main configuration screw for MATSim, often just referred to as `config file` or as `config.xml`. Also see config. xxxi, xxxii, 11–14, 16, 18, 19, 21, 35–40, 44, 48, 49, 55, 56, 58, 79, 86, 226, 227, 241, 268, 284, 290, 294, 298, 504, 549, 550, 554

Configuration object The object in the MATSim code containing configuration options. It can be modified by the config file, but also by other mains, in particular by scripts-in-Java. 549

Contribution An extension contributed by the MATSim community and hosted in the MATSim repository. See `http://matsim.org/extensions`. xxxi, 48, 49, 92, 95, 121, 135, 140, 146, 157, 159, 226, 259, 272, 274, 281, 293, 294, 296, 437, 443, 471, 528

Eclipse The standard integrated development environment (IDE) used by the MATSim developers. 295, 309

Equilibrium A system state where are competing forces are balanced. 7

Event Small pieces of information reported by the mobsim, describing a simulation object action at a specific time. xxxii, 17, 127, 298, 301–304

Extension Core MATSim uses only a config file, population file and network file, corresponding to the book's part I. An extension is any code that extends this core MATSim, corresponding to this book's part II. They hook to MATSim via the extension points described in Chapter 45. xxxi, xxxii, 39, 48, 49, 290, 294, 298, 299, 311, 549

Facility An optional element in MATSim to further specify an activity location. 57, 221, 549

Framework A software concept, providing generic functionality and application-specific software. It is selectively changed by user code. MATSim is currently a framework, but is developing towards also being useful as a library/toolbox. xix, xx, 4, 188, 302, 303

Geocoding Adding geographic coordinates to locations identified by addresses. 501

Git A free and open source distributed version control system. 292, 295

GitHub A web-based Git repository hosting service, see `https://github.com/`. 9, 292–294

Google Earth Google's virtual globe. 13

Identifier A name that labels an object in a unique way. 12, 77, 78, 299

Iteration Numerical equilibrium search methods, such as MATSim, are iterative. A MATSim run is thus composed of a configurable number of iterations. xix, 11, 16, 550

Java A modern, object-oriented, cross-platform programming language run in virtual machines. xxxii, 4, 11, 35, 37, 38, 42, 48–50, 56, 59, 61, 62, 78, 120, 210, 219, 220, 225, 226, 231, 254, 255, 267, 285, 290, 292, 293, 298, 299, 303, 309–311, 409, 510, 539, 549–551

Javadoc Source code documentation compiled from javadoc annotations in the source files. 88, 173, 295

Jenkins A software tool for continuous integration. 295

Large-scale Denoting large, extended simulation scenarios, often modeling complete cities, or even countries. 3, 140, 313, 461, 502

Leg A plan element, part of a trip performed with a specific mode. In transport planning, this is often called a stage. 15, 302, 551, 552

Library A set of routines providing services to independent programs. Usually not executable on its own. 10

Link A network component representing streets. 14

Linux A unix-like operating system released by Linus Torvalds at the end of 1991. 10, 309

Logsum The Expected Maximum Utility (EMU) for a user that has several options. Computed as the logarithm of the sum of exponential functions. 241, 244

Mac OS The operating system by Apple Inc. developed for their Macintosh computer systems. 10, 309

MATSim run A configurable number set of iterations, typically ending with an equilibrium solution of transport supply and demand. 4, 10, 16, 36, 298, 550

Maven A build automation tool tailored to Java. 10, 66, 259, 292, 294, 295, 297

Microsimulation The modeling of the temporal development of a real-world system, or process, by explicitly considering the interactions of micro units such as individuals or vehicles. For concise definitions and further information, see e.g., Miller (1996, Section 2) or Banks J (2001, p. 3), Bossel (2004) or Orcutt (1957), who is often referred to as the inventor of microsimulation. xxxi, 176, 309, 413, 414, 424, 541, 550, 551

Model A universal concept reducing a real system to the aspects relevant for understanding or solving a specific problem. 327, 551

Module According to Merriam-Webster (`http://www.merriam-webster.com`), a module is "one of a set of parts that can be connected or combined to build or complete something" or more specifically "a part of a computer or computer program that does a particular job". That is, "module" is not a very specific term, and, in consequence, modules exist in MATSim at many levels. xxxii, 57, 221, 304, 311

Multimodal Combining different means of transport. 38, 49, 79, 106, 112, 135, 136, 140, 416, 453, 477, 486, 523, 524

NAVTEQ A geographical information system data provider, particularly for navigation maps. 67, 117, 374, 380

Node An element of a MATSim network representing intersections. Note that intersections are not modeled explicitly in MATSim, i.e., cars do not interact at intersections. 14

Objective function A central element in optimization problems, among others. An objective function, sometimes also called loss or cost function, is mapping of candidate solutions onto a real number. 165, 551, 552

OmniTRANS A transport Modeling Software Platform. 115

Osmosis Command line Java application for processing OSM data. See `http://wiki.openstreetmap.org/wiki/Osmosis`. 62

Plan The agent's day schedule and, after run completion, an associated score. xix, 4, 15, 301

QSim The standard MATSim mobsim. 6, 19, 36–38, 42, 44, 77–79, 106, 135, 191–193, 195, 196, 263–268, 298, 347, 350, 351, 542

Replanning The stage when agents modify their plans. xix, 4, 12, 15, 297, 301, 304, 479, 501, 504

Scenario In MATSim context, a scenario is defined as: the combination of specific agent populations, their initial plans and activity locations (home, work, education), the network and facilities where, and on which, they compete in time-space for their slots and modules, i.e., behavioral dimensions, which they can adjust during their search for equilibrium. xix, xxxi, xxxii, 6, 9, 11, 303

Score After execution in the infrastructure, the agents' day plans are evaluated through an individual objective function, the MATSim scoring function. Also see utility. xxxii, 4, 24, 304, 551, 552

Scoring see score. 304

Senozon AG A spin-off company founded by two core developers of MATSim. xxii, 110, 219, 290, 291, 295, 392, 443, 467, 497, 503, 552

Simulation Evaluating a model capturing the temporal development of a real-world system or process. 327

Stage A stage is part of a trip, performed with a single mode. In MATSim called leg. 550

Study The basic organizational unit of research in empirical science. Comparable to the experiment in natural sciences. 4

SustainCity A project addressing the modeling and computational issues of integrating modern mobility simulations with the latest microsimulation land use models, see `http://www.sustaincity.org`. 405

Teleportation Moving vehicles from origin to destination, at a predefined speed, without considering interactions in the network. 16, 42, 44, 79, 106, 135, 136, 419, 505

Teleported see teleportation. 105, 374, 375, 380, 386, 525

Traffic assignment Traditionally, this is the last step of the four-step model, calculating trip distribution over the different routes between O-D-pairs. Dynamic traffic assignment may additionally consider departure time choices. xxxii

Trip The connection between two activities, composed of multiple legs. 17, 550

UrbanSim Software-based simulation system for supporting planning and analysis of urban development. See `http://www.urbansim.org`. 283, 284, 405, 495, 540

Utility A central economic concept representing satisfaction through goods consumption. The MATSim score can be interpreted in utility units. xxxii, 24, 551

Utility function A MATSim agent's objective function. See also score. xix

Via The Senozon AG visualizer. 19, 47, 292, 392, 443, 445, 446, 450, 483, 526

Windows An operating system developed by Microsoft, first released in 1985. 10, 309, 310

Symbols and Typographic Conventions

Symbols

Variables

c	monetary costs
d	distance
t	time
U	utility variable ($V + \varepsilon$)
V	systematic component of utility variable
S	score (= the un-interpreted MATSim value)
β	utility function coefficient
$\hat{\beta}$	estimated utility function coefficient
ε	random component of utility variable
φ	replanning share
μ	scale parameter of the multinomial logit model

Indices and Subscripts

i	index of plans
k	index of iterations
n	index of agents
q	index of plan activities
ℓ	index of activity locations/facilities

Typographic Conventions

The `listing-format` is used for text that you typically see when you run MATSim, i.e., program snippets, commands, on-screen computer output, input and output file names and content, and configurations to be specified in the config file. Larger snippets are shown as complete listings.

```
... main( ... ) {
    // construct the config object:
    Config config = ConfigUtils.xxx(...) ;
    config.xxx().setYyy(...) ;
    ...
}
```

Important passages are *emphasized.*

Vertical bar | is a separator for mutually exclusive items. For example: "`KeepLastSelected | BestScore | SelectExpBeta`"

Math mode, e.g., $x = 42$ is used for mathematical terms.

Acronyms are given with the abbreviation and the description following in parenthesis (e.g., MATSim (Multi-Agent Transport Simulation)) on first occurrence, later only the abbreviation (e.g., MATSim) is given.

Bibliography[6]

Abdulaal, M. and L. J. LeBlanc (1979) Methods for combining modal split and equilibrium assignment models, *Transportation Science*, **13** (4) 292–314.

Abedin, Z. U. and R. A. Waraich (2014) Modelling inductive charging of battery electric vehicles using an agent-based approach, *Journal of Sustainable Development of Energy, Water and Environment Systems*, **2** (3) 219–233.

Agarwal, A., G. Lämmel and K. Nagel (2015a) Modelling of backward travelling holes in mixed traffic conditions, in: *Traffic and Granular Flow '15*, Delft. Also VSP WP 15-15.

Agarwal, A., M. Zilske, K. Rao and K. Nagel (2013) Person-based dynamic traffic assignment for mixed traffic conditions, in: *Conference on Agent-Based Modeling in Transportation Planning and Operations*, Blacksburg, VA. Also VSP WP 12-11.

Agarwal, A., M. Zilske, K. Rao and K. Nagel (2015b) An elegant and computationally efficient approach for heterogeneous traffic modelling using agent based simulation, *Procedia Computer Science*, **52** (C) 962–967, doi:10.1016/j.procs.2015.05.173.

Aguirre, H., A. Oyama and K. Tanaka (2013) Adaptive ϵ-sampling and ϵ-hood for evolutionary many-objective optimization, in: *7th International Conference on Evolutionary Multi-Criterion Optimization (EMO 2013)*, Sheffield, March 2013.

Ahlheim, M. and M. Rose (1989) *Messung individueller Wohlfahrt*, Springer, Heidelberg.

Ahn, K. and H. Rakha (2008) The effects of route choice decisions on vehicle energy consumption and emissions, *Transportation Research Part D: Transport and Environment*, **13** (3) 151–167.

Ahrens, G.-A., F. Ließke, R. Wittwer and S. Hubrich (2009) Endbericht zur Verkehrserhebung 'Mobiität in Städten - SrV 2008' und Auswertungen zum SrV-Städtepegel, TU Dresden, Lehrstuhl Verkehrs- und Infrastrukturplanung, `http://www.tu-dresden.de/srv/`.

Amt für Verkehr, Volkswirtschaftsdirektion Kanton Zürich (2011) Gesamtverkehrsmodell, webpage, March 2011, `http://www.afv.zh.ch/internet/volkswirtschaftsdirektion/afv/de/home.html`.

André, M. and M. Rapone (2009) Analysis and modelling of the pollutant emissions from European cars regarding the driving characteristics and test cycles, *Atmospheric Environment*, **43** (5) 986–995.

[6] IVT working papers are available from `http://www.ivt.ethz.ch/vpl/publications/reports`. Swiss Transport Research Conference (STRC) papers are available from `http://www.strc.ch`. VSP working papers (VSP WP) are available from `http://www.vsp.tu-berlin.de/publications`.

Arentze, T. A., M. Kowald and K. W. Axhausen (2012) A method to model population-wide social networks for large scale activity-travel micro-simulation, in: *91st Annual Meeting of the Transportation Research Board*, Washington, D.C., January 2012.

Arentze, T. A., M. Kowald and K. W. Axhausen (2013) An agent-based random-utility-maximization model to generate social networks with transitivity in geographic space, *Social Networks*, **35** (3) 451–459.

Arentze, T. A. and H. J. P. Timmermans (2000) *ALBATROSS: A Learning-Based Transportation Oriented Simulation*, EIRASS, Eindhoven.

Arentze, T. A. and H. J. P. Timmermans (2004) A learning-based transportation oriented simulation system, *Transportation Research Part B: Methodological*, **38** (7) 613–633.

Arentze, T. A. and H. J. P. Timmermans (2009) A need-based model of multi-day, multi-person activity generation, *Transportation Research Part B: Methodological*, **43** (2) 251–265.

Arentze, T. A., H. J. P. Timmermans, D. Janssens and G. Wets (2006) Modeling short-term dynamics in activity-travel patterns: From Aurora to Feathers, in: *Innovations in Travel Demand Modeling (ITM'06)*, Austin, May 2006.

Armas, R., H. Aguirre and K. Tanaka (2014) Effects of mutation and crossover operators in the optimization of traffic signal parameters, in: *The Tenth International Conference on Simulated Evolution And Learning (SEAL 2014)*, Otago, December 2014.

Arnott, R., A. de Palma and R. Lindsey (1990) Economics of a bottleneck, *Journal of Urban Economics*, **27** (1) 111–130, doi:10.1016/0094-1190(90)90028-L.

Arnott, R., A. de Palma and R. Lindsey (1993) A structural model of peak-period congestion: A traffic bottleneck with elastic demand, *The American Economic Review*, **83** (1) 161–179.

Arthur, B. (1994) Inductive reasoning, bounded rationality, and the bar problem, *American Economic Review (Papers and Proceedings)*, **84**, 406–411.

ASTRA (2006) Swiss federal roads authority, webpage, `http://www.astra.admin.ch/`.

Autodor (2013) *Recommendations for forecasting the intensity of traffic on the toll road sections of the state company "Avtodor" and the proceeds of their exploitation*, Basic Books, Moscow.

Axelrod, R. (1984) *The Evolution of Cooperation*, Basic Books, New York.

Axhausen, K. W. (1989) Eine ereignisorientierte Simulation von Aktivitätenketten zur Parkstandswahl (A Simultaneous Simulation of Activity Chains), *Schriftenreihe*, **40**, Institut für Verkehrswesen, University of Karlsruhe, Karlsruhe.

Axhausen, K. W. (1990) A simultaneous simulation of activity chains and traffic flow, in: P. M. Jones (ed.) *Developments in Dynamic and Activity-Based Approaches to Travel Analysis*, 206–225, Avebury, Aldershot.

Axhausen, K. W. (2005) Social networks and travel: Some hypotheses, in: K. P. Donaghy, S. Poppelreuter and G. Rudinger (eds.) *Social Dimensions of Sustainable Transport: Transatlantic Perspectives*, Ashgate, Aldershot.

Axhausen, K. W. (2006) Neue Modellansätze der Verkehrsnachfragesimulation: Entwicklungslinien, Stand der Forschung, Forschungsperspektiven, *Stadt Region Land*, **81**, 149–164.

Axhausen, K. W. (2009) Update on MATSim, presentation, the 88th Annual Meeting of the Transportation Research Board, Washington, D.C., January.

Axhausen, K. W. (2014) MATSim platform and applications in Europe and elsewhere, presentation, Demonstration of an Integrated Dynamic Policy Sensitive Model of Travel Demand for the Mega-Region of New York, New York, May 2014.

Axhausen, K. W., I. Dimitropoulos and E. Dimitrakopolou (1995) Adapting to change: some evidence from a simple learning model, in: *23rd European Transport Forum*, Warwick.

Axhausen, K. W. and P. B. Goodwin (1991) Eurotopp: Towards a dynamic and activity-based modelling framework, in: *Advanced Telematics in Road Transport*, 1021–1039, Amsterdam.

Axhausen, K. W. and R. Herz (1989) Simulating activity chains: German approach, *Journal of Transportation Engineering*, **115** (3) 316–325.

Axhausen, K. W. and J. W. Polak (1989) *The Role of Parking Search Strategies in Understanding Parking Behaviour*, Transport Studies Unit, Oxford University, Oxford.

Baker, J., D. Grewel and A. Parasuraman (1994) The influence of store environment on quality inferences and store image, *Journal of the Academy of Marketing Science*, **22** (4) 328–339.

Balac, M., F. Ciari and K. W. Axhausen (2015) Carsharing demand estimation: Case study of Zurich area, in: *94th Annual Meeting of the Transportation Research Board*, Washington, D.C., January 2015.

Balijepalli, N., D. Watling and R. Liu (2007) Doubly dynamic traffic assignment - simulation modeling framework and experimental results, *Transportation Research Record*, **2029**, 39–48.

Balmer, M. (2007) Travel demand modeling for multi-agent traffic simulations: Algorithms and systems, Ph.D. Thesis, ETH Zurich, Zurich, May 2007.

Balmer, M. (2014) Coupling ABD Model of NY with MATSim, presentation, Zürich Meets New York, New York, May.

Balmer, M., M. Bernard and K. W. Axhausen (2005a) Matching geo-coded graphs, in: *5th Swiss Transport Research Conference*, Ascona, March 2005.

Balmer, M., A. Horni, K. Meister, F. Ciari, D. Charypar and K. W. Axhausen (2009a) Wirkungen der Westumfahrung Zürich: Eine Analyse mit einer Agenten-basierten Mikrosimulation, *Final Report*, Baudirektion Kanton Zurich, IVT, ETH Zurich, Zurich, February 2009.

Balmer, M., K. Meister, M. Rieser, K. Nagel and K. Axhausen (2008) Agent-based simulation of travel demand: Structure and computational performance of MATSim-T, in: *Innovations in Travel Modeling (ITM) '08*, Portland, OR, June 2008. Also VSP WP 08-07.

Balmer, M., K. Meister, R. A. Waraich, A. Horni, F. Ciari and K. W. Axhausen (2010) Agenten-basierte Simulation für location based services, *Final Report*, F&E Förderung: Science to Market, **KTI 8443.1 ESPP-ES**, Datapuls AG, IVT, ETH Zurich, Zurich, February 2010.

Balmer, M., B. Raney and K. Nagel (2005b) Adjustment of activity timing and duration in an agent-based traffic flow simulation, in: H. Timmermans (ed.) *Progress in activity-based analysis*, 91–114, Elsevier, Oxford.

Balmer, M., M. Rieser, K. Meister, D. Charypar, N. Lefebvre, K. Nagel and K. Axhausen (2009b) MATSim-T: Architecture and simulation times, in: A. Bazzan and F. Klügl (eds.) *Multi-Agent Systems for Traffic and Transportation*, 57–78, IGI Global.

Banks, J. (2001) *Discrete-event system simulation*, Prentice Hall, Upper Saddle River.

Bar-Gera, H. (2013) Transportation network test problems, webpage, `http://www.bgu.ac.il/~bargera/tntp/`. Webpage under construction accessed on 22/08/2013.

Bar-Gera, H., K. Konduri, B. Sana, X. Ye and R. M. Pendyala (2009) Estimating survey weights with multiple constraints using entropy optimization methods, in: *88th Annual Meeting of the Transportation Research Board*, Washington, D.C., January 2009.

Barcelo, J., H. Grzybowska and S. Pardo (2007) Vehicle routing and scheduling models, simulation and city logistics, 163–195, Springer, NY.

Batty, M. (2009) Accessibility: In search of a unified theory, *Environment and Palanning B: Planning and Design*, **36**, 191–194.

BAV (2013) Faktenblatt FABI: So erfolgt die Finanzierung, *Technical Report*, Bundesamt für Verkehr. See `http://www.bav.admin.ch/fabi/`, accessed 16.02.2015.

Beazley, D., P. Lomdahl, N. Gronbech-Jensen, R. Giles and P. Tamayo (1995) Parallel algorithms for short-range molecular dynamics, in: D. Stauffer (ed.) *Annual reviews of computational physics III*, 119–176, World Scientific, Singapore.

Becker, R., R. Caceres, K. Hanson, S. Isaacman, J. Loh, M. Martonosi, J. Rowland, S. Urbanek, A. Varshavsky and C. Volinsky (2013) Human mobility characterization from cellular network data, *Communications of the ACM*, **56** (1).

Beckx, C., T. A. Arentze, L. Int Panis, D. Janssens, J. Vankerkorn and G. Wets (2009) An integrated activity-based modelling framework to assess vehicle emissions: Approach and application, *Environment and Planning B*, **36** (6) 1086–1102.

Bekhor, S., C. Dobler and K. W. Axhausen (2011) Integration of activity-based with agent-based models: an example from the Tel Aviv model and MATSim, in: *90th Annual Meeting of the Transportation Research Board*, Washington, D.C., January 2011.

Bell, M., V. Trozzi, S. Hosseinloo, G. Gentile and A. Fonzone (2012) Time-dependent Hyperstar algorithm for robust vehicle navigation, *Transportation Research Part A: Policy and Practice*, **46** (5) 790–800.

Bell, M. G. H. (2009) Hyperstar: A multi-path astar algorithm for risk averse vehicle navigation, *Transportation Research Part B: Methodological*, **43** (1) 97–107.

Bemetz, V. and M. Hohenfellner (2014) Implementation of a parking choice model to analyse the effect of parking prices on electric vehicles, presentation, Bachelor Thesis, IVT, ETH Zürich.

Ben-Akiva, M. and M. Bierlaire (1999) Discrete choice methods and their applications to short-term travel decisions, in: R. Hall (ed.) *Handbook of Transportation Science*, 5–34, Kluwer.

Ben-Akiva, M. E. and B. Boccara (1995) Discrete choice models with latent choice sets, *International Journal of Research in Marketing*, **12** (1) 9–24.

Ben-Akiva, M. E., J. L. Bowman and D. Gopinath (1996) Travel demand model system for the information era, *Transportation*, **23** (3) 241–266.

Ben-Akiva, M. E. and S. R. Lerman (1985) *Discrete Choice Analysis: Theory and Application to Travel Demand*, MIT Press, Cambridge.

Benfield, S. S., J. Hendrickson and D. Galanti (2006) Making a strong business case for multiagent technology, in: *Proceedings of Autonomous Agents and Multi-Agent Systems (AAMAS)*, 10–15.

Bhat, C., J. Guo, S. Srinivasan and A. Sivakumar (2008) *CEMDAP User's Manual*, University of Texas at Austin, AustinTX.

Bhat, C. R. and F. S. Koppelman (2003) Activity-based modeling of travel demand, in: R. W. Hall (ed.) *Handbook of Transportation Science*, 39–65, Springer, New York.

Bickel, P., R. Friedrich, A. Burgess, P. Fagiani, A. Hunt, G. de Jong, J. Laird, C. Lieb, G. Lindberg, P. Mackie, S. Navrud, T. Odgaard, A. Ricci, J. Shires and L. Tavasszy (2006) Proposal for Harmonised Guidelines, deliverable 5 of HEATCO [harmonised european approaches for transport costing and project assessment] commissioned by the European Union.

Bischoff, J. (2010) Verkehrsabhängige Lichtsignalanlagensteuerung – Vergleich und simulationsbasierte Evaluation, Bachelor's thesis, TU Berlin, Institute for Land and Sea Transport Systems, Berlin, December 2010.

Bischoff, J. (2013) Agentenbasierte Simulation elektrifizierter Taxiflotten, Master's thesis, TU Berlin, Institute for Land and Sea Transport Systems, Berlin, 09 2013.

Bischoff, J. and M. Maciejewski (2014) Agent-based simulation of electric taxicab fleets, *Transportation Research Procedia*, **4**, 191–198, doi:10.1016/j.trpro.2014.11.015. Also VSP WP 14-10.

Bockemühl, F. (2014) MORBAMS - setting up a regional MATSim model, *Technical Report*, Hasselt University, Hasselt.

Boesch, P. M. (2014) Calibration of Generic Base Scenarios, presentation, MATSim Conceptual Meeting, Wiepersdorf, August 2014.

Boesch, P. M. and F. Ciari (2014) Climate Change Influence on Swiss Transport, Tourism and Energy – A Stakeholders Perspective, in: *14th Swiss Transport Research Conference*, Ascona, May 2014.

Boesch, P. M., F. Ciari and A. Perrels (2014) Overview of system responsiveness to climate change impacts in energy, transport and tourist sectors, *ToPDAd Deliverable*, **2.3**, ToPDAd Consortium Partners, S.I.

Böhme, S. and L. Eigenhüller (2006) Pendlerbericht Bayern, *Technical Report*, IAB: Institut für Arbeitsmarkt- und Berufsforschung der Bundesagentur für Arbeit, Nürnberg.

Bordini, R. H., J. F. Hübner and M. Wooldridge (2007) *Programming Multi-agent Systems in AgentSpeak Using Jason*, Wiley, New York.

Börjesson, M. and J. Eliasson (2014) Experiences from the Swedish value of time study, *Transportation Research Part A: Policy and Practice*, **59**, 144–158, doi:10.1016/j.tra.2013.10.022.

Bossel, H. (2004) *Systeme, Dynamik, Simulation: Modellbildung, Analyse und Simulation komplexer Systeme*, Books on Demand, www.bod.de.

Bouman, P., L. Kroon, G. Maróti and P. Vervest (2013) Capacity, comfort and crowdedness: The El Farol train game, in: *5th International Seminar on Railway Operations Modelling and Analysis - RailCopenhagen, Copenhagen, May 2013.*

Bouman, P., M. Lovric, T. Li, E. 'van der Hurk, L. Kroon and P. Vervest (2012) Recognizing demand patterns from smart card data for agent-based micro-simulation of public transport, in: *11th International Joint Conference on Autonomous Agents and Multiagent Systems (AAMAS)*, Valencia, June 2012.

Bowman, J. and M. Ben-Akiva (1998) Activity based travel demand model systems, in: P. Marcotte and S. Nguyen (eds.) *Equilibrium and advanced transportation modelling*, 27–46, Kluwer, Dordrecht.

Bowman, J., M. Bradley, Y. Shiftan, T. Lawton and M. Ben-Akiva (1998) Demonstration of an activity-based model for Portland, in: *World Transport Research: Selected Proceedings of the 8th World Conference on Transport Research 1998*, vol. 3, 171–184, Elsevier, Oxford.

Bowman, J. L. (2009a) Historical development of activity based model theory and practice (part 1), *Traffic Engineering and Control*, **50** (2) 59–62.

Bowman, J. L. (2009b) Historical development of activity based model theory and practice (part 2), *Traffic Engineering and Control*, **50** (7) 314–318.

Bowman, J. L. and M. E. Ben-Akiva (1996) Activity-based travel forecasting, in: *Activity-Based Travel Forecasting Conference*, New Orleans, June 1996.

Bowman, J. L. and M. E. Ben-Akiva (2001) Activity-based disaggregate travel demand model system with activity schedules, Transportation Research Part A: Policy and Practice, 35 (1) 1–28.

Bradley, M. A. and J. L. Bowman (2006) Design features of activity-based microsimulation models for U.S. metropolitan planning organizations, in: *Innovations in Travel Demand Modeling (ITM'06)*, Austin, May 2006.

Bradley, M. A., J. L. Bowman and B. Griesenbeck (2010) SACSIM: An applied activity-based model system with fine-level spatial and temporal resolution, *Journal of Choice Modelling*, **3** (1) 5–31.

Bradley, M. A. and P. Vovsha (2005) A model for joint choice of daily activity pattern types of household members, *Transportation*, **32** (5) 545–571.

Braubach, L., A. Pokahr and W. Lamersdorf (2005) Jadex: A BDI agent system combining middleware and reasoning, in: *Software Agent-Based Applications, Platforms and Development Kits*, 143–168.

Bundesagentur für Arbeit (2010) Pendlerstatistik 2010, Nürnberg, CD-ROM.

Bundesinstitut für Bau-, Stadt- und Raumforschung (accessed March 2015) Erreichbarkeitsmodell des BBSR, `http://www.bbsr.bund.de`.

Burnett, K. P. (1980) Spatial constraints-oriented modelling as an alternative approach to movement: Microeconomic theory and urban policy, *Urban Geography*, **1** (1) 53–67.

Burnett, K. P. and S. Hanson (1979) Rationale for an alternative mathematical approach to movement as complex behavior, *Transportation Research Record*, **723**, 11–24.

Büttner, B., J. Keller and G. Wulfhorst (2010) Ein Erreichbarkeitsatlas für die europäische Metropolregion München, *Schlussbericht für ein Projekt im Auftrag von Europäische Metropolregion München e. V.*, Technische Universität München, München.

Button, K. J. and E. T. Verhoef (eds.) (1998) *Road Pricing, Traffic Congestion and the Environment: Issues of Efficiency and Social Feasibility*, Edward Elgar, Cheltenham.

BVG (2012) Berliner verkehrsbetriebe, zahlenspiegel 2013, online, 12 2012, http://www.bvg.de.

BVU Beratergruppe Verkehr + Umwelt GmbH und Intraplan Consult GmbH (2007) Prognose der deutschlandweiten Verkehrsverflechtungen 2025, *Technical Report*, ITP and BVU, München und Freiburg.

Cabrita, I., S. Gayda, R. Hurtubia, D. Efthymiou, I. Thomas, D. Peeters, J. Jones, C. Cotteels, K. Nagel, T. Nicolai and D. Röder (2015) Integrated land use and transport microsimulation for Brussels, in: M. Bierlaire, A. de Palma, R. Hurtubia and P. Waddell (eds.) *Integrated Transport and Land Use Modeling for Sustainable Cities*, 373-412, EPFL press, Lausanne.

Caliper (accessed 2015) TransModeler web site, http://www.caliper.com/transmodeler/.

Carrasco, J. A. and K. M. N. Habib (2009) Understanding the social embeddedness of activity-travel participation: The case of frequency and duration of social activities, in: *12th International Conference on Travel Behaviour Research (IATBR)*, Jaipur, December 2009.

Carrasco, J. A., E. J. Miller and B. Wellman (2008) How far and with whom do people socialize? empirical evidence about the distance between social network members, *Transportation Research Record*, **2076**, 114–122.

Cascetta, E. (1989) A stochastic process approach to the analysis of temporal dynamics in transportation networks, *Transportation Research Part B*, **23** (1) 1–17.

Cascetta, E. and G. Cantarella (1991) A day-to-day and within-day dynamic stochastic assignment model, *Transportation Research Part A*, **25** (5) 277–291.

Cascetta, E., A. Nuzzolo, F. Russo and A. Vitetta (1996) A modified logit route choice model overcoming path overlapping problems. Specification and some calibration results for interurban networks., in: *Proceedings of the 13th International Symposium on Transportation and Traffic Theory*, 697–711, Lyon, July 1996.

Certicky, M., M. Jakob, R. Pibil and Z. Moler (2014) Agent-based simulation testbed for on-demand transport services, in: *Proceedings of the 2014 International Conference on Autonomous Agents and Multi-agent Systems*, AAMAS '14, 1671–1672, Richland, May 2014.

Cetin, N. (2005) Large-scale parallel graph-based simulations, Ph.D. Thesis, ETH Zurich, Zurich.

Cetin, N., A. Burri and K. Nagel (2003) A large-scale agent-based traffic microsimulation based on queue model, in: *Swiss Transport Research Conference (STRC)*, Monte Verita. See http://www.strc.ch.

Chakirov, A. and A. Erath (2011) Use of public transport smart card fare payment data for travel behaviour analysis in singapore, in: *16th international conference of Hong Kong Society for Transportation Studies*, Hong Kong, December 2011.

Chakirov, A. and A. Erath (2012) Activity identification and primary location modelling based on smart card payment data for public transport, in: *13th International Conference on Travel Behaviour Research (IATBR)*, Toronto, July 2012.

Chakirov, A. and P. J. Fourie (2014) Enriched Sioux Falls Scenario with Dynamic and Disaggregate Demand, *Working Paper*, Future Cities Laboratory, Singapore-ETH Centre (SEC), Singapore.

Chan, C. (2007) The state of the art of electric, hybrid, and fuel cell vehicles, *PIEEE*, **95** (4) 704–718.

Chang, G.-L. and H. S. Mahmassani (1989) The dynamics of commuting decision behaviour in urban transportation networks, in: B. Gerardin (ed.) *Travel Behaviour Research*, 15{–26, Gower, Aldershot.

Charypar, D. (2008) Efficient algorithms for the microsimualtion of travel behavior in very large scenarios, Ph.D. Thesis, ETH Zurich, Zurich.

Charypar, D., K. Axhausen and K. Nagel (2007a) An event-driven parallel queue-based microsimulation for large scale traffic scenarios, in: *World Conference on Transport Research (WCTR'07)*, Berkeley, CA. Also VSP WP 07-03.

Charypar, D., K. Axhausen and K. Nagel (2007b) Event-driven queue-based traffic flow microsimulation, *Transportation Research Record*, **2003**, 35–40, doi:10.3141/2003-05.

Charypar, D., M. Balmer and K. W. Axhausen (2009) High-performance traffic flow microsimulation for large problems, in: *88th Annual Meeting of the Transportation Research Board*, Washington, D.C., January 2009.

Charypar, D. and K. Nagel (2005) Generating complete all-day activity plans with genetic algorithms, *Transportation*, **32** (4) 369–397, doi:10.1007/s11116-004-8287-y.

Choi, C. C. and R. Toh (2010) Household interview surveys from 1997 to 2008: A decade of changing travel behaviours, *Journeys*, **5**, 52–61.

Chow, J. Y. and W. W. Recker (2012) Inverse optimization with endogenous arrival time constraints to calibrate the household activity pattern problem, *Transportation Research Part B: Methodological*, **46** (3) 463–479.

Chu, Y. L. (2003) Empirical analysis of commute stop-making behavior, *Transportation Research Record*, **1831**, 106–113.

Ciari, F. (2012) Sharing as a key to rethink urban mobility: Investigating and modelling innovative transport systems, Ph.D. Thesis, ETH Zurich, Zurich.

Ciari, F., M. Balmer and K. W. Axhausen (2007) Mobility tool ownership and mode choice decision processes in multi-agent transportation simulation, in: *7th Swiss Transport Research Conference*, Ascona, September 2007.

Ciari, F., M. Balmer and K. W. Axhausen (2008) A new mode choice model for a multi-agent transport simulation, in: *8th Swiss Transport Research Conference*, Ascona, October 2008.

Ciari, F., B. Bock and M. Balmer (2014) Modeling station-based and free-floating carsharing demand: test case study for Berlin, *Transportation Research Record*, **2416**, 37–47.

Ciari, F., B. Bock and M. Balmer (forthcoming) Modeling the effect of different pricing schemes on free-floating carsharing travel demand: test case study for Zurich, Switzerland, *Transportation*.

Ciari, F. and C. Weis (forthcoming) Carsharing membership in Switzerland: modeling the influence of socio-demographics and accessibility, *EURO Journal on Transportation and Logistics*.

Clarke, J.-P., T. Melconian, E. Bly and F. Rabbani (2007) MEANS–MIT extensible air network simulation, *Simulation*, **83** (5) 385–399, 05 2007, doi:10.1177/0037549707063766.

Cools, M., B. Kochan, T. Bellemans, D. Janssens and G. Wets (2011) Assessment of the effect of micro-simulation error on key travel indices: Evidence from the activity-based model FEATHERS, in: *90th Annual Meeting of the Transportation Research Board*, Washington, D.C., January 2011.

Cools, M., E. Moons and G. Wets (2010) Calibrating activity-based models with external origin-destination information: Overview of different possibilities, *Transportation Research Record*, **2175**, 98–110.

Cornélis, E., J.-P. Hubert, P. Huynen, K. Lebrun, G. Patriarche, A. de Witte, L. Creemers, K. Declercq, D. Janssens, M. Castaigne, L. Hollaert and F. Walle (2012) Belgian daily mobility—BELDAM: Enquête sur la mobilité quotidienne des Belges, *Research Report*, Politique scientifique fédérale - Programme AGORA, SPF Mobilité & Transports, FUNDP, GRT, VUB, MOSI-T, UHasselt, IMOB, Brussels.

Corthout, R., G. Flötteröd, F. Viti and C. Tampere (2012) Non-unique flows in macroscopic first-order intersection models, *Transportation Research Part B*, **46** (3) 343–359.

Creutzig, F. and D. He (2009) Climate change mitigation and co-benefits of feasible transport demand policies in Beijing, *Transportation Research Part D: Transport and Environment*, **14** (2) 120–131.

Curtis, C., J. Scheurer and M. Burke (2013) Using new accessibility tools to guide policy innovation, *Built Environment*, **39** (4) 454–472.

Daganzo, C. (1998) Queue spillovers in transportation networks with a route choice, *Transportation Science*, **32** (1) 3–11.

Daganzo, C. F. (1994) The cell transmission model: A dynamic representation of highway traffic consistent with the hydrodynamic theory, *Transportation Research Part B*, **28** (4) 269–287.

Daganzo, C. F. and Y. Sheffi (1977) On stochastic models of traffic assignment, *Transportation Science*, **11** (3) 253–274.

Dagsvik, J. K. and A. Karlström (2005) Compensating variation and hicksian choice probabilities in random utility models that are nonlinear in income, *The Review of Economic Studies*, **72** (1) 57–76.

Daly, A. (2013) The role of choice modelling in travel demand forecasting, in: *2nd Annual Meeting of the European Association for Research in Transportation (hEART)*, Stockholm.

Daly, A., G. de Jong, N. Ibáñez, R. Batley and M. de Bok (2008) Welfare measures from discrete choice models in the presence of income effect, in: *European Transport Conference (ETC)*.

de Jong, G., A. Daly, E. Kroes and T. van der Hoorn (2006) Using the logsum in project appraisal, in: *11th Conference on Travel Behavior Research (IATBR)*, Kyoto, Japan.

de Jong, G., A. Daly, M. Pieters and T. van der Hoorn (2007) The logsum as an evaluation measure: Review of the literature and new results, *Transportation Research Part A: Policy and Practice*, **41** (9) 874–889, doi:10.1016/j.tra.2006.10.002.

de Palma, A. and F. Marchal (2002) Real case applications of the fully dynamic METROPOLIS tool-box: An advocacy for large-scale mesoscopic transportation systems, *Networks and Spatial Economics*, **2** (4) 347–369.

Deflorio, F. P. (2011) Simulation of requests in demand responsive transport systems, *Intelligent Transport Systems, IET*, **5** (3) 159–167.

Del Rosario, B. T. (2011) Effects of tropical depression (shanshan), *Technical Report*, **Update Sitrep No. 13 re**, NDRRMC, Quezon City.

DeSerpa, A. (1971) A theory of the economics of time, *Economic Journal*, **81**, 828–846.

Dewals, B. J., A. Mustafa and M. Bruwier (2015) Landuse change and future flood risk: an integrated and multi-scale approach, in: *36th IAHR World Congress*, Delft, June 2015.

Dial, R. (1971) A probabilistic multipath traffic assignment model which obviates the need for path enumeration, *Transportation Research*, **5** (2) 83–111.

Dibbelt, J., T. Pajor, B. Strasser and D. Wagner (2013) Intriguingly simple and fast transit routing, in: V. Bonifaci, C. Demetrescu and A. Marchetti-Spaccamela (eds.) *Experimental Algorithms*, vol. 7933 of *Lecture Notes in Computer Science*, 43–54, Springer, Heidelberg.

Dijkstra, E. W. (1959) A note on two problems in connexion with graphs, *Numerische Mathematik*, **1**, 269–271.

DIW (2014) Verkehr in Zahlen, Berlin.

DLR (2012) Luftverkehrsbericht 2011 – Daten und Kommentierungen des deutschen und weltweiten Luftverkehrs, *Technical Report*, Deutsches Zentrum für Luft- und Raumfahrt e.V. – DLR, Köln.

Dobler, C. (2009) Implementations of within day replanning in MATSim-T, *Working Paper*, **598**, IVT, ETH Zurich, Zurich.

Dobler, C. (2010) Implementation of a time step based parallel queue simulation in MATSim, in: *10th Swiss Transport Research Conference*, Ascona, September 2010.

Dobler, C. (2013) Travel behaviour modelling for scenarios with exceptional events - methods and implementations, Ph.D. Thesis, ETH Zurich, Zurich.

Dobler, C. and K. W. Axhausen (2011) Design and implementation of a parallel queue-based traffic flow simulation, *Working Paper*, **732**, IVT, ETH Zurich, Zurich.

Dobler, C., A. Horni and K. W. Axhausen (2014) Integration of activity-based and agent-based models: Recent developments for tel aviv, israel, *Working Paper*, **1027**, IVT, ETH Zurich, Zurich.

Dobler, C., M. Kowald, N. Schüssler and K. W. Axhausen (2012) Within-day replanning of exceptional events, *Transportation Research Record*, **2302**, 138–147.

Dobler, C. and G. Lämmel (2012) A framework for large-scale multi-modal microscopic evacuation simulations, in: *International Conference on Evacuation Modeling and Management*, Northwestern University, Evanston.

Dobler, C. and G. Lämmel (2014) Integration of a multi-modal simulation module into a framework for large-scale transport systems simulation, in: U. Weidmann, U. Kirsch and M. Schreckenberg (eds.) *Pedestrian and Evacuation Dynamics 2012*, 739–754, Springer, Berlin.

Dressler, D., G. Flötteröd, G. Lämmel, K. Nagel and M. Skutella (2011) Optimal evacuation solutions for large-scale scenarios, in: B. Hu, K. Morasch, S. Pickl and M. Siegle (eds.) *Operations Research Proceedings 2010*, 239–244, Springer Berlin Heidelberg, doi:10.1007/978-3-642-20009-0_38.

Dubernet, T. and K. W. Axhausen (2013) Including joint decision mechanisms in a multiagent transport simulation, *Transportation Letters*, **5** (4) 175–183.

Dubernet, T. and K. W. Axhausen (2014) Solution Concepts for the Simulation of Household-Level Joint Decision Making in Multi-Agent Travel Simulation Tools, in: *14th Swiss Transport Research Conference*, Ascona, May 2014.

Dubernet, T. and K. W. Axhausen (forthcoming) Implementing a household joint activity-travel multi-agent simulation tool: First results, *Transportation*.

Durst, D., G. Lämmel and H. Klüpfel (2014) Large-scale multi-modal evacuation analysis with an application to Hamburg, in: U. Weidmann, U. Kirsch and M. Schreckenberg (eds.) *Pedestrian and Evacuation Dynamics 2012*, 361–369, Springer, Berlin.

DynusT (accessed August 2014) Dynamic urban systems for Transportation, `http://dynust.net`.

Eiben, A. E. and J. E. Smith (eds.) (2003) *Introduction to Evolutionary Computing*, Springer, Berlin.

Elalouf, A. (2012) Efficient routing of emergency vehicles under uncertain urban traffic conditions, *Journal of Service Science and Management*, **5** (3) 23–36.

Emmerink, R. H. M., K. W. Axhausen, P. Nijkamp and P. Rietveld (1995) The potential of information provision in a simulated road transport network with non-recurrent congestion, *Transportation Research Part C: Emerging Technologies*, **3** (5) 293–309.

Erath, A., A. Chakirov, P. J. Fourie, S. A. Ordóñez Medina, M. Shah, M. A. B. van Eggermond and K. W. Axhausen (2012) A large-scale agent-based transport travel demand model for Singapore: The implementation of MATSim, *Working Paper*, Future Cities Laboratory, Singapore-ETH Centre (SEC), Singapore.

Erath, A., P. J. Fourie and S. Ordonez (in preparation) Using smartcard data for agent-based transport simulation: the case of Singapore, in: J.-D. Schmoecker and F. Kurauchi (eds.) *Public Transport Planning with Smart Card Data*, Taylor & Francis.

Erath, A., M. A. B. van Eggermond, P. J. Fourie and A. Chakirov (2013) Decision support tool for large scale, agent-based transport demand model, in: *10th International Conference of Eastern Asia Society for Transportation Studies*, Taipei, September 2013.

Eroglu, S. and G. D. Harrell (1986) Retail crowding: Theoretical and strategic implications, *Journal of Retailing*, **62** (4) 346–363.

Eroglu, S. and K. A. Machleit (1990) An empirical study of retail crowding: Antecedents and consequences, *Journal of Retailing*, **66** (2) 201–221.

Eroglu, S., K. A. Machleit and T. Feldman Barr (2005) Perceived retail crowding and shopping satisfaction: The role of shopping values, *Journal of Business Research*, **58** (8) 1146–1153.

ESRI (1998) *ESRI Shapefile Technical Description*, Environmental Systems Research Institute (ESRI) Inc., `http://www.esri.com/library/whitepapers/pdfs/shapefile.pdf`.

ESRI (2011) ArcGIS Desktop: Release 9.3, software, `http://www.esri.com/software/arcgis/arcinfo/index.html`.

Ettema, D., G. Tamminga, H. Timmermans and T. Arentze (2003) A micro-simulation model system of departure time and route choice under travel time uncertainty, in: *meeting of the International Association for Travel Behavior Research (IATBR)*, Lucerne. See www.ivt.baug.ethz.ch.

Ettema, D. F. and H. J. P. Timmermans (eds.) (1997) *Activity-Based Approaches to Travel Analysis*, Pergamon, Oxford.

European Environment Agency (2011) Corine Land Cover 2006 seamless vector data, 08 2011. Version 15.

Farooq, B., M. Bierlaire, R. Hurtubia and G. Flötteröd (2013) Simulation based population synthesis, *Transportation Research Part B: Methodological*, **58** (4) 243–263, Dec 2013, doi:10.1016/j.trb.2013.09.012.

Faura, I. (2012) Accessibility evaluation of Kungsmässan shopping centre with traffic simulation, Master Thesis, Chalmers University of Technology, Göteborg.

Federal Highway Administration (2013) Urban Congestion Report, Website, March 2013. `http://www.ops.fhwa.dot.gov/perf_measurement/ucr/reports/fy2013_q2.htm`.

Feil, M. (2010) Choosing the daily schedule: Expanding activity-based travel demand modelling, Ph.D. Thesis, ETH Zurich, Zurich.

Ferber, J. (1999) *Multi-agent systems. An Introduction to distributed artificial intelligence*, Addison-Wesley, New York.

FGSV (2009) *Handbuch für die Bemessung von Strassenverkehrsanlagen: HBS*, Forschungsgesellschaft für Straßen- und Verkehrswesen, Cologne.

FHWA (2013) TRANSIMS Background, webpage, `http://www.fhwa.dot.gov/planning/tmip/resources/transims/`.

Ficici, S. G., O. Melnik and J. B. Pollack (2005) A game-theoretic and dynamical-systems analysis of selection method in coevolution, *IEEE Transactions on Evolutionary Computation*, **9** (6) 580–602.

Flötteröd, G. (2008) Traffic state estimation with multi-agent simulations, Ph.D. Thesis, TU Berlin, Berlin.

Flötteröd, G. (2010) Cadyts – Calibration of dynamic traffic simulations – Version 1.1.0 manual, `http://people.kth.se/~gunnarfl/cadyts.html`.

Flötteröd, G. (accessed 2015) Cadyts web site, `http://people.kth.se/~gunnarfl/cadyts.html`.

Flötteröd, G. and M. Bierlaire (2013) Metropolis-Hastings sampling of paths, *Transportation Research Part B*, **48**, 53–66.

Flötteröd, G., M. Bierlaire and K. Nagel (2011) Bayesian demand calibration for dynamic traffic simulations, *Transportation Science*, **45** (4) 541–561.

Flötteröd, G., Y. Chen and K. Nagel (2011a) Behavioral calibration and analysis of a large-scale travel microsimulation, *Networks and Spatial Economics*, **12** (4) 481–502, doi:10.1007/s11067-011-9164-9.

Flötteröd, G., Y. Chen and K. Nagel (2011b) Behavioral calibration and analysis of a large-scale travel microsimulation, presentation, 3rd MATSim User Meeting, Berlin, April 2011.

Flötteröd, G., Y. Chen and K. Nagel (2012) Choice model refinement from network data, in: *13th Conference on Travel Behaviour Research (IATBR)*, Toronto, Canada. Also VSP WP 12-08.

Flötteröd, G. and G. Lämmel (2010) Evacuation simulation with limited capacity sinks – an evolutionary approach to solve the shelter allocation and capacity assignment problem in a multi-agent evacuation simulation., in: J. Filipe and J. Kacprzyk (eds.) *IJCCI (ICEC)*, 249–254, SciTePress.

Flötteröd, G. and G. Lämmel (2015) Bidirectional pedestrian fundamental diagram, *Transportation Research Part B: Methodological*, **71** (1) 194–212.

Flötteröd, G. and J. Rohde (2011) Operational macroscopic modeling of complex urban road intersections, *Transportation Research Part B*, **45** (6) 903–922.

Flügel, S. and J. Kern (2014) Workshop on activity-based traffic simulations, presentation, Workshop at Hasselt University, Hasselt.

Flügel, S., N. G. Voll and F. Bockemühl (2014) Prøves for vegnettverket i Trondheim: Ny og bedre metode for å beregne trafikkavvikling, press release, June 2014, `http://samferdsel.toi.no/nr-06/ny-og-bedre-metode-for-a-beregne-trafikkavvikling-article32606-1462.html`.

Follmer, R., D. Gruschwitz, B. Jesske, S. Quandt, B. Lenz, C. Nobis, K. Köhler and M. Mehlin (2010) Mobilit{ä}t in Deutschland - Ergebnisbericht.

Follmer, R., U. Kunert, J. Kloas and H. Kuhfeld (2004) Mobilität in Deutschland – Ergebnisbericht, *Technical Report*, infas/DIW.

Ford, L. and D. Fulkerson (1962) *Flows in Networks*, Princeton University Press.

Fourie, P. (2009) An initial implementation of a multi-agent transport simulator for South Africa, Master Thesis, University of Pretoria, Pretoria.

Fourie, P. J. (2010) Agent-based transport simulation versus equilibrium assignment for private vehicle traffic in Gauteng, in: *29th Annual Southern African Transport Conference*, Pretoria, July 2010.

Fourie, P. J. (2014) Reconstructing bus vehicle trajectories from transit smart-card data, *Working Paper*, Future Cities Laboratory, Singapore-ETH Centre (SEC), Singapore.

Fourie, P. J., J. Illenberger and K. Nagel (2013) Increased convergence rates in multi-agent transport simulations with pseudo-simulation, *Transportation Research Record*, **2343**, 68–76.

Fourie, P. J. and J. W. Joubert (2009) The first agent steps in agent-based transport planning, in: *28th Annual Southern African Transport Conference*, Pretoria, July 2009.

Fowkes, A. (2010) The value of travel time savings, *Applied Transport Economics: A Management and Policy Perspective*, 547–569.

Fowler, M. (2004) *Refactoring: Improving the design of existing code*, Addison-Wesley.

Frei, A. (2012) Networks, geography and travel: Travel between infrastructure and social structure, Ph.D. Thesis, ETH Zurich, Zurich.

Frejinger, E. and M. Bierlaire (2007) Capturing correlation with subnetworks in route choice models, *Transportation Research Part B: Methodological*, **41** (3) 363–378.

Frejinger, E. and M. Bierlaire (2010) On path generation algorithms for route choice models, in: *Choice modelling: the state-of-the-art and the state-of-practice*, 307–315, Emerald Group Publishing Limited.

Frejinger, E., M. Bierlaire and M. Ben-Akiva (2009a) Sampling of alternatives for route choice modeling, *Transportation Research Part B*, **43** (10) 984–994.

Frejinger, E., M. Bierlaire and M. E. Ben-Akiva (2009b) Sampling of alternatives for route choice modeling, *Transportation Research Part B: Methodological*, **43** (10) 984–994.

Fujiyama, T. (2004) Evaluation of accessible design of public transport facilities, *Research Progress Report*, University College London.

Galus, M. D. and G. Andersson (2011) Balancing renewable energy source with vehicle to grid services from a large fleet of plug-in hybrid electric vehicles controlled in a metropolitan area distribution network, in: *Cigré 2011 Bologna Symposium*, Bologna, September 2011.

Galus, M. D., G. Georges and R. A. Waraich (2012a) Abschlussbericht des Projekts ARTEMIS (Abating Road Emissions Through Efficient (electric) Mobility - Interactions with the electric System), *Final Report*, Elektrizitätswerk der Stadt Zürich (EWZ), IVT, ETH Zurich, Zurich.

Galus, M. D., R. A. Waraich, M. Balmer, G. Andersson and K. W. Axhausen (2009) A framework for investigating the impact of PHEVs, in: *International Advanced Mobility Forum 2009*, Geneva, March 2009.

Galus, M. D., R. A. Waraich, F. Noembrini, K. Steurs, G. Georges, K. Boulouchos, K. W. Axhausen and G. Andersson (2012b) Integrating power systems, transport systems and vehicle technology for electric mobility impact analysis and efficient control, *IEEE Transactions on Smart Grid*, **3** (2) 934–949.

Gan, L. P. and W. W. Recker (2008) A mathematical programming formulation of the household activity rescheduling problem, *Transportation Research Part B: Methodological*, **42** (6) 571–606.

Gao, W., M. Balmer and E. J. Miller (2010) Comparison of MATSim and EMME/2 on Greater Toronto and Hamilton Area network, Canada, *Transportation Research Record*, **2197**, 118–128.

Garcia, A., D. Reaume and R. Smith (2000) Fictitious play for finding system optimal routings in dynamic traffic networks, *Transportation Research Part B*, **34** (2) 147–156.

Gartner, N., C. Messer, A. Rathi, C. on Traffic Flow Theory and Characteristics (2001) *Traffic Flow Theory: A State-of-the-art Report*, Committe on Traffic Flow Theory and Characteristics, National Academy of Science, Washington, D.C.

Gawron, C. (1998) An iterative algorithm to determine the dynamic user equilibrium in a traffic simulation model, *International Journal of Modern Physics C*, **9** (3) 393–408.

Geotools (accessed 2015) The open source Java GIS toolkit, webpage. `http://www.geotools.org`.

Geroliminis, N. and C. F. Daganzo (2007) Macroscopic modeling of traffic in cities, in: *86th Annual Meeting of the Transportation Research Board*, Washington, D.C., January 2007.

Geroliminis, N. and C. F. Daganzo (2008) Existence of urban-scale macroscopic fundamental diagrams: Some experimental findings, *Transportation Research Part B: Methodological*, **42** (9) 759–770.

Geroliminis, N. and J. Sun (2011) Hysteresis phenomena of a macroscopic fundamental diagram in freeway networks, *Transportation Research Part A: Policy and Practice*, **45** (9) 966–979.

Geurs, K., M. de Bok and B. Zondag (2012a) Accessibility benefits of integrated land use and public transport policy plans in the Netherlands, in: K. Geurs, K. Krizek and A. Reggiani (eds.) *Accessibility Analysis and Transport Planning*, 135–153, Edward Elgar, Cheltenham.

Geurs, K., K. Krizek and A. Reggiani (2012b) Accessibility analysis and transport planning: an introduction, in: K. Geurs, K. Krizek and A. Reggiani (eds.) *Accessibility Analysis and Transport Planning*, 1–12, Edward Elgar.

Geurs, K. T. and B. van Wee (2004) Accessibility evaluation of land-use and transport strategies: review and research directions, *Journal of Transport Geography*, **12** (2) 127–140.

GitHub (2015) Web-based git repository hosting service, webpage, `https://github.com`.

Giuliano, G. and J. Dargay (2006) Car Ownership, Travel and Land Use: A Comparison of the US and Great Britain, *Transportation Research Part A: Policy and Practice*, **40** (2) 106–124.

GKS (2010) National population census, volume 1: The size and distribution of the population, section 5, webpage, `http://www.gks.ru/free_doc/new_site/perepis2010/croc/perepis_itogi1612.htm`.

Gliebe, J., M. A. Bradley, N. Ferdous, M. Outwater, H. Lin and J. Chen (2014) Transfer of activity-based model parameters from Sacramento, California, to Jacksonville, and to Tampa, Florida, in: *93rd Annual Meeting of the Transportation Research Board*, Washington, D.C., January 2014.

Gliebe, J. P. and F. S. Koppelman (2002) A model of joint activity participation, *Transportation*, **29** (1) 49–72.

Gliebe, J. P. and F. S. Koppelman (2005) Modeling household activity-travel interactions as parallel constrained choices, *Transportation*, **32** (5) 449–471.

Gloor, C. and K. Nagel (2005) A message-based framework for real-world mobility simulations, in: F. Klügl, A. Bazzan and S. Ossowski (eds.) *Applications of Agent Technology in Traffic and Transportation*, Whitestein Series in Software Agent Technologies and Autonomic Computing, 193–209, Birkhäuser, Basel, doi:10.1007/3-7643-7363-6_13.

Goldberg, D. E. (1989) *Genetic Algorithms in Search, Optimization and Machine Learning*, Addison-Wesley, Reading.

Google Developers (2015) Protocol Buffers, webpage, December 2015, `https://developers.google.com/protocol-buffers/`.

Goseberg, N., G. Lämmel, H. Taubenböck, N. Setiadi, J. Birkmann and T. Schlurmann (2013) The last-mile evacuation project: A multi-disciplinary approach to evacuation planning and risk reduction in tsunami-threatend coastal areas, in: F. Wenzel and J. Zschau (eds.) *Early Warning for Geological Disasters: Scientific Methods and Current Practice*, chap. 11, 205–224, Springer, doi:10.1007/978-3-642-12233-0_11.

Goseberg, N. and T. Schlurmann (2009) Enhanced hazard mapping on a medium-resolved numerical grid for the city of Padang, West Sumatra, *Journal of Ship Technology*, **5** (2) 13–21, July 2009.

Gottardi, G. and S. Bürgler (1999) Güterverkehrsmodell Kanton Zürich, *Strasse und Verkehr*, **85** (2) 76–79.

Gourieroux, C., A. Monfort and E. Renault (1993) Indirect Inference, *Journal of Applied Econometrics*, **8** (S) 85–118.

Gradoteka (2015) Exciting statistics Samara city in infographics, webpage, `http://gradoteka.ru/city/samara/detail/transport-c/avtomobilizaciya`.

Graf, A. (2013) Die Bewertung der Qualität des Schülerverkehrs unter Anwedung der Multiagentensimulation MATSim, Diplomarbeit (Diploma Thesis), TU Berlin, Institute for Land and Sea Transport Systems, Berlin, 05 2013.

Grant-Muller, S. M., P. Mackie, J. Nellthorp and A. Pearman (2001) Economic appraisal of European transport projects: The state-of-the-art revisited, *Transport Reviews*, **21** (2) 237–261, doi:10.1080/01441640119423.

Grether, D., J. Bischoff and K. Nagel (2011a) Traffic-actuated signal control: Simulation of the user benefits in a big event real-world scenario, in: *2nd International Conference on Models and Technologies for ITS, Leuven, Belgium*. Also VSP WP 11-12.

Grether, D., Y. Chen, M. Rieser, U. Beuck and K. Nagel (2008) Emergent effects in multi-agent simulations of road pricing, in: *Annual Meeting of the European Regional Science Association (ERSA)*. Also VSP WP 08-08.

Grether, D., Y. Chen, M. Rieser and K. Nagel (2009) Effects of a simple mode choice model in a large-scale agent-based transport simulation, in: A. Reggiani and P. Nijkamp (eds.) *Complexity and Spatial Networks. In Search of Simplicity*, Advances in Spatial Science, 167–186, Springer, doi:10.1007/978-3-642-01554-0.

Grether, D., S. Fürbas and K. Nagel (2013) Agent-based modelling and simulation of air transport technology, *Procedia Computer Science*, **19**, 821–828, doi:10.1016/j.procs.2013.06.109.

Grether, D., B. Kickhöfer and K. Nagel (2010) Policy evaluation in multi-agent transport simulations, *Transportation Research Record*, **2175**, 10–18, doi:10.3141/2175-02.

Grether, D., A. Neumann and K. Nagel (2011b) Traffic light control in multi-agent transport simulations, *VSP Working Paper*, **11-08**, TU Berlin, Transport Systems Planning and Transport Telematics, Berlin.

Grether, D., A. Neumann and K. Nagel (2012) Simulation of urban traffic control: A queue model approach, *Procedia Computer Science*, **10**, 808–814, doi:10.1016/j.procs.2012.06.104.

Grether, D. S. (2014) Extension of a multi-agent transport simulation for traffic signal control and air transport systems, Ph.D. Thesis, TU Berlin, Berlin.

Gühnemann, A., A. Pearman and J. Laird (2011) National Secondary Roads Needs Study MCA Methodology Report, *Technical Report*, National Roads Authority, April 2011.

Gulhan, G., H. Ceylan, O. Baskan and Ceylan (2014) Using Potential Accessibility Measure for Urban Public Transportation Planning: A Case Study of Denizli, Turkey, *Promet Traffic and Transportation*, **26** (2) 129–139.

Guo, J. Y. and C. R. Bhat (2007) Population synthesis for microsimulating travel behavior, *Transportation Research Record*, **2014** (12) 92–101.

Gurram, S., A. L. Stuart and A. R. Pinjari (2015) Impacts of travel activity and urbanicity on exposures to ambient oxides of nitrogen and on exposure disparities, *Air Quality, Atmosphere & Health*, **8** (1) 97–114.

Habib, K. M. N. and J. A. Carrasco (2011) Investigating the role of social networks in start time and duration of activities: Trivariate simultaneous econometric model, *Transportation Research Record*, **2230**, 1–8.

Hackney, J. K. (2009) Integration of social networks in a large-scale travel behavior microsimulation, Ph.D. Thesis, ETH Zurich, Zurich.

Hagberg, A. A., D. A. Schult and P. J. Swart (2008) Exploring network structure, dynamics, and function using networkx, in: *Proceedings of the 7th Python in Science Conference (SciPy2008)*, 11–15.

Hägerstrand, T. (1970) What about people in Regional Science?, *Papers in Regional Science*, **24**, 6–21, doi:10.1007/BF01936872.

Hall, R. (1993) Non-recurrent congestion: how big is the problem? Are traveler information systems the solution?, *Transportation Research Part C*, **1** (1) 89–103.

Han, Q., T. A. Arentze, H. J. P. Timmermans, D. Janssens and G. Wets (2011) The effects of social networks on choice set dynamics: Results of numerical simulations using an agent-based approach, *Transportation Research Part A*, **45** (4) 310–322.

Hansen, W. (1959) How accessibility shapes land use, *Journal of the American Planning Association*, **25** (2) 73–76.

Harrell, G. D., M. D. Hutt and J. C. Anderson (1980) Path analysis of buyer behavior under conditions of crowding, *Journal of Marketing Research*, **17** (1) 45–51.

Hastings, W. (1970) Monte Carlo sampling methods using Markov chains and their applications, *Biometrika*, **57** (1) 97–109.

Hatcher, S. G. and H. S. Mahmassani (1992) Daily variability of route and trip scheduling decisions for the evening commute, *Transportation Research Record*, **1357**, 72–81.

Hatzopoulou, M. and E. J. Miller (2010) Linking an activity-based travel demand model with traffic emission and dispersion models: Transport's contribution to air pollution in Toronto, *Transportation Research Part D: Transport and Environment*, **15** (6) 315–325.

Herman, R., E. Montroll, R. Potts and R. Rothery (1959) Traffic dynamics: Analysis of stability in car following, *Operations Research*, **7** (1) 86–106.

Herriges, J. and C. Kling (1999) Nonlinear income effects in random utility models, *The Review of Economics and Statistics*, **81** (1) 62–72.

Heyndrickx, C., J. Purwanto, F. Ciari, P. M. Boesch and A. Perrels (2014) The impact of extreme weather events on urban mobility in Switzerland: combining a traffic micro-simulation with an economic macro-model, *Working Paper*, **1012**, IVT, ETH Zurich, Zurich.

Heyndrickx, C., F. Rodric, P. M. Bösch and F. Ciari (2016) Benefits of informing travellers in case of extreme precipitation events: A model based case study for Zurich using MATSim, presentation, the 95th Annual Meeting of the Transportation Research Board, Washington, D.C., January.

Hirschmann, K., M. Zallinger, M. Fellendorf and S. Hausberger (2010) A new method to calculate emissions with simulated traffic conditions, in: *Intelligent Transportation Systems Conference (ITSC)*, Madeira, September 2010.

Ho, C. and C. Mulley (2013) Tour-based mode choice of joint household travel patterns on weekend and weekday, *Transportation*, **40** (4) 789–811.

Hofbauer, J. and K. Sigmund (1998) *Evolutionary games and replicator dynamics*, Cambridge University Press, Cambridge.

Holland, J. H. (1992) *Adaptation in Natural and Artificial Systems: An Introductory Analysis with Applications to Biology, Control, and Artificial Intelligence*, MIT Press, Cambridge.

Horni, A. (2013) Destination choice modeling of discretionary activities in transport microsimulations, Ph.D. Thesis, ETH Zurich, Zurich.

Horni, A. (2016) MATSim destination choice documentation, webpage, http://matsim.org/docs/extensions/destination-choice.

Horni, A. and K. W. Axhausen (2012a) Distribution of benefits and losses from roadpricing illustrated in a microsimulation scenario, *Working Paper*, **974**, IVT, ETH Zurich, Zurich.

Horni, A. and K. W. Axhausen (2012b) MATSim agent heterogeneity and week scenario, *Working Paper*, **836**, IVT, ETH Zurich, Zurich.

Horni, A., D. Charypar and K. W. Axhausen (2011a) Variability in transport microsimulations investigated with the multi-agent transport simulation MATSim, *Working Paper*, **692**, IVT, ETH Zurich, Zurich.

Horni, A., F. Ciari and K. W. Axhausen (2012a) Coupling customers' destination choice and retailers' location choice in MATSim, *Working Paper*, **808**, IVT, ETH Zurich, Zurich.

Horni, A. and D. Grether (2007) Counts, internal presentation, MATSim-T Workshop, IVT, ETH Zurich, Castasegna, October 2007.

Horni, A., L. Montini and K. W. Axhausen (2013a) An Agent-Based Cellular Automaton Cruising-For-Parking Simulation, *Transportation Letters*, **5** (4) 167–175.

Horni, A., L. Montini and K. W. Axhausen (2013b) An Agent-Based Cellular Automaton Cruising-For-Parking Simulation, *IATBR Special Issue of Transportation Letters*.

Horni, A., K. Nagel and K. W. Axhausen (2012b) High-resolution destination choice in agent-based models, *Annual Meeting Preprint*, **12-1988**, Transportation Research Board, Washington, D.C. Also VSP WP 11-17.

Horni, A., D. M. Scott, M. Balmer and K. W. Axhausen (2009) Location choice modeling for shopping and leisure activities with MATSim: Combining micro-simulation and time geography, *Transportation Research Record*, **2135**, 87–95.

Horni, A., B. J. Vitins and K. W. Axhausen (2011b) The Zurich scenario: A technical overview, *Working Paper*, **687**, IVT, ETH Zurich, Zurich.

Hraber, P., T. Jones and S. Forrest (1994) The ecology of Echo, *Artificial Life*, **3** (3) 165–190.

Hu, T.-Y. and H. Mahmassani (1997) Day-to-day evolution of network flows under real-time information and reactive signal control, *Transportation Research Part C*, **5** (1) 51–69, doi:10.1016/S0968-090X(96)00026-5.

Hugenbusch, D. (2012) Simulation of Evacuation Scenarios Using MATSim. Case Study on Hamburg, B.S. Thesis, University of Kiel, Kiel.

Hui, M. K. and J. E. G. Bateson (1991) Perceived control and the effects of crowding and consumer choice on the service experience, *Journal of Consumer Research*, **18** (2) 174–184.

Hülsmann, F., R. Gerike, B. Kickhöfer, K. Nagel and R. Luz (2011) Towards a multi-agent based modeling approach for air pollutants in urban regions, in: *Conference on "Luftqualität an Straßen"*, 144–166. Also VSP WP 10-15.

Hülsmann, F., B. Kickhöfer and R. Gerike (2013) Air pollution hotspots in urban areas – how effective are pricing strategies to comply with the EU limits for NO_2?, in: R. Gerike, K. Roller and F. Hülsmann (eds.) *Strategies for Sustainable Mobilities: Opportunities and Challenges*, 105–128, Ashgate Publishing Ltd.

Illenberger, J. (2012) Social networks and cooperative travel behaviour, Ph.D. Thesis, TU Berlin, Berlin.

Illenberger, J., M. Kowald, K. W. Axhausen and K. Nagel (2011) Insights into a spatially embedded social network from a large-scale snowball-sample, *European Physical Journal B*, **84** (4) 549–561, doi:10.1140/epjb/e2011-10872-0.

INRO (2015) EMME, webpage, `http://www.inro.ca/en/products/emme/index.php`.

Isaacman, S., R. Becker, R. Caceres, S. Kobourov, M. Martonosi, J. Rowland and A. Varshavsky (2011) Identifying important places in people's lives from cellular network data, *Lecture Notes in Computer Science*, **6696**, 133–151.

Istanbul Chamber of Industry (2012) Istanbul Chamber of Industry, webpage, `http://www.iso.org.tr/`.

Ito, M., D. Fukuda and J. Ma (2015) Development of hyperpath-based risk-averse route guidance system and its verification, in: *51st Conference of Infrastructure Planning and Management, Japan Society for Civil Engineers (in Japanese)* (in Japanese).

Jacob, R. R., M. V. Marathe and K. Nagel (1999) A computational study of routing algorithms for realistic transportation networks, *ACM Journal of Experimental Algorithms*, **4** (1999es, Article No. 6), doi:10.1145/347792.347814.

Jara-Díaz, S. (2007) *Transport Economic Theory*, Emerald Group Publishing Limited, Bingley.

Jara-Díaz, S. and C. Guevara (2003) Behind the subjective value of travel time savings, *Journal of Transport Economics and Policy*, **37** (1) 29–46.

Jara-Díaz, S., M. Munizaga, P. Greeven, R. Guerra and K. W. Axhausen (2008) Estimating the value of leisure from a time allocation model, *Transportation Research B*, **42** (10) 946–957, doi:10.1016/j.trb.2008.03.001.

Jara-Díaz, S. and J. Videla (1989) Detection of income effect in mode choice: Theory and application, *Transportation Research Part B: Methodological*, **23** (6) 393–400, doi:10.1016/0191-2615(89)90040-4.

Joh, C.-H. (2004) Measuring and predicting adaptation in multidimensional activity-travel patterns, Ph.D. Thesis, Technical University Eindhoven, Eindhoven.

Jones, P. M. (1979) New approaches to understand travel behaviour: the human activity approach, in: D. A. Hensher and P. R. Stopher (eds.) *Behavioural Travel Modelling*, 55–80, Croom Helm Ltd, Kent.

Jones, P. M., M. C. Dix, M. I. Clarke and I. G. Heggie (1983) *Understanding Travel Behaviour*, Gower, Aldershot.

JOSM (2014) Java OpenStreetMap Editor, `https://josm.openstreetmap.de/`.

Joubert, J., D. Ziemke and K. Nagel (2015) Accessibility in a post-apartheid city: Comparison of two approaches for the computation of accessibility indicators, in: *55th ERSA Congress*. Also VSP WP 15-17.

Joubert, J. W. and K. W. Axhausen (2011) Inferring commercial vehicle activities in Gauteng, South Africa, *Journal of Transport Geography*, **19** (1) 115–124.

Joubert, J. W., P. J. Fourie and K. W. Axhausen (2010) Large-scale agent-based combined traffic simulation of private cars and commercial vehicles, *Transportation Research Record*, **2168**, 24–32.

Jourquin, B. and S. Limbourg (2006) Equilibrium traffic assignment on large virtual networks: Implementation issues and limits for multi-modal freight transport, *European Journal of Transport and Infrastructure Research*, **6** (3) 205–228.

Kaiser, C. and A. Pozdnouhkov (2013) Enabling real-time city sensing with kernel stream oracles and MapReduce, *Pervasive and Mobile Computing*, **7** (5) 708–721.

Kato, H. and M. Matsumoto (2009) Intra-household interaction in a nuclear family: A utility-maximizing approach, *Transportation Research Part B: Methodological*, **43** (2) 191–203.

Kaufman, D., K. Wunderlich and R. Smith (1991) An iterative routing / assignment method for anticipatory real-time route guidance, in: *Vehicle Navigation and Information Systems Conference*, vol. 2, 693–700, doi:10.1109/VNIS.1991.205814.

Kempton, W. and J. Tomic (2005) Vehicle-to-grid power fundamentals: Calculating capacity and net revenue, *Journal of Power Sources*, **144** (1) 268–279.

Kickhöfer, B. (2009) Die Methodik der ökonomischen Bewertung von Verkehrsmaßnahmen in Multiagentensimulationen, Master Thesis, Technische Universität Berlin, Berlin.

Kickhöfer, B. (2014) Economic policy appraisal and heterogeneous users, Ph.D. Thesis, TU Berlin, Berlin.

Kickhöfer, B., D. Grether and K. Nagel (2011) Income-contingent user preferences in policy evaluation: application and discussion based on multi-agent transport simulations, *Transportation*, **38** (6) 849–870, doi:10.1007/s11116-011-9357-6.

Kickhöfer, B., D. Hosse, K. Turner and A. Tirachini (2016) Creating an open MATSim scenario from open data: The case of Santiago de Chile, *VSP Working Paper*, **16-02**, TU Berlin, Transport Systems Planning and Transport Telematics.

Kickhöfer, B., F. H. Hülsmann, R. Gerike and K. Nagel (2013) Rising car user costs: Comparing aggregated and geo-spatial impacts on travel demand and air pollutant emissions, in: T. Vanoutrive and A. Verhetsel (eds.) *Smart Transport Networks: Decision Making, Sustainabil-*

ity and Market structure, NECTAR Series on Transportation and Communications Networks Research, 180–207, Edward Elgar Publishing Ltd, Cheltenham.

Kickhöfer, B. and J. Kern (2015) Pricing local emission exposure of road traffic: An agent-based approach, *Transportation Research Part D: Transport and Environment*, **37** (1) 14–28, doi:10.1016/j.trd.2015.04.019.

Kickhöfer, B. and K. Nagel (2011) Mapping emissions to individuals – new insights with multi-agent transport simulations, in: *12th Conference on Computers in Urban Planning and Urban Management (CUPUM)*, Lake Louise, Canada, July 2011. Also VSP WP 11-02.

Kickhöfer, B. and K. Nagel (2013) Towards high-resolution first-best air pollution tolls, *Networks and Spatial Economics*, 1–24, doi:10.1007/s11067-013-9204-8.

Kickhöfer, B., M. Zilske and K. Nagel (2010) Income dependent economic evaluation and public acceptance of road user pricing, in: *Kuhmo Nectar Conference on Transportation Economics*, Valencia, July 2010. Also VSP WP 10-03.

Kim, J. Y., Y. S. Yu, S. J. Lee, H. Hu and J. G. Sung (2012) Application of multi-agent transport simulation for urban road network operation in incident case, *International Journal of Highway Engineering*, **14** (4) 163–163.

Knight, F. H. (1924) Some fallacies in the interpretation of social cost, *Quarterly Journal of Economics*, **38** (4) 582–606.

Köhler, E. and M. Strehler (2010) Traffic signal optimization using cyclically expanded networks, in: *10th Workshop on Algorithmic Approaches for Transportation Modelling, Optimization, and Systems*, no. 14 in: OpenAccess Series in Informatics (OASIcs), 114–129, Dagstuhl, doi:http://dx.doi.org/10.4230/OASIcs.ATMOS.2010.114.

Kohli, S. and A. Daly (2006) The use of logsums in welfare estimation: Application in PRISM, in: *European Transport Conference, Strasbourg*.

Kowald, M. and K. W. Axhausen (2012) Focusing on connected personal leisure networks: Selected results from a snowball sample, *Environment and Planning A*, **44** (5) 1085–1100.

Kowald, M., A. Frei, J. K. Hackney, J. Illenberger and K. W. Axhausen (2009) Collecting data on leisure travel: The link between leisure acquaintances and social interactions, in: *Applications of Social Network Analysis*, Zurich, August 2009.

Kowald, M., P. van den Berg, A. Frei, J. A. Carrasco, T. A. Arentze, K. W. Axhausen, D. Mok, H. J. P. Timmermans and B. Wellman (2013) Distance patterns of personal networks in four countries: a comparative study, *Journal of Transport Geography*, **31** (3) 236–248.

Kraschl-Hirschmann, K., M. Zallinger, R. Luz, M. Fellendorf and S. Hausberger (2011) A method for emission estimation for microscopic traffic flow simulation, in: *IEEE Forum on Integrated and Sustainable Transportation Systems*, Vienna, June 2011.

Kucirek, P. (2012) Comparison between matsim & emme: Developing a dynamic, activity-based micro-simulation transit assignment model for toronto, Master Thesis, University of Toronto, Toronto.

Kühnel, N. (2014) Grafisches Editieren von Straßen- und ÖV-Netzmodellen, Master's thesis, TU Berlin, Institute for Land and Sea Transport Systems, Berlin. Also VSP WP 15-10 .

Kwan, M.-P. (1998) Space-time and integral measures of individual accessibility: A comparative analysis using a point-based framework, *Geographical Analysis*, **30** (3) 191–216, doi:10.1111/j.1538-4632.1998.tb00396.x.

Lämmel, G. (2011) Escaping the tsunami: Evacuation strategies for large urban areas. concepts and implementation of a multi-agent based approach, Ph.D. Thesis, TU Berlin, Berlin.

Lämmel, G., M. Chraibi, A. Kemloh Wagoum and B. Steffen (2016) Hybrid multi- and inter-modal transport simulation: A case study on large-scale evacuation planning, *Annual Meeting Preprint*, **16-2259**, Transportation Research Board, Washington, D.C.

Lämmel, G. and G. Flötteröd (2009) Towards system optimum: Finding optimal routing strategies in time-tependent networks for large-scale evacuation problems, in: B. Mertsching, M. Hund and Z. Aziz (eds.) *KI 2009: Advances in Artificial Intelligence*, vol. 5803 of *LNCS (LNAI)*, 532–539, Springer, Berlin Heidelberg, doi:10.1007/978-3-642-04617-9_67.

Lämmel, G., D. Grether and K. Nagel (2010) The representation and implementation of time-dependent inundation in large-scale microscopic evacuation simulations, *Transportation Research Part C: Emerging Technologies*, **18** (1) 84–98, doi:10.1016/j.trc.2009.04.020.

Lämmel, G. and H. Klüpfel (2012) Slower is faster: the influence of departure time distribution on the overall evacuation performance, in: *International Conference on Evacuation Modeling and Management*, Northwestern University, Evanston, Illinois, USA.

Lämmel, G., H. Klüpfel and K. Nagel (2009) The MATSim network flow model for traffic simulation adapted to large-scale emergency egress and an application to the evacuation of the Indonesian city of Padang in case of a tsunami warning, in: H. Timmermans (ed.) *Pedestrian Behavior*, chap. 11, 245–265, Emerald Group Publishing Limited.

Lämmel, G., H. Klüpfel and K. Nagel (2011) Risk minimizing evacuation strategies under uncertainty, in: R. Peacock, E. Kuligowski and J. Averill (eds.) *Pedestrian and Evacuation Dynamics*, 287–296, Springer, New York, doi:10.1007/978-1-4419-9725-8_26.

Lämmel, G. and M. Plaue (2014) Getting out of the way: collision avoiding pedestrian models compared to the real world, in: U. Weidmann, U. Kirsch and M. Schreckenberg (eds.) *Pedestrian and Evacuation Dynamics 2012*, 1275–1289, Springer, Berlin.

Lämmer, S. and D. Helbing (2010) Self-stabilizing decentralized signal control of realistic, saturated network traffic, *Working Paper*, **10-09-019**, Santa Fe Institute, Santa Fe.

Land Transport Authority (2009) Update of Economic Evaluation Parameters, *Final Report (unpublished)*, Land Transport Authority, Singapore.

Larsen, J., K. W. Axhausen and J. Urry (2006) Geographies of social networks: Meetings, travel and communications, *Mobilities*, **1** (2) 261–283.

Lebacque, J. (1996) The Godunov scheme and what it means for first order traffic flow models, in: *Proceedings of the 13th International Symposium on Transportation and Traffic Theory*, 647–677, Lyon, July 1996.

Lebacque, J. and M. Khoshyaran (2005) First-order macroscopic traffic flow models: intersection modeling, network modeling, in: *Proceedings of the 16th International Symposium on Transportation and Traffic Theory*, 365–386, Maryland, July 2005.

LeBlanc, L. J., E. K. Morlok and W. P. Pierskalla (1975) An efficient approach to solving the road network equilibrium traffic assignment problem, *Transportation Research*, **9** (5) 309–318.

Lee, S. J. and A. Ali (2014) Analyzing subway origin-destination flows by utilizing transit smart card data: Case of seoul metropolitan area, in: *1st International Workshop on Utilizing Transit Smart Card Data for Service Planning*, Gifu, July 2014.

Lefebvre, N. and M. Balmer (2007) Fast shortest path computation in time-dependent traffic networks, presentation, The 7th Swiss Transport Research Conference (STRC), Ascona, September.

Leutzbach, W. (1972) *Einführung in die Theorie des Verkehrsflusses*, Springer, Berlin Heidelberg.

Leutzbach, W. and R. Wiedemann (1986) Development and applications of traffic simulation models at the Karlsruhe Institut für Verkehrswesen, *Traffic Engineering and Control*, **27** (5) 270–278.

Li, S., I. V. Kolmanovsky and A. G. Ulsoy (2011) Battery swapping modularity design for plug-in HEVs using the augmented lagrangian decomposition method, in: *American Control Conference*, San Francisco, July 2011.

Liao, L., T. A. Arentze and H. J. P. Timmermans (2013) Multi-state supernetwork framework for the two-person joint travel problem, *Transportation*, **40** (4) 813–826.

Liao, T.-Y., T.-Y. Hu and D.-J. Chen (2008) Object-oriented evaluation framework for dynamic vehicle routing problems under real-time information, *Annual Meeting Preprint*, **08-2222**, Transportation Research Board, Washington, D.C., January 2008.

Lighthill, M. and J. Witham (1955) On kinematic waves II. a theory of traffic flow on long crowded roads, *Proceedings of the Royal Society A*, **229**, 317–345.

Liu, H., X. He and B. He (2007) Method of successive weighted averages (MSWA) and self-regulated averaging schemes for solving stochastic user equilibrium problem, *Networks and Spatial Economics*, **9** (4) 485–503.

Liu, R., D. van Vliet and D. P. Watling (2006) Microsimulation models incorporating both demand and supply dynamcis, *Transportation Research Part A: Policy and Practice*, **40** (2) 125–150.

Liu, S. and X. Zhu (2004) Accessibility analyst: An integrated GIS tool for accessibility analysis in urban transportation planning, *Environment and Planning B: Planning and Design*, **31** (2) 105–124.

Lovric, M., T. Li and P. Vervest (2013) Sustainable revenue management: A smart card enabled agent-based modeling approach, *Decision Support Systems*, **54** (2 - Special Issue: Rapid Modeling for Sustainability) 1587–1601.

Lu, C.-C., H. Mahmassani and X. Zhou (2009) Equivalent gap function-based reformulation and solution algorithm for the dynamic user equilibrium problem, *Transportation Research Part B: Methodological*, **43** (3) 345–364, doi:10.1016/j.trb.2008.07.005.

Lu, Q., B. George and S. Shekhar (2005) Capacity constrained routing algorithms for evacuation planning: A summary of results, *Lecture Notes in Computer Science*, **3633**, 291–307.

Lu, Y., M. Adnan, K. Basak, F. C. Pereira, C. Carrion, V. Hamishagi Saber, H. Loganathan and M. E. Ben-Akiva (2015) SimMobility mid-term simulator: A state of the art integrated agent based demand and supply model, *Annual Meeting Preprint*, **15-3937**, Transportation Research Board, Washington, D.C., January 2015.

Luce, R. D. and P. Suppes (1965) Preference, utility and subjective probability, in: R. D. Luce, R. R. Bush and E. H. Galanter (eds.) *Handbook of Mathematical Psychology Vol. 3*, 249–410, Wiley, New York.

Ma, H., T. A. Arentze and H. J. P. Timmermans (2012) Incorporating selfishness and altruism int dynamic joint activity-travel scheduling, in: *13th International Conference on Travel Behaviour Research (IATBR)*, Toronto, July 2012.

Ma, H., N. Ronald, T. A. Arentze and H. J. P. Timmermans (2011) New credit mechanism for semi-cooperative agent-mediated joint activity-travel scheduling, *Transportation Research Record*, **2230**, 104–110.

Ma, J., D. Fukuda and J.-D. Schmöcker (2013) Faster hyperpath generating algorithms for vehicle navigation, *Transportmetrica A: Transport Science*, **9** (10) 925–948.

Maciejewski, M. (2014) Online taxi dispatching via exact offline optimization, *Logistyka*, **3**, 2133–2142.

Maciejewski, M. and K. Nagel (2012) Towards multi-agent simulation of the dynamic vehicle routing problem in MATSim, in: R. Wyrzykowski et al (ed.) *Parallel Processing and Applied Mathematics (PPAM), Revised Selected Papers, Part II*, Lecture Notes in Computer Science, Springer, Berlin, doi:10.1007/978-3-642-31500-8_57.

Maciejewski, M. and K. Nagel (2013a) The influence of multi-agent cooperation on the efficiency of taxi dispatching, in: *10th International Conference Parallel Processing and Applied Mathematics (PPAM)*, Warsaw, Poland, September 2013. Also VSP WP 13-10.

Maciejewski, M. and K. Nagel (2013b) A microscopic simulation approach for optimization of taxi services, in: T. Albrecht, B. Jaekel and M. Lehnert (eds.) *3rd International Conference on Models and Technologies for Intelligent Transportation Systems 2013*, 1–10, TUDpress. Also VSP WP 13-12.

Maciejewski, M. and K. Nagel (2013c) Simulation and dynamic optimization of taxi services in MATSim, *VSP Working Paper*, **13-05**, TU Berlin, Transport Systems Planning and Transport Telematics.

Maciejewski, M., B. Piatkowski and W. Walerjanczyk (2014) Od makroskopowego modelu popytu na podroze do calodobowej mikroskopowej symulacji przeplywu ruchu, *Przeglad Komunikacyjny*, **2**, 27–31.

Mackie, P., S. Jara-Díaz and A. Fowkes (2001) The value of travel time savings in evaluation, *Transportation Research Part E: Logistics and Transportation Review*, **37** (2) 91–106, doi:10.1016/S1366-5545(00)00013-2.

Mackie, P. and T. Worlsley (2013) International Comparisons of Transport Appraisal Practice, *Technical Report*, ITS Leeds.

Mahmassani, H. S. and Y.-H. Liu (1999) Dynamics of commuting decision behaviour under advanced traveller information systems, *Transportation Research Part C: Emerging Technologies*, **7** (2–3) 91–107.

Manski, C. F. (1977) The structure of random utility models, *Theory and Decision*, **8** (3) 229–254.

Marchal, F. and K. Nagel (2005) Modeling location choice of secondary activities with a social network of cooperative agents, *Transportation Research Record*, **1935**, 141–146, doi:10.3141/1935-16.

Märki, F., D. Charypar and K. W. Axhausen (2014) Agent-based model for continuous activity planning with an open planning horizon, *Transportation*, **41** (4) 905–922.

MATSim (2016) Multi-Agent Transportation Simulation, webpage, `http://www.matsim.org`.

McArdle, G., E. Furey, A. Lawlor and A. Pozdnoukhov (2012) City-scale traffic simulation from digital footprints, in: *International Workshop on Urban Computing*, Beijing, August 2012.

McArdle, G., E. Furey, A. Lawlor and A. Pozdnoukhov (2014) Using digital footprints for a city-scale traffic simulation, *ACM Transactions on Intelligent Systems and Technology*, **5** (3) 1–3.

McCloskey, J., D. Lange, F. Tilmann, S. Nalbant, A. Bell, D. Natawidjaja and A. Rietbrock (2010) The September 2009 Padang earthquake, *Nature Geoscience*, **3** (1) 70–72.

McFadden, D. (1975) The revealed preferences of a government bureaucracy: Theory, *The Bell Journal of Economics*, **6** (2) 401–416. Earlier unpublished version with same title in 1967.

McGrath, R. and A. Pozdnoukhov (2014) A generative model of urban activities: simulating a population, in: *ACM SIGKDD International Workshop on Urban Computing*, New York, August 2014.

Meister, K. (2008) Erstellung von MATSim Facilities für das Schweiz-Szenario, *Working Paper*, **541**, IVT, ETH Zurich.

Meister, K. (2011) Contribution to agent-based demand optimization in a multi-agent transport simulation, Ph.D. Thesis, ETH Zurich, Zurich.

Meister, K., M. Balmer, K. Axhausen and K. Nagel (2006) planomat: A comprehensive scheduler for a large-scale multi-agent transportation simulation, in: *11th Conference on Travel Behavior Research (IATBR)*, Kyoto, Japan. Also VSP WP 06-07.

Meister, K., M. Balmer, F. Ciari, A. Horni, M. Rieser, R. A. Waraich and K. W. Axhausen (2010) Large-scale agent-based travel demand optimization applied to Switzerland, including mode choice, in: *12th World Conference on Transportation Research*, Lisbon, July 2010.

Meister, K., M. Frick and K. W. Axhausen (2005) A GA-based household scheduler, *Transportation*, **32** (5) 473–494.

Meister, K., M. Rieser, F. Ciari, A. Horni, M. Balmer and K. W. Axhausen (2009) Anwendung eines agentenbasierten Modells der Verkehrsnachfrage auf die Schweiz, *Straßenverkehrstechnik*, **53** (5) 269–280.

Meneguzzer, C. (1997) Review of models combining traffic assignment and signal control, *Journal of Transportation Engineering*, **123** (2) 148–155, doi:10.1061/(ASCE)0733-947X(1997)123:2(148).

Metropolis, N., A. Rosembluth, M. Rosembluth and A. Teller (1953) Equation of state calculations by fast computing machines, *Journal of Chemical Physics*, **21**, 1087–1092.

Michalewicz, Z. and D. B. Fogel (2004) *How to Solve It: Modern Heuristics*, Springer, Heidelberg.

Michiels, H., I. Mayers, L. Int Pais, L. De Nocker, F. Deutsch and W. Lefebvre (2012) *$PM_{2.5}$* and *NO_x* from traffic — human health impacts, external costs and policy implications from the Belgian perspective, *Transportation Research Part D: Transport and Environment*, **17** (8) 569–577.

Mikheeva, T. (2008) *Structural-parametric synthesis of intelligent transport systems*, Samara scientific center of RAS, Samara.

Mikheeva, T., S. Mikheev and O. Saprykin (2012) *Neural models of heterogeneous spatial coordinated data*, vol. 3 of *Programmnye producty i sistemy*, Factor i K, Tver.

Miller, E. and M. Roorda (2003) A prototype model of household activity/travel scheduling, *Transportation Research Record*, **1831**, 114–121.

Miller, E. J. (1996) Microsimulation and activity-based forecasting, in: *Activity-Based Travel Forecasting Conference*, New Orleans, June 1996.

Miller, E. J., M. J. Roorda and J. A. Carrasco (2005) A tour-based model of travel mode choice, *Transportation*, **32** (4) 399–422.

Monderer, D. and L. Shapley (1996) Fictitious play property for games with identical interests, *Journal of Economic Theory*, **68** (1) 258–265.

Moore, J., J. A. Carrasco and A. Tudela (2013) Exploring the links between personal networks, time use, and the spatial distribution of social contacts, *Transportation*, **40** (4) 773–788.

Morlok, E. K., J. L. Schofer, W. P. Pierskalla, R. Marsten, S. K. Agarwal, J. W. Stoner, J. L. Edwards, L. J. LeBlanc and D. T. Spacek (1973) Development and application of a highway network design model, volumes 1 and 2, *Final Report*, Contract Number DOT-PH-11, Federal Highway Administration, Northwestern University, Evanston.

Morris, J., P. Dumble and M. Wigan (1979) Accessibility indicators for transport planning, *Transportation Research Part A*, **13A** (2) 91–109.

Moyo Oliveros, M. (2013) Calibration of public transit routing for multi-agent simulation, Ph.D. Thesis, TU Berlin, Berlin.

Moyo Oliveros, M. and K. Nagel (2012) Automatic calibration of microscopic, activity-based demand for a public transit line, *Annual Meeting Preprint*, **12-3279**, Transportation Research Board, Washington, D.C., January 2012. Also VSP WP 11-13.

Moyo Oliveros, M. and K. Nagel (in press) Automatic calibration of agent-based public transit assignment path choice to count data, *Transportation Research Part C*. Also VSP WP 13-13.

Müller, K. (2011a) IPF within multiple domains: Generating a synthetic population for Switzerland, in: *11th Swiss Transport Research Conference*, Ascona, May 2011.

Müller, K. (2011b) Occam's Razor and some randomness: Generating a synthetic population for Switzerland, in: *European Transport Conference*, Glasgow, October 2011.

Müller, K. (2012) Using the Swiss PUS to generate a synthetic population for Switzerland, in: *13th International Conference on Travel Behavior Research*, Toronto, July 2012.

Müller, K. and K. W. Axhausen (2011) Population synthesis for microsimulation: State of the art, in: *90th Annual Meeting of the Transportation Research Board*, Washington, D.C., January 2011.

Müller, K. and K. W. Axhausen (2012) Multi-level fitting algorithms for population synthesis, in: *1st European Symposium on Quantitative Methods in Transportation Systems*, Lausanne, September 2012.

Müller, K. and K. W. Axhausen (2013) Using survey calibration and statistical matching to reweight and distribute activity schedules, *Working Paper*, **948**, IVT, ETH Zurich, Zurich.

Müller, K. and G. Flötteröd (2014) Population synthesis with regression trees, in: *3rd Symposium of the European Association for Research in Transportation (hEART 2014)*, Leeds, September 2014.

Munizaga, M., S. Jara-Díaz, P. Greeven and C. Bhat (2008) Econometric calibration of the joint time assignment-mode choice model, *Transportation Science*, **42** (2) 208–219, May 2008, doi:10.1287/trsc.1080.0231.

Muñoz, J. C., M. Batarce and D. Hidalgo (2014) Transantiago, five years after its launch, *Research in Transportation Economics*, **48**, 184–193.

Mustafa, A., I. Saadi, M. Cools and J. Teller (2014) Measuring the effect of stochastic perturbation component in Cellular Automata urban growth model, *Procedia Environmental Sciences*, **22**, 156–168.

MySQL (accessed 2014) MySQL wikipedia article, See `http://en.wikipedia.org/wiki/MySQL`.

Nagel, K. (1995) High-speed microsimulations of traffic flow, Ph.D. Thesis, University of Cologne, Cologne.

Nagel, K. (1996) Individual adaptation in a path-based simulation of the freeway network of Northrhine-Westfalia, *International Journal of Modern Physics C*, **7** (6) 883.

Nagel, K. (1999) From particle hopping models to traffic flow theory, *Transportation Research Record*, **1644**, 1–9.

Nagel, K. and C. Barrett (1997) Using microsimulation feedback for trip adaptation for realistic traffic in Dallas, *International Journal of Modern Physics C*, **8** (3) 505–526.

Nagel, K. and G. Flötteröd (2012) Agent-based traffic assignment: Going from trips to behavioural travelers, in: R. Pendyala and C. Bhat (eds.) *Travel Behaviour Research in an Evolving World – Selected papers from the 12th international conference on travel behaviour research*, 261–294, International Association for Travel Behaviour Research.

Nagel, K., B. Kickhöfer and J. W. Joubert (2014) Heterogeneous tolls and values of time in multi-agent transport simulation, *Procedia Computer Science*, **32**, 762–768, doi:10.1016/j.procs.2014.05.488.

Nagel, K. and M. Rickert (2001) Parallel implementation of the TRANSIMS micro-simulation, *Parallel Computing*, **27** (12) 1611–1639.

Nagel, K., M. Rickert, P. M. Simon and M. Pieck (2000) The dynamics of iterated transportation simulations, see `http://www.arXiv.org/abs/nlin/0002040`. Earlier version in: Proceedings of 3rd TRIannual Symposium on Transportation ANalysis (TRISTAN-III) 1998, San Juan, Puerto Rico.

Nagel, K. and A. Schleicher (1994) Microscopic traffic modeling on parallel high performance computers, *Parallel Computing*, **20**, 125–146, doi:10.1016/0167-8191(94)90117-1.

Nagel, K. and M. Schreckenberg (1992) A cellular automaton model for freeway traffic, *Journal de Physique I France*, **2**, 2221–2229.

Nagel, K., P. Stretz, M. Pieck, S. Leckey, R. Donnelly and C. L. Barrett (1997) TRANSIMS traffic flow characteristics, *Los Alamos Unclassified Report (LA-UR)*, **97-3531**, Los Alamos National Laboratory, Los Alamos, NM.

Nagel, K., D. Wolf, P. Wagner and P. M. Simon (1998) Two-lane traffic rules for cellular automata: A systematic approach, *Physical Review E*, **58** (2) 1425–1437.

Nagurney, A. and J. Dong (2002) *Supernetworks: Decision-Making for the Information Age*, Edward Elgar, London.

NAVTEQ (2011) NAVTEQ Maps and Traffic, webpage, `http://www.navteq.com`.

Neumann, A. (2008) Modellierung und Evaluation von Lichtsignalanlagen in Queue-Simulationen, Diplomarbeit (Diploma Thesis), TU Berlin, Institute for Land and Sea Transport Systems, Berlin. Also VSP WP 08-24.

Neumann, A. (2014) A paratransit-inspired evolutionary process for public transit network design, Ph.D. Thesis, TU Berlin, Berlin.

Neumann, A., M. Balmer and M. Rieser (2014) Converting a static trip-based model into a dynamic activity-based model to analyze public transport demand in Berlin, in: M. Roorda and E. Miller (eds.) *Travel Behaviour Research: Current Foundations, Future Prospects*, 151–176, International Association for Travel Behaviour Research (IATBR).

Neumann, A., I. Kaddoura and K. Nagel (2013) Mind the gap – passenger arrival patterns in multi-agent simulations, in: *Conference on Agent-Based Modeling in Transportation Planning and Operations 2013*, Blacksburg, VA. Also VSP WP 13-14.

Neumann, A., D. Röder and J. W. Joubert (2015) Towards a simulation of minibuses in South Africa, *Journal of Transport and Land Use*, **8** (1) 137–154, 05 2015, doi:10.5198/jtlu.2015.390.

Newell, G. (1993) A simplified theory of kinematic waves in highway traffic, part I: General theory, *Transportation Research Part B*, **27** (4) 281–287.

Ng, C. F. and K. A. Small (2012) Tradeoffs among free-flow speed, capacity, cost, and environmental footprint in highway design, *Transportation*, **39** (6) 1259–1280.

Nicolai, T. W. (2013) MATSim for UrbanSim: Integrating an urban simulation model with a travel model, Ph.D. Thesis, TU Berlin, Berlin.

Nicolai, T. W. and K. Nagel (2014) High resolution accessibility computations, in: A. Condeço, A. Reggiani and J. Gutiérrez (eds.) *Accessibility and Spatial Interaction*, 62–91, Edward Elgar.

Nicolai, T. W. and K. Nagel (2015) Integration of agent-based transport and land use models, in: M. Bierlaire, A. de Palma, R. Hurtubia and P. Waddell (eds.) *Integrated Transport and Land Use Modeling for Sustainable Cities*, 333–354, EPFL press, Lausanne.

Nicolai, T. W., L. Wang, K. Nagel and P. Waddell (2011) Coupling an urban simulation model with a travel model – A first sensitivity test, in: *12th Conference on Computers in Urban Planning and Urban Management (CUPUM)*, Lake Louise, Canada, July 2011. Also VSP WP 11-07.

Nurminski, E., I. Pugachev and N. Shamray (2014) *Modeling Car Correspondence Regional Transport System (on Example of the Irkutsk Area)*, vol. 4, Vestnik TOGU, Vladivostok.

OECD (2006) *Cost-benefit Analysis and the Environment: Recent Developments*, Organisation for Economic Cooperation and Development.

Oguchi, T., T. Satoh, M. Katakura and S. Shikata (2003) Analysis of traffic congestion and route choice behavior influenced by traffic information, *International Journal of ITS Research*, **1** (1) 75–82.

Openshaw, S. (1984) *The Modifiable Areal Unit Problem, vol. 38 of Concepts and techniques in modern geography*, Geo Books, Norwhich, England.

OpenStreetMap (2014) Map data © OpenStreetMap contributors, CC BY-SA.

OpenStreetMap (2015) The Free Wiki World Map, webpage, `http://www.openstreetmap.org`.

ORACLE www page (accessed 2005) Oracle database server, `http://www.oracle.com/products`.

Orcutt, G. H. (1957) A new type of socio-economic system, Review of Economics and Statistics, 39 (2) 116–123.

Ordóñez Medina, S. A. (2011a) Adding bus lanes using matsim network editing tool, `http://vimeo.com/37719740`.

Ordóñez Medina, S. A. (2011b) Demonstration of semi-automatic tool for bus route map matching, `http://vimeo.com/27137889`.

Ordóñez Medina, S. A. and A. Erath (2011) Semi-automatic tool for map-matching bus routes on high-resolution navigation networks, in: *16th international conference of Hong Kong Society for Transportation Studies*, Hong Kong, December 2011.

Ordóñez Medina, S. A. and A. Erath (2013a) Estimating dynamic workplace capacities by means of public transport smart card data and household travel survey in Singapore, *Transportation Research Record*, **2344**, 20–30.

Ordóñez Medina, S. A. and A. Erath (2013b) New dynamic events-based public transport router for agent-based simulations, *Working Paper*, **921**, IVT, ETH Zurich, Zurich.

Ortúzar, J. d. D. and L. G. Willumsen (2011) *Modelling Transport*, 4th edn., John Wiley & Sons, Chichester.

Osorio, C. and G. Flötteröd (published online in *Articles in Advance*) Capturing dependency among link boundaries in a stochastic dynamic network loading model, *Transportation Science*.

Osorio, C., G. Flötteröd and M. Bierlaire (2011) Dynamic network loading: a stochastic differentiable model that derives link state distributions, *Transportation Research Part B*, **45** (9) 1410–1423.

Padgham, L., K. Nagel, D. Singh and Q. Chen (2014) Integrating BDI Agents into a MATSim Simulation, *Frontiers in Artificial Intelligence and Applications*, **263** (ECAI 2014) 681–686.

Padgham, L. and Winikoff (2004) *Developing Intelligent Agent Systems: A Practical Guide*, Wiley, New York.

Pagliara, F. and H. J. P. Timmermans (2009) Choice set generation in spatial contexts: A review, *Transportation Letters*, **1** (1) 181–196.

Palmer, R. G., B. W. Arthur, J. H. Holland, B. Lebaron and P. Tayler (1994) Artificial economic life: A simple model of a stockmarket, *Physica D: Nonlinear Phenomena*, **75** (1–3) 264–274.

Parkin, J. and J. Rotheram (2010) Design speeds and acceleration characteristics of bicycle traffic for use in planning, design and appraisal, *Transport Policy*, **17** (5) 335–341.

Parsons Brinckerhoff (2005) Transportation models and data initiative - new york best practice model (nybpm), *Final Report*, New York Metropolitan Transportation Council (NYMTC), Parsons Brinckerhoff Quade & Douglas, Inc., New York.

Parsons Brinckerhoff (2009) New York Best Practice Model (NYBPM) For Regional Travel Demand Forecasting - NYBPM User Documentation BPM 2005 Update, *Final Report*, New York Metropolitan Transportation Council (NYMTC), Parsons Brinckerhoff Quade & Douglas, Inc., New York.

Pawlak, J., A. Sivakumar and J. W. Polak (2011) The consequences of the productive use of travel time: Revisiting the goods-leisure trade-off in the era of pervasive ICT, in: *2nd International Choice Modelling Conference*, Leeds, July 2011.

Peeta, S. and A. Ziliaskopoulos (2001) Foundations of dynamic traffic assignment: the past, the present and the future, *Networks and Spatial Economics*, **1** (3/4) 233–265.

Pendyala, R. M., K. P. Christian and K. C. Konduri (2011) *PopGen 1.1 User's Guide*, Lulu Publishers, Raleigh.

Piatkowski, B. and M. Maciejewski (2012) Zastosowanie map OSM w budowie modelu sieci aglomeracji poznanskiej dla symulacji w MATSim, *Zeszyty Naukowo-Techniczne Stowarzyszenia Inzynierow i Technikow Komunikacji Rzeczypospolitej Polskiej, Oddzial w Krakowie*, **98** (2) 163–177.

Piatkowski, B. and M. Maciejewski (2013) Zastosowanie danych GIS o zagospodarowaniu przestrzennym w mikroskopowej symulacji ruchu, *Prace Naukowe Politechniki Warszawskiej - Transport*, **95**, 411–419.

Piatkowski, B., M. Maciejewski, W. Walerjanczyk and A. Szarata (2014) A 24-hour microscopic traffic flow simulation in the Poznan agglomeration, *Logistyka*, **4**, 3145–3155.

Pigou, A. C. (1920) *The Economics of Welfare*, Macmillan and Co., London.

Pilli-Sihvola, K., V. Nurmi, A. Perrels, A. Harjanne, P. M. Boesch and F. Ciari (forthcoming) Innovations in weather services as a crucial building block for climate change adaptation in road transport, *European Journal of Transport and Infrastructure Research*.

Poeck, M. and D. Zumkeller (1978) Simulation of the effects of energy shortage in shortage in regional transport systems, in: *Planning and Transport, Research and Computation (PTRC) Summer Meeting*, Warwick, July 1978.

Polak, J. W. and F. Oladeine (2002) An empirical model of travelers' day-to-day learning in the presence of uncertain travel times, *Working Paper*, Centre for Transport Studies, Imperial College London, London.

Pons, F., M. Laroche and M. Mourali (2006) Consumer reactions to crowded retail settings: Cross-cultural differences between North America and the Middle East, *Psychology & Marketing*, **23** (7) 555–572.

Popovici, E., R. P. Wiegand and E. D. De Jong (2012) Coevolutionary principles, in: G. Rozenberg, T. Bäck and J. N. Kok (eds.) *Handbook of Natural Computing*, 987–1033, Springer, Heidelberg.

Prato, C. (2009) Route choice modeling: past, present and future research directions, *Journal of Choice Modelling*, **2** (1) 65–100, doi:10.1016/S1755-5345(13)70005-8.

Prechelt, L. (1999) Technical opinion: comparing Java vs. c/c++ efficiency differences to interpersonal differences, *Communications of the ACM*, **42** (10) 109–112, Oct 1999, doi:10.1145/317665.317683.

Prefeitura Municipal de Joinville (PMJ) (2015) `https://www.joinville.sc.gov.br`.

Protocol Buffers web page (accessed 2015) See `https://developers.google.com/protocol-buffers/`.

PTV (2009) Ptv, webpage, `http://www.ptvag.com`.

PTV (2011) *VISUM 12 - New features at a glance*, PTV, Karlsruhe.

PTV (2013) PTV AG: PTV Planung Transport Verkehr AG, `http://www.ptvgroup.com`.

PTV AG (accessed 2015) VISSIM web site, http://vision-traffic.ptvgroup.com/en-us/products/ptv-vissim/.

Quadstone Paramics Ltd. (accessed 2015) Paramics web site, http://www.paramics-online.com.

Quick, T. (2012) Modellierung und Simulation des deutschlandweiten Eisenbahnverkehrs, Bachelor's thesis, TU Berlin, Institute for Land and Sea Transport Systems, Berlin, August 2012.

Ramos, B. T. (2011) Effects of tropical storm (washi), *Technical Report*, **Update Sitrep No. 9 re**, NDRRMC, Quezon City.

Raney, B. (2005) Learning framework for large-scale multi-agent simulations, Ph.D. Thesis, ETH Zurich, Zurich.

Raney, B. and K. Nagel (2004) Iterative route planning for large-scale modular transportation simulations, *Future Generation Computer Systems*, **20** (7) 1101–1118, doi:10.1016/j.future.2003.11.001.

Raney, B. and K. Nagel (2006) An improved framework for large-scale multi-agent simulations of travel behaviour, in: P. Rietveld, B. Jourquin and K. Westin (eds.) *Towards better performing European Transportation Systems*, 305–347, Routledge, London.

Recker, W. (2001) A bridge between travel demand modeling and activity-based travel analysis, *Transportation Research Part B*, **35** (5) 481–506.

Recker, W. W. (1995) The household activity pattern problem: General formulation and solution, *Transportation Research Part B: Methodological*, **29** (1) 61–77.

Redmond, L. S. and G. Mokhtarian (2001) The positive utility of the commute: Modeling ideal commute time and relative desired commute amount, *Transportation*, **28** (2) 179–205.

Regan, A., H. Mahmassani and P. Jaillet (1998) Evaluation of dynamic fleet management systems: Simulation framework, *Transportation Research Record*, **1645**, 176–184.

Reinhold, T. (2006) Konzept zur integrierten Optimierung des Berliner Nahverkehrs, in: *Öffentlicher Personennahverkehr*, 131–146, Springer, Berlin, doi:10.1007/3-540-34209-5_8.

Richards, P. (1956) Shock waves on highways, *Operations Research*, **4**, 42–51.

Rickert, M. and K. Nagel (2001) Dynamic traffic assignment on parallel computers in TRANSIMS, *Future Generation Computer Systems*, **17** (5) 637–648.

Rieser, M. (2010) Adding transit to an agent-based transportation simulation concepts and implementation, Ph.D. Thesis, TU Berlin, Berlin.

Rieser, M., U. Beuck, M. Balmer and K. Nagel (2008) Modelling and simulation of a morning reaction to an evening toll, in: *Innovations in Travel Modeling (ITM) '08*, Portland, OR, June 2008. Also VSP WP 08-01.

Rieser, M., U. Beuck and K. Nagel (2007a) Researching the influence of time-dependent tolls with a multi-agent traffic simulation, in: *European Transport Conference (ETC)*, Leiden. Also VSP WP 07-17.

Rieser, M., D. Grether and K. Nagel (2009) Adding mode choice to a multi-agent transport simulation, *Transportation Research Record*, **2132**, 50–58, doi:10.3141/2132-06.

Rieser, M. and K. Nagel (2008) Network breakdown "at the edge of chaos" in multi-agent traffic simulations, *European Journal of Physics*, **63** (3) 321–327, doi:10.1140/epjb/e2008-00153-6.

Rieser, M., K. Nagel, U. Beuck, M. Balmer and J. Rümenapp (2007b) Truly agent-oriented coupling of an activity-based demand generation with a multi-agent traffic simulation, *Transportation Research Record*, **2021**, 10–17, doi:10.3141/2021-02.

Rizzi, L. and S. Steinmetz (2015) On using standard values of time in project appraisal: Income equity vs. preference equity, *Technical Report*, Pontificia Universidad Católica de Chile, Santiago de Chile.

Röder, D. (2013) Simulation of South African minibus taxis, Master's thesis, TU Berlin, Institute for Land and Sea Transport Systems, Berlin. Also VSP WP 13-11.

Röder, D., I. Cabrita and K. Nagel (2013) Simulation-based sketch planning, part III: Calibration of a MATSim-model for the greater Brussels area and investigation of a cordon pricing for the highway ring, *VSP working paper*, **13-16**, TU Berlin, Berlin.

Rodríguez, H., E. Quarantelli and R. Dyness (eds.) (2006) *Handbook of Disaster Research*, Series: Handbooks of Sociology and Social Research, Springer, Berlin.

Ronald, N. (2014) iMoD: Intelligent Mobility on Demand, webpage, `http://imod-au.info/`. Last accessed 6 Nov 2014.

Ronald, N., T. A. Arentze and H. J. P. Timmermans (2012) Modelling social interactions between individuals for joint activity scheduling, *Transportation Research Part B: Methodological*, **46** (2) 276–290.

Ronald, N., R. G. Thompson and S. Winter (2014) Simulating demand-responsive transportation: A review of agent-based approaches, *submitted to Transport Reviews*.

Ronald, N., R. G. Thompson and S. Winter (2015) A comparison of constrained and ad-hoc demand-responsive transportation systems, in: *94th Transportation Research Board Annual Meeting, Washington, D.C., January 2015*.

Ross, S. (2006) *Simulation*, fourth edn., Elsevier, Oxford.

Russel, S. J. and P. Norvig (2010) *Artificial Intelligence – A Modern Approach*, Pearson Education, Upper Saddle River.

Saadi, I., H. Eftekhar, A. Mustafa, J. Teller and M. Cools (2014) An agent-based micro-simulation framework to assess the impact of river floods on transportation systems: implementation trajectory for an assessment in the brussels metropolitan area, in: *International Conference for Traffic and Transport Engineering*, Belgrade, November 2014.

Sanders, P. and D. Schultes (2017) Engineering fast route planning algorithms, *Lecture Notes in Computer Science*, **4525**, 23–36.

Saprykina, O. and O. Saprykin (2014) *The detection algorithm of concentration points of traffic flows on the road network in GIS*, vol. 1 of *IT&Transport*, Inteltrans, Samara.

Saprykina, O. and O. S. T. Mikheeva (2012) *Realization calculation model of matrices of correspondence for transport network, paper presented at The 14th International Workshop on Computer Science and Information Technologies (CSIT'2012)*, USATU, Moscow.

SATURN (2014) Simulation and Assignment of Traffic to Urban Road Networks, webpage, `http://www.saturnsoftware.co.uk/`. Accessed on 07/05/2014.

Schieffer, S. V. (2011) Decentralized charging decisions for the smart grid, Master Thesis, IVT, ETH Zurich, Zurich.

Schlich, R., S. Schönfelder, S. Hanson and K. W. Axhausen (2004) Structures of leisure travel: Temporal and spatial variability, *Transport Reviews*, **24** (2) 219–237.

Schneider, S. (2011) A methodology for the extrapolation of trip chain data, Ph.D. Thesis, TU Berlin, Berlin.

Schnittger, S. and D. Zumkeller (2004) Longitudinal microsimulation as a tool to merge transport planning and traffic engineering models - the MobiTopp model, in: *European Transport Conference*, Strasbourg, October 2004.

Schröder, S., M. Zilske, G. Liedtke and K. Nagel (2012) Towards a multi-agent logistics and commercial transport model: The transport service provider's view, *Procedia Social and Behavioral Sciences*, **39**, 649–663, doi:10.1016/j.sbspro.2012.03.137.

Schummer, J. and R. V. Vohra (2007) Mechanism design without money, in: N. Nisan, T. Roughgarden, É. Tardos and V. V. Vazirani (eds.) *Algorithmic Game Theory*, 243–266, Cambridge University Press, Cambridge.

Schüssler, N. (2010) Accounting for similarities between alternatives in discrete choice models based on high-resolution observations of transport behaviour, Ph.D. Thesis, ETH Zurich, Zurich.

Schüssler, N. and K. W. Axhausen (2009) Map-matching of GPS traces on high-resolution navigation networks using the Multiple Hypothesis Technique (MHT), *Working Paper*, **568**, IVT, ETH Zurich, Zurich.

Schwerdtfeger, T. (1984) DYNEMO: A model for the simulation of traffic flow in motorway networks, in: J. Volmuller and R. Hamerslag (eds.) *Proceedings of the Ninth International Symposium on Transportation and Traffic Theory*, 65–87, VNU Science Press, Utrecht.

SECTRA (2014) Actualización y recolección de información del sistema de transporte urbano, Etapa IX, Encuesta Origen Destino Santiago 2012, *Technical Report*, Informe Final, Observatorio Social Universidad Alberto Hurtado.

Seddon, P. (1972) Program for simulating dispersion of platoons in road traffic, *Simulation*, **18** (3) 81–90.

Senozon (2013) Senozon AG: understanding mobility, http://www.senozon.com, `http://www.senozon.com`.

senozon AG (2015) via – visualization and analysis tool, webpage, `http://via.senozon.com`.

Shah, M. (2010) Activity-based travel demand modelling including freight and cross-border traffic with transit simulation, *Working Paper*, **654**, IVT, ETH Zurich, Zurich.

Sheffi, Y. (1985) *Urban Transportation Networks: Equilibrium Analysis with Mathematical Programming Methods*, Prentice Hall, Englewood Cliffs.

Shoup, D. (2004) The ideal source of local public revenue, *Regional Science and Urban Economics*, **34** (6) 753–784.

Shoup, D. (2005) *The High Cost of Free Parking*, Planners Press, Chicago.

Shvetsov, V. (2003) *Mathematical Modeling of Traffic Flows*, Kluwer Academic Publishers-Plenum Publishers, Dordrecht.

Sigua, R. G. (2008) *Fundamentals of Traffic Engineering*, University of the Philippines Press, Quezon City.

Simini, F., M. C. González, A. Maritan and A.-L. Barabási (2012) A universal model for mobility and migration patterns, *Nature*, **484**, 96–100.

Simon, P. M., J. Esser and K. Nagel (1999) Simple queueing model applied to the city of Portland, *International Journal of Modern Physics C*, **10** (5) 941–960.

Small, K. and H. Rosen (1981) Applied welfare economics with discrete choice models, *Econometrica*, **49** (1) 105–130.

Small, K. A. (2012) Valuation of travel time, *Economics of Transportation*, **1**, 2–14, doi:10.1016/j.ecotra.2012.09.002.

Smith, L., R. Beckman, D. Anson, K. Nagel and M. Williams (1995) TRANSIMS: TRansportation ANalysis and SIMulation System, in: *5th National Transportation Planning Methods Applications Conference*, Seattle, WA.

Song, G., L. Yu and Y. Zhang (2012) Applicability of traffic microsimulation models in vehicle emissions estimates, *Transportation Research Record*, **2270**, 132–141.

Sparmann, U. and W. Leutzbach (1980) ORIENT: Ein verhaltensorientiertes Simulationsmodell zur Verkehrsprognose, *Working Paper*, **20**, Institut für Verkehrswesen, University of Karlsruhe.

Speckman, P., D. Sun and K. Vaughn (1998) Synthesizing activity-travel patterns: A resampling approach, *Working Paper*, **1**, National Institute of Statistical Sciences (NISS), Research Triangle Park, NC.

Spiess, H. and M. Florian (1989) Optimal strategies: A new assignment model for transit networks, *Transportation Research Part B: Methodological*, **23B** (2) 82–102.

Stadt Zürich, Dienstabteilung Verkehr (2008) Grünzeiten der stadtzürcherischen Verkehrsregelungsanlagen im September 2007, unpublished, Traffic Division, Zurich Police Departement, Zurich.

Steinmeyer, I. and T. Wagner (2005) Verwendung der 'Kraftfahrzeugverkehr in Deutschland' (KiD 2002) für städtische bzw. regionale Fragestellungen, in: *Wirtschaftsverkehr 2005 - Trends - Modelle - Konzepte.*

Strippgen, D. (2009) Investigating the technical possibilities of real-time interaction with simulations of mobile intelligent particles, Ph.D. Thesis, TU Berlin, Berlin.

Strippgen, D. and K. Nagel (2009a) Multi-agent traffic simulation with CUDA, in: *High Performance Computing & Simulation (HPCS)*, Leipzig, doi:10.1109/HPCSIM.2009.5192895.

Strippgen, D. and K. Nagel (2009b) Using common graphics hardware for multi-agent traffic simulation with CUDA, in: *2nd International Conference on Simulation Tools and Techniques*, Rome, Italy, doi:10.4108/ICST.SIMUTOOLS2009.5666.

Sudret, B. (2012) Meta-models for structural reliability and uncertainty quantification, in: *5th Asian-Pacific Symposium on Structural Reliability and its Applications*, Singapore, May 2012.

Sumalee, A. and F. Kurauchi (2006) Network capacity reliability analysis considering traffic regulation after a major disaster, *Networks and Spatial Economics*, **6** (3) 205–219.

Sun, L., J. G. Jin, D.-H. Lee and K. W. Axhausen (2014a) Characterizing multimodal transfer time using smart card data: the effect of time, passenger age, crowdedness and collective pressure, *Working Paper*, **1013**, IVT, ETH Zurich, Zurich.

Sun, L., A. Tirachini, K. W. Axhausen, A. Erath and D.-H. Lee (2014b) Models of bus boarding and alighting dynamics, *Transportation Research Part A: Policy and Practice*, **69**, 447–460.

Swiderski, D. (1983) A model for simulating spatially and temporally coordinated activity sequences on the basis of mental maps, in: S. M. Carpenter and P. M. Jones (eds.) *Recent Advances in Travel Demand Analysis*, 313–334, Gower, Aldershot.

Swiss Federal Statistical Office (BFS) (2000) Eidgenössische Volkszählung 2000, http://www.bfs.admin.ch/bfs/portal/de/index/infothek/erhebungen__quellen/blank/blank/vz/uebersicht.html.

Swiss Federal Statistical Office (BFS) (2001) *Eidgenössische Betriebszählung 2001 - Sektoren 2 und 3, GEOSTAT Datenbeschreibung*, Swiss Federal Statistical Office (BFS), GEOSTAT, Neuchatel, http://www.bfs.admin.ch/bfs/portal/de/index/dienstleistungen/geostat/datenbeschreibung/betriebszaehlung05.Document.111424.pdf.

Swiss Federal Statistical Office (BFS) (2006) *Ergebnisse des Mikrozensus 2005 zum Verkehrsverhalten*, Swiss Federal Statistical Office (BFS), Neuchatel.

Tableau Software (2013) Tableau desktop, software, http://www.tableausoftware.com/products/desktop.

Tampere, C., R. Corthout, D. Cattrysse and L. Immers (2011) A generic class of first order node models for dynamic macroscopic simulations of traffic flows, *Transportation Research Part B*, **45** (1) 289–309.

Taubenböck, H., N. Goseberg, G. Lämmel, N. Setiadi, T. Schlurmann, K. Nagel, F. Siegert, J. Birkmann, K.-P. Traub, S. Dech, V. Keuck, F. Lehmann, G. Strunz and H. Klüpfel (2013) Risk reduction at the "Last-Mile": an attempt to turn science into action by the example of Padang, Indonesia, *Natural Hazards*, **65** (1) 915–1945, doi:10.1007/s11069-012-0377-0.

Teknomo, K. (2008) Modeling mobile traffic agents on network simulation, in: *16th Annual Conference of the Transportation Science Society of the Philippines (TSSP)*, Metro Manila, Philippines, September 2008.

Tele Atlas MultiNet (2010) MultiNet: The most powerful map database ever built, webpage, January 2010, http://www.teleatlas.com/OurProducts/MapData/Multinet/index.htm.

Thill, J.-C. (1992) Choice set formation for destination choice modelling, *Progress in Human Geography*, **16** (3) 361–382.

Tian, X., M. Mahut, M. Jha and M. Florian (2007) Dynameq application to evaluating the impact of freeway reconstruction, in: *Proceedings of the 86. Annual Meeting of the Transportation Research Board*, Washington, DC, January 2007.

Tirachini, A., D. A. Hensher and J. M. Rose (2014) Multimodal pricing and optimal design of urban public transport: The interplay between traffic congestion and bus crowding, *Transportation Research Part B: Methodological*, **61** (0) 33–54.

Train, K. E. (2003) *Discrete Choice Methods with Simulation*, Cambridge University Press, New York.

TRANSIMS Open Source (2013) Getting Started with TRANSIMS, webpage, `http://code.google.com/p/transims/wiki/GettingStarted`.

Transportation Research Board (2010) *Highway Capacity Manual*, Transportation Research Board, Washington, D.C., December 2010.

TRIPP, iTrans and VKS (2009) Comprehensive mobility plan for Patna urban agglomeration area, *Technical Report*, Department of Urban Development. Government of Bihar.

TSS Transport Simulation Systems (accessed 2015) AIMSUN web site, http://www.aimsun.com.

Turkish Statistical Institute (2011) Turkish Statistical Institute, webpage, `http://www.turkstat.gov.tr/`.

Turnbull, K. F. (1990) High-occupancy vehicle project case studies history and institutional arrangements, Website.

URA (2008) The planning act masterplan written statement, *Technical Report*, Urban Redevelopement Authority, Singapore.

U.S. Bureau of Public Roads (1964) *Traffic Assignment Manual*, U.S. Department of Commerce, Washington, D.C.

van der Zijpp, N. and C. Lindveld (2001) Estimation of origin-destination demand for dynamic assignment with simultaneous route and departure time choice, *Transportation Research Record*, **1771**, 75–82.

van Eggermond, M. A. B., A. Erath and K. W. Axhausen (2012) Vehicle ownership in Singapore using revealed-preference data and spatial variables, in: *13th International Conference on Travel Behaviour Research (IATBR)*, Toronto, July 2012.

van Zuylen, H. and L. Willumsen (1980) The most likely trip matrix estimated from traffic counts, *Transportation Research*, **14B**, 281–293.

Vickrey, W. S. (1969) Congestion theory and transport investment, *The American Economic Review*, **59** (2) 251–260.

Vovsha, P. and S. Gupta (2013) A model for work activity schedules with synchronization for multiple-worker households, *Transportation*, **40** (4) 827–845.

Vovsha, P., E. Petersen and R. Donnelly (2002) Microsimulation in travel demand modeling: Lessons learned from the New York best practice model, *Transportation Research Record*, **1805**, 68–77.

Vrtic, M. and P. Fröhlich (2010) Nationales Personenverkehrsmodell des UVEK, Basismodell 2005, *Research Report*, Swiss Federal Department for Environment, Transport, Energy and Communication, Swiss Federal Office for Spatial Development (ARE), Swiss Federal Roads Authority and Swiss Federal Office of Transport, IVT, ETH Zurich, Emch und Berger and Institute for Transportation Planning and Traffic, Technical University Dresden, Berne.

Vrtic, M., P. Fröhlich and K. W. Axhausen (2003) Schweizerische Netzmodelle für Strassen- und Schienenverkehr, in: T. Bieger, C. Lässer and R. Maggi (eds.) *Jahrbuch 2002/2003 Schweizerische Verkehrswirtschaft*, 119–140, Schweizerische Verkehrswissenschaftliche Gesellschaft (SVWG), St. Gallen.

Vrtic, M., N. Schüssler, A. Erath, M. Bürgle, K. W. Axhausen, E. Frejinger, M. Bierlaire, R. Rudel, S. Scagnolari and R. Maggi (2008) Einbezug der Reisekosten bei der Modellierung des Mobilitätsverhaltens, *Schriftenreihe*, **1191**, Bundesamt für Strassen, UVEK, Bern, CH. Final Report for Project SVI 2005/004.

Vrtic, M., N. Schüssler, A. Erath, K. Meister and K. W. Axhausen (2007) Tageszeitliche Fahrtenmatrizen im Personenverkehr an Werktagen im Jahr 2000, *Research Report*, Swiss Federal Department for Environment, Transport, Energy and Communication, Swiss Federal Office for Spatial Development (ARE), Swiss Federal Roads Authority and Swiss Federal Office of Transport, IVT, ETH Zurich, Zurich.

VSP (2012) Transport Systems Planning and Transport Telematics, Technische Universität Berlin (TU Berlin), `http://www.vsp.tu-berlin.de`.

Waddell, P., A. Borning, M. Noth, N. Freier, M. Becke and G. Ulfarsson (2003) Microsimulation of urban development and location choices: Design and implementation of UrbanSim, *Networks and Spatial Economics*, **3** (1) 43–67.

Waraich, R. A. (2012a) A Framework for Modeling the Electricity Demand by Plug-in Electric Vehicles, presentation, 25th European Conference on Operational Research, Vilnius, July 2012.

Waraich, R. A. (2012b) Impact of vehicle charging on the electric grid in Zurich, presentation at the 4th MATSim User Meeting, Berlin, March 2012.

Waraich, R. A. (2013) Agent-based simulation of electric vehicles: Design and implementation of a framework, Ph.D. Thesis, ETH Zurich, Zurich.

Waraich, R. A. (2014) Can Pseudo-Simulation be used for Modelling Parking Search?, presentation, Brown Bag, Zurich, April 2014.

Waraich, R. A. and K. W. Axhausen (2012) Agent-Based Parking Choice Model, *Transportation Research Record*, **2319**, 39–46.

Waraich, R. A. and K. W. Axhausen (2013) Integration of optimal charging locations into a transportation energy simulation framework, in: *8th Conference on Sustainable Development of Energy, Water and Environment Systems*, Dubrovnik, September 2013.

Waraich, R. A., D. Charypar, M. Balmer and K. W. Axhausen (2009) Performance improvements for large scale traffic simulation in MATSim, in: *9th Swiss Transport Research Conference*, Ascona, September 2009.

Waraich, R. A., D. Charypar, M. Balmer and K. W. Axhausen (2015) Performance improvements for large-scale traffic simulation in matsim, in: M. Helbich, J. J. Arsanjani and M. Leitner (eds.) *Computational Approaches for Urban Environments*, vol. 13 of *Geotechnologies and the Environment*, 211–233, Springer, Cham.

Waraich, R. A., C. Dobler and K. W. Axhausen (2012) Modelling Parking Search Behaviour with an Agent-Based Approach, presentation, 13th International Conference on Travel Behaviour Research, Toronto, July 2012.

Waraich, R. A., C. Dobler and K. W. Axhausen (2013a) A parking search strategy equilibrium model, *Working Paper*, **914**, IVT, ETH Zurich, Zurich.

Waraich, R. A., C. Dobler and K. W. Axhausen (2013b) Simulating parking search, in: *13th Swiss Transport Research Conference*, Ascona, April 2013.

Waraich, R. A., C. Dobler, C. Weis and K. W. Axhausen (2013c) Optimizing Parking Prices Using an Agent Based Approach, poster presentation, 92nd Annual Meeting of the Transportation Research Board, Washington, D.C., January 2013.

Waraich, R. A., M. D. Galus, C. Dobler, M. Balmer, G. Andersson and K. W. Axhausen (2013d) Plug-in hybrid electric vehicles and smart grids: Investigations based on a microsimulation, *Transportation Research Part C: Emerging Technologies*, **28**, 74–86.

Waraich, R. A., G. Georges, M. D. Galus and K. W. Axhausen (2014a) Adding electric vehicle modeling capability to an, agent-based transport simulation, in: D. Janssens, A.-U.-H. Yasar and L. Knapen (eds.) *Data Science and Simulation in Transportation Research*, 282–318, IGI Global, Hershey.

Waraich, R. A., S. Ranganathan and K. W. Axhausen (2014b) The parking game, in: *14th Swiss Transport Research Conference*, Ascona, May 2014.

Wardrop, J. G. (1952) Some theoretical aspects of road traffic research, *Proceedings of the Institution of Civil Engineers*, **1** (3) 325–362.

Watling, D. and M. Hazelton (2003) The dynamics and equilibria of day-to-day assignment models, *Networks and Spatial Economics*, **3** (3) 349–370.

Weidmann, U. (1992) Transporttechnik der Fussgänger - Transporttechnische Eigenschaften des Fussgängerverkehrs, Literaturauswertung, *Schriftenreihe*, **90**, IVT, ETH Zurich, Zurich.

Weilenmann, M., J.-Y. Favez and R. Alvarez (2009) Cold-start emissions of modern passenger cars at different low ambient temperatures and their evolution over vehicle legislation categories, *Atmospheric Environment*, **43** (15) 2419–2429.

Weis, C. (2012) Activity oriented modelling of short- and long-term dynamics of travel behaviour, Ph.D. Thesis, ETH Zurich, Zurich.

Weiss, A., M. Mahmoud, P. Kucirek and K. N. Habib (2014) Merging transit schedule information with a planning network to perform dynamic multimodal assignment: Lessons from a case study of the Greater Toronto and Hamilton area, *Canadian Journal of Civil Engineering*, **41** (10) 900–908.

Wiedemann, R. (1974) Simulation des Verkehrsflusses, Ph.D. Thesis, University of Karlsruhe, Karlsruhe.

Wiethölter, D., D. Bogai and J. Carstensen (2010) Pendlerbericht Berlin-Brandenburg 2009, *IAB-Regional*, **3/2010**, Institut für Arbeitsmarkt- und Berufsforschung. See `http://www.iab.de/238/section.aspx/Publikation/k100921n04`.

Wilson, A. (1971) A family of spatial interaction models, and associated developments, *Environment and Planning*, **3** (1) 1–32.

Wilson, P. (2008) Venezuela: Land of 12-cent gas, *Businees Week*, May 2008.

Winikoff, M. (2005) Jack intelligent agents: An industrial strength platform, in: R. H. Bordini, M. Dastani, J. Dix and A. E. Fallah-Seghrouchni (eds.) *Multi-Agent Programming: Languages, Platforms and Applications*, vol. 15 of *Multiagent Systems, Artificial Societies, and Simulated Organizations*, 175–193, Springer, Heidelberg.

Wismans, L., R. van den Brink, L. Brederode, K. Zantema and E. van Berkum (2013) Comparison of estimation of emissions based on static and dynamic traffic assignment models, in: *92nd Annual Meeting of the Transportation Research Board*, Washington, D.C., January 2013.

WolframAlpha (2012) Wolfram|alpha: Computational knowledge engine, webpage, `www.wolframalpha.com`.

Wooldridge, M. (2009) *An Introduction to MultiAgent Systems*, John Wiley & Sons, Chichester.

Yamada, K., J. Ma and D. Fukuda (2013) Simulation analysis of the market diffusion effects of risk-averse route guidance on network traffic, *Procedia Computer Science*, **19**, 874–881.

Yamagata, Y., D. Murakami, K. Minami, N. Arizumi, S. Kuroda, T. Tanjo and H. Maruyama (2015) A comparative study of clustering algorithms for electricity self-sufficient community extraction, in: *7th International Conference on Applied Energy*, Abu Dhabi, March 2015.

Yamagata, Y. and H. Seya (2013) Simulating a future smart city: An integrated land use-energy model, *Applied Energy*, **112** (12) 1466–1474.

Yamagata, Y. and H. Seya (2015) Proposal for a local electricity-sharing system: A case study of Yokohama city, Japan, *Intelligent Transport Systems, IET*, **9** (1) 38–49.

Yamagata, Y., H. Seya and S. Kuroda (2014) Energy resilient smart community: Sharing green electricity using V2C technology, *Energy Procedia*, **61**, 84–87.

Yamagata, Y., H. Seya and K. Nakamichi (2013) Creation of future urban environmental scenarios using a geographically explicit land-use model: a case study of Tokyo, *Annals of GIS*, **19** (3) 153–168.

Yandex Company (2013) Traffic jams in Russian cities, webpage, `https://company.yandex.ru/researches/reports/2013/city_jams/city_jams_2013.xml`.

Yperman, I. (2007) The link transmission model for dynamic network loading, Ph.D. Thesis, Katolieke Universiteit Leuven.

Yperman, I., S. Logghe, C. Tampere and B. Immers (2006) The multi-commodity link transmission model for dynamic network loading, in: *Proceedings of the 85. Annual Meeting of the Transportation Research Board*, Washington, DC.

Zhang, J., H. J. P. Timmermans and A. W. J. Borgers (2005) A model of household task allocation and time use, *Transportation Research Part B: Methodological*, **39** (1) 81–95.

Zhang, J., H. J. P. Timmermans and A. W. J. Borgers (2007) Utility-maximizing model of household time use for independent, shared, and allocated activities incorporating group decision mechanisms, *Transportation Research Record*, **1807**, 1–8.

Zhang, K., H. Mahmassani and C.-C. Lu (2008) Probit-based time-dependent stochstic user equilibrium traffic assignment model, *Transportation Research Record*, **2085**, 86–94.

Zhang, L., W. Yang, J. Wang and Q. Rao (2014) Large-scale agent-based transport simulation in Shanghai, China, *Transportation Research Record*, **2399**, 34–43.

Zheng, J. and J. Y. Guo (2008) Destination choice model incorporating choice set formation, in: *87th Annual Meeting of the Transportation Research Board*, Washington, D.C., January 2008.

Zhou, X. and J. Taylor (2014) DTALite: A queue-based mesoscopic traffic simulator for fast model evaluation and calibration, *Cogent Engineering*, **1** (1).

Zhu, S., D. Levinson and L. Zhang (2008) Agent-based route choice with learning and exchange of information, *Annual Meeting Preprint*, **08-2152**, Transportation Research Board, Washington, D.C., January 2008.

Zhuge, C. (2014) Dynamic evolution mechanism of urban transport-land use based on self-organizing theory, Ph.D. Thesis, Beijing Jiaotong University, Beijing.

Ziemke, D., K. Nagel and C. Bhat (2015) Integrating CEMDAP and MATSim to increase the transferability of transport demand models, *Transportation Research Record*, **2493**, 117–125, doi:10.3141/2493-13.

Zilske, M. and K. Nagel (2015) A simulation-based approach for constructing all-day travel chains from mobile phone data, *Procedia Computer Science*, **52**, 468–475, doi:10.1016/j.procs.2015.05.017.

Zilske, M., S. Schröder, K. Nagel and G. Liedtke (2012) Adding freight traffic to MATSim, *VSP Working Paper*, **12-02**, TU Berlin, Transport Systems Planning and Transport Telematics.

Zöllig Renner, C. (2014) The role of real estate developers in the context of land use development and transport, Ph.D. Thesis, ETH Zurich, Zurich.

www.ingramcontent.com/pod-product-compliance
Lightning Source LLC
Chambersburg PA
CBHW061929190326
41458CB00009B/2695

* 9 7 8 1 9 0 9 1 8 8 7 5 4 *